AF893464

Nitrogen-fixing Actinorhizal Symbioses

Nitrogen Fixation: Origins, Applications, and Research Progress

VOLUME 6

The titles published in this series are listed at the end of this volume.

Nitrogen-fixing Actinorhizal Symbioses

Edited by

Katharina Pawlowski
Stockholm University,
Sweden

and

William E. Newton
Virginia Polytechnic Institute and State University,
Blacksburg, VA, USA

Springer

A C.I.P. Catalogue record for this book is available from the Library of Congress.

ISBN 978-1-4020-3540-1
ISBN 978-1-4020-3547-0 (eBook)

Published by Springer,
P.O. Box 17, 3300 AA Dordrecht, The Netherlands.

www.springer.com

Background figure caption:
"A seed crop of clover (*Trifolium hirtum*) in flower near Moora, Western Australia. Photograph courtesy of Mike Davies, Senior Technical Officer, Pasture Research Group of Agriculture WA and reproduced with permission."

Vol. 6-specific figure caption:
"Scanning electron micrograph of a sporulating *Frankia* Cc13 culture grown on agar (upper left triangle; courtesy of David R. Benson, University of Connecticut, USA, and reproduced with permission) and a light microscopy picture of the infection zone near the tip of the lobe of a *Casuarina glauca* nodule, showing fully infected cells at bottom left and a line of three cells undergoing hyphal invasion (lower right triangle; courtesy of Kirill Demchenko, Komarov Botanical Institute, Russian Academy of Sciences, St. Petersburg, Russia, and reproduced with permission) ."

Printed on acid-free paper

All Rights Reserved
© 2008 Springer
No part of this work may be reproduced, stored in a retrieval system, or transmitted in any form or by any means, electronic, mechanical, photocopying, microfilming, recording or otherwise, without written permission from the Publisher, with the exception of any material supplied specifically for the purpose of being entered and executed on a computer system, for exclusive use by the purchaser of the work.

TABLE OF CONTENTS

PREFACE TO THE SERIES

Nitrogen Fixation: Origins, Applications, and Research Progress

Nitrogen fixation, along with photosynthesis as the energy supplier, is the basis of all life on Earth (and maybe elsewhere too!). Nitrogen fixation provides the basic component, fixed nitrogen as ammonia, of two major groups of macromolecules, namely nucleic acids and proteins. Fixed nitrogen is required for the N-containing heterocycles (or bases) that constitute the essential coding entities of deoxyribonucleic acids (DNA) and ribonucleic acids (RNA), which are responsible for the high-fidelity storage and transfer of genetic information, respectively. It is also required for the amino-acid residues of the proteins, which are encoded by the DNA and that actually do the work in living cells. At the turn of the millennium, it seemed to me that now was as good a time as any (and maybe better than most) to look back, particularly over the last 100 years or so, and ponder just what had been achieved. What is the state of our knowledge of nitrogen fixation, both biological and abiological? How has this knowledge been used and what are its impacts on humanity?

In an attempt to answer these questions and to capture the essence of our current knowledge, I devised a seven-volume series, which was designed to cover all aspects of nitrogen-fixation research. I then approached my long-time contact at Kluwer Academic Publishers, Ad Plaizier, with the idea. I had worked with Ad for many years on the publication of the Proceedings of most of the International Congresses on Nitrogen Fixation. My personal belief is that congresses, symposia, and workshops must not be closed shops and that those of us unable to attend should have access to the material presented. My solution is to capture the material in print in the form of proceedings. So it was quite natural for me to turn to the printed word for this detailed review of nitrogen fixation. Ad's immediate affirmation of the project encouraged me to share my initial design with many of my current co-editors and, with their assistance, to develop the detailed contents of each of the seven volumes and to enlist prospective authors for each chapter.

There are many ways in which the subject matter could be divided. Our decision was to break it down as follows: nitrogenases, commercial processes, and relevant chemical models; genetics and regulation; genomes and genomics; associative, endophytic, and cyanobacterial systems; actinorhizal associations; leguminous symbioses; and agriculture, forestry, ecology, and the environment. I feel very fortunate to have been able to recruit some outstanding researchers as co-editors for this project. My co-editors were Mike Dilworth, Claudine Elmerich, John Gallon, Euan James, Werner Klipp, Bernd Masepohl, Rafael Palacios, Katharina Pawlowski, Ray Richards, Barry Smith, Janet Sprent, and Dietrich Werner. They worked very hard and ably and were most willing to keep the volumes moving along reasonably close to our initial timetable. All have been a pleasure to work with and I thank them all for their support and unflagging interest.

Nitrogen-fixation research and its application to agriculture have been ongoing for many centuries – from even before it was recognized as nitrogen fixation. The Romans developed the crop-rotation system over 2000 years ago for maintaining and improving soil fertility with nitrogen-fixing legumes as an integral component. Even though crop rotation and the use of legumes was practiced widely but intermittently since then, it wasn't until 1800 years later that insight came as to how legumes produced their beneficial effect. Now, we know that bacteria are harbored within nodules on the legumes' roots and that they are responsible for fixing N_2 and providing these plants with much of the fixed nitrogen required for healthy growth. Because some of the fixed nitrogen remains in the unharvested parts of the crop, its release to the soil by mineralization of the residue explains the follow-up beneficial impact of legumes. With this realization, and over the next 100 years or so, commercial inoculants, which ensured successful bacterial nodulation of legume crops, became available. Then, in the early 1900's, abiological sources of fixed nitrogen were developed, most notable of these was the Haber-Bosch process. Because fixed nitrogen is almost always the limiting nutrient in agriculture, the resulting massive increase in synthetic fixed-nitrogen available for fertilizer has enabled the enormous increase in food production over the second half of the 20th century, particularly when coupled with the new "green revolution" crop varieties. Never before in human history has the global population enjoyed such a substantial supply of food.

Unfortunately, this bright shiny coin has a slightly tarnished side! The abundance of nitrogen fertilizer has removed the necessity to plant forage legumes and to return animal manures to fields to replenish their fertility. The result is a continuing loss of soil organic matter, which decreases the soil's tilth, its water-holding capacity, and its ability to support microbial populations. Nowadays, farms do not operate as self-contained recycling units for crop nutrients; fertilizers are trucked in and meat and food crops are trucked out. And if it's not recycled, how do we dispose of all of the animal waste, which is rich in fixed nitrogen, coming from feedlots, broiler houses, and pig farms? And what is the environmental impact of its disposal? This problem is compounded by inappropriate agricultural practice in many countries, where the plentiful supply of cheap commercial nitrogen fertilizer, plus farm subsidies, has encouraged high (and increasing) application rates. In these circumstances, only about half (at best) of the applied nitrogen reaches the crop plant for which it was intended; the rest leaches and "runs off" into streams, rivers, lakes, and finally into coastal waters. The resulting eutrophication can be detrimental to marine life. If it encroaches on drinking-water supplies, a human health hazard is possible. Furthermore, oxidation of urea and ammonium fertilizers to nitrate progressively acidifies the soil – a major problem in many agricultural areas of the world. A related problem is the emission of nitrogen oxides (NO_x) from the soil by the action of microorganisms on the applied fertilizer and, if fertilizer is surface broadcast, a large proportion may be volatilized and lost as ammonia. For urea in rice paddies, an extreme example, as much as 50% is volatilized and lost to the atmosphere. And what goes up must come down; in the case of fertilizer nitrogen, it returns to Earth in the rain, often acidic in nature. This uncontrolled deposition has unpredictable environmental effects, especially in pristine environments like forests, and may also affect biodiversity.

Some of these problems may be overcome by more efficient use of the applied fertilizer nitrogen. A tried and tested approach (that should be used more often) is to ensure that a balanced supply of nutrients (and not simply applying more and more) is applied at the right time (maybe in several separate applications) and in the correct place (under the soil surface and not broadcast). An entirely different approach that could slow the loss of fertilizer nitrogen is through the use of nitrification inhibitors, which would slow the rate of conversion of the applied ammonia into nitrate, and so decrease its loss through leaching. A third approach to ameliorating the problems outlined above is through the expanded use of biological nitrogen fixation. It's not likely that we shall soon have plants, which are capable of fixing N_2 without associated microbes, available for agricultural use. But the discovery of N_2-fixing endophytes within the tissues of our major crops, like rice, maize, and sugarcane, and their obvious benefit to the crop, shows that real progress is being made. Moreover, with new techniques and experimental approaches, such as those provided by the advent of genomics, we have reasons to renew our belief that both bacteria and plants may be engineered to improve biological nitrogen fixation, possibly through developing new symbiotic systems involving the major cereal and tuber crops.

In the meantime, the major impact might be through agricultural sustainability involving the wider use of legumes, reintroduction of crop-rotation cycles, and incorporation of crop residues into the soil. But even these practices will have to be performed judiciously because, if legumes are used only as cover crops and are not used for grazing, their growth could impact the amount of cultivatable land available for food crops. Even so, the dietary preferences of developed countries (who eats beans when steak is available?) and current agricultural practices make it unlikely that the fixed-nitrogen input by rhizobia in agricultural soils will change much in the near-term future. A significant positive input could accrue, however, from matching rhizobial strains more judiciously with their host legumes and from introducing "new" legume species, particularly into currently marginal land. In the longer term, it may be possible to engineer crops in general, but cereals in particular, to use the applied fertilizer more efficiently. That would be a giant step the right direction. We shall have to wait and see what the ingenuity of mankind can do when "the chips are down" as they will be sometime in the future as food security becomes a priority for many nations. At the moment, there is no doubt that commercially synthesized fertilizer nitrogen will continue to provide the key component for the protein required by the next generation or two.

So, even as we continue the discussion about the benefits, drawbacks, and likely outcomes of each of these approaches, including our hopes and fears for the future, the time has arrived to close this effort to delineate what we know about nitrogen fixation and what we have achieved with that knowledge. It now remains for me to thank personally all the authors for their interest and commitment to this project. Their efforts, massaged gently by the editorial team, have produced an indispensable reference work. The content is my responsibility and I apologize upfront for any omissions and oversights. Even so, I remain confident that these volumes will serve well the many scientists researching nitrogen fixation and related fields, students considering the nitrogen-fixation challenge, and administrators wanting to either become acquainted with or remain current in this field. I also acknowledge the many scientists who were not direct contributors to this series of books, but whose contributions to the field are documented in their pages. It would be remiss of me not to acknowledge also the patience and assistance of the several members of the Kluwer staff who have assisted me along the way. Since my initial dealings with Ad Plaizier, I have had the pleasure of working with Arno Flier, Jacco Flipsen, Frans van Dunne, and Claire van Heukelom; all of whom provided encouragement and good advice – and there were times when I needed both!

It took more years than I care to remember from the first planning discussions with Ad Plaizier to the completion of the first volumes in this series. Although the editorial team shared some fun times and a sense of achievement as volumes were completed, we also had our darker moments. Two members of our editorial team died during this period. Both Werner Klipp (1953-2002) and John Gallon (1944-2003) had been working on Volume II of the series, *Genetics and Regulation of Nitrogen-Fixing Bacteria,* and that volume is dedicated to their memory. Other major contributors to the field were also lost in this time period: Barbara Burgess, whose influence reached beyond the nitrogenase arena into the field of iron-sulfur cluster biochemistry; Johanna Döbereiner, who was the discoverer and acknowledged leader in nitrogen-fixing associations with grasses; Lu Jiaxi, whose "string bag" model of the FeMo-cofactor prosthetic group of Mo-nitrogenase might well describe its mode of action; Nikolai L'vov, who was involved with the early studies of molybdenum-containing cofactors; Dick Miller, whose work produced new insights into MgATP binding to nitrogenase; Richard Pau, who influenced our understanding of alternative nitrogenases and how molybdenum is taken up and transported; and Dieter Sellmann, who was a synthetic inorganic chemistry with a deep interest in how N_2 is activated on metal sites. I hope these volumes will in some way help both preserve their scientific contributions and reflect their enthusiasm for science. I remember them all fondly.

Only the reactions and interest of you, the reader, will determine if we have been successful in capturing the essence and excitement of the many sterling achievements and exciting discoveries in the research and application efforts of our predecessors and current colleagues over the past 150 years or so. I sincerely hope you enjoy reading these volumes as much as I've enjoyed producing them.

William E. Newton
Blacksburg, February 2004

PREFACE

Nitrogen-fixing Actinorhizal Symbioses

This book is part of a seven-volume series that was launched in 2004 and covers all aspects of nitrogen fixation from the biological systems to the industrial processes. Volume 6 covers nitrogen-fixing actinorhizal symbioses, which occur between soil actinomycetes of the genus *Frankia* and a diverse group of dicotyledonous plants, collectively called actinorhizal plants. These symbioses play vital roles in native ecosystems as well as important components in both forestry and land reclamation.

The volume is divided into 11 chapters, all authored by well-known scientists in the field. As in previous volumes of this series, the first chapter presents an historical perspective and describes the development of actinorhizal research with its focus on the period after the first reproducible isolation of the responsible microorganism by John Torrey's group in 1978.

Very early on, the initial attempts to characterize the bacterium taxonomically had considered this endosymbiont as an obligate symbiotic bacterium and used its ability to form root nodules and its morphological characteristics within root-nodule cells as discriminative criteria to distinguish it from other actinomycetes. These efforts led to the emendation of the family *Frankiaceae* with the type genus *Frankia* and also to the definition of host-specificity groups based on inoculation experiments using crushed nodules. However, after *Frankia* strains were isolated from nodules and pure cultures became available, many of these early results had to be discarded. Chapter 2 describes the techniques used to obtain phenotypic, genotypic and phylogenetic information on the members of the genus *Frankia*.

In contrast to most rhizobia (see Volume 7, *Nitrogen-fixing Leguminous Symbioses*), *Frankia* strains can fix N_2 in the free-living state which improves their ability to survive in soil and *Frankia* strains have been found in soils devoid of actinorhizal plants. Chapter 3 covers the recent advances in knowledge of *Frankia* strains as soil microorganisms and their relationship to other soil microorganisms.

Actinorhizal symbioses occur with dicotyledonous plants from eight different families, *i.e.*, in quite a diverse group of plants. Chapter 4 deals with the phylogeny of both the host plants and the *Frankia* endosymbionts. Host and endosymbiont phylogeny are then compared in an effort to address the question of whether the partners are an example of co-evolution.

Aerobic organisms, like *Frankia*, suffer from the so-called 'oxygen (O_2) dilemma of nitrogen fixation'. Nitrogenase (see Volume 1, *Catalysts for Nitrogen Fixation*) is highly sensitive to O_2 and can only function in an O_2-free environment, so a high respiratory O_2 flux has to exist adjacent to a vanishingly low O_2 tension at the sites of nitrogen fixation. Generally, strategies combine external O_2 barriers with high O_2 utilization at the nitrogenase site to maintain a steep O_2 gradient. Because, in contrast to the rhizobia, *Frankia* can provide its own O_2-protection system by forming specialized vesicles with restricted O_2 access, O_2-protection mechanisms in actinorhizal symbioses involve more than modification of nodule structure and physiology. Chapter 5 covers the diverse forms in which the O_2 dilemma is solved in different actinorhizal symbioses.

The different ways by which *Frankia* strains can enter the plant root and induce organogenesis, as well as mechanisms for autoregulation of nodule formation, are reviewed in Chapter 6. The N_2 fixed by the resulting actinorhizal plants is often the major nitrogen input to many terrestrial ecosystems and Chapter 7 deals with the interplay of carbon and fixed-nitrogen metabolism in actinorhizal nodules of different plant species.

Although actinorhizal plants play important functional roles in native uncultivated ecosystems and are usually the first species to colonize devastated land, relatively little is known about ecological constraints on their capacity for nitrogen fixation because most studies of actinorhizal associations have been conducted in laboratories, growth chambers, and greenhouses. The impact of ecological effects is addressed in Chapter 8, which reviews studies of actinorhizal symbioses in a variety of natural situations.

The molecular-level analysis of actinorhizal symbioses has lagged behind similar studies of legume symbioses for several good reasons and not for the lack of effort! Actinorhizal plants are, with one exception, woody plants, trees or shrubs, and have a long generation time, which together renders them recalcitrant to molecular-genetic analysis. In the last decade, however, molecular-level studies have been initiated with several actinorhizal species, aided by the development of transformation procedures for actinorhizal trees of the Casuarinaceae family. Chapter 9 describes the contribution of plant molecular-biology approaches to our understanding of actinorhizal symbioses.

We've known, since 1995, that all plant species, which are able to enter into root-nodule symbiosis with nitrogen-fixing soil bacteria, *i.e.*, legumes, *Parasponia* sp., and actinorhizal plants, belong to a single clade that also contains many non-symbiotic plant species. It appears that the common ancestor of this clade (called Rosid I) had acquired a property based upon which a root-nodule symbiosis could and, in some cases did, develop. If this property could be identified, it would offer the possibility of transferring the capacity to form nitrogen-fixing root-nodule symbioses to plant species outside the Rosid I clade. As a preliminary in the search for such a property, comparative analysis of different symbiotic systems should allow the identification of common features *versus* system-specific adaptations. Hence, a comparison of actinorhizal and legume symbioses is presented in Chapter 10. Finally, Chapter 11 discusses the prospects for the future of actinorhizal research.

It's been nearly five years from its inception to the completion of this volume and we would like to sincerely thank all of the contributors for their efforts and patience. We dedicate this volume to Antoon Akkermans, a long-time proponent of this area of research, who died suddenly in 2006. We remember him fondly.

Katharina Pawlowski
Stockholm, April, 2007

William E. Newton
Blacksburg, April, 2007

LIST OF CONTRIBUTORS

Antoon D. L. AKKERMANS
Laboratory of Microbiology,
Wageningen University,
Wageningen, The Netherlands.
Email: Antoon.Akkermans@wur.nl

Florence AUGUY
Equipe Rhizogenèse, UMR 1098,
Institut de Recherche pour le
Développement (IRD), BP 64501,
911 Avenue Agropolis, 34394
Montpellier Cedex 5, FRANCE.
Email: Florence.Auguy@mpl.ird.fr

David R. BENSON
Department of Molecular and Cell
Biology, University of Connecticut,
Storrs, CT 06269-3044, USA.
Email: dbenson@uconnvm.uconn.edu

R. Howard BERG
Donald Danforth Plant Science
Center, 975 N. Warson Road,
St. Louis, MO 634132, USA.
Email: RHBerg@danforthcenter.org

Alison M. BERRY
Department of Environmental
Horticulture, University of California
One Shields Avenue, Davis,
CA 95616, USA.
Email: amberry@ucdavis.edu

Didier BOGUSZ
Equipe Rhizogenèse, UMR 1098,
Institut de Recherche pour le
Développement (IRD), BP 64501,
911 Avenue Agropolis, 34394
Montpellier Cedex 5, FRANCE.
Email: Didier.Bogusz@mpl.ird.fr

Jeffrey O. DAWSON
Department of Natural Resources and
Environmental Sciences and Plant
Biology, University of Illinois,
Urbana, IL 61801, USA.
Email: jdawson2@uiuc.edu

Claudine FRANCHE
Equipe Rhizogenèse, UMR 1098,
Institut de Recherche pour le
Développement (IRD), BP 64501,
911 Avenue Agropolis, 34394
Montpellier Cedex 5, FRANCE.
Email: Claudine.Franche@mpl.ird.fr

Dittmar HAHN
Department of Biology,
Texas State University,
601 University Drive,
San Marcos, TX 78666, USA.
Email: dh49@txstate.edu

Kerstin HUSS-DANELL
Department of Agricultural Research
for Northern Sweden, Crop Science
Section, Swedish University of
Agricultural Sciences,
SE-904 03 Umeå, SWEDEN.
Email: Kerstin.Huss-Danell@njv.slu.se

Laurent LAPLAZE
Equipe Rhizogenèse, UMR 1098,
Institut de Recherche pour le
Développement (IRD), BP 64501,
911 Avenue Agropolis, 34394
Montpellier Cedex 5, FRANCE.
Email: Laurent.Laplaze@mpl.ird.fr

Beth C. MULLIN
Department of Botany,
The University of Tennessee,
Knoxville, TN 37996-1100, USA.
Email: bmullin@utk.edu

William E. NEWTON
Department of Biochemistry,
Virginia Polytechnic Institute and
State University,
Blacksburg, VA 24061, USA.
Email: wenewton@vt.edu

Philippe NORMAND
Department of Microbial Ecology,
CNRS, University Claude Bernard,
Lyon, FRANCE.
Email:
normand@biomserv.univ-lyon1.fr

Katharina PAWLOWSKI
Department of Botany, Stockholm
University, 10691 Stockholm,
SWEDEN.
Email:
Katharina.Pawlowski@botan.su.se

Carole SANTI
Equipe Rhizogenèse, UMR 1098,
Institut de Recherche pour le
Développement (IRD), BP 64501,
911 Avenue Agropolis, 34394
Montpellier Cedex 5, FRANCE.
Email: Carole.Santi@mpl.ird.fr

Christa R. SCHWINTZER
Department of Biological Sciences,
University of Maine, Orono,
ME 04469-5722, USA.
Email: christas@maine.edu

Warwick B. SILVESTER
Department of Biological Sciences,
University of Waikato, Private Bag
3105, Hamilton,
NEW ZEALAND.
Email: w.silvester@waikato.ac.nz

Janet I. SPRENT
32 Birkhill Avenue, Wormit, Fife,
DD6 8PW, Scotland, UK.
Email: JISprent@aol.com

Sergio SVISTOONOFF
Equipe Rhizogenèse, UMR 1098,
Institut de Recherche pour le
Développement (IRD), BP 64501,
911 Avenue Agropolis, 34394
Montpellier Cedex 5, FRANCE.
Email:
Sergio.Svistoonoff@mpl.ird.fr

Susan M. SWENSEN
Department of Biology, Ithaca
College, Ithaca, NY14850, USA.
Email: sswensen@ithaca.edu

John D. TJEPKEMA
Department of Biological Sciences,
University of Maine, Orono,
ME 04469-5722, USA.
Email: tjepkema@maine.edu

Maria VALDÉS
Instituto Politécnico Nacional,
Escuela de Ciencias Biológicas,
Departamento de Microbiología,
11340 México, DF, MEXICO.
Email: mvaldes@encb.ipn.mx

Claudio VALVERDE
Programa de Investigación sobre Interacciones Biológicas, Departamento de Ciencia y Tecnología, Universidad Nacional de Quilmes, Roque Saenz Peña 180, Bernal, B1876BXD, Argentina.
Email:
Claudio.Valverde@imf.unil.ch

Luis G. WALL
Programa de Investigación sobre Interacciones Biológicas, Departamento de Ciencia y Tecnología, Universidad Nacional de Quilmes, Roque Saenz Peña 180, Bernal, B1876BXD, Argentina.
Email: lgwall@unq.edu.ar

C. T. WHEELER
Plant Science Group, Bower Building, IBLS, University of Glasgow, Glasgow G12 8QQ, Scotland, UK.
Email:
wheeler@glas55.freeserve.co.uk

Antoon D. L. Akkermans
(1940-2006)

This book is dedicated to the memory of Antoon Akkermans, who was Associate Professor at Wageningen University from 1987 to 2003. His early research concerned symbiotic nitrogen fixation between *Frankia* and actinorhizal plants, particularly the symbiosis of black elder. Later, he focused more on the molecular ecology of bacterial communities, examining bacterial interactions in the mammalian (human, pig and mouse) gastrointestinal tract, in grassland soil, and in anaerobic sludge. *Akkermansia muciniphila,* a human intestinal mucin-degrading bacterium, was named for him to honor his contributions to microbial ecology. Actinorhizal symbioses were Antoon's first love and he maintained his interest in them even after his switch to microbial ecology. He co-authored the first chapter in this volume, a review of the history of actinorhizal research, a history in which he played an important role.

Antoon Akkermans had many interests besides science, most particularly music and art. In fact, after his retirement, he turned to painting with the proceeds of the sale of his oil and acryl paintings of bacterial communities going to the Akkermansia Foundation, which supports young scientists (< 65 year) working in the field of microbial ecology. Antoon's experience, wisdom and sense of humor will be missed by all who knew him.

Chapter 1

FRANKIA AND ACTINORHIZAL PLANTS: A HISTORICAL PERSPECTIVE

C. T. WHEELER[1], A. D. L. AKKERMANS[2] AND A. M. BERRY[3]
[1]Plant Science Group, Bower Building, IBLS, University of Glasgow, Glasgow G12 8QQ, Scotland, UK
[2]Laboratory of Microbiology, Wageningen University, Wageningen, The Netherlands
[3]University of California, Department of Environmental Horticulture, One Shields Avenue, Davis, CA 95616, U.S.A.

1. INTRODUCTION

> *"...when in the wide estuaries the mangroves have in due time reclaimed the swampy land from the water, the Casuarina tree plants itself and in its turn settles, solidifies and fertilizes the soil till it is ripe for a more varied and luxuriant growth"*
>
> *W. Somerset Maugham, 1926.*

The origins of the two organismal names in the title of this chapter are of particular historical interest. First, the generic name *Frankia* was proposed initially by Brunchorst (1886-1888), in honor of Professor Frank, to describe the endophyte of root-nodulated non-legumes. The name was adopted by Becking (1970) in his early taxonomic study and is now universally accepted. *Frankia* is at present the only confirmed member of the family Frankiaceae. Prior to 1979, the term "non-leguminous plant symbioses" was in common usage to describe not only what are now known as "actinorhizal plants" but all nitrogen-fixing associations between microorganisms and plants that do not belong to the Leguminosae (Fabaceae). In an excellent review in 1965, Allen and Allen considered the history of research not only of root-nodulated non-leguminous members of the Angiospermae but also of the Cycadaceae colonized by cyanobacteria. The term "actinorhizal" was proposed

K. Pawlowski and W. E. Newton (eds.), Nitrogen-fixing Actinorhizal Symbioses, 1–24.
© 2008 *Springer*.

at the first international meeting on "Symbiotic Nitrogen Fixation in Actinomycete-nodulated Plants", held at Harvard Forest, to provide a convenient and more positive designation for the field than the term "non-legume" (Torrey and Tjepkema, 1979).

Following the first reproducible isolation of the microorganism by the John Torrey's group (Callaham *et al.,* 1978), which was a watershed event in the development of *Frankia*-actinorhizal research, there was a rapid increase internationally in the number of scientists interested in *Frankia* and an exponential increase in the number of isolates in culture. This focused attention on the taxonomy of *Frankia* and necessitated the establishment of a classification system. At the international conference on the "Biology of *Frankia*", held in Wisconsin in 1982, it was agreed that criteria were lacking for a system based on species names, and so a system for numbering *Frankia* strains was proposed in the first "Catalog of *Frankia* strains" (Lechevalier, 1983; 1986), which is still in use. This system employs three letters to designate the collection and up to ten numbers to reflect the origins of the strain. Thus CpI1, the first *Frankia* strain to be isolated at Harvard Forest from *Comptonia peregrina*, is designated HFP070101. The first two numbers encode the host genus from which the strain was isolated and the third and fourth number the species. Thus, one hundred and fifty years after Meyen (1829) first published a description of alder root nodules, the designation and acceptance by the scientific community of a unique and distinguishing nomenclature accorded proper and full recognition of the scientific and economic importance of these symbiotic relationships. By definition, the term "actinorhizal" excludes *Parasponia* of the Ulmaceae, the only non-legume shown definitively to bear rhizobial nodules.

Actinorhizal plants are perennial, woody species, with the exception of *Datisca* spp., which are herbaceous perennials. Most genera fruit heavily or spread vegetatively and show relative shade intolerance, which are characteristics of species of early- to mid-succession in plant community development (Dawson, 1990). The primary colonization of deglaciated areas in Alaska, first by *Dryas drummondii* and then by *Alnus sinuata*, both actinorhizal species, provided a benchmark chronosequence that has served as a paradigm for the role of actinorhizal plants as pioneer species and facilitators of fixed-nitrogen accretion in nutrient-poor habitats (Crocker and Major, 1955; Lawrence *et al.,* 1967; Bormann and Sidle, 1990; Chapin *et al.,* 1994). The invasion of young volcanic fields by *Myrica faya*, after its introduction in the Hawaiian Islands, illustrates the successful domination of this ecosystem (to the exclusion of native trees) due to its characteristics of nitrogen fixation, wind pollination, prolific fruiting, and ready seed dispersal by birds (Vitousek, 1989).

In this chapter, key advances made in the period prior to 1978 will be discussed briefly because excellent accounts, including historical aspects, are available in the literature (McKee, 1962; Allen and Allen, 1965; Bond, 1973; Akkermans and van Dijk, 1981; Quispel, 1990; Quispel *et al.,* 1993). Further, because the pace of research has increased rapidly since 1978, not all of the major advances can be considered in this description of the historical development of the area. The authors extend their apologies to any who feel that their contributions to the history of the area have been neglected. However, due accord is given to the work of many other major contributors in the specialist chapters in this volume. Earlier scientific progress in this area is covered in "The Biology of *Frankia* and Actinorhizal Plants" edited by Schwintzer and Tjepkema (1990) and has also been the subject of many reviews in recent years, e.g., Normand and Lalonde, 1986; Tjepkema *et al.,* 1986; Benson and Silvester, 1993; Berry, 1994; Mullin and Dobritsa, 1996; Pawloski and Bisseling, 1996; Dommergues, 1997; Huss-Danell, 1997; Franche *et al.,* 1998; Benson and Clawson, 2000; Laplaze *et al.,* 2000; Wall 2000; Schwencke and Caru, 2001. The history of research on casuarinas is admirably documented in a specialist volume (Subbarao and Rodriguez-Barrueco, 1995).

2. THE EARLY YEARS

This section draws primarily on the review of Quispel (1990), who divided the efforts of scientists prior to 1978 into three periods. By the end of the nineteenth century, good experimental evidence had been obtained for the utilisation of atmospheric N_2 by nodulated *Alnus glutinosa* plants (Hiltner, 1895). Anatomical studies showed the presence of intercellular hyphae in nodules (Woronin, 1866), but the nature of the microorganism was in much dispute, being described variously as either a parasitic fungus or a myxomycete. The next 50 years were characterised by research that consolidated and extended these earlier findings. Notably, careful cytological studies provided strong evidence that the organisms in nodules of several genera of plants, such as *Alnus*, *Myrica*, *Casuarina*, and *Elaeagnus*, were actinomycetes (reviewed in Schaede, 1962), although definitive proof was not obtained due to failure to isolate the organism reproducibly. Cross-inoculation experiments with crushed nodules indicated that there were specific differences between the microorganisms that nodulated different plant genera.

In the period from 1950 to 1978, interest in actinorhizal root nodules greatly increased, promoted by the efforts of several influential scientists. Notable among these were George Bond and Anton Quispel. Bond and his co-workers conducted many classical experiments to demonstrate an increase in total nitrogen in nodulated actinorhizal plants. Their experiments showed how environmental factors, such as pH, combined nitrogen, O_2, and light periodicity, affected nodulation and N_2 fixation. Importantly, in experiments that paralleled those of Virtanen *et al.* (1954), ^{15}N methodology was used to prove unequivocally that actinorhizal nodules fix atmospheric N_2 (Bond, 1955; 1956). Biochemical studies confirmed the importance of citrulline as the main product of N_2 fixed in alder and its role in nitrogen transport from root nodules (Leaf *et al.,* 1958). One of Bond's other important contributions was as coordinator of that part of the International Biological Programme of the International Council of Scientific Unions concerning the biological importance of actinorhizal plants; this activity focused on a world-wide survey of root nodulation of these plants involving some 50 collaborators from 30 countries (Bond, 1976). The ubiquitous nodulation of *Alnus* species and the irregularity of nodulation of some other genera, notably *Casuarina* and *Dryas*, were noted. A further 36 species in genera known to be nodulated were recorded and *Colletia* was described as a new nodulated genus in the Rhamnaceae. In addition to the important scientific contributions, the IBP programme served to arouse interest in actinorhizal-plant biology and to provide added stimulus to all aspects of nitrogen-fixation research. Bond died in 1988 and a fascinating summary of his life and achievements is found in the Memoirs of the Royal Society of London, to which he was elected as Fellow in 1972 (Nutman, 1990).

Quispel also had significant interests in the host plant, but particularly with respect to the processes of infection and nodulation (Quispel, 1974). He had a long-standing personal interest in the endophyte and its cultivation *in vitro*. Early publications by others of their attempts to grow *Frankia* in pure culture were highly controversial because the experiments either did not fulfill Koch's postulates or were not reproduced. In 1959, Pommer published his study on the isolation of the micro-symbiont of alder, the characteristics of which suggest that his attempt was almost certainly successful (Pommer, 1959). Unfortunately, his cultures were lost. In the early nineteen fifties, Quispel designed a series of experiments to help lay the foundations for the isolation and culture of the endophyte. He demonstrated that surface-sterilized root-nodule tissue of *Alnus glutinosa* remained infective during storage on nutrient agar and that infectivity increased in media to which an alcoholic extract of alder nodules had been added (Quispel, 1954; 1955). Actinomycete hyphae were evident at the border of the infective nodule pieces (Quispel, 1960). Following his appointment as Professor in Experimental Botany at the University of Leiden, it was another eight years before Quispel and Teun Tak returned to research on actinorhizal plants and demonstrated that there was a close correlation between *Frankia* hyphal biomass and infectivity (Quispel and Tak, 1968). A further twenty years were to elapse before dipterocarpol, the compound in alcoholic extracts that enhanced the infectivity of nodule fragments, was isolated and identified (Quispel *et al.,* 1989). Quispel supervised many doctoral students, several of whom have contributed with distinction to our knowledge of actinorhizal plants. Quispel's involvement in the early work of his students is often not apparent for he encouraged their publication of doctoral work as the sole author. Quispel's contributions are highlighted further in Akkermans *et al.* (1983).

The availability of the electron microscope facilitated the resolution of important questions concerning the pleoiomorphic nature of the endophyte. The bacteria-like cells seen in sections of many actinorhizal nodules were identified as spores contained within sporangia (van Dijk and Merkus, 1976). The ultrastructure of the vesicles of several genera was described and histochemical techniques coupled with light microscopy identified them as the locus for nitrogenase (Akkermans, 1971), observations since confirmed using immunocytological methods (Huss-Danell and Bergman, 1990). Vesicles are absent from nodules of *Casuarina* and here nitrogenase is located in the hyphal tips (Berg and McDowell, 1987).

The discovery by Dilworth, Burris and co-workers in 1966 (see an historical account by Turner and Gibson, 1980; and volume 1 of this series, *Catalysts for Nitrogen Fixation: Nitrogenases, Relevant Chemical Models and Commercial Processes*) that nitrogenase reduces acetylene to ethylene was soon utilised as an assay for nitrogenase activity in field studies of nitrogen fixation by actinorhizal plants (Stewart *et al.,* 1967). The technique has been used frequently in both field and laboratory assays of nitrogenase activity since, but many problems arise in the quantitative extrapolation of acetylene-reduction data to nitrogen-fixation rates. These include an acetylene-induced decline in nitrogenase activity, inactivation of hydrogenase in some actinorhizal nodules, and effects of water stress on the acetylene-to-$^{15}N_2$ conversion ratio (Winship *et al.,* 1987; Schwintzer and Tjepkema, 1994; Johnson *et al.,* 1997). Although nitrogenase has not been purified from actinorhizal nodules, studies with nodule homogenates and cell-free extracts show that actinorhizal nitrogenase is similar to that of rhizobia (Benson *et al.,* 1979; Roelofsen and Akkermans, 1979). Further, molecular-genetic techniques show that *nifHDK,* the structural genes for the Fe and MoFe proteins, of the actinorhizal nitrogenase have extensive sequence similarity with those of other nitrogen-fixing organisms (Ruvkun and Ausubel, 1980) and, in several *Frankia* strains, are clustered on the chromosome as occurs in other bacteria (Simonet *et al.,* 1990).

3. TWO DECADES TO THE NEW MILLENNIUM

By 1978, a battery of new techniques was becoming available to study actinorhizal symbioses and, as indicated above, the advent and application of molecular techniques was providing new ways to tackle problems previously thought to be impossible to resolve. However, the number of scientists involved and the funding available was much less than for research on legumes, so that progress with actinorhizal plants has been slower. Although several actinorhizal genera contain species that are economically important in forestry and land regeneration and all are of ecological importance, only a few e.g. *Hippophae* berries, are used for food (Wheeler and Miller, 1990). Because of their slower growth and generation time, and their high phenolic content, woody plants in general are less amenable to physiological and molecular analysis than herbaceous plants. Nevertheless, micropropagation techniques offer a partial solution to these last problems and have been devised for species such as *Alnus* (Perinet and Lalonde, 1983; Tremblay and Lalonde, 1984); *Myrica* (Tavares *et al.,* 1998), *Hippophae* (Yao, 1995); *Datisca* (Wang and Berry, 1996), and casuarinas (Duhoux *et al.,* 1996).

The availability of such methodology has facilitated efforts to genetically transform casuarinas by the group of Emile Duhoux, Claudine Franche and Didier Bogusz. Following inoculation with *Frankia,* active nitrogen-fixing nodules were formed on a high proportion of transformed plants of *Allocasuarina verticillata,* which had been regenerated from calli induced on wounded embryos co-cultivated with *Agrobacterium tumefaciens*. Integration of the transgenes into the *Allocasuarina* genome did not interfere with the nodulation process (Franche *et al.,* 1999a; 1999b).

Although the actinomycetous micro-symbiont is slow-growing and more difficult to culture and use than the single-celled rhizobia, there have been striking advances in recent years in our knowledge of *Frankia* and its symbioses. In the next section, we have attempted to place into a historical context some of the most important research findings or "milestones" that have influenced the direction of actinorhizal research in the last quarter century.

3.1. Isolation, Culture and Taxonomy of Frankia

The repeated early failures to isolate the endophyte and to obtain proof of re-infectivity (see Baker and Torrey, 1979) led many to suggest that either an obligate association with the host or a synergistic interaction among one or more different microorganisms was necessary for endophyte growth and host-plant infection. As late as 1970, there were suggestions that the endophyte forms associations akin to that of an obligate parasite (Becking, 1970). Such a possibility was effectively eliminated when *Frankia* strain CpI1 was isolated in 1978 (Callaham *et al.,* 1978). The medium used was of a relatively complex composition and a range of media of differing composition have been used subsequently (Lechevalier and Lechevalier, 1990). Because of the relatively slow growth rate of the organism in culture, successful isolation from nodules required careful application of procedures both to surface-sterilise nodules and to remove as far as possible contaminating organisms and inhibitory plant compounds, such as phenolics (Lechevalier and Lechevalier, 1990). However, even today and despite numerous attempts since 1978, *Frankia* strains, which have been identified as a single clade by 16S-rDNA amplification from actinorhizal root nodules and infective only on rosaceous hosts, *Ceanothus* spp., *Datisca* and *Coriaria* spp. (Benson and Clawson, 2000), have still not been isolated successfully in pure culture.

In the mid–1970s, Torrey branched out into actinorhizal-plant research from his major and internationally distinguished contributions to root biology, root tissue culture, and legume-nodule physiology. Initially, his special interest was *Casuarina* because of its importance in the Tropics. He developed his interests further during a visit with Bond, which encouraged him to begin studies on the initiation and development of *Casuarina* root nodules (Torrey, 1976). It is remarkable that Torrey's group achieved the first reproducible isolation of the endophyte, the objective of many researchers previously, so soon after establishing his research program on actinorhizal plants. This important advance attracted many researchers to his laboratory and many of these scientists have gone on to distinguished careers and to make major contributions to the subject. John Torrey retired from active research in 1992, when he, along with Yvon Dommergues and Mary Lechevalier, were honoured at the International Conference on *Frankia* and Actinorhizal Plants, held in Lyon, France in September 1991 (Normand *et al.,* 1992). Sadly, John Torrey died only a few months later. The meeting on *Frankia* and actinorhizal plants, held in Ohakune, New Zealand in 1993, was dedicated to his memory and contains a tribute to his life and work (Baker and Berry, 1994).

Although actinorhizal-plant research had been included since 1976 in both the International Symposia on Nitrogen Fixation and the Symposia on Nitrogen Fixation with Non-Legumes, these symposia were largely focused on legume and on non-symbiotic fixation, respectively. The increased interest and activity in actinorhizal symbioses that was stimulated by the work of the Harvard group led to the organisation of a series of meetings devoted to *Frankia* and actinorhizal symbioses, which have helped to coordinate scientific activities (see Table 1). In addition, progress has been reviewed at specialist meetings, for example on *Casuarina* in Canberra (1981), Cairo (1990), and Da Nang, Vietnam (Pinyopusarerk

Table 1. Proceedings of International Conferences on Frankia *and Actinorhizal Plants.*

Number	Year	Place	Proceedings
1	1978	Petersham, MA, USA	Torrey and Tjepkema (1979)
2	1979	Corvallis, OR, USA	Gordon *et al.* (1979)
3	1982	Madison, WI, USA	Torrey and Tjepkema (1983)
4	1983	Wageningen, The Netherlands	Akkermans *et al.* (1984)
5	1984	Québec, Canada	Lalonde *et al.* (1985)
6	1986	Umeå, Sweden	Huss-Danell and Wheeler (1987)
7	1989	Storrs, CT, USA	Winship and Benson (1989)
8	1991	Lyon, France	Normand *et al.* (1992)
9	1993	Okahune, New Zealand	Harris and Silvester (1994)
10	1995	Davis, CA, USA	Berry and Myrold (1997)
11	1998	Champaign, IL, USA	Dawson *et al.* (1999)
12	2001	Carry-le-Rouet, France	Normand *et al.* (2003)

et al., 1996), and at many local meetings, where some presentations were published in journals such as *Acta Botanica Gallica* (Duhoux and Diem, 1996).

Biochemical and physiological techniques have continued to play an important role in unravelling the taxonomy of both *Frankia* and the host plant. Research in the Lalonde laboratory showed early on that *Frankia* is characterised by the presence of 2-*O*-methyl–D-mannose, a sugar not found in other actinomycetes (Mort *et al.,* 1983). Biochemical and serological study of different strains initially suggested a preliminary division of the genus into two groups: one group that nodulates *Alnus* and *Myrica,* and a second, more heterogeneous group (Baker *et al.,* 1981; Lechevalier, 1984). However, cross-inoculation experiments suggested that strains fell into at least three or four host-compatibility groups (Normand and Lalonde, 1986; Baker, 1987).

A more definitive analysis of *Frankia* taxonomy became available as molecular techniques for analysing *Frankia* DNA were developed. Analyses of DNA–DNA relatedness and of the DNA, which encodes the universal, slowly evolving and functionally conserved 16S rRNA sequence, have been particularly important. The groups of Normand and Fernandez and Dobritsa (Akimov and Dobritsa, 1992) played significant roles in the early application of these techniques. Fernandez *et al.* (1989) applied DNA-hybridisation techniques to 43 isolates of *Frankia* and differentiated nine genomic species, including three among strains compatible with *Alnus* species, five among strains compatible with Elaeagnaceae, and one among strains compatible with Casuarinaceae. Nazaret *et al.* (1991) determined phylogenetic relationships among eight of the genomic species by amplification and sequencing of 16S rDNA. They first showed that strains in the *Alnus* and *Casuarina* infectivity groups were closely related, but well separated from those in the *Elaeagnus* infectivity group, which also included atypical strains isolated from *Casuarina.* Three cohesive *Frankia* clades with distinct host-specificity ranges have now been defined by DNA analysis, as well as a fourth clade of non-nodulating, non-nitrogen-fixing *Frankia* relatives (Benson and Clawson, 2000).

Reverse-transcriptase sequencing of 16S rRNA led Hahn *et al.* (1989) to suggest a close phylogenetic relationship between *Frankia* and *Geodermatophilus*. Both organisms have multilocular sporangia. However, Maréchal *et al.* (2000) re-sequenced the *rrs* gene and the *recA* gene of *Acidothermus cellulolyticus* to show an even closer proximity of this actinomycete to *Frankia*, compared to the morphologically more similar *Geodermatophilus*. Further, both *Acidothermus* and *Frankia* contained high levels of hopanoid lipids, which had been found earlier to be abundant in *Frankia* cells and in nodules (Berry *et al.*, 1991).

3.2. Taxonomy and Evolution of the Host Plant and New Nodulating Genera

There are eight Angiosperm families known to be nodulated by *Frankia*. Until 1979, only seven families were commonly known to be actinorhizal hosts, when Chaudhary (1979) reported that *Datisca* (Datiscaceae) also forms *Frankia* symbioses. Interestingly, *Datisca* was first described as a nodulated plant by Severini (1922), but this report had gone relatively unnoticed until Chaudhary's rediscovery. Three new genera of the Casuarinaceae were defined by dividing the former genus *Casuarina* into *Casuarina, Allocasuarina, Gymnostoma* and *Ceuthostoma* (Johnson, 1980; 1982; 1988). Nodulation of species in the first three of these genera has been observed regularly.

New nodulated genera in the Rhamnaceae (*Colletia, Trevoa, Talguenea and Retanilla),* which are native to South America, have also been discovered and *Frankia* strains from these shrubs characterised (Caru, 1993), whereas nodulation of *Rubus* has now been discounted (Stowers, 1985). Nodulation of genera in new families, e.g. *Atriplex* of the Chenopodiaceae (Caucas and Abril, 1996), always requires careful, independent confirmation. "Nodules", often called tubercles in older literature, which are produced by mycorrhizal or other forms of microbial infection, may easily be confused with *Frankia* nodulation. Arbuscular mycorrhizal nodules, which like actinorhizas are modified lateral roots, have been reported recently for *Gymnostoma* (Duhoux *et al.*, 2001).

Classical taxonomy, based on morphology and floristics, suggested that the distribution of actinorhizal plants through eight families was characterised by taxonomic unrelatedness (Bond, 1983), thus introducing the possibility that actinorhizal and legume-nodule symbioses have arisen in taxonomically unrelated plant groups. This concept has been effectively eliminated by the findings of Swensen and Mullin, who used molecular techniques rather than morphological criteria to study the phylogeny of actinorhizal plants. Working with collaborators in the USA and Australia, they used chloroplast-gene sequence data (*rbcL*) to show that representatives of all eight actinorhizal-plant families, together with representatives of the rhizobia-nodulated families, occurred in a single "nitrogen-fixing clade", interspersed with non-nodulating genera (Soltis *et al.*, 1995). Additional molecular data, together with phylogenetic trees constructed from *Frankia nifH*-gene and 16S-rDNA sequences, have given insights into the co-evolution of actinorhizal symbioses and suggests that actinorhizal symbioses originated three times after the divergence of the large plant clade (Swensen, 1996;

Swensen and Mullin, 1997a; 1997b; Jeong *et al.,* 1999, Benson and Clawson, 2000). Here, as in virtually all areas of biology, the "molecular revolution" has transformed our ability to answer questions that previously either could not be tackled or could be studied only with the expenditure of much time and effort. The impact on the ecology of actinorhizal plants is considered below.

3.3. Infection and Nodule Development

The application of electron microscopy facilitated further detailed observations of infection pathways and nodule development (Berry and Sunell, 1990). One of the most important developments was the recognition of two different infection pathways used by *Frankia* hyphae. The "traditional" pathway, which occurs in genera such as *Alnus*, *Myrica* and *Casuarina*, involved penetration of deformed root hairs, followed by intracellular penetration of cells of the root cortex. The "alternative" pathway, which was first recognised in *Elaeagnus* (Miller and Baker, 1985), involved epidermal penetration, followed by intercellular colonisation of the cortex, before intracellular penetration of mature cortical cells and ultimately the host cells of the developing nodule. Furthermore, whereas hyphae in intracellular infections are encapsulated in a host-derived matrix of polysaccharides, cellulose, hemicellulose and pectin (Berg, 1990), during intercellular colonisation, the hyphae are not encapsulated until they penetrate the host cells.

The molecular signals that initiate infection remain unknown. No convincing evidence of either *nod* genes or Nod-factor homologs has been demonstrated in *Frankia* (Cérémonie *et al.,* 1998a; 1998b), although a root hair-deforming factor (or Had factor) is produced constitutively by some *Frankia* strains (van Ghelue *et al.,* 1997; Cérémonie *et al.,* 1998b). Because of their role as signal molecules in legume symbioses, flavonoids excreted by the host plant have been examined, but clear evidence for their involvement in nodulation has not been obtained (Benoit and Berry, 1997; Hughes *et al.,* 1999).

The possibility that plant hormones may regulate nodule development has led to the observation of elevated levels of cytokinins, auxins and gibberellins in nodules (Wheeler *et al.,* 1979) and changes in abscisic acid and polyamines in relation to dormancy and the supply of mineral nitrogen, respectively, have been reported (Watts *et al.,* 1987; Wheeler *et al.,* 1994). Auxins and cytokinins are known to be secreted by *Frankia* both *in vitro* (Wheeler *et al.,* 1984; Stevens and Berry, 1988; Berry *et al.,* 1989) and in nodules (Hammad *et al.,* 2003), but direct evidence of their involvement in nodule development is lacking. It has been suggested that the failure of most transgenic plants of *Allocasuarina verticillata* to nodulate could be due to effects of the auxin genes of the transforming organism, *Agrobacterium rhizogenes,* on host-plant hormone balance (Franche *et al.,* 1999a). The availability of genetic transformation systems will undoubtedly facilitate resolution of the role of hormones in the nodulation process and the nature of the molecular signalling systems that must regulate interactions between *Frankia* and the host plant.

Progress is being made in identifying and determining the expression of nodule-specific genes – actinorhizal nodulin genes – and their gene products. Séguin and

Lalonde (1993) used two-dimensional polyacrylamide gel electrophoresis to detect several nodule-specific polypeptides in developing actinorhizal nodules. This research area has been developed further, particularly by Katharina Pawlowski, together with Ton Bisseling and Antoon Akkermans, and also independently by Beth Mullin, Didier Bogusz and Claudine Franche, Alison Berry, Ann Hirsch, and Chung Sun An. Screens of cDNA libraries from nodules of *Alnus glutinosa* and *Datisca glomerata* revealed a number of genes that are expressed during early nodulation and the products of some are known, *e.g.*, *ag12* encodes a subtilisin-type protease (Ribeiro *et al.,* 1995), whereas *Dg93* shares sequence homology with an early nodulin gene from legumes (Okubara *et al.,* 2000). After expression *in vitro*, proteins, which are encoded by two nodule-specific cDNAs isolated from *A. glutinosa* nodules, have been purified and characterised (Gupta *et al.,* 2002). These proteins represent a new class of plant metal-binding proteins, which have potential use in bioremediation. Cell-specific expression of chitinase genes in nodules of *Elaeagnus umbellate* has also been described (Kim and An, 2002). Urgent challenges for future research include matching gene structure and protein function with the processes involved in nodule developmental and metabolism.

3.4. Life with Oxygen (O_2)

Nitrogen fixation by *Frankia* in both the free-living and symbiotic state is supported by aerobic metabolism and is maximal at about atmospheric O_2 partial pressures, so special mechanisms must be in place to protect nitrogenase from inactivation by O_2. Early identification of vesicles as the probable site of nitrogen fixation in cultured *Frankia* was confirmed by immunogold labelling of nitrogenase (Tjepkema *et al.,* 1981; Meesters *et al.,* 1987). The individual research programs of Tjepkema, Silvester, and Huss-Danell and Berry have been instrumental in resolving the complexity and diversity of protective mechanisms that operate in free-living and symbiotic *Frankia*. Actinorhizal nodules are well aerated with large numbers of air spaces and do not have the diffusion-resistant "nodule endodermis" that restricts O_2 diffusion in legume nodules. In the cultured organism, the layered walls of the vesicles, which contain large amounts of hopanoid lipids (Berry *et al.,* 1993), vary in thickness in response to changes in pO_2, thus regulating O_2 diffusion (Parsons *et al.,* 1987). Similarly, at least in the *Alnus* symbiosis, where nitrogenase is located in vesicles, increased O_2 concentration results in both an increase in vesicle envelope thickness and changes in the relative proportions of hopanoid lipids present (Silvester *et al.,* 1988; Huss-Danell, 1990; Kleemann *et al.,* 1994).

The production of vesicles in actinorhizal nodules of different species is highly variable. At one end of the spectrum, vesicles are not produced in *Casuarina* nodules, most probably because they are only ever exposed to low pO_2. Experimental proof (Tjepkema, 1979) of the earlier observation of O_2-transporting hemoglobins in *Casuarina* nodules (Davenport, 1960) provided the basis of this explanation because low pO_2 in the infected cell areas is consistent with the presence of a functional O_2 transporter. Later, Murry *et al.* (1985) showed that, under conditions of very low pO_2, vesicles do not form in cultured *Frankia* but the hyphae still develop nitrogenase activity. In contrast, nodules of *Alnus* and some other genera form well-defined vesicles, but contain only a low haemoglobin concentration (Suharjo and Tjepkema, 1995).

For *Casuarina glauca*, hemoglobin is synthesised early in young infected nodule cells to prepare the environment for functional nitrogenase. Hemoglobin cDNA has been cloned and induction of the hemoglobin gene prior to the detection of *Frankia nifH* mRNA has been demonstrated by *in situ* hybridisation (Gherbi *et al.,* 1997). Interestingly, cultured *Frankia* produces haemoglobin, raising important questions concerning the source and regulation of hemoglobins *in vivo* (Beckwith *et al.,* 2002). An additional method of O_2 protection was found in nodules of *Coriaria,* which has thin vesicles that radiate inwards towards a central cell vacuole. Here, mitochondria are distributed around the vesicles and also around the intercellular spaces that penetrate the nodule, indicating that pO_2 could be kept low by respiratory scavenging of O_2 (Silvester *et al.*, 1999). It should be noted that the shape of the *Frankia* vesicle *in vivo* is host-plant dependent (Tjepkema *et al.,* 1986).

Nitrogenase in actinorhizal nodules seems also to be protected by "oxygen transients" in which nitrogenase switches off rapidly in response to an increase in pO_2, possibly by conformational protection, and then recovers when O_2 levels return to normal (Silvester and Winship, 1990). Free radical scavenging systems, such as superoxide dismutase, have also been detected in cultured and symbiotic *Frankia* and may complement the battery of defenses that help prevent O_2 inactivation (Steele and Stowers, 1986; Puppo *et al.,* 1989; Alskog and Huss-Danell, 1997). Interestingly, expression of the *Frankia* gene for Fe superoxide dismutase, *sodF,* is induced by host-root exudates (Hammad *et al.,* 2001), suggesting it plays an additional earlier role in the symbiosis (Maréchal *et al.,* 2003).

3.5. Metabolism, Nitrogen Cycling, and the Regulation of Metabolism

Frankia in most actinorhizal nodules has a particularly active H_2-uptake system, catalysed by uptake hydrogenase (Hup), which metabolises H_2 evolved by nitrogenase during nitrogen fixation (Schubert and Evans, 1976; Roelofsen and Akkermans, 1979). Respiratory activity eventually leads to donates of the electrons from H_2 oxidation to O_2 and may thus contribute to both energy conservation through ATP generation and help prevent O_2 inactivation of nitrogenase. Support for these suggestions came from studies of *Alnus incana* in symbiosis with a Hup(–) *Frankia* strain, which showed lower nitrogen fixation than plants inoculated with Hup(+) *Frankia* (Sellstedt *et al.,* 1986). Further, nitrogenase activity of *Frankia* increased when cultured in a gas mix with elevated pO_2 and pH_2 (Murry and Lopez, 1989). Immunological studies by Anita Sellstedt have shown that the hydrogenases of *Frankia* are located primarily in vesicles and to a lesser extent in hyphae (Lindblad and Sellstedt, 1989; Sellstedt and Lindblad, 1990) and are similar to membrane-bound [NiFe] hydrogenases (Mattson *et al.,* 2001).

The composition of different media used to culture *Frankia* shows that the organism can use a wide range of carbon substrates *in vitro*, such as amino acids, pyruvate, propionate, and glucose. However, the nature of the carbon substrate that is transported into symbiotic *Frankia* to support endophyte growth and nitrogen fixation is still unknown despite comprehensive physiological studies by Kerstin Huss-Danell's group, who developed assay techniques for symbiotic *Frankia* cells (Lundquist and Huss-Danell 1992) that continue to be used today.

The likely first step in the assimilation of the ammonia produced by nitrogenase activity is the formation of glutamine. Extensive biochemical studies by David Benson's laboratory showed that this is true for cultured *Frankia,* which possesses two glutamine synthetases. GS-I is present during growth on either NH_4^+ or N_2 and is similar to the classical bacterial glutamine synthetase, being regulated by adenylation. GS-II is derepressed when cultures are starved of combined nitrogen and accounts for most of the glutamine-synthtase activity in such cultures (Edmands *et al.,* 1987). However, glutamine synthetase is thought not to be present in *Frankia* in symbiosis (Lundquist and Huss-Danell, 1992). The NH_4^+ from fixation is exported to the plant-cell cytosol for assimilation into an organic form and exported to the host plant as either citrulline (alders and casuarinas) or amides (most other actinorhizal genera). The cells of the nodule pericycle of *Alnus* show high levels of expression of genes that code for enzymes, such as sucrose synthase and glutamine synthase, and of several other nodulin genes, different from those of the root pericycle. These observations suggest that the *Alnus* nodule pericycle may play a special role in the exchange of metabolites between the stele and the nodule cortex (Pawlowski and Bisseling, 1996).

Photosynthesis is the main source of carbon and provides translocated sucrose to drive nodule metabolism, with lesser amounts coming from CO_2 fixation through the action of phosphenol pyruvate carboxylase and, in alders, ornithine carbamyl phosphate synthase (McClure *et al.,* 1983). It was thought originally that either the supply of carbohydrates or nitrite inhibition following uptake of nitrate were primary regulators of nitrogenase activity. However, Parsons, Raven, Sprent and co-workers proposed that the concentration of nitrogen-containing compounds (amino acids, amides, ureides) in phloem sap inversely regulates the rates of both nitrogen fixation and nodule growth (Parsons *et al.,* 1993). This mechanism has gained favour as the primary metabolic signal that regulates nodule growth and activity in actinorhizal nodules (Baker *et al.,* 1997a; 1997b; Parsons and Sunley, 2001) and further work is in progress to identify the sensor mechanisms that detect fluctuations in supply of combined nitrogen to nodules.

3.6. *Ecology and Applications*

The utilisation of actinorhizal plants in land reclamation and forestry continues to be researched (Dawson, 1983; 1986; Dommergues, 1997; Gordon and Wheeler, 1983; Schwenke and Caru, 2001; Tjepkema and Schwintzer, 1990). Traditional techniques of ecological physiology were employed in the Himalayas to determine the contributions of *Alnus nepalensis* to nutrient cycling and primary production in agroforestry, both in different aged plantations and in naturally regenerated landslip sites (Sharma *et al.,* 1998). These studies showed clearly how uptake of recycled mineral nitrogen replaced the high-energy processes of both nitrogen fixation and nodule production as soil-nitrogen concentration increased with stand age (Sharma and Ambasht, 1988).

New molecular techniques have facilitated investigation of questions such as the persistence and competitiveness of introduced *Frankia* strains in managed environments and provided approaches to determine the contributions of indigenous and introduced actinorhizal plants to the nitrogen economy of particular ecosystems. The groups of Antoon Akkermans, Philippe Normand, Dittmar Hahn, David Myrold, and David Benson have been particularly successful in developing molecular techniques to study the ecology of *Frankia* populations. These techniques include both utilising probes that will hybridise specifically with *Frankia* 16S or 23S rRNA (Hahn *et al.,* 1990; Akkermans *et al.,* 1994; Hönerlage *et al.,* 1994) and sequence analysis of PCR-amplified ribosomal DNA, *nif* genes or intergenic spacers (Hahn *et al.,* 1999; see chapter by Hahn in this volume). These last authors note that, whereas studies to date have focused on the analysis of *Frankia* populations in nodules and to a lesser extent in soil and the rhizosphere, the establishment of more sophisticated methods should allow detailed studies of the environmental dynamics of *Frankia* populations.

3.6.1. Sporulation in Frankia

The capacity of molecular techniques to provide unequivocal answers to previously insoluble questions is well illustrated by research on the sporulation of symbiotic *Frankia* in natural plant communities. In the mid-seventies, microscopic studies of the endophyte of *Alnus glutinosa* nodules led van Dijk and Merkus (1976) to propose that the term "spore" should be used to replace "bacteroids", which was in common use to describe the "granulated bodies" seen in some infected cells. Further analysis of the root nodules of different *Alnus glutinosa* populations showed two types of nodules, one containing spores and the other from which spores were absent (van Dijk, 1978). The distribution of these two types showed considerable clustering and, in due course, led to the terminology "spore(+)" and "spore(–)" to describe the two nodule types. These nodules types are not confined to alders and, in general, the spore(–) nodule type seems to be less infective but more effective than spore(+) nodules (VandenBosch and Torrey, 1984; Holman and Schwintzer, 1987). Cross-inoculation experiments suggested that the ability to form sporangia in the symbiotic state is controlled by the endophyte (van Dijk, 1978; Vanden Bosch and Torrey, 1985). Proof of this was finally obtained by molecular characterisation through PCR amplification and sequencing of 16S-rRNA sequences, which showed specific differences in the DNA of endophytic *Frankia* from spore(+) and spore(–) nodules (Simonet *et al.,* 1994).

3.6.2. Irregular Nodulation

As mentioned earlier, there are many reports of irregular nodulation among different host species. Most frequent causes are growth in either unfavourable environmental/soil conditions or the occurrence in soil of non-infective, poorly effective or ineffective strains of *Frankia.* A study of alders growing in a wet soil in the Netherlands showed that these strains can form a significant proportion of the total soil population (van Dijk and Sluimer-Stolk, 1990). A comparison with effective nodules showed that ineffective nodules contain higher amounts of polyphenols. Hybridisation with a *Frankia* antisense 16S-rRNA probe showed that *Frankia* is degraded at an early stage in the development of infected cells in the ineffective nodules (Mirza *et al.,* 1991; Guan *et al.,* 1996; Wolters *et al.,* 1997). The infective, but ineffective, strains are of particular ecological importance because they may either compete for nodulation with effective strains in the soil (van Dijk and Sluimer-Stolk, 1990) or provide complementary signalling molecules during nodulation (see chapter by Wall and Berry in this volume).

3.6.3. Responses to Environmental Changes or Other Microorganisms

An area of continuing interest involves the responses of the host plant to different environments and to the presence of other microorganisms. For example, the formation of cluster, or proteoid, roots of *Casuarina* spp. has been shown to alleviate poor mineral nutrition, particularly of phosphorus or iron (Diem and Arahou, 1996), even though *Frankia* itself excretes siderophores, particularly under conditions of iron deficiency – however, it is unclear how important this process is in soil (Aronson and Boyer, 1993; Boyer *et al.,* 1997).

Mycorrhizal associations are important, particularly for phosphorus nutrition. The most detailed studies of these associations have concerned the economically important alders and casuarinas. A historical list of observations on the occurrence of ecto- and arbuscular mycorrhizal associations in actinorhizal nodules is given by Daft *et al.* (1985). Arbuscular mycorrhizal fungi are generally considered to be of particular importance in lighter, sandy soil, which are often more typical of warmer climates. Ectomycorrhizas assume importance on heavier soils (Gardner, 1986). The occurrence and positive effects of these associations on actinorhizal plant growth are well documented (Rose, 1980; Molina, 1981; Berliner and Torrey, 1988; Diem and Dommergues, 1990; Sempavalan *et al.,* 1995; Diem, 1996). The highest growth of *Alnus incana* seedlings was achieved when inoculated with three symbionts, namely *Frankia*, an ecto- and an arbuscular mycorrhizal fungus (Chatarpaul *et al.,* 1989).

The range of mutualistic organisms that form beneficial associations with actinorhizal plants presents a daunting task for those who wish to develop and apply inoculation practices to improve the growth of actinorhizal plants in the field. In addition to *Frankia* and mycorrhizal organisms, positive effects on host-plant growth of inoculation with some rhizobacteria have been reported (Probanza *et al.,* 1996; Garcia *et al.,* 2003). Although the availability of molecular techniques has rendered monitoring the persistence of introduced organisms an easier task, enormous problems remain in interpreting field experiments because of potential interactive effects of indigenous microbial populations. This is particularly problematical in areas of high rainfall, where water movement may carry contaminants across barriers and wide boundary areas between plots into the experimental areas. For example, experiments with *Alnus rubra*, inoculated as seedlings with specific *Frankia* strains and planted out on mine spoil in a high rain fall area of Scotland, showed that massive movement of *Frankia* occurred across

carefully constructed barriers. Furthermore, the strains used for seedling inoculation were shown to be poorly competitive with other *Frankia* invading the plantation area (McEwan *et al.,* 1996).

Despite practical experimental problems, there is no doubt that nodulation and mycorrhization will improve host-plant growth in seed beds and, later, in nutrient-poor soils (lsopi *et al.,* 1994; Lumini *et al.,* 1994). Further examples are given in reviews concerning applications (Schwintzer and Tjepkema, 1990). Mycorrhizal inocula are available commercially but, although large-scale production of *Frankia* has been attempted (Perinet *et al.,* 1985; Diem and Dommergues, 1990), it appears not to be financially viable at present to offer *Frankia* inocula for purchase.

3.6.4. Measurements of Nitrogen-Fixation Rates in the Field

Quispel (1990) listed a number of early attempts to determine rates of nitrogen fixation in the field using nitrogen difference, fixation of N_2 gas, or ^{15}N isotope-dilution methodology. These methods have been compared for assessment of nitrogen fixation by *Casuarina* (Gauthier *et al.,* 1985), but they are laborious and time consuming, which has encouraged researchers both to take advantage of mass spectrometers with increased sensitivity and to utilise assays of ^{15}N natural abundance ($\delta^{15}N$) to detect and to measure nitrogen fixation (Shearer and Kohl, 1993). Application of this last technique in field situations is subject to an array of problems (Handley and Raven, 1992; Hogberg, 1997), but it is particularly attractive for application to nitrogen-fixing tree systems, with their attendant problems of tree size, difficulties in assessing nodulation status, and the requirement for integrating changing rates of nitrogen fixation over long periods of time. Even so, the data obtained must be viewed with caution.

Mariotti *et al.* (1992) found good correspondence between estimates of nitrogen fixation obtained using $\delta^{15}N$ assays and the total nitrogen difference method for inoculated and uninoculated three-year-old *Casuarina equisetifolia* trees planted on sand-dune soil in Senegal. The $\delta^{15}N$ technique was used subsequently by Anne-Marie Domenach and coworkers to provide much useful ecological data, *e.g.*, to estimate nitrogen fixation in an alder forest (Domenach *et al.,* 1989). It continues to be of value in defined circumstances. Thus, Kohls *et al.* (1994) concluded that *Dryas octopetala* and *D. integrifolia*, which were colonising the foreland of a glacier in Canada, did not fix N_2 because the mean $\delta^{15}N$ values were similar to those of non-nitrogen-fixing vegetation; in contrast, *D. drummondii* did fix N_2 but not until the mid-to-late stages of the sere. Hurd *et al.* (2001) found that foliar $\delta^{15}N$ of *Alnus incana* ssp. *Rugosa*, which was growing in wetlands in the Adirondack Mountains, did not differ from that of alders grown in fixed-nitrogen-free water culture and estimated that 85–100 % of their nitrogen was obtained from fixation.

4. PERSPECTIVES

Examination of the many publications on the actinorhizal symbiosis show clearly how far we have moved; from an era where a very few scientists carried out their work independently or with a few students to the current position, where successful pursuit of experiments involving specialized biochemical and molecular techniques requires collaboration between among researchers with different expertise across several laboratories, often in different countries. We are fortunate that this is a field which, although competitive, has always thrived on collaboration and which has as its hallmark the encouragement of a friendly and co-operative spirit between its disciples.

> *"Where runs the river? Who can say*
> *Who hath not followed all the way*
> *By alders green and sedges grey?"*
> *André Marty, 1931.*

REFERENCES

Akimov, V. N., and Dobritsa, S. V. (1992). Grouping of *Frankia* strains on the basis of DNA relatedness. *Syst. Appl. Bacteriol., 15,* 372-379.

Akkermans, A. D. L. (1971). *Nitrogen fixation and nodulation of Alnus and Hippophaë under natural conditions.* Ph.D. thesis, University of Leiden, The Netherlands, 85 pp. Dordrecht, The Netherlands: Publisher Dordtse Sociale Werkplaats.

Akkermans, A. D. L., and van Dijk, C. (1981). Non-leguminous root-nodule symbioses with actinomycetes and *Rhizobium.* In W. J. Broughton (Ed.), *Nitrogen fixation. Vol. 1. Ecology* (pp. 57-103). Oxford, UK: Clarendon Press.

Akkermans, A. D. L., Mirza, M. S., Harmsen, H J. M., Blok, H. J. Sessitsch, A., and Akkermans, W. M. (1994). Molecular ecology of microbes: A review of promises, pitfalls and true progress. *FEMS Microbiol. Rev., 15,* 185-194.

Akkermans, A. D. L., Baker, D., Huss-Danell, K., and Tjepkema, J. D. (1984). *Frankia* symbioses. Proceedings of Workshop. *Plant Soil, 78,* 1-2.

Akkermans, A. D. L., van Dijk, C., and Oremus, P. A. I. (1983). *Frankia*: Tool and toy of the microbiologist. Bio-essays to Dr. A. Quispel. University of Leiden, 5-10.

Allen, E. K., and Allen, O. N. (1965). Nonleguminous plant symbioses. In C. M. Gilmour, and O. N. Allen (Eds.), *Microbiology and soil fertility* (pp. 77-106). Corvallis, OR: Oregon State University Press.

Alskog, G., and Huss-Danell, K. (1997). Superoxide dismutase, catalase and nitrogenase activities of symbiotic *Frankia* (*Alnus incana*) in response to different oxygen tensions. *Physiol. Plant.*, 99, 286-292.

Aronson, D. B., and Boyer, G. L. (1993). *Frankia* produces a hydroxamate siderophores under iron limitation. *J. Plant Nutr., 16,* 2193-2201.

Baker, A., Hill, G. F., and Parsons, R. (1997a). Alteration of N nutrition in *Myrica gale* induces changes in nodule growth, nodule activity and amino acid composition. *Physiol. Plant., 99,* 632-639.

Baker, A., Hill, G. F., and Parsons, R. (1997b). Evidence for N feedback regulation of N_2 fixation in *Alnus glutinosa. J. Exp. Bot., 48,* 67-74.

Baker, D. D. (1987). Relationships among pure cultured strains of *Frankia* based on host specificity. *Physiol. Plant., 70,* 245-248.

Baker, D. D., and Berry, A. (1994). A tribute to John G. Torrey 1921-1993. *Soil Biol. Biochem., 26,* vii-viii.

Baker, D. D., and Torrey, J. G. (1979). The isolation and cultivation of actinomycetous root nodule endophytes. In J. C. Gordon, D. A. Perry, and C. T. Wheeler (Eds.), *Symbiotic nitrogen fixation in the management of temperate forests* (pp. 38-56). Corvallis, OR: Oregon State University Press.

Baker, D. D., Pengelly, W. L., and Torrey, J. G. (1981). Immunological analysis of relationships among the isolated frankiae (Actinomycetales). *Int. J. Syst. Bacteriol., 31,* 148-151.

Becking, J. H. (1970). *Frankiaceae* fam. Nov. (Actinomycetales) with one new combination and six new species of the genus *Frankia* Brunchorst 1886. *Int. J. System. Bacteriol., 20,* 201-220.

Beckwith, J., Tjepkema, J. D., Cashon, R. E., Schwintzer, C. R., and Tisa, L. S. (2002). Hemoglobin in five genetically diverse *Frankia* strains. *Can. J. Microbiol., 48,* 1048-1055.

Benoit, L. F., and Berry, A. M. (1997). Flavonoid-like compounds from seeds of red alder (*Alnus rubra*) influence host nodulation by *Frankia* (Actinomycetales). *Physiol. Plant., 99,* 588-593

Benson, D. R., Arp. D. J., and Burris, R. H. (1980). Cell free nitrogenase and hydrogenase from actinorhizal root nodules. *Science, 205,* 688-689.

Benson, D. R., and Clawson, M. L. (2000). Evolution of the actinorhizal plant symbioses. In E. W. Triplett (Ed.), *Prokaryotic nitrogen fixation: A model system for analysis of a biological process* (pp. 207-224). Wymondham, UK: Horizon Scientific Press.

Benson, D. R., and Silvester, W. B. (1993). Biology of *Frankia* strains, actinomycete symbionts of actinorhizal plants. *Microbiol. Rev., 57,* 293-319.

Berg, R. H. (1990). Cellulose and xylans in the interface capsule in symbiotic cells of actinorhizae. *Protoplasma, 159,* 35-43.

Berg, R. H., and McDowell, L. (1987). Endophyte differentiation in *Casuarina* actinorhizae. *Protoplasma, 136,* 104-117.

Berliner, R., and Torrey, J. G. (1988). On tripartite *Frankia*-mycorrhizal associations in the Myricaceae. *Can. J. Bot., 67,* 1708-1712.

Berry, A. M. (1994). Recent developments in the actinorhizal symbiosis. *Plant Soil, 161,* 135-145.

Berry A. M., and Myrold, D. D. (Eds.) (1997). Proceedings of the 10th international conference on *Frankia* and actinorhizal plants. *Physiol. Plant., 99,* 564-731.

Berry, A. M., and Sunell, L. A. (1990). The infection process and nodule development. In C. R. Schwintzer, and J. D. Tjepkema (Eds.), *The biology of* Frankia *and actinorhizal plants* (pp. 61-82). San Diego, CA: Academic Press.

Berry, A. M., Kahn, R. K. S., and Booth, M. C. (1989). Identification of indole compounds secreted by *Frankia* HFPArI3 in defined culture conditions. *Plant Soil, 118,* 205-209.

Berry, A. M, Moreau, R. A., and Jones, A. D. (1991). Bacteriohopanetetrol: Abundant lipid in *Frankia* cells and in nitrogen-fixing nodule tissue. *Plant Physiol., 95,* 111-115.

Berry, A. M., Harriot, O. T., Moreau, R. A., Osman, S. F., Benson, D. R., and Jones, A. D. (1993). Hopanoid lipids compose the *Frankia* vesicle envelope, presumptive barrier of oxygen diffusion to nitrogenase. *Proc. Nat. Acad. Sci. U.S.A., 90,* 6091-6094.

Bond, G. (1955). An isotopic study of the fixation of nitrogen associated with nodulated plants of *Alnus, Myrica* and *Hippophae*. *J. Exp. Bot., 6,* 303-311.

Bond, G. (1956). Evidence for fixation of nitrogen by root nodules of alder (*Alnus*) under field conditions. *New Phytol., 55,* 147-153.

Bond, G. (1973) Root nodule symbioses with actinomycete-like organisms. In A. Quispel (Ed.), *The biology of nitrogen fixation* (pp. 342-378). Amsterdam, The Netherlands: North Holland Publishing Company.

Bond, G. (1976). The results of the IBP survey of root-nodule formation in non-leguminous angiosperms. In P.S. Nutman (Ed.), *Symbiotic nitrogen fixation in plants* (pp. 443-474). Cambridge, UK: Cambridge University Press.

Bond, G. (1983). Taxonomy and distribution of non-legume nitrogen-fixing systems. In J.C. Gordon and C. T. Wheeler (Eds.), *Biological nitrogen fixation in forest ecosystems: Foundations and applications* (pp. 55-88). The Hague, The Netherlands: Martinus Nijhoff/Dr. W. Junk Publishers.

Bormann B. T., and Sidle, R. C. (1990). Changes in productivity and distrubtion of nutrients in a chronosequence at Glacier Bay National Park, Alaska. *J. Ecol., 78,* 561-578.

Boyer, G. L., Kane, S. A., Alexander, J. A., and Aronson, D. B. (1999). Siderophore formation in iron-limited cultures of *Frankia* sp. strain 52065 and *Frankia* sp. strain CeSI5. *Can. J. Bot., 77,* 1316-1320.

Brunchorst, J. (1886-1888). Über einige wurzelanschwellungen, besonders diejenigen von *Alnus* und den Elaeagnaceen. *Untersuch. Bot. Instit. Tübingen, 2,* 150-177.

Callaham, D., Del Tredici, P., and Torrey, J. G. (1978). Isolation and cultivation *in vitro* of the actinomycetes causing root nodulation in *Comptonia. Science*, 199, 899-902.

Caucas, V., and Abril, A. (1996). *Frankia* sp. infects *Atriplex cordobensis*. Cross-inoculation assays and symbiotic efficiency. *Phyton, 59,* 103-110.

Cérémonie, H., Cournoyer, B., Maillet, F., Normand, P., and Fernandez, M. P. (1998). Genetic complementation of rhizobial *nod* mutants with *Frankia* DNA: Artifact or reality? *Mol. Gen. Genet., 260,* 115-119.

Cérémonie, H., Debelle, F., and Fernandez, M. P. (1999). Structural and functional comparison of *Frankia* root hair deforming factor and rhizobial Nod factor. *Can. J. Bot., 77,* 1293-1301.
Chapin, F. S. I., Walker, L. R., Fastie, C. L., and Sharman, L. C. (1994). Mechanisms of primary succession following deglaciation at Glacier Bay, Alaska. *Ecol. Monogr., 64,* 149-175.
Chatarpaul, L., Chakravarty, P., and Subramaniam, P. (1989). Studies in tetrapartite symbioses. I. Role of ecto- and endomycorrhizal fungi and *Frankia* on the growth performance of *Alnus incana. Plant Soil, 118,* 145-150.
Chaudhary, A. H. (1979). Nitrogen-fixing root nodules in *Datisca cannabina* L. *Plant Soil, 51,* 163-165.
Crocker, R. L., and Major, J. (1955). Soil development in relation to vegetation and surface age at Glacier Bay, Alaska. *J. Ecol., 43,* 427-448.
Daft, M. J., Clelland, D. M., and Gardner, I. C. (1985). Symbiosis with endomycorrhizas and nitrogen-fixing organisms. *Roy. Soc. Edinburgh B., 85,* 283-298.
Davenport, H. E. (1960). Haemoglobin in the root nodules of *Casuarina cunninghamiana. Nature, 186,* 653-654.
Dawson, J. O. (1983). Dinitrogen fixation in forest ecosystems. *Can. J. Microbiol., 29,* 979-992.
Dawson, J. O. (1986). Actinorhizal plants – their use in forestry and agriculture. *Outlook Agric., 15,* 202-298.
Dawson, J. O. (1990). Interactions among actinorhizal and associated plant species. In C. R. Schwintzer, and J. D. Tjepkema (Eds.), *The biology of* Frankia *and actinorhizal plants* (pp. 299-316). San Diego, CA: Academic Press.
Dawson, J. O, Berg, R. H., Paschke, M. W., and Wheeler, C. T. (1999). The 11th international conference on *Frankia* and actinorhizal plants at Champaign. *Can. J. Bot., 77,* 1203-1400.
Diem, H. G. (1996). Les mycorrhizes des plantes actinorhiziennes. *Acta Bot. Gall., 143,* 581-592.
Diem, H. G., and Arahou, M. (1996). A review of cluster root formation: A primary strategy of Casuarinaceae to overcome soil nutrient deficiency. In K. Pinyopusarek, J. W. Turnbull, and S. J. Midgley (Eds.), *Recent Casuarina research and development, Proceedings of the 3rd international* Casuarina *workshop, Da Nang* (pp. 51-58). Canberra, Australia: CSIRO.
Diem, H. G., and Dommergues, Y. R. (1990). Current and potential uses and management of Casuarinaceae in the Tropics. In C. R. Schwintzer, and J. D. Tjepkema (Eds.), *The biology of* Frankia *and actinorhizal plants* (pp. 317-342). San Diego, CA: Academic Press.
Diem, H. G., Duhoux, E., and Arahou, M. (2000). Cluster roots in Casuarinaceae: Role and relationship to soil nutrient factors. *Ann. Bot., 85,* 929-936.
Domenach, A. M., Kurdali, F., and Bardin, R. (1989). Estimation of symbiotic dinitrogen fixation in alder forest by the method based on natural ^{15}N abundance. *Plant Soil, 118,* 51-59.
Dommergues, Y. R. (1997). Contribution of actinorhizal plants to tropical soil productivity and rehabilitation. *Soil Biol. Biochem., 29,* 931-941.
Duhoux, E., and Diem, H. G. (Eds.) (1996). *Les plantes actinorhiziennes. Acta Bot. Gall., 143,* 563-679.
Duhoux, E., Franche, C., Bogusz, D., Diouf, D., Le, V.Q., Gherbi, H., Sougoufara, B., Le Roux, C., and Dommergues, Y. R. (1996). *Casuarina* and *Allocasuarina* species. In Y. P. S Bajaj (Ed.), *Trees IV.* (pp. 76-94). Berlin: Springer-Verlag.
Duhoux, E., Rinaudo, G., Diem, H. G., Auguy, F., Fernandez, D., Bogusz, D., Franche, C., Dommergues, Y., and Huguenin, B. (2001). Angiosperm *Gymnostoma* trees produce root nodules colonized by arbuscular mycorrhizal fungi related to *Glomus. New Phytol., 149,* 115-125.
Edmands, J., Noridge, N. A., and Benson, D. R. (1987). The actinorhizal root-nodule symbiont *Frankia* sp. CpI1 has two glutamine synthetases. *Proc. Natl. Acad. Sci. U.S.A., 84,* 6126-6130.
Fernandez, M. P., Meugnier, H., Grimont, P. A. D., and Bardin, R. (1989). Deoxyribonucleic acid relatedness among members of the genus *Frankia. Int. J. Syst. Bacteriol., 39,* 424-429.
Franche, C., Laplaze, L. Duhoux, E., and Bogusz, D. (1998). Actinorhizal symbioses: Recent advances in plant molecular and genetic transformation studies. *Crit. Rev. Plant Sci., 17,* 1-28.
Franche, C., N'Diaye, A., Gobe, C., Alloneu, C. Bogusz, D., and Duhoux E. (1999a). Genetic transformation of *Allocasuarina verticillata.* In Y. P. S Bajaj (Ed.), *Biotechnology in agriculture and forestry, 44. Transgenic trees* (pp. 1-14). Berlin, Germany: Springer-Verlag.
Franche, C., Bogusz, D., Smouni, A., Diouf, D., and Duhoux E. (1999b). Genetic transformation of *Casuarina glauca.* In Y. P. S Bajaj (Ed.), *Biotechnology in agriculture and forestry, 44. Transgenic trees* (pp. 15-28). Berlin, Germany: Springer-Verlag.
Gardner, I. C. (1986). Mycorrhizae of actinorhizal plants. *MIRCEN J., 2,* 147-160.

Gauthier, D., Diem, H. G., and Dommergues, Y. R. (1985). Assessment of N_2 fixation by *Casuarina equisetifolia* inoculated with *Frankia* ORS021001 using ^{15}N methods. *Soil Biol. Biochem., 17,* 375-379.

Gherbi, H., Duhoux, E., Franche, C., Pawlowski, K., Nassar, A., Berry, A. M., and Bogusz, D. (1997). Cloning of a full-length hemoglobin cDNA and *in situ* localization of the corresponding mRNA in *Casuarina glauca* root nodule. *Physiol. Plant., 99,* 608-616.

Gordon, J. C., Perry, D. A., and Wheeler, C. T. (Eds.) (1979). *Symbiotic nitrogen fixation in the management of temperate forests* (501 pp.). Corvallis, OR: Oregon State University Press.

Guan, C., Wolters, D. J., van Dijk, C., Akkermans, A. D. L., van Kammen, A., Bisseling, T., and Pawlowski, K. (1996). Gene expression in ineffective actinorhizal nodules of *Alnus glutinosa. Acta Bot. Gall., 143,* 613-620.

Gupta, R. K., Dobritsa, S. V., Stiles, C. A., Essington, M. E., Liu, Z. Y., Chen, C.-H., Serpersu, E. H., and Mullin, B. C. (2002). Metallohistins: A new class of plant metal-binding proteins. *J. Protein Chem., 21,* 529-536.

Hahn, D., Lechevalier, M. P., Fischer, A., and Stackebrandt, E. (1989). Evidence for a close phylogenetic relationship between members of the genera *Frankia, Geodermatophilus* and "*Blastococcus*" and emendation of the family Frankiaceae. *Syst. Appl. Microbiol., 11,* 236-242.

Hahn, D., Starrenburg, M. J. C., and Akkermans, A. D. L. (1990). Oligonucleotide probes that hybridise with rRNA as a tool to study *Frankia* strains in root nodules. *Appl. Env. Microbiol., 56,* 1342-1346.

Hahn, D., Nickel, A., and Dawson, J. (1999). Assessing *Frankia* populations in plants and soil using molecular methods. *FEMS Microbiol. Ecol., 29,* 215-227.

Hammad, Y., Maréchal, J., Cournoyer, B., Normand, P., and Domenach, A. M. (2001). Modification of the protein expression pattern induced in the nitrogen-fixing actinomycete *Frankia* sp. strain ACN14a-tsr by root exudates of its symbiotic host *Alnus glutinosa* and cloning of the *sodF* gene. *Can. J. Microbiol., 47,* 541-547.

Hammad, Y., Nalin, R., Maréchal, J., Fiasson, K., Pepin, R., Berry, A.M., Normand, P., and Domenach, A. M. (2003). A possible role for phenyl acetic acid (PAA) on *Alnus glutinosa* nodulation by *Frankia. Plant Soil, 254,* 193-205.

Handley, L. L., and Raven, J. A. (1992). The use of natural abundance of nitrogen isotopes in plant physiology and ecology. *Plant Cell Environ., 15,* 965-985.

Harris S. L., and Silvester, W. B. (Eds.) (1994). *Frankia* and actinorhizal plants: 9th international conference. *Soil Biol. Biochem., 26,* 525-661.

Hiltner, L. (1895). Über die Bedeutung der Wurzelknöllchen von *Alnus glutinosa* für die Stickstoffernährung dieser Pflanze. *Landwirtsch. Vers.-Stn., 46,* 153-161.

Hogberg, P. (1997). ^{15}N natural abundance in soil-plant systems. *New Phytol., 137,* 179-203.

Holman, R. M., and Schwintzer, C. R. (1987). Distribution of spore-positive and spore-negative nodules of *Alnus incana* ssp. *rugosa* in Maine, USA. *Plant Soil, 104,* 103-111.

Hönerlage, W., Hahn, D., Zepp, K. Zeyer, J., and Normand, P. (1994). A hypervariable 23S rRNA region provides a discriminating target for specific characterisation of uncultured and cultured *Frankia. Syst. Appl. Microbiol., 17,* 433-443.

Hughes, M., Donnelly, C., Crozier, A., and Wheeler, C. T. (1999). Effects of the exposure of roots of *Alnus glutinosa* to light on flavonoids and nodulation. *Can. J. Bot., 77,* 1311-1315.

Hurd, T. M., Raynal, D. J., and Schwintzer, C. R. (2001). Symbiotic N_2 fixation of *Alnus incana* ssp. *rugosa* in shrub wetlands of the Adirondack Mountains, New York. *Oecologia, 126,* 94-103.

Huss-Danell, K. (1990). The physiology of actinorhizal nodules. In C. R. Schwintzer and J. D. Tjepkema (Eds.), *The biology of* Frankia *and actinorhizal plants* (pp. 61-82). San Diego, CA: Academic Press.

Huss-Danell, K. (1997). Actinorhizal symbioses and their N_2 fixation. *New Phytol., 136,* 375-405.

Huss-Danell, K., and Bergman, B. (1990). Nitrogenase in *Frankia* from root nodules of *Alnus incana* (L.) Moench: Immunolocalization of the Fe- and MoFe- proteins during vesicle differentiation. *New Phytol., 116,* 443-455.

Huss-Danell, K., and Wheeler, C. T. (Eds.) (1987). *Frankia* and actinorhizal plants. Proceedings of the international meeting, Umeå, Sweden. *Physiol. Plant., 70,* 235-377.

Isopi, R., Lumini, E., Frattegiani, M., Puppi, P., Bosco, M., Favilli, F., and Buresti, E. (1994). Inoculation of *Alnus cordata* with selected microsymbionts: Effects of *Frankia* and *Glomus* spp. on seedling growth and development. *Symbiosis, 17,* 237-245.

Jeong, S. C., Ritchie N. J., and Myrold, D. D. (1999). Molecular phylogenies of plants and *Frankia* support multiple origins of actinorhizal symbioses. *Mol. Phylogen. Evol., 13,* 493-503.

Johnson, L. A. S. (1980). Notes on Casuarinaceae (I). *Telopea, 2,* 83-84.

Johnson, L. A. S. (1980). Notes on Casuarinaceae (II) *J. Adelaide Bot. Gard., 6,* 73-87.

Johnson, L. A. S. (1980). Notes on Casuarinaceae (III) *Telopea, 3,* 133-137.

Johnson, G. V., Schwintzer, C. R., and Tjepkema, J. D. (1997). The acetylene induced decline in nitrogenase activity in root nodules of *Elaeagnus angustifolia. Plant Soil, 191,* 157-161.

Kim, H. B., and An, C. S. (2002). Differential expression patterns of an acidic chitinase and a basic chitinase in the root nodule of *Elaeagnus umbellata. Mol. Plant-Microbe Interact., 15, 209-215.*

Kleemann, G., Alskog, G., Berry, A. M., and Huss-Danell, K. (1994). Lipid-composition and nitrogenase activity of symbiotic *Frankia* (*Alnus incana*) in response to different oxygen concentrations. *Protoplasma, 183,* 107-115.

Kohls, S. J., van Kessel, C., Baker, D. D., Grigal, D. F., and Lawrence, D. B. (1994). Assessment of N_2 fixation and N cycling by *Dryas* along a chronosequence within the forelands of the Athabaca Glacier, Canada. *Soil Biol. Biochem., 26,* 623-632.

Lalonde, M., and Camire, C. (Eds.) (1985). *Frankia* and actinorhizal plants. Proceedings of the international meeting, Quebec, Canada. *Plant Soil, 87,* 1-208.

Laplaze, L., Bon, M-C., Sy, M.O., Smouni, A., Alloneu, C., Auguy, F., Thierry, F., Rio, M., Guermache, F., Duhoux E., Franche, C., and Bogusz, D. (2000). Molecular biology of tropical nitrogen-fixing trees in the Casuarinaceae family. In S. M. Jain and S. C. Minoicha (Eds.), *Molecular biology of woody plants, I.* (pp. 269-285). Dordrecht, The Netherlands: Kluwer.

Lawrence, D. B., Schoenike, R. E., Quispel, A., and Bong, G. (1967). The role of *Dryas drummondii* in vegetation development following ice recession at Glacier Bay, Alaska, with special reference to its nitrogen fixation by root nodules. *J. Ecol., 55,* 793-813.

Leaf, G., Gardner, I. C., and Bond, G. (1958). Observations on the composition and metabolism of the nitrogen-fixing root nodules of *Alnus. J. Exp. Bot., 9,* 320-331.

Lechevalier, M. P. (1983). Cataloging *Frankia* strains. *Can. J. Bot., 61,* 2964-2967.

Lechevalier, M. P. (1984). The taxonomy of the genus *Frankia. Plant Soil, 78,* 1-6.

Lechevalier, M. P. (1986). Catalog of *Frankia* strains. *Actinomycetes, 19,* 131-162.

Lechevalier, M. P., and Lechevalier, H. A. (1990). Systematics, isolation and culture of *Frankia.* In C. R. Schwintzer and J. D. Tjepkema (Eds.), *The biology of* Frankia *and actinorhizal plants* (pp. 35-60). San Diego, CA: Academic Press.

Lindblad, P., and Sellstedt, A. (1989). Immunogold localization of hydrogenase in free-living *Frankia* CpI1. *FEMS Microbiol. Lett., 60,* 311-316.

Lumini, E., Bosco, M., Puppi, G., Isopi, R., Frattegiani, M., Buresti, E., and Favilli, F. (1994). Field performance of *Alnus cordata* Loisel (Italian alder) inoculated with *Frankia* and VA-mycorrhizal strains in mine-spoil afforestation plots. *Soil Biol. Biochem., 26,* 659-661.

Lundquist, P.-O., and Huss-Danell, K. (1992). Immunological studies of glutamine synthetase in *Frankia-Alnus incana* symbioses. *FEMS Microbiol. Lett., 91,* 141-146.

Maréchal, J., Clement, B., Nalin, R., Gandon, C., Orso, S., Cvejic, J. H., Bruneteau, M., Berry, A., and Normand, P. (2000). A *recA* gene phylogenetic analysis confirms the close proximity of *Frankia* to *Acidothermus. Int. J. Syst. Evol. Microbiol., 50,* 781-785.

Maréchal, J., Santos, R., Hammad, Y., Alloisio, N., Domenach, A. M., and Normand, P. (2003). Characterization of the *sodF* gene region of *Frankia* sp. strain ACN14a and complementation of *Escherichia coli sod* mutant. *Can J. Microbiol., 49,* 294-300.

Marriotti, A., Sougoufara, B., and Dommergues, Y. R. (1992). Estimation de la fixation d'azote atmosphérique par le tracage isotopique naturel dans une plantation de *Casuarina equisetifolia* (Forst.). *Soil Biol. Biochem., 24,* 647-653.

Mattsson, U., Johansson, L., Sandström, G., and Sellstedt, A. (2001). *Frankia* KB5 possesses a hydrogenase immunologically related to membrane-bound [NiFe]-hydrogenases. *Curr. Microbiol., 42,* 438-441.

McClure, P. R., Coker, G. T., and Schubert, K. R. (1983). CO_2 fixation in roots and nodules of *Alnus glutinosa*: Role of PEP carboxylase and carbamyl phosphate synthetase in dark CO_2 fixation, citrulline biosynthesis and N_2 fixation. *Plant Physiol., 71,* 652-657.

McEwan, N. R., Gould, E. M. O., Lumini, E., McNeill, J. D., and Wheeler, C. T. (1996). Competitivity and persistence of *Frankia* in a managed environment. In K. A. Malik, M. S. Mirza, and J. K. Ladha (Eds.), *Nitrogen fixation with non-legumes. Proceedings of the* 7th *international symposium, Faisalabad* (pp. 315-326). Dordrecht, The Netherlands: Kluwer.

McKee, H. S. (1962). Fixation of free atmospheric nitrogen. In H. S. McKee (Ed.), *Nitrogen metabolism in plants* (pp. 39-102). Oxford, UK: Clarendon Press.

Meesters, T. M., van Vliet, W. M., and Akkermans, A. D. L. (1987). Nitrogenase is restricted to the vesicles in *Frankia* strain EAN1pec. *Physiol. Plant., 70,* 267-271.

Meyen, J. (1829). Über das hervorwachsen parasitischer gebilde aus den wurzeln anderer pflanzen. *Flora (Jena), 12,* 49-64.

Miller, I. M., and Baker, D. D. (1985). Initiation, development and structure of root nodules in *Elaeagnus angustifolia* L. (Elaeagnaceae). *Protoplasma, 128,* 107-119.

Molina, R. (1981). Ectomycorrhizal specificity in the genus *Alnus*. *Can. J. Bot., 59,* 325-334.

Mort, A., Normand, P., and Lalonde, M. (1983). 2-*O*-methyl-D-mannose, a key sugar in the taxonomy of *Frankia*. *Can. J. Microbiol., 29,* 993-1002.

Mullin, B. C., and Dobritsa, S. V. (1996). Molecular analysis of actinorhizal symbiotic systems: Progress to date. *Plant Soil, 186,* 9-20.

Murry, M. A., and Lopez, M. F. (1989). Interaction between hydrogenase, nitrogenase and respiratory activities in a *Frankia* isolate from *Alnus rubra*. *Can. J. Microbiol., 35,* 636-641.

Murry, M. A., Zhang, Z., and Torrey, J. G. (1985). Effect of O_2 on vesicle formation, acetylene reduction and O_2-uptake kinetics in *Frankia* sp. HFPCcI3 isolated from *Casuarina cunninghamiana*. *Can. J. Microbiol., 31,* 804-809.

Nazaret, S., Cournoyer, B., Normand, P., and Simonet, P. (1991). Phylogenetic relationships among *Frankia* genomic species determined by use of amplified 16S rDNA sequences. *J. Bacteriol., 173,* 4072-4078.

Normand, P., and Lalonde, M. (1986). The genetics of actinorhizal *Frankia*; a review. In F. A. Skinner and P. Uomala (Eds.), *Nitrogen fixation with non-legumes* (pp. 429-454). Dordrecht, The Netherlands: Martinus Nijhoff Publishers.

Normand, P., Fernandez, M., Simonet, P., and Domenach, A. M. (1992). *Frankia* and actinorhizal plants. Proceedings of the 8th international conference, Lyon, France. *Acta Oecol., 13,* 1-516.

Normand, P., Pawlowski, K., and Dawson, J. O. (Eds.) (2003). *Frankia* symbiosis. Proceeding of the 12th meeting on *Frankia* and actinorhizal plants, Carry-le-Rouet, France, June 2001. *Plant Soil, 254,* 1-244.

Nutman, P. S. (1990). George Bond. *Biograph. Mem. Fellows Roy. Soc., 36,* 61-82.

Okubara, P. A., Fujishige, N. A., Hirsch, A. M., and Berry, A. M. (2000). Dg93, a nodule abundant mRNA *of Datisca glomerata* with homology to a soybean early nodulin gene. *Plant Physiol.*, 122, 1073-1079.

Parsons, R., and Sunley, R. J. (2001). Nitrogen nutrition and the role of root-shoot nitrogen signalling particularly in symbiotic systems. *J. Exp. Bot., 52,* 435-443.

Parsons, R., Stanforth, A., Raven, J. A., and Sprent, J. I. (1993). Nodule growth and activity may be regulated by a feedback mechanism involving phloem nitrogen. *Plant Cell Environ., 16,* 125-136.

Parsons, R., Silvester, W. B., Harris, S., Gruitjers, W. T. M., and Bullivant, S. (1987). *Frankia* vesicles provide inducible and absolute oxygen protection for nitrogenase. *Plant Physiol., 83,* 728-731.

Pawlowski K., and Bisseling, T. (1996). Rhizobial and actinorhizal symbioses: What are the shared features? *Plant Cell, 8,* 1899-1913.

Perinet, P., and Lalonde, M. (1983). Axenic nodulation of *in vitro* propagated *Alnus glutinosa* plantlets by *Frankia* strains. *Can. J. Bot., 61,* 2883-2888.

Perinet, P., Brouillette, J. G., Fortin, J. A., and Lalonde, M. (1985). Large scale inoculation of actinorhizal plants with *Frankia*. *Plant Soil, 87,* 175-184.

Pinyopusarek, K., Yurnbull, J. W., and Midgley, S. J. (1996). *Recent* Casuarina *research and development, Proceedings of the 3rd international* Casuarina *workshop, Da Nang* (pp. 247). Canberra, Australia: CSIRO.

Pommer, E. H. (1959). Über die isolierung des endophyten aus den wurzelknöllchen von *Alnus glutinosa* Gaertn. und über erfolgreiche reinfektionsversuche. *Ber. Dtsch. Bot. Ges., 72,* 138-150.

Probanza, A., Lucas, J. A., Acero, N., and Gutierrez Manero, F. J. (1996). The influence of native rhizobacteria on European alder (*Alnus glutinosa* (L.) Gaertn.) growth. *Plant Soil, 182,* 59-66.

Puppo, A., Dimitrijevic, L., and Rigaud, J. (1989). Superoxide dismutase and catalase activities in purified *Frankia* vesicles. *Physiol. Plant., 77,* 308-311.

Quispel, A. (1954). Symbiotic nitrogen fixation in non-leguminous plants. 2. The influence of inoculation density and external factors on the nodulation of *Alnus glutinosa* and its importance to our understanding of the mechanism of infection. *Acta Bot. Neerl., 3,* 495-511.

Quispel, A. (1955). Symbiotic nitrogen fixation in non-leguminous plants. 3. Experiments on the growth *in vitro* of the endophyte of *Alnus glutinosa*. *Acta Bot. Neerl., 4,* 671-689.
Quispel, A. (1960). Symbiotic nitrogen fixation in non-leguminous plants. 5. The growth requirements of the endophyte of *Alnus glutinosa*. *Acta Bot. Neerl., 9,* 380-396.
Quispel, A. (1974). The endophytes of the root nodules in non-leguminous plants. In A. Quispel (Ed.), *The biology of nitrogen fixation* (pp. 499-520). Amsterdam, The Netherlands: North Holland Publishing Company.
Quispel, A. (1990). Discoveries, discussions and trends in research on actinorhizal root nodule symbioses before 1978. In C. R. Schwintzer and J. D. Tjepkema (Eds.), *The biology of* Frankia *and actinorhizal plants* (pp. 15-33). San Diego, CA: Academic Press.
Quispel, A., and Tak, T. (1968). Studies on the growth of the endophyte of *Alnus glutinosa* (L.) Vill. in nutrient solutions. *New Phytol., 81,* 587-600.
Quispel, A., Svendsen A. B., Schripsems, J., Baas, W. J., Erkelens, C., and Lugtenberg, B. J. (1989). Identification of dipterocarpol as isolation factor for the induction of primary isolation of *Frankia* from root nodules of *Alnus glutinosa* (L.) Gaertner. *Mol. Plant-Microbe Interact., 2,* 107-112.
Quispel A., Rodriguez-Barrueco, C., and Subbarao, N. S. (1993) Some general considerations on symbioses of nitrogen-fixing trees. In N. S. Subbarao, and C. Rodriguez-Barrueco (Eds.), *Casuarinas* (pp. 1-32). New Delhi, India: Oxford and IBH Publishing Co.
Racette, S., Torrey, J. G., and Berg, R. H. (1991). Sporulation in root nodules of actinorhizal plants inoculated with pure cultured strains of *Frankia*. *Can. J. Bot., 69,* 1471-1476.
Ramos, B., Garcia, J. A. L., Probanzaa, A., Domenach, J., and Manero, F. J. G. (2003). Influence of an indigenous European alder (*Alnus glutinosa* (L.) Gaertn.) rhizobacterium (*Bacillus pumilis*) on the growth of alder and its rhizosphere microbial community structure in two soils. *New Forests, 25,* 149-159.
Ribeiro, A., Akkermans, A. D. L., van Kammen, A., Bisseling, T., and Pawlowski, K. (1995). A nodule-specific gene encoding a subtilisin-like protease is expressed in early stages of actinorhizal nodule development. *Plant Cell, 7,* 785-794.
Roelofsen, W., and Akkermans, A. D. L. (1979). Uptake and evolution of H_2 and reduction of C_2H_2 by root nodules and homogenates of *Alnus glutinosa*. *Plant Soil, 52,* 571-578.
Rose, S. L. (1980). Mycorrhizal associations of some actinomycete nodulated nitrogen fixing plants. *Can. J. Bot., 58,* 1449-1454.
Ruvkun, G. B., and Ausubel, F. M. (1980). Interspecies homology of nitrogenase genes. *Proc. Natl. Acad. Sci. U.S.A., 77,* 191-194.
Schaede, R. (1962). *Die pflanzen symbiosen*, 3rd edition (pp. 42-54). Stuttgart: Fischer.
Schubert, K. R., and Evans, H. J. (1976). Hydrogen evolution: a major factor affecting the efficiency of nitrogen fixation in nodulated symbionts. *Proc. Natl. Acad. Sci. U.S.A., 73,* 1207-1211.
Schwencke, J., and Carú, M. (2001). Advances in actinorhizal symbiosis: Host plant-*Frankia* interactions, biology, and applications in arid land reclamation. A review. *Arid Land Res. Manag., 15,* 285-327.
Schwintzer, C. R., and Tjepkema, J. D. (Eds.) (1990). *The biology of* Frankia *and actinorhizal plants* (408 pp.). San Diego, CA: Academic Press.
Schwintzer, C. R., and Tjepkema, J. D. (1990). Factors affecting the acetylene to $^{15}N_2$ conversion ratio in root nodules of *Myrica gale* (L). *Plant Physiol., 106,* 1041-1047.
Sellstedt, A., Huss-Danell, K., and Ahlqvist, A.-S. (1986). Nitrogen fixation and biomass production in symbiosis between *Alnus incana* and *Frankia* strains with different hydrogen metabolism. *Physiol. Plant., 66,* 99-107.
Sellstedt, A., and Lindblad, P. (1990). Activities, occurrence, and localization of hydrogenase in free-living and symbiotic *Frankia*. *Plant Physiol.*, 92, 809-815.
Sempavalan, J., Wheeler, C. T., and Hooker, J. E. (1995). Lack of competition between *Frankia* and *Glomus* for infection and colonisation of roots of *Casuarina equisetifolia*. *New Phytol., 136,* 429-436.
Sequin, A., and Lalonde, M. (1993). Modification of polypeptide patterns during nodule development in the *Frankia-Alnus* symbiosis. *Symbiosis, 15,* 135-149.
Severini, G. (1922). Sui tubercoli radicali di *Datisca cannabina* L. *Ann. Bot. (Roma), 15,* 29-51.
Sharma, E., and Ambasht, R. S. (1988). Nitrogen accretion and its energetics in the Himalayan alder. *Funct. Ecol., 2,* 229-235.

Sharma, E., Sharma, R., and Pradhan, M. (1998). Ecology of Himalayan alder (*Alnus nepalensis* D. Don) *PINSA B64, 1,* 59-78.

Shearer, G., and Kohl, D. H. (1993). Natural abundance of ^{15}N: Fractional contribution of two sources to a common sink and use of isotope discrimination. In R. Knowles, and T. H. Blackburn (Eds.), *Nitrogen isotope techniques* (pp. 89-126). San Diego, CA: Academic Press.

Silvester, W. B., and Winship, L. J. (1990). Transient response of nitrogenase to acetylene and oxygen by actinorhizal nodules and cultured *Frankia. Plant Physiol., 92,* 480-486.

Silvester, W. B., Silvester, J. K., and Torrey, J. G. (1988). Adaptations of nitrogenase to varying oxygen tension and the role of the vesicle in root nodules of *Alnus incana* ssp. *rugosa. Can. J. Bot., 66,* 1772-1779.

Silvester, W. B., Langenstein, B., and Berg, R. H. (1999). Do mitochondria provide the oxygen diffusion barrier in root nodules of *Coriaria* and *Datisca*? *Can. J. Bot., 77,* 1358-1366.

Simonet, P., Normand, P., Hirsch, A. M., and Akkermans, A. D. L. (1990). The genetics of the *Frankia* actinorhizal symbiosis. In P. M. Gresshoff (Ed.), *The molecular biology of symbiotic nitrogen fixation* (pp. 70-109). Boca Raton, FA: CRC Press, Inc.

Simonet, P., Bosco, M., Chapelon, C., Moiroud, A., and Normand, P. (1994). Molecular characterisation of *Frankia* microsymbionts from spore-positive and spore-negative nodules in a natural alder stand. *Appl. Env. Microbiol., 60,* 1335-1341.

Subbarao, N. S., and Rodriguez-Barrueco, C. (1995). *Casuarinas* (240 pp.). New Delhi, India: Oxford and IBH Publishing Co.

Suharjo, U. K. J., and Tjepkema, J. D. (1995). Occurrence of hemoglobin in the nitrogen-fixing root nodules of *Alnus glutinosa. Physiol. Plant., 95,* 247-252.

Soltis, D. E., Soltis, P. S., Morgan, D. R., Swensen, S. M., Mullin, B. C., Dowd, J. M., and Martin, P. G. (1995). Chloroplast gene sequence data suggest a single origin of the predisposition for symbiotic nitrogen fixation in angiosperms. *Proc. Natl. Acad. Sci. U.S.A., 92,* 2647-2651.

Steele, D. B., and Stowers, M. D. (1986). Superoxide dismutase and catalase in *Frankia. Can. J. Microbiol., 32,* 409-413.

Stewart, W. D. P., Fitzgerald, G. P., and Burris, R. H. (1967). *In situ* studies on N_2 fixation using the acetylene reduction technique. *Proc. Natl. Acad. Sci. U.S.A.*, 58, 2071-2078.

Stevens, G. A., and Berry, A. M. (1988). Cytokinin secretion by *Frankia* sp. HFPArI3 in defined medium. *Plant Physiol., 87,* 15-16.

Stowers, M. D. (1985). Further studies on the nodulating potential of *Rubus ellipticus* by the actinomycete *Frankia*. In H. J. Evans, P. J. Bottomley, and W. E. Newton (Eds.), *Nitrogen fixation research progress* (p. 702). The Hague, The Netherlands: Martinus Nijhoff.

Swensen, S. M. (1996). The evolution of actinorhizal symbioses: Evidence for multiple origins of the symbiotic association. *Amer. J. Bot., 83,* 1503-1512.

Swensen, S. M., and Mullin, B. C. (1997a). Phylogenetic relationships among actinorhizal plants. The impact of molecular systematics and implications for the evolution of actinorhizal symbioses. *Physiol. Plant., 99,* 565-573.

Swensen, S. M., and Mullin, B. C. (1997b). The impact of molecular systematics on hypotheses for the evolution of root nodule symbioses and implications for expanding symbioses to new host plant genera. *Plant Soil, 194,* 185-192.

Tavares, F., Abreu, I., and Salema, R. (1998) Regeneration of the actinorhizal plant *Myrica gale* L. from epicotyl explants. *Plant Sci., 135,* 203-210.

Tjepkema, J. D. (1979). Oxygen relations in leguminous and actinorhizal nodules. In J. C. Gordon, D. A. Perry and C. T. Wheeler (Eds.), *Symbiotic nitrogen fixation in the management of temperate forests* (pp. 175-186). Corvallis, OR: Oregon State University Press.

Tjepkema, J. D., Ormerod, W., and Torrey, J. G. (1981). Factors affecting vesicle formation and acetylene reduction (nitrogenase activity) in *Frankia* sp. CpI1. *Can. J. Microbiol., 27,* 815-823.

Tjepkema, J. D., Schwintzer, C. R., and Benson, D. R. (1986). Physiology of actinorhizal nodules. *Ann. Rev. Plant Physiol., 37,* 209-232.

Torrey, J. G. (1976). Initiation and development of root nodules of *Casuarina* (Casuarinaceae). *Amer. J. Bot., 63,* 335-344.

Torrey, J. G., and Tjepkema, J. D. (1979). Symbiotic nitrogen fixation in actinomycete nodulated trees: Preface. *Bot. Gaz. 140 (Suppl.),* i-ii.

Torrey, J. G., and Tjepkema, J. D. (1983). International conference on the biology of *Frankia. Can. J. Bot., 61,* 2765-2967.

Tremblay, F. M., and Lalonde, M. (1984). Requirements for *in vitro* propagation of seven nitrogen-fixing *Alnus* species. *Plant Cell Tisssue Organ Culture, 3,* 189-199.

Turner, G. L., and Gibson, A. H. (1980). Measurement of nitrogen fixation by indirect means. In F. J. Bergersen (Ed.), *Methods for evaluating biological nitrogen fixation* (pp. 111-138). Chichester, UK: John Wiley and Sons.

VandenBosch, K. A., and Torrey, J.G. (1984a). Consequences of sporangial development for nodule function in root nodules of *Comptonia peregrina* and *Myrica gale. Plant Physiol., 76,* 556-560.

VandenBosch, K. A., and Torrey, J.G. (1984b). Development of endophytic *Frankia* sporangia in field- and laboratory-grown nodules of *Comptonia peregrina* and *Myrica gale. Amer. J. Bot.*, 72, 99-108.

van Dijk, C. (1978). Spore formation and endophyte diversity in root nodules of *Alnus glutinosa* (L.) Vill. *New Phytol., 81,* 601-615,

van Dijk, C., and Merkus, E. (1976). A microscopical study of the development of a spore-like stage in the life cycle of the root nodule endophyte of *Alnus glutinosa* (L.) Gaertn. *New Phytol., 77,* 73-91.

van Dijk, C., and Sluimer-Stolk, A. (1990). An ineffective strain type of *Frankia* in the soil of natural stands of *Alnus glutinosa* (L.) Gaertner. *Plant Soil, 127,* 107-121.

Van Ghelue, M., Lovaas, E., Ringo, E., and Solheim, B. (1997). Early interactions between *Alnus glutinosa* and *Frankia* strain ArI3. Production and specificity of root hair deformation factor(s). *Physiol. Plant., 99,* 579-587.

Virtanen, A. L., Moisio, Y., Allison, R. M., and Burris, R. H. (1954). Fixation of molecular nitrogen by excised nodules of the alder. *Acta Chem. Scand., 8,* 1730-1731.

Vitousek, P. M. (1989). Biological invasion by *Myrica faya* in Hawaii: Plant demography, nitrogen fixation, ecosystem effects. *Ecol. Monogr., 59,* 247-265.

Wall, L. G. (2000). The actinorhizal symbiosis. *J. Plant Growth Reg., 19,* 167-182.

Wang, H.-Y., and Berry, A. M. (1996). Plant regeneration from leaf segments of *Datisca glomerata. Acta Bot. Gall., 143,* 609-612.

Watts, S. H., Wheeler, C. T., and Hillman, J. R. (1987). Abscisic acid and dormancy in root-nodulated *Alnus glutinosa. New Phytol., 105,* 459-468.

Wheeler, C. T., Crozier, A., and Sandberg, G. (1984). The biosynthesis of indole-3-acetic acid by *Frankia. Plant Soil, 77,* 99-104.

Wheeler, C. T., and Miller, I. M. (1990) Current and potential uses of aactinorhizal plants in Europe. In C. R. Schwintzer, and J. D. Tjepkema (Eds.), *The biology of* Frankia *and actinorhizal plants* (pp. 365-390). San Diego, CA: Academic Press.

Wheeler, C. T., Henson, I. E., and McLaughlin, M. E. (1979). Hormones in plants bearing nitrogen fixing root nodules. *Bot. Gaz., 140 (Suppl.),* S52-S57.

Wheeler, C. T., Tonin, G. S., and Sutcliffe, A. (1994). Polyamines of *Frankia* in relation to nitrogen nutrition. *Soil Biol. Biochem., 26,* 577-581.

Winship, L. J., and Benson, D. R. (1989). Proceedings of the 7th international conference on *Frankia* and actinorhizal plants. *Plant Soil, 118,* 1-247.

Winship, L. J., Martin, K. J., and Sellstedt, A. (1987). The acetylene reduction assay inactivates root nodule uptake hydrogenase in some actinorhizal plants. *Physiol. Plant., 70,* 361-366.

Wolters, D. J., van Dijk, C., Zoetendal, E. G., and Akkermans, A. D. L. (1997). Phylogenetic characterisation of ineffective *Frankia* in *Alnus glutinosa* (L.) Gaertn. nodules from wetland soil inoculants. *Mol. Ecol., 6,* 971-981.

Woronin, M. (1885). Bemerkungen zu dem aufsatze von Herrn H. Müller über *Plasmodiophora Alni. Ber. Dtsch. Bot. Ges., 5,* 50-58.

Yao, Y. (1995). Micropropagation of sea buckthorn (*Hippophae rhamnoides* L.). *Agric. Sci. Finland, 4,* 503-512.

Chapter 2

POLYPHASIC TAXONOMY OF THE GENUS *FRANKIA*

D. HAHN

Department of Chemical Engineering, New Jersey Institute of Technology (NJIT), University Heights, Newark, NJ 07102, and Department of Biological Sciences, Rutgers University, 101 Warren Street, Smith Hall 135, Newark, NJ 07102, U.S.A.

1. INTRODUCTION

Members of the genus *Frankia* are generally described as nitrogen-fixing actinomycetes that form root nodules in symbiosis with more than 200 species of non-leguminous woody plants in 25 genera of angiosperms (Benson and Silvester, 1993; Huss-Danell, 1997; Schwintzer and Tjepkema, 1990). Although knowledge about the interaction of woody plants and *Frankia* has a long history, pure cultures of this actinomycete have been obtained only relatively recently. Initial attempts to characterize the bacterium taxonomically referred to the microsymbiont as an obligate symbiotic bacterium and used its ability to form root nodules and morphological characteristics, such as filament and vesicle formation in cells of these root nodules, as discriminative criteria with respect to other actinomycetes (Becking, 1970). These criteria led to the emendation of the family *Frankiaceae* with the type genus *Frankia* (Becking, 1970). Ten species were proposed within the genus *Frankia* based on the formation of specific host-plant infection groups after inoculation with crushed nodules (Becking, 1970). Later investigations with pure cultures, however, demonstrated that the host-plant infection groups were quite different (Baker, 1987; Normand and Lalonde, 1986).

The first successful isolation of *Frankia* occurred less than 30 years ago (Callaham *et al.,* 1978). Since then, hundreds of *Frankia* isolates have been obtained from root nodules, a natural locale of enrichment for this organism, using different isolation techniques (Baker, 1990; Baker *et al.,* 1979a; 1979b; Baker and O'Keefe, 1984; Benson, 1982; Berry and Torrey, 1979; Burggraaf, 1984; Diem and Dommergues, 1983; Hiyoshi *et al.,* 1988; Kim *et al.*, 1993). However, no general isolation protocols have been developed and only a small percentage of isolation

K. Pawlowski and W. E. Newton (eds.), Nitrogen-fixing Actinorhizal Symbioses, 25–47.
© 2008 *Springer*.

attempts succeeds (Rosbrook *et al.,* 1989; St-Laurent and Lalonde, 1987). Successful attempts to isolate frankiae from soil, its second ecological niche, have been reported only once (Baker and O'Keefe, 1984).

Problems with the isolation of frankiae have been attributed to long generation times, the requirement of special isolation factors (Quispel *et al.*, 1989), the lack of specific nutritional requirements of pure cultures obtained so far (Akkermans *et al.,* 1992), and the large variability of isolates combined in the genus *Frankia* (Akimov and Dobritsa, 1992; Fernandez *et al.,* 1989; Lechevalier, 1994). Isolation and cultivation, however, are prerequisites for the characterization of *Frankia* by traditional techniques that focus on the analysis of protein and isoenzyme patterns, susceptibility and resistance to antibiotics, sugar or fatty acid composition, or other criteria, such as pigment production (Benson and Hanna, 1983; Benson *et al.,* 1984; Dobritsa, 1998; Gardes *et al.,* 1987; Gardes and Lalonde, 1987; Girgis and Schwencke, 1993; Hafeez *et al.,* 1984; Igual *et al.*, 2001; Lalonde *et al.*, 1988; Lechevalier and Ruan, 1984; Mirza *et al.,* 1991; Mort *et al.,* 1983; Simon *et al.,* 1989; St-Laurent *et al.,* 1987; Wheeler *et al.,* 1986).

Many isolates can clearly be distinguished from other bacterial genera on the basis of morphological, cytochemical, and physiological characteristics (Baker, 1990). Distinctive morphological characteristics of isolates include filamentous growth, the formation of multilocular sporangia containing non-motile spores, and the differentiation of thick-walled vesicular structures at the tips of filaments (vesicles), which are the principal locus of nitrogenase and, thus, of nitrogen fixation (Lechevalier, 1994). Cells are characterized by a type-III cell wall that contains *meso*-diaminopimelic acid, glutamic acid, alanine, glucosamine and muramic acid, type PI phospholipids with phosphatidylinositol, phosphatidylinositol mannosides and diphosphatidylglycerol, and 2-*O*-methyl-D-mannose as a diagnostic sugar (Lechevalier, 1994). The most prominent physiological characteristics are their ability to fix N_2 under nitrogen-limited conditions, even in pure culture, and the formation of root nodules with specific host plants (Lechevalier, 1994).

This combination of phenotypic characteristics has frequently proven useful for the identification of frankiae, even though single features might be shared by other organisms. In certain morphological features, such as cells dividing in more than one plane, *Frankia* strains resemble *Geodermatophilus obscurus* and *Dermatophilus congolensis*, but differ from these in the possession of sporangia. Still, morphological features and cell-wall type of both *Frankia* and *Geodermatophilus* were the basis for a classification in a taxon "multilocular sporangia" (Goodfellow, 1986). The expression and thus the presence of many of the characteristic phenotypic features of *Frankia*, however, might vary with respect to the isolate and its growth conditions. The extent of sporangia formation, for instance, varies from isolate to isolate and might be totally suppressed under certain growth conditions. Although many *Frankia* strains produce typical terminal sporangia in liquid culture, the extent to which these sporangia are formed can differ significantly between strains (Figure 1). Others do not form terminal sporangia under the same growth conditions, but display intercalary sporangia only or no sporangia formation at all

(Figure 1). The latter two might form typical terminal sporangia under different growth conditions. Spores of frankiae are also found in root nodules of several actinorhizal plants (*e.g.*, *Alnus* sp.) (van Dijk, 1984; van Dijk *et al.,* 1988; Weber, 1986). However, isolates producing spores in root nodules (spore (+) type) have not yet been obtained in pure culture, thus, all *Frankia* strains are of the spore (–) type.

Vesicles, the sites of nitrogen fixation, are generally formed under nitrogen-limited growth conditions, but may be formed, though inactive, in the presence of fixed-nitrogen resources (Figure 2). In addition to the typical diazotrophic frankiae, some non-N_2-fixing (atypical) isolates have been obtained from actinorhizal nodules (Baker *et al.,* 1980; Hahn *et al.,* 1988; Mirza *et al.,* 1992). Recently, these atypical *Frankia* strains have been shown to occur quite commonly in natural habitats (Wolters *et al.,* 1997a; 1997b). Atypical isolates produce hyphae and terminal sporangia, but no vesicles in either pure culture or root nodules (Figure 2); consequently, they do not grow in the absence of fixed-nitrogen resources (Hahn *et al.*, 1988).

Root-nodule formation by atypical isolates has been observed, but might be suppressed by the plant (Hahn *et al.*, 1988; van Dijk and Sluimer, 1994), thus, limiting the validity of root-nodule formation as a characteristic trait of frankiae. Root-nodule formation on specific host plants has been used as a differentiating characteristic between groups of frankiae. Most isolates obtained to date can be classified into four host-infection groups; those that nodulate i) *Alnus* and *Myrica*,

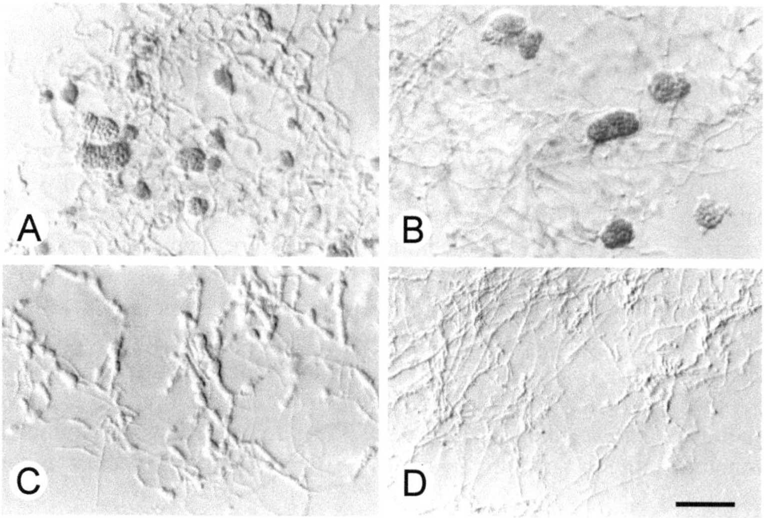

Figure 1. Frankia *strains AgGS'84/44 (A), AgGS'84/45 (B), AgKG'84/4 (C), and AgGA'84/2 (D) grown in liquid P+N medium. Although AgGS'84/44 and AgGS'84/45 form typical terminal sporangia, but to different extents, strain AgKG'84/4 forms intercalary sporangia only and AgGA'84/2 exhibits no sporangia formation on this medium. Bar represents 50 μm.*

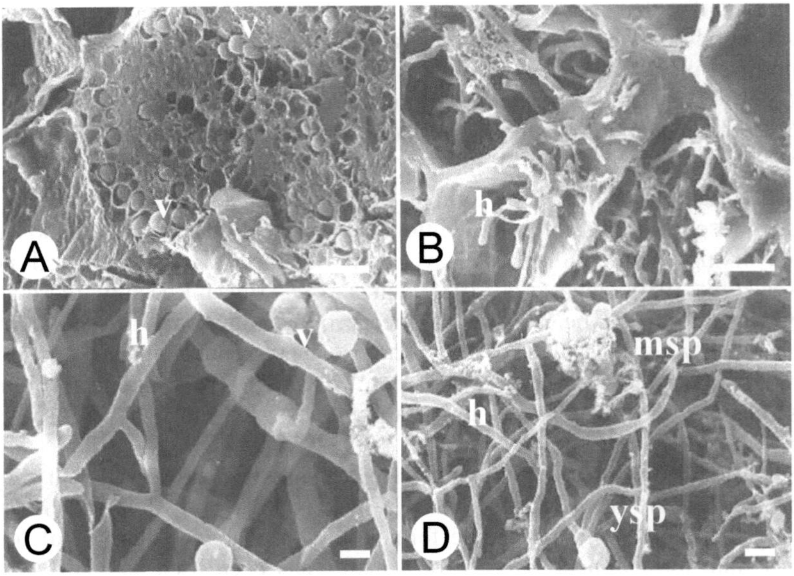

Figure 2. Root nodules formed by the N_2-fixing Frankia *strain Ag45/Mut15 (A) display typical vesicle (v) clusters in plant cells, whereas those formed by the non-N_2-fixing (atypical)* Frankia *strain AgB1.9 (B) only harbor hyphae (h) but not vesicles. The N_2-fixing strain develops vesicles in pure culture, but not sporangia (C). Sporangia (young (ysp) and mature (msp) sporangia), but no vesicles, are produced by the non-N_2-fixing strain (D). Bars represent 10 µm (A, B) and 1 µm (C, D), respectively (modified from Hahn et al., 1988).*

ii) *Casuarina* and *Myrica*, iii) *Myrica* and *Elaeagnus*, and iv) members of the *Elaeagnaceae* (*Elaeagnus*, *Hippophaë*, *Shepherdia*) (Baker, 1987). However, this classification is incomplete because isolates from several actinorhizal plant taxa (*e.g.*, *Coriaria*, *Datisca*) are still lacking. Furthermore, some isolates cross the boundaries between host-infection groups (Bosco *et al.*, 1992), whereas others show variable compatibility to plant species of the same genus (Hahn *et al.*, 1988; Huss-Danell and Myrold, 1994; Weber, 1990). Thus, the usefulness of "unique" morphological and physiological features for the identification of frankiae is limited. Consequently, a reliable and more objective classification and identification of frankiae and *Frankia*-like bacteria requires the introduction and application of new taxonomic methods, notably chemical, numerical and molecular taxonomic procedures.

Recently, molecular taxonomic procedures have increasingly been used as alternatives to traditional techniques that are limited by successful isolation of *Frankia*. Whole plasmids (Simonet *et al.*, 1988), *nif* genes (Baker and Mullin, 1994; Rouvier *et al.*, 1996; Mirza *et al.*, 1994; Normand *et al.*, 1992; Simonet *et al.*, 1990; 1991), intergenic spacer regions (IGS) (Bosco *et al.*, 1996; Lumini and Bosco, 1996; Lumini *et al.*, 1996; Navarro *et al.*, 1997; Rouvier *et al.*, 1992; Simonet *et al.*, 1991), glutamine synthetase genes (*gln*) (Cournoyer and Lavire, 1999; Hosted *et al.*,

1993; Rochefort and Benson, 1990), the *rec*A gene (Marechal *et al.*, 2000) and ribosomal RNAs (Bosco *et al.*, 1992; Hahn *et al.*, 1989a; Hahn *et al.*, 1997; Hönerlage *et al.*, 1994; Mirza *et al.*, 1992; Mirza *et al.*, 1994; Navarro *et al.*, 1997; Nazaret *et al.*, 1989; 1991; Rouvier *et al.*, 1992; 1996; Simonet *et al.*, 1994; Zepp *et al.*, 1997) have all been used successfully as specific targets for the characterization of isolates as well as for the identification of uncultured endophytes in root nodules. These new procedures reveal much about the genetic diversity and distribution of *Frankia*, and have refined and expanded knowledge about endophyte-host specificities. The most prominent advancement in studies on the taxonomy of *Frankia* has resulted from the use of polyphasic taxonomy approaches integrating information retrieved by a wide range of techniques on different levels of taxonomic resolution (Murray *et al.*, 1990).

2. POLYPHASIC TAXONOMY APPROACH

Polyphasic taxonomy approaches integrate phenotypic, genotypic, and phylogenetic information retrieved by a relatively large set of techniques with different resolving power (Figure 3; see Rossello-Mora and Amann, 2001; Vandamme *et al.*, 1996 for review). Methods that retrieve phenotypic information comprise all methods not directed toward DNA or RNA, including chemotaxonomic techniques (Vandamme *et al.*, 1996). Examples are API or BIOLOG substrate-utilization screening, cell wall-structure analysis, fatty-acid methyl-ester analysis (FAME), whole-cell protein profiling, both multilocus- and iso-enzyme electrophoresis (MLEE, IEE), and serological and phage-typing techniques (Figure 3).

Genotypic information is obtained by methods that use either complete or partial features of the genome. The former include DNA base-composition analysis (%G+C), DNA-DNA relatedness or re-association studies, and fingerprinting techniques, such as ribo-typing, low-frequency restriction-fragment analysis (LFRFA), pulsed-field gel electrophoresis (PFGE), and restriction fragment length polymorphism (RFLP) analysis. Partial genomic-typing techniques use the potential of the polymerase chain reaction (PCR) that enables thermocyclic amplification of target sequences on DNA by extension of specifically annealed primers by a thermostable DNA polymerase (e.g. *Taq* polymerase) (Mullis and Faloona, 1987; Saiki *et al.*, 1988). PCR-based typing techniques include arbitrary primed PCR (AP-PCR), randomly amplified polymorphic DNA analysis (RAPD), amplified ribosomal DNA restriction analysis (ARDRA), and genomic fingerprinting that targets endogenous interspersed repetitive sequences (rep-PCR). The introduction of PCR and the shift in focus from the complete genome to specific genes has opened possibilities to characterize uncultured organisms. PCR-assisted retrieval of specific genes, e.g. ribosomal RNA genes that carry phylogenetically significant sequence information, might then be used as basis for comparative sequence or probe analyses (Ludwig *et al.*, 1998).

A reliable classification should focus on different levels of taxonomic resolution. At higher taxonomic levels, comparative 16S-rRNA sequence analysis is most appropriate because large differences in sequence between organisms can be expected. Comparative 16S-rRNA sequence analysis is reliably used for

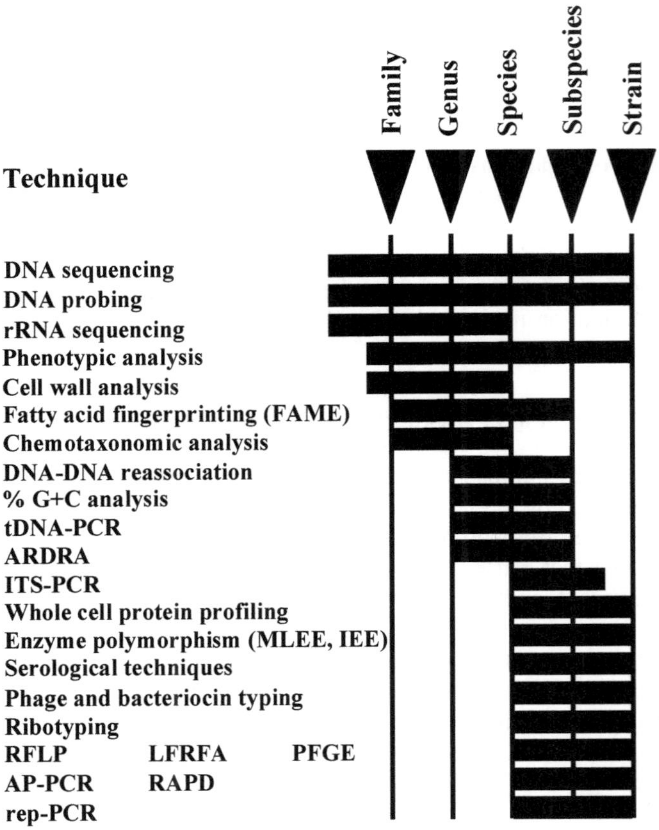

Figure 3. Taxonomic resolution of some currently used techniques (modified from Vandamme et al., 1996).

assignments from the domain level down to the genus level as characterized by a 95% sequence similarity borderline (Ludwig *et al.*, 1998). Further assignment at the species level is indicated by a 97% similarity borderline of the 16S rRNA, however, although generally confirmatory (Rossello-Mora and Amann, 2001; Stackebrandt and Goebel, 1994), there are several examples of different species with identical or nearly identical 16S-rRNA sequences (Fox, Wisotzkey, and Jurtshuk, 1992; Martinez-Murcia, Benlloch, and Collins, 1992). Furthermore, both micro-heterogeneity of the 16S-rRNA genes within a single species (Bennasar, Rossello-Mora, Lalucat, and Moore, 1996) and single organisms with two or more 16S rRNA genes with relatively high sequence divergence (Mylvaganam and Dennis, 1992; Nübel *et al.*, 1996) have been reported.

The current standard for species delineation is DNA-DNA relatedness because a high degree of correlation was found between DNA-DNA relatedness and chemotaxonomic, genomic, serological and numerical similarity (Rossello-Mora and Amann, 2001). A species is defined as a group of strains, including the type strain, sharing 70% or greater whole genomic DNA-DNA relatedness. A further division of the species into subspecies, the lowest official taxonomic rank in nomenclature, could be based on different typing techniques, indicating minor but consistent pheno- or genotypic differences between strains of the species (Rossello-Mora and Amann, 2001). However, consistency is often not achieved and, thus, typing techniques often result in groupings of strains without formal nomenclatural standing.

For *Frankia*-like isolates or uncultured endophytes in root nodules, a classification or identification should incorporate information obtained in a polyphasic top-to-bottom approach of taxonomic resolution. Focussing on nucleic-acid analyses, an initial comparative 16S-rRNA sequence analysis provides information on their phylogenetic position and permits an assignment on the genus level. Subsequent characterization on the species level through DNA-DNA relatedness studies is followed up by sub-species level analyses provided by either comparative sequence analysis of variable regions on rRNAs or by fingerprinting techniques.

3. 16S-rRNA SEQUENCE DATA

Because of their ubiquity and genetic stability, ribosomal RNA (rRNA) sequences have been used intensively to investigate quantitative evolutionary relationships among numerous bacteria (Ludwig *et al.*, 1998; Woese, 1987; Woese, Kandler, and Wheelis, 1990). Phylogenetic characterization has been performed using sequence comparison of all three rRNAs (5S, 16S and 23S rRNA) (Fox and Stackebrandt, 1987; Ludwig *et al.*, 1992; Ludwig *et al.*, 1998; Woese *et al.*, 1990; Winker and Woese, 1991), even though only 16S and 23S rRNA molecules are of appropriate size for broad phylogenetic analyses. At the moment, the most elaborated databases of primary structures of informative macromolecules have been compiled for 16S-rRNA molecules (Maidak *et al.*, 1997).

Initially, oligonucleotide catalogs obtained by partial sequence analysis, using oligonucleotide fragments produced by digestion with ribonuclease T_1, were used to establish distant phylogenetic relationships (Fox and Stackebrandt, 1987; Stackebrandt *et al.*, 1985). Oligonucleotide catalogs of the 16S rRNA of *Frankia* strain AirI1 and *Geodermatophilus obscurus* indicated a high degree of phylogenetic relationship between the organisms, which led to considerations of assigning both genera to the family *Frankiaceae*. This informal proposal was confirmed when a combination of oligonucleotide cataloguing and reverse-transcriptase sequencing of 16S rRNA (Embley *et al.*, 1988; Lane *et al.*, 1985; Lane *et al.*, 1988) was used to refine the phylogenetic position of two *Frankia* strains, AirI1 and Ag45/Mut15 (Hahn *et al.*, 1989b). This analysis led to the clustering of *Frankia*, "*Blastococcus*" and *Geodermatophilus* into one family *Frankiaceae*, which, however, excluded *Dermatophilus* (Hahn *et al.*, 1989b).

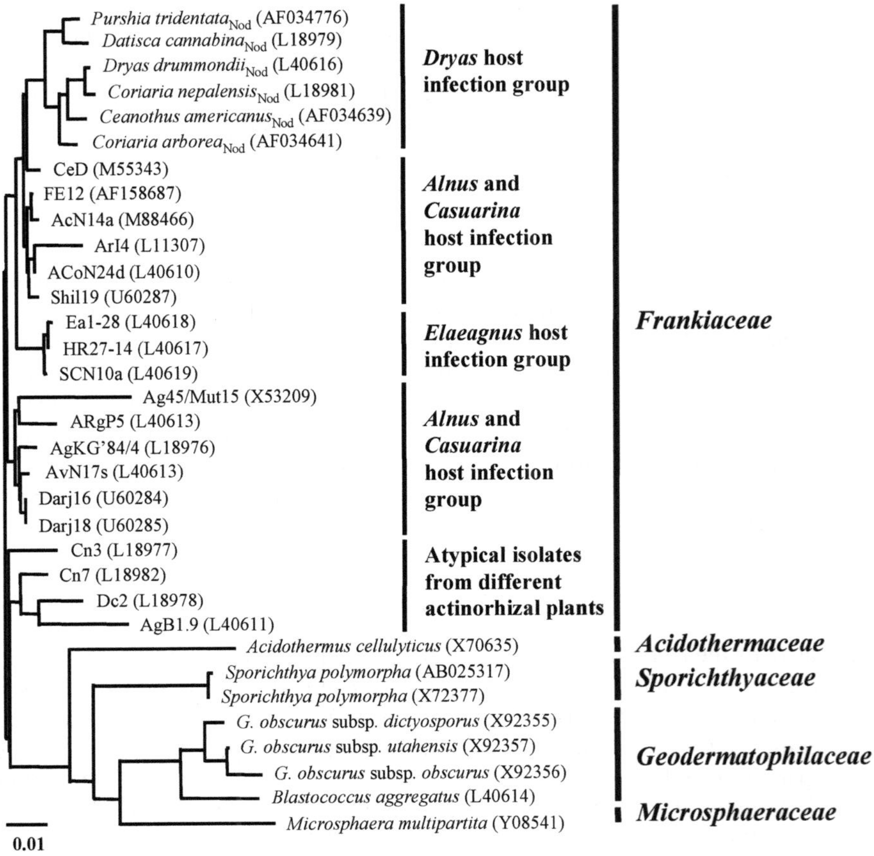

Figure 4. Neighbor-joining tree of aligned, almost complete 16S rRNA sequences of selected members of the suborder Frankineae *within the order Actinomycetales.*

The emendation of the family *Frankiaceae* was refined again when rapid sequencing protocols and large databases became available. Comparative sequence analysis of almost complete PCR-amplified 16S rDNAs of pure cultures of *Frankia*, as well as of uncultured endophytes in root nodules, led to the emendation of the family *Frankiaceae* to contain only the genus *Frankia* (Normand *et al.*, 1996). Taxa related to *Frankia* were the genera *Geodermatophilus*, "*Blastococcus*", *Sporichthya*, *Acidothermus*, and *Actinoplanes*. Similar results were obtained when a new hierarchical classification system, which was solely based on 16S rRNA and rDNA, was proposed for the taxa between the levels of genus and class for the actinomycete line of descent (Stackebrandt et al., 1997). The genus *Frankia* was the only genus within the family *Frankiaceae*. Together with the families *Acidothermaceae*, *Geodermatophilaceae*, *Microsphaeraceae* and *Sporichthyaceae*, the family *Frankiaceae* were assigned to the suborder *Frankineae*, one of ten suborders within the order *Actinomycetales* (Stackebrandt *et al.*, 1997).

With the exception of the *Geodermatophilaceae*, each family is represented by a single genus; *Microsphaera*, *Acidothermus*, *Sporichthya*, *Geodermatophilus* and *Blastococcus*, and *Frankia*, respectively (Figure 4). Within the genus *Frankia*, four main subdivisions were proposed (Normand *et al.*, 1996). These superseded an earlier separation of closely related strains that belonged to the *Alnus/Casuarina* host-infection group from strains of the *Elaeagnus* host-infection group (Nazaret *et al.*, 1991). These new subdivisions included: (i) a large group mainly comprised of typical N_2-fixing strains belonging to the *Alnus* and the *Casuarina* host-infection groups; (ii) uncultured endophytes of *Dryas*, *Coriaria*, and *Datisca* species; (iii) strains of the *Elaeagnus* host-infection group; and (iv) atypical non-N_2-fixing strains (Figures 4 and 5) (Normand *et al.*, 1996). Although based on a relatively small number of sequences, the proposed subdivisions are still valid for sequences obtained from isolates or uncultured nodule populations, all of which cluster into the respective groups, even though many sequences are only short and partial (Figure 5).

Isolates and/or uncultured endophytes of nodules from *Gymnostoma* (Navarro *et al.*, 1997) and from *Colletia hystrix*, *Talguenea quinquenervia*, *Trevoa trinervis*, *Retanilla ephedra*, *Discaria serratifolia*, and *D. toumatou*, for example, were closely related to frankiae of the *Elaeagnus* host-infection group (Figure 5) (Benson *et al.,* 1996). Further studies suggested that frankiae that infect members of the *Elaeagnaceae* might share a common ancestor with uncultured endophytes in root nodules of *Ceanothus* spp. (Murry *et al.,* 1997). Endophytes in nodules of *C. americanus* (Clawson *et al.*, 1998), *C. griseus* (Benson *et al.*, 1996) and *C. caerulens* (Ramirez-Saad *et al.*, 1998) were assigned to the *Dryas* host-infection group (Figures 4 and 5). The uncultured endophytes in root nodules of *Purshia tridentata* (Benson *et al.*, 1996), *Coriaria arborea* (Nick et al., 1992), *Coriaria plumosa* (Benson *et al.*, 1996), as well as the closely related endophytes in nodules of *Datisca cannabina* (Mirza *et al.*, 1994), also clustered with the *Dryas* group. Non-N_2-fixing and non-nodulating isolates from nodules of *C. caerulens* (Ramirez-Saad *et al.*, 1998), *D. cannabina* and *Coriaria nepalensis* (Mirza *et al.*, 1994) were not related to this group, however (Normand *et al.*, 1996), and clustered with the atypical frankiae isolates derived from *A. glutinosa* nodules (Figure 4). Uncultured non-N_2-fixing endophytes in nodules of *A. glutinosa* were found to belong to this hitherto undescribed cluster within the genus *Frankia* (Wolters *et al.*, 1997b).

Although 16S rRNA sequences enabled differentiation between frankiae of different host-infection groups as well as between uncultured endophytes of the spore(+) and spore(–) types in nodules of *A. incana* (Simonet *et al.*, 1994), differences are generally relatively small. Therefore, 16S-rRNA sequence analysis is the method of choice for the classification of frankiae at the genus level, but is of limited value for further classification within the genus (Hahn *et al.*, 1989a; Nazaret *et al.*, 1991), even though some correlation between groupings based on the analysis of variable regions of 16S rRNA (Nazaret *et al.*, 1991) and DNA-DNA relatedness studies (Fernandez *et al.*, 1989) has been observed.

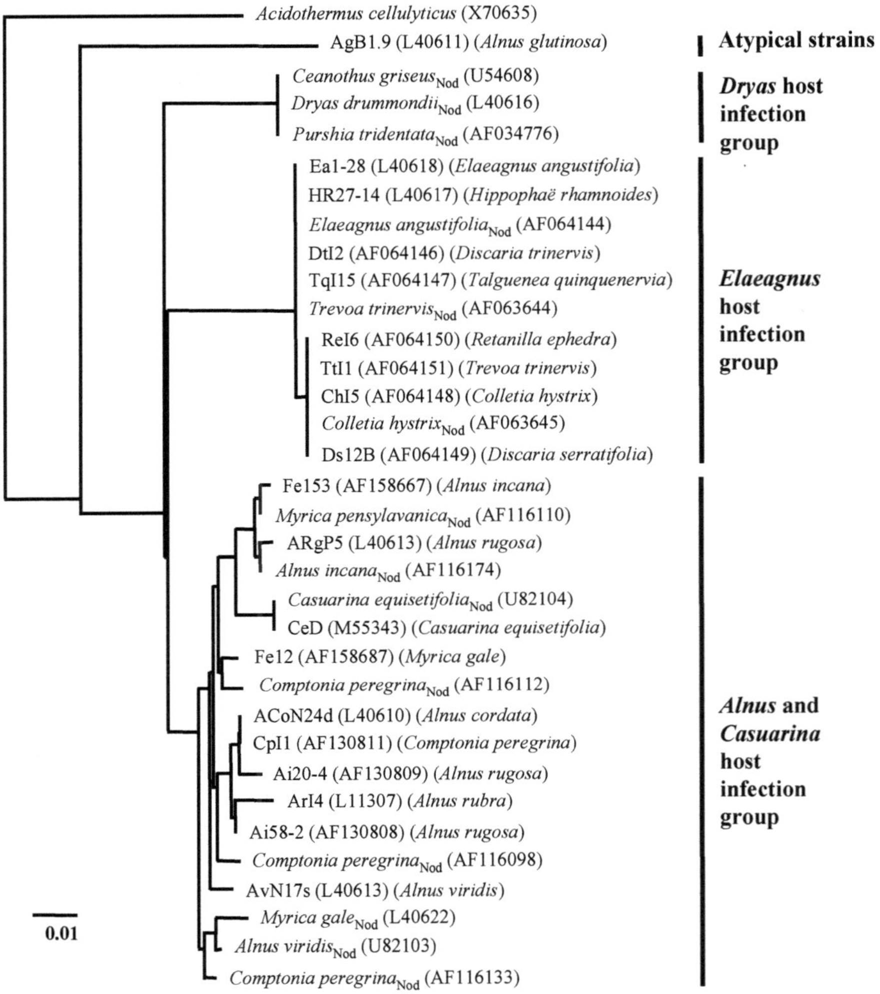

Figure 5. Neighbor-joining tree of aligned, mainly partial 16S rRNA sequences of selected Frankia *strains (with original host plant in brackets) and of uncultured nodule populations (host plant$_{Nod}$).*

4. DNA-DNA RELATEDNESS

The species in bacterial taxonomy is defined as a group of strains, including the type strain, that share 70% or greater whole genomic DNA-DNA relatedness with a ΔT_m of 5°C or less (Stackebrandt and Goebel, 1994; Wayne *et al.*, 1987). T_m is the

Strains analyzed	Genomic species (Original No.)	*Reference*	Genomic species (Proposed No.)
***Alnus* and *Casuarina* host infection group**			
ArI4, ArbN4b, ACN1, AGN1g, CpI1, AGN1$_{EXO}^{AG}$, MpI1, AirI1, AvcI1	A1	(An *et al.*, 1985)	1
ACoN24d, Ar24H3, Ar24O2, Ag24$_{251}$, ArI3, ARgN22d, ACN1, CpI1, A2J	1	(Fernandez *et al.*, 1989)	1
AvcI1, RBR162013, RBR162008, RBR162010, RBR162014	B1	(Bloom *et al.*, 1989)	1
CpI1, ArI3, ArI4, AvcI1, AvcI1.R1, AirI1, An2.1, An2.24, AgPM2.8, A7, A43, A51, A153	P1	(Akimov and Dobritsa, 1992)	1
ArI4	S2	(Shi and Ruan, 1992)	1
AV22c, ANV17o, Ac23$_{40}$, AI43$_1$	2	(Fernandez *et al.*, 1989)	2
ARgP5^{AG}	3	(Fernandez *et al.*, 1989)	3
Ai1c, Ai15a, Ag9b	P2	(Akimov and Dobritsa, 1992)	?
Ai6b	P3	(Akimov and Dobritsa, 1992)	?
AirI2	P5	(Akimov and Dobritsa, 1992)	?
CeD, CcI3, CeF, AllI1, ORS020608, ORS020609, CjI82, ORS022602	9	(Fernandez *et al.*, 1989)	4
C1, C2, C3, C4, C5	S1	(Shi and Ruan, 1992)	?
***Elaeagnus* host infection group**			
Ea1.12, Ea1$_2$, Ea2$_6$, Ea3$_3$, Eacm5$_1$, HR27$_{14}$	4	(Fernandez *et al.*, 1989)	5
TX31e^{HR}, EAN1$_{pec}$, HRX401a	5	(Fernandez *et al.*, 1989)	6
EUN1f	6	(Fernandez *et al.*, 1989)	7
HRN18a	7	(Fernandez *et al.*, 1989)	8
Ea50$_1$	8	(Fernandez *et al.*, 1989)	9
EaI10, EaI11, EaI8	P6	(Akimov and Dobritsa, 1992)	?
Cc1.17	P7	(Akimov and Dobritsa, 1992)	?
S15, S14, S13, Hr5-o, H109	P8	(Akimov and Dobritsa, 1992)	?
E1, E13, E15	10	(Lumini *et al.*, 1996)	?
2.1.7, HrI1	11	(Lumini *et al.*, 1996)	?
EuI1b	P9	(Akimov and Dobritsa, 1992)	?
G2, D11**	A2	(An *et al.*, 1985)	?
Atypical isolates from different actinorhizal plants			
AgI5	P4	(Akimov and Dobritsa, 1992)	10
PtI1	S3	(Shi and Ruan, 1992)	11

Figure 6. Data on DNA-DNA relatedness support the differentiation into host-infection groups and indicate a high diversity of the genus Frankia *with at least eleven genomic species reliably separated and several others to be confirmed (as indicated by the question marks).*

melting temperature of the hybrid determined by stepwise denaturation, and ΔT_m is the difference in T_m in degrees Celsius between the homologous and the heterologous hybrids formed under standard conditions (Vandamme *et al.*, 1996). Attempts to define species within the genus *Frankia* were initially based on host-infection groups as the major criterion. This criterion resulted in the proposal of two species, *F. elaeagni* and *F. alni*, with the latter composed of two subspecies, *vandijkii* and *pommerii*, based on the ability or not to form spores in root nodules (Lalonde *et al.*, 1988). DNA-DNA relatedness was used to differentiate strains from different host-infection groups with ratios of homology between 67% and 94% within one host-infection group or levels lower than 50% between different host-infection groups (Akimov and Dobritsa, 1992; An *et al.*, 1985; Bloom *et al.*, 1989; Fernandez *et al.*, 1989; Shi and Ruan, 1992).

These results have led to the emendation of *F. alni* (Fernandez *et al.*, 1989), but have also indicated a much higher diversity of the genus *Frankia* than initially suggested (Lalonde *et al.*, 1988). At least four genomic species were proposed in the *Alnus* and *Casuarina* host-infection group and five genomic species in the *Elaeagnus* host-infection group (Fernandez *et al.*, 1989). The large diversity in DNA-DNA relatedness of *Frankia* strains was confirmed by other studies that distinguished five genomic species in the *Alnus* and *Casuarina* host-infection group and four within the *Elaeagnus* host-infection group (Akimov and Dobritsa, 1992). To date, data on DNA-DNA relatedness are available for about 80 strains (Figure 6). Unfortunately, virtually all data sets were obtained with different strains so making comparative analysis and delineation of distinct genomic species highly speculative. Although the assignment of 19 genomic species has been indicated among *Frankia* strains (Dobritsa, 1998), eleven genomic species might be reliably separated with several others to be confirmed (Figure 6). This confirmation and further delineation of distinct genomic species, however, requires comprehensive studies on DNA-DNA relatedness that include the respective reference strains and additional phenotypic characterization (Wayne *et al.*, 1987).

5. 23S-rRNA SEQUENCE DATA

In contrast to 16S rRNA, 23S rRNA generally shows a greater variation of both sequence and length due to the presence or absence of additional sequence stretches compared to the 23S-rRNA gene sequence of *E. coli* (Höpfl *et al.*, 1989). Domain III of the 23S rRNA of gram-positive bacteria with a high DNA G+C content contains a large insertion specific for these organisms (Roller et al., 1992). This region exhibits considerable sequence differences between organisms of different genera (Roller *et al.*, 1992). A sequence comparison of the whole ribosomal operon for two *Frankia* strains, one belonging to the *Alnus* and the other to the *Casuarina* host-infection group, showed that the large variability of this region was only exceeded by that of the intergenic spacer regions (Hahn *et al.*, 1997).

Remarkable variation in this region was observed within the genus *Frankia*, when sequences of *Frankia* strains belonging to the *Alnus* host-infection group (Hönerlage *et al.*, 1994; Maunuksela *et al.*, 1999) were compared to those of *Frankia* belonging to other host-infection groups (Hönerlage *et al.*, 1994). The results confirmed the classification of the strains into host-infection groups (Hönerlage *et al.*, 1994) as proposed by 16S-rRNA sequence analysis (Normand *et al.*, 1996). In addition, strains belonging to the *Alnus* host-infection group could be separated roughly into four subgroups, three containing typical N_2-fixing strains and a fourth exclusively of non-N_2-fixing strains (Hönerlage *et al.*, 1994). Recently, a further subgroup containing typical N_2-fixing *Frankia* strains of the *Alnus* host-infection group was proposed (Maunuksela *et al.*, 1999).

Using sequence differences in the entire insertion as well as in target sites for existing probes as criteria for the design of new subgroups, a further differentiation could be made. Based on sequence divergence between *Frankia* strains of subgroup IIIb, for example, three subgroups represented by the original reference strain AgB32 (IIIb), as well as by strain ARgP5 (IIId) and by strain AgKG'84/4 (IIIe) might be designed within this *Alnus* host-infection subgroup (Figure 7).

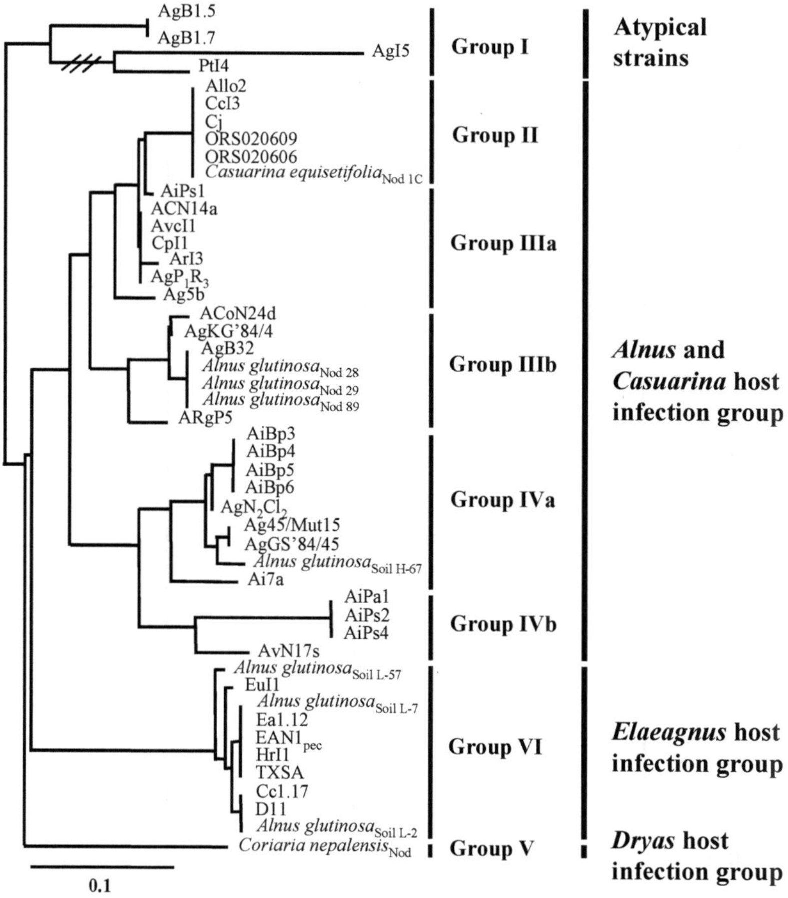

Figure 7. Neighbor-joining tree based on comparative sequence analysis of an actinomycetes-specific insertion in Domain III of the 23S rRNA of different Frankia *strains and uncultured populations in root nodules.*

Similar considerations could be made for subgroup I which comprises non-N_2-fixing *Frankia* strains. However, because members of the currently established subgroups I and IIIb were rarely encountered in root nodules of different origin (Maunuksela *et al.,* 1999; 2000; Nickel *et al.*, 1999; 2001; Zepp *et al.*, 1997), it is questionable whether the designation of further subgroups within these groups will

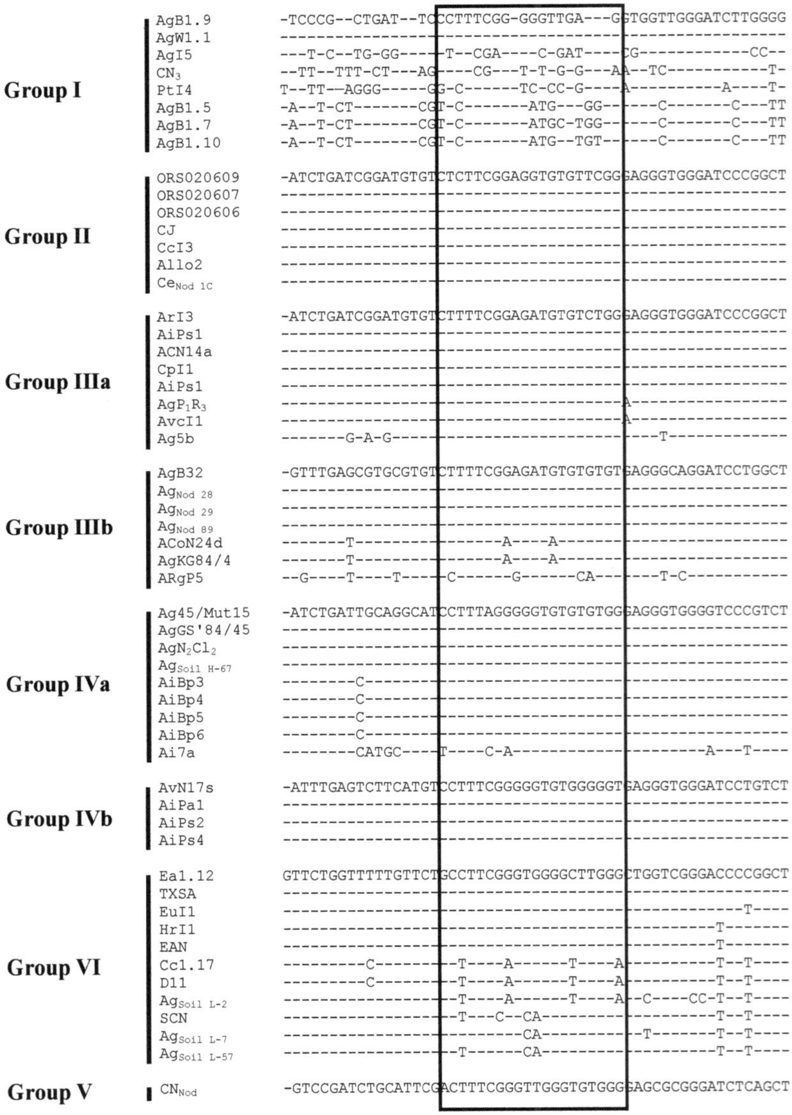

Figure 8. Potential signature sequences on the insertion in Domain III of the 23S rRNA that characterize pure cultures of Frankia *or uncultured endophytes in root nodules or soil that belong to specific host-infection groups and sub-groups.*

have any future implications on ecological studies. This might be different for the *Elaeagnus* host-infection group, where large sequence differences in the insertion of strains Cc1.17, Ea1.12 and SCN10a (up to 9 differences) might be the basis for the design of ecologically significant subgroups of frankiae.

The correlation of sequences of the highly variable insertion in Domain III of the 23S rRNA with host infection groups is surprising (Figure 8). Although this region is not supposed to have any phylogenetic significance (Roller *et al.*, 1992), these results confirm comparable data from the variable regions of 16S rRNA (Nazaret *et al.*, 1991b). The use of sequences from this region as targets for the detection and identification of frankiae is, therefore, limited to sequences retrieved from pure cultures and uncultured populations in root nodules, which allows a clear assignment of sequence information to members of the genus *Frankia* from a certain host-infection group. Based on this sequence information, attempts to detect frankiae in more complex environments, such as soils, become feasible. However, sequence information retrieved from such environments need to have its origin confirmed, even though high similarity values between the sequence retrieved and existing sequences that originated from confirmed frankiae might exist. An application of this approach for the analysis of diversity of *Frankia* populations in soils is, therefore, limited.

6. GENOMIC FINGERPRINTING

Fingerprinting or typing techniques can be used to discriminate frankiae at lower taxonomic levels. Restriction fragment length polymorphism (RFLP) analysis on DNA, either extracted from individual nodule lobes (Baker and Mullin, 1994) or on PCR-amplified nucleic acids, such as the 16S rDNA (Huguet *et al.*, 2001; McEwan *et al.,* 1994), the *gln*II gene (Cournoyer and Normand, 1994), the IGS of the 16S-23S rRNA operon (Gauthier et al., 1999; Maggia *et al.,* 1992; Navarro *et al.*, 1999; Ritchie and Myrold, 1999), the IGS of *nif*H-D (Cournoyer and Normand, 1994), or the IGS of *nif*D-K (Jamann *et al.,* 1993), have permitted discrimination among different *Frankia* genomic groups (Lumini *et al.*, 1996; McEwan *et al.*, 1994; Nalin *et al.,* 1997). However, the discriminative ability of this technique might be affected by the target nucleic acid and the restriction enzyme (or the combination of restriction enzymes) used (Normand and Chapelon, 1997).

A combination of RFLP analysis of 16S-23S rDNA IGS amplicons, which were digested with the restriction enzyme *Nci*I, with hybridization to specific oligonucleotide probes of PCR amplified *nif*H-D revealed a close relationship between 60 *Frankia* isolates obtained from root nodules of *Casuarina equisetifolia* and the reference strain CeD (Maggia *et al.*, 1992), This result suggested a small genetic diversity among *Frankia* strains that infect members of the Casuarinacaeae.

In contrast, five distinct genetic groups were recognized, using a set of different restriction enzymes, when amplicons of the 16S-23S rDNA IGS and the *nif*D-K IGS of pure cultures of *Frankia* belonging to the *Casuarina* host-infection group and uncultured populations in nodules on two *Casuarina* and two *Allocasuarina* species were analyzed (Rouvier *et al.*, 1996). Again, all pure cultures had similar patterns and were assigned to a group along with most of the uncultured endophytes in nodules from *C. equisetifolia.* A second group consisted of two uncultured endophytes from *C. equisetifolia*, whereas three further groups were comprised only of uncultured endophytes from *C. cunninghamiana*, *A. torulosa*, and *A. littoralis*, respectively (Rouvier *et al.*, 1996).

At the sub-species level, fingerprinting protocols, which are based on PCR with arbitrary primers, provide an effective means to distinguish closely related *Frankia* strains (Sellstedt *et al.,* 1992). Differentiation to the species, subspecies and strain level can also be achieved by PCR-assisted fingerprinting techniques that target consensus motifs of repetitive elements common to prokaryotic genomes (rep-PCR) (de Bruijn, 1992; Lupski and Weinstock, 1992; Nick and Lindstrom, 1994; Versalovic *et al.,* 1991; 1994) (Figure 9).

Rep-PCR genomic fingerprinting covers three PCR-based fingerprinting protocols (REP-PCR, ERIC-PCR, and BOX-PCR genomic fingerprinting, respectively), in which distinct genomic regions that are located between the

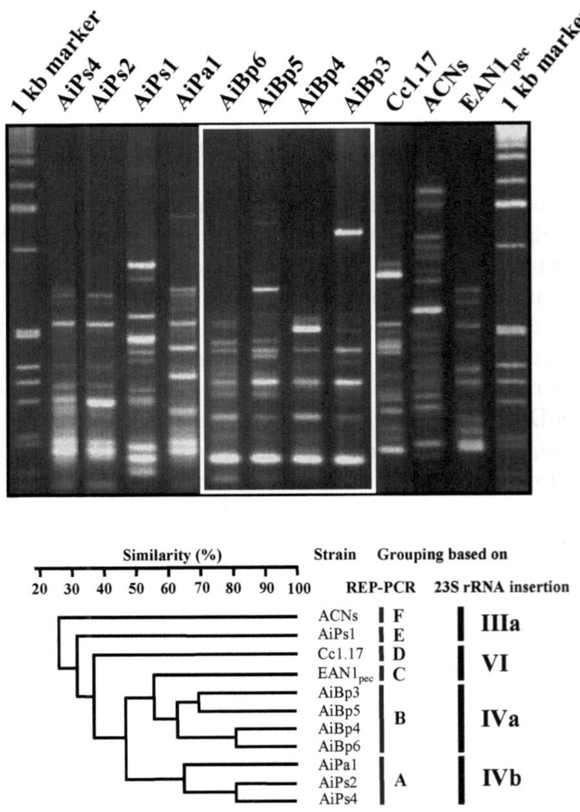

Figure 9. Rep-PCR fingerprint pattern of genomic DNA from Frankia *strains, using BOXA1R primer and the corresponding UPGMA dendrogram (modified from Maunuksela et al., 1999).*

repetitive extragenic palindromic (REP) sequence (Stern et al., 1984), the enterobacterial repetitive intergenic consensus (ERIC) sequence (Hulton *et al.*, 1991), or the BOX element (Martin *et al.*, 1992), respectively, are selectively amplified. The resulting patterns of amplification products (fingerprints) generally correlate to the classification based on other methods (Louws *et al.*, 1994; Versalovic *et al.*, 1994). For *Frankia* strains, a good correlation between degrees of relatedness has been shown by comparative 16S-rRNA sequence analysis and rep-PCR fingerprinting (Murry *et al.*, 1995). Using rep-PCR, the genomic fingerprints of DNA from root nodules of several *Ceanothus* spp. differentiated twelve distinct groups, indicating considerable genetic diversity of *Frankia* in the nodules sampled (Murry *et al.*, 1997). A similar high degree of diversity was obtained for isolates from nodules of *A. incana* (Maunuksela *et al.*, 1999; 2000) as well as for endophytes in nodule homogenates (Maunuksela *et al.*, 2000). These could not be differentiated by comparative sequence analysis of a 23S-rRNA insertion but all exhibited different rep-PCR patterns. In nodule lobe homogenates, the complexity of the fingerprint patterns was found to be similar to the patterns obtained using DNA from *Frankia* isolates (Jeong and Myrold, 1999; Maunuksela *et al.*, 1999), suggesting no significant impact of either plant or additional microbial DNA on the generation of rep-PCR fingerprint patterns.

7. CONCLUSIONS

Polyphasic taxonomy approaches that integrate information retrieved by a wide range of techniques on different levels of taxonomic resolution are revealing much about the genetic diversity and distribution of *Frankia*, and are refining and expanding knowledge of endophyte-host specificities. PCR-based approaches have been used to unravel the phylogenetic relationships of isolates, as well as of uncultured endophytes in root nodules of many actinorhizal plants from which no isolates have been obtained. Furthermore, these analyses have led to the emendation of the family *Frankiaceae* to contain a single genus *Frankia* with four main subdivisions: (i) a large group mainly comprising *Frankia alni* and other typical N_2-fixing strains belonging to the *Alnus* and the *Casuarina* host infection groups, respectively; (ii) uncultured endophytes of *Dryas*, *Coriaria*, and *Datisca* species; (iii) strains of the *Elaeagnus* host-infection group; and (iv) atypical non-N_2-fixing strains. The differentiation into host-infection groups is generally supported by data on DNA-DNA relatedness of pure cultures of *Frankia* and has led to the emendation of *F. alni* and indicated a large diversity within the genus *Frankia.* Considerable diversity among both cultured *Frankia* strains and uncultured endophytes in nodules has been confirmed by fingerprinting techniques. The growing databases of discriminative target sequences open the door to more sophisticated studies of the distribution and fate of specific *Frankia* populations in the environment and, thus, may lead to advancements in the management of actinorhizal plants and *Frankia* for human benefit.

REFERENCES

Akimov, V. N., and Dobritsa, S. V. (1992). Grouping of *Frankia* strains on the basis of DNA relatedness. *Syst. Appl. Microbiol., 15*, 372-379.

Akkermans, A. D. L., Hahn, D. and Baker, D. D. (1991). The family *Frankiaceae*. In A. Balows, H. G. Trüper, M. Dworkin, W. Harder and K.-H. Schleifer (Eds.), *The Prokaryotes* (pp. 1069-1084). Heidelberg, Germany: Springer-Verlag.

An, C. S., Riggsby, W. S. and Mullin, B. C. (1985). Relationships of *Frankia* isolates based on deoxyribonucleic acid homology studies. *Int. J. Syst. Bacteriol., 35*, 140-146.

Baker, D. D. (1990). Methods for the isolation, culture and characterization of the *Frankiaceae*: Soil actinomycetes and symbionts of actinorhizal plants. In H. D. Labeda (Ed.), *Isolation of biotechnological organisms from nature* (pp. 213-236). New York, USA: McGraw Hill, Inc.

Baker, D. D., Kidd, G. H., and Torrey, J. G. (1979a). Separation of actinomycete nodule endophytes from crushed nodule suspensions by Sephadex fractionation. *Bot. Gaz., 140 (Suppl.),* S49-S51.

Baker, D. D., Newcomb, W., and Torrey, J. G. (1980). Characterization of an ineffective actinorhizal microsymbiont, *Frankia* EuI1 (Actinomycetales). *Can. J. Microbiol., 26,* 1072-1089.

Baker, D. D., and O'Keefe, D. (1984). A modified sucrose fractionation procedure for the isolation of frankiae from actinorhizal root nodules and soil samples. *Plant Soil, 78,* 23-28.

Baker, D. D., Torrey, J. G., and Kidd, G. H. (1979b). Isolation by sucrose-density fractionation and cultivation *in vitro* of actinomycetes from nitrogen-fixing root nodules. *Nature, 281*, 76-78.

Baker, D. D. (1987). Relationships among pure cultured strains of *Frankia* based on host specificity. *Physiol. Plant., 70*, 245-248.

Baker, D. D., and Mullin, B. C. (1994). Diversity of *Frankia* nodule endophytes of the actinorhizal shrub *Ceanothus* as assessed by RFLP patterns from single nodule lobes. *Soil Biol. Biochem., 26*, 547-552.

Becking, J. H. (1970). *Frankiaceae* fam. nov (*Actinomycetales*) with one new combination and six new species of the genus *Frankia* Brunschorst 1886, 174. *Int. J. Syst. Bacteriol., 20*, 201-220.

Bennasar, A., Rossello-Mora, R., Lalucat, J., and Moore, E. R. B. (1996). 16S rRNA gene sequence analysis relative to genomovars of *Pseudomonas stutzeri* and proposal of *Pseudomonas balearica* sp. nov. *Int. J. Syst. Bacteriol., 46*, 200-205.

Benson, D. R. (1982). Isolation of *Frankia* strains from alder actinorhizal root nodules. *Appl. Environ. Microbiol., 44*, 461-465.

Benson, D. R., Buchholz, S. E., and Hanna, D. G. (1984). Identification of *Frankia* strains by two-dimensional polyacrylamide gel electrophoresis. *Appl. Environ. Microbiol., 47*, 489-494.

Benson, D. R., and Hanna, D. G. (1983). *Frankia* diversity in an alder stand as estimated by sodium dodecyl sulfate-polyacrylamide gel electrophoresis of whole-cell proteins. *Can. J. Bot., 61*, 2919-2923.

Benson, D. R., and Silvester, W. B. (1993). Biology of *Frankia* strains, actinomycete symbionts of actinorhizal plants. *Microbiol. Rev., 57*, 293-319.

Benson, D. R., Stephens, D. W., Clawson, M. L., and Silvester, W. B. (1996). Amplification of 16S rRNA genes from *Frankia* strains in root nodules of *Ceanothus griseus, Coriaria arborea, Coriaria plumosa, Discaria toumatou,* and *Purshia tridentata. Appl. Environ. Microbiol., 62*, 2904-2909.

Berry, A., and Torrey, J. G. (1979). Isolation and characterization *in vivo* and *in vitro* of an actinomycetous endophyte from *Alnus rubra* Bong. In J. C. Gordon, C. T. Wheeler and D. A. Perry (Eds.), *Symbiotic nitrogen fixation in the management of temperate forests* (pp. 69-83). Corvallis, OR: Oregon State University.

Bloom, R. A., Mullin, B. C., and Tate, III, R. L. (1989). DNA restriction patterns and DNA-DNA solution hybridization studies of *Frankia* isolates from *Myrica pennsylvanica* (Bayberry). *Appl. Environ. Microbiol., 55*, 2155-2160.

Bosco, M., Fernandez, M. P., Simonet, P., Materassi, R., and Normand, P. (1992). Evidence that some *Frankia* sp. strains are able to cross boundaries between *Alnus* and *Elaeagnus* host specificity groups. *Appl. Environ. Microbiol., 58*, 1569-1576.

Bosco, M., Lumini, E., and Normand, P. (1996). PCR-RFLP direct fingerprinting of uncultured *Frankia* microsymbiont from *Dryas drummondii* nodules. *Ann. Microbiol. Enzim., 46*, 115-123.

Burggraaf, A. J. P. (1984). *Isolation, cultivation and characterization of Frankia strains from actinorhizal root nodules.* Ph.D. Thesis, University of Leiden, The Netherlands.

Callaham, D., Del Tredici, P., and Torrey, J. G. (1978). Isolation and cultivation *in vitro* of the actinomycete causing root nodulation in *Comptonia. Science, 199*, 899-902.

Clawson, M. L., Caru, M., and Benson, D. R. (1998). Diversity of *Frankia* sp. strains in root nodules of plants from the *Elaeagnaceae* and *Rhamnaceae*. *Appl. Environ. Microbiol., 64*, 3539-3543.

Cournoyer, B., and Lavire, C. (1999). Analysis of *Frankia* evolutionary radiation using *glnII* sequences. *FEMS Microbiol. Lett., 177*, 29-34.

Cournoyer, B., and Normand, P. (1994). Characterization of a spontaneous thiostrepton-resistant *Frankia alni* infective isolate using PCR-RFLP of *nif* and *gln*II genes. *Soil Biol. Biochem., 26*, 553-559.

de Bruijn, F .J. (1992). Use of repetitive (repetitive extragenic palindromic and enterobacterial repetitive intergenic consensus) sequences and the polymerase chain reaction to fingerprint the genomes of *Rhizobium meliloti* isolates and other soil bacteria. *Appl. Environ. Microbiol., 58*, 2180-2187.

Diem, H. G., and Dommergues, Y. R. (1983). The isolation of *Frankia* from nodules of *Casuarina*. *Can. J. Bot., 61*, 2822-2825.

Dobritsa, S. V. (1998). Grouping of *Frankia* strains on the basis of susceptibility to antibiotics, pigment production and host specificity. *Int. J. Syst. Bacteriol., 48*, 1265-1275.

Embley, T. M., Smida, J., and Stackebrandt, E. (1988). Reverse transcriptase sequencing of 16S ribosomal RNA from *Faenia rectivirgula*, *Pseudonocardia thermophila* and *Saccharopolyspora hirsuta*, three Wall type IV actinomycetes which lack mycolic acids. *J. Gen. Microbiol., 134*, 961-966.

Fernandez, M. P., Meugnier, H., Grimont, P. A. D., and Bardin, R. (1989). Deoxyribonucleic acid relatedness among members of the genus *Frankia*. *Int. J. Syst.Bacteriol., 39*, 424-429.

Fox, G. E., and Stackebrandt, E. (1987). The application of 16S rRNA cataloguing and 5S rRNA sequencing in microbial systematics. *Method. Microbiol., 19* , 405-458.

Fox, G. E., Wisotzkey, J. D., and Jurtshuk, P .J. (1992). How close is close: 16S rRNA sequence identiy may not be sufficient to guarantee species identity. *Int. J. Syst. Bacteriol., 42*, 166-170.

Gardes, M., Bousquet, J., and Lalonde, M. (1987). Isozyme variation among 40 *Frankia* strains. *Appl. Environ. Microbiol., 53*, 1596-1603.

Gardes, M., and Lalonde, M. (1987). Identification and subgrouping of *Frankia* strains using sodium dodecyl sulfate-polyacrylamide gel electrophoresis. *Physiol. Plant., 70*, 237-244.

Gauthier, D., Jaffré, T., and Prin, Y. (1999). Occurrence of both *Casuarina*-infective and *Elaeagnus*-infective *Frankia* strains within actinorhizae of *Casuarina collina*, endemic to New Caledonia. *Eur. J. Soil Biol., 35*, 9-15.

Girgis, M. G. Z., and Schwencke, J. (1993). Differentiation of *Frankia* strains by their electrophoretic patterns of intracellular esterases and aminopeptidases. *J. Gen. Microbiol., 139*, 2225-2232.

Goodfellow, M. (1986). Actinomycete systematics: Present and future prospects. In G. Szabo, S. Biro and M. Goodfellow (Eds.), *Biological, biochemical and biomedical aspects of actinomycetes*. (pp. 487-496). Budapest, Hungary: Akademiai Kiado.

Hafeez, F., Akkermans, A. D. L., and Chaudhary, A. H. (1984). Morphology, physiology and infectivity of two *Frankia* isolates An 1 and An 2 from root nodules of *Alnus nitida*. *Plant Soil, 78*, 45-59.

Hahn, D., Dorsch, M., Stackebrandt, E., and Akkermans, A. D. L. (1989a). Synthetic oligonucleotide probes for identification of *Frankia* strains. *Plant Soil, 118*, 211-219.

Hahn, D., Lechevalier, M. P., Fischer, A., and Stackebrandt, E. (1989b). Evidence for a close relationship between members of the genera *Frankia*, *Geodermatophilus* and *Blastococcus* and emendation of the family *Frankiaceae*. *Syst. Appl. Microbiol., 11*, 236-242.

Hahn, D., Starrenburg, M. J. C., and Akkermans, A. D. L. (1988). Variable compatibility of cloned *Alnus glutinosa* ecotypes against ineffective *Frankia* strains. *Plant Soil, 107*, 233-243.

Hahn, D., Zepp, K., and Zeyer, J. (1997). Whole cell hybridization as a tool to study *Frankia* populations in root nodules. *Physiol. Plant., 99*, 696-706.

Hiyoshi, T., Sasakawa, H., and Yatazawa, M. (1988). Isolation of *Frankia* strains from root nodules of *Myrica rubra*. *Soil Sci. Plant Nutr., 34*, 107-116.

Hönerlage, W., Hahn, D., Zepp, K., Zeyer, J., and Normand, P. (1994). A hypervariable 23S rRNA region provides a discriminating target for specific characterization of uncultured and cultured *Frankia*. *Syst. Appl. Microbiol., 17*, 433-443.

Höpfl, P., Ludwig, W., Schleifer, K.-H., and Larsen, N. (1989). The 23S ribosomal RNA higher-order structure of *Pseudomonas cepacia* and other procaryotes. *Eur. J. Biochem., 185*, 355-364.

Hosted, T. J., Rochefort, D. A., and Benson, D. R. (1993). Close linkage of genes encoding glutamine synthetases I and II in *Frankia alni* CpI1. *J. Bacteriol., 175*, 3679-3684.

Huguet, V., McCray Batzli, J., Zimpfer, J. F., Normand, P., Dawson, J. O., and Fernandez, M. P. (2001). Diversity and specificity of *Frankia* strains in nodules of sympatric *Myrica gale*, *Alnus incana*, and *Shepherdia canadensis* determined by *rrs* gene polymorphism. *Appl. Environ. Microbiol., 67*, 2116-2122.

Hulton, C. S., Higgins, C. F., and Sharp, P. M. (1991). ERIC sequences: A novel family of repetitive elements in the genomes of *Escherichia coli*, *Salmonella typhimurium* and other enterobacteria. *Mol. Microbiol., 5*, 825-834.

Huss-Danell, K. (1997). Actinorhizal symbioses and their N_2-fixation. *New Phytol., 136*, 375-405.

Huss-Danell, K., and Myrold, D. D. (1994). Intragenic variation in nodulation of *Alnus*: Consequences for quantifying *Frankia* nodulation units in soil. *Soil Biol. Biochem., 26*, 525-531.

Igual, J. M., Velazquez, E., Mateos, P. F., Rodriguez-Barrueco, C., Cervantes, E., and Martinez-Molina, E. (2001). Cellulase isoenzyme profiles in *Frankia* strains belonging to different cross-inoculation groups. *Plant Soil, 229*, 35-39.

Jamann, S., Fernandez, M. P., and Normand, P. (1993). Typing method for N_2-fixing bacteria based on PCR-RFLP-application to the characterization of *Frankia* strains. *Mol. Ecol., 2*, 17-26.

Jeong, S.-C., and Myrold, D. D. (1999). Genomic fingerprinting of *Frankia* microsymbionts from *Ceanothus* copopulations using repetitive sequences and polymerase chain reaction. *Can. J. Bot., 77*, 1220-1230.

Kim, S. C., Ku, C. D., Park, M. C., Kim, C. H., Song, S. D., and An, C. S. (1993). Isolation of symbiotic *Frankia* Eulk1 strain from root nodule of *Elaeagnus umbellata. Korean J. Bot., 36*, 177-182.

Lalonde, M., Simon, L., Bousquet, J., and Seguin, A. (1988). Advances in the taxonomy of *Frankia*: Recognition of species *alni* and *elaeagni* and novel subspecies *pommerii* and *vandijkii*. In H. Bothe, F. J. de Bruijn and W. E. Newton (Eds.), *Nitrogen fixation: Hundred years after* (pp. 671-680). Stuttgart, Germany: Gustav Fischer Verlag.

Lane, D. J., Field, K. G., Olsen, G. J., and Pace, N. R. (1988). Reverse transcriptase sequencing of ribosomal RNA for phylogenetic analysis. *Method. Enzymol., 167*, 138-144.

Lane, D. J., Pace, B., Olsen, G. J., Stahl, D. A., Sogin, M. L., and Pace, N. R. (1985). Rapid determination of 16S ribosomal RNA sequences for phylogenetic analyses. *Proc. Natl. Acad. Sci. USA, 82*, 6955-6959.

Lechevalier, M. P. (1994). Taxonomy of the genus *Frankia* (Actinomycetales). *Int. J. Syst. Bacteriol., 44*, 1-8.

Lechevalier, M. P., and Ruan, J. S. (1984). Physiology and chemical diversity of *Frankia* spp. isolated from nodules of *Comptonia peregrina* (L.) Coult. and *Ceanothus americanus* L. *Plant Soil, 78*, 15-22.

Louws, F. J., Fulbright, D. W., Stephens, C. T., and de Bruijn, F. J. (1994). Specific genomic fingerprints of phytopathogenic *Xanthomonas* and *Pseudomonas* pathovars and strains generated with repetitive sequences and PCR. *Appl. Environ. Microbiol., 60*, 2286-2295.

Ludwig, L., Kirchhof, G., Klugbauer, N., Weizenegger, M., Betzl, D., Ehrmann, M., Hertel, C., Jilg, S., Tatzel, R., Zitzelsberger, H., Liebl, S., Hochberger, M., Shah, J., Lane, D., Wallnofer, P. R., and Schleifer, K.-H. (1992). Complete 23S ribosomal RNA sequences of gram-positive bacteria with a low DNA G+C content. *Syst. Appl. Microbiol., 15*, 487-501.

Ludwig, W., Strunk, O., Klugbauer, N., Weizenegger, M., Neumaier, J., Bachleitner, M., and Schleifer, K.-H. (1998). Bacterial phylogeny based on comparative sequence analysis. *Electrophoresis, 19*, 554-568.

Lumini, E., and Bosco, M. (1996). PCR-restriction fragment length polymorphism identification and host range of single-spore isolates of the flexible *Frankia* sp. strain UFI 132715. *Appl. Environ. Microbiol., 62*, 3026-3029.

Lumini, E., Bosco, M., and Fernandez, M. P. (1996). PCR-RFLP and total DNA homology revealed three related genomic species among broad-host-range *Frankia* strains. *FEMS Microbiol. Ecol., 21*, 303-311.

Lupski, J. R., and Weinstock, G. M. (1992). Short, interspersed repetitive DNA sequences in prokaryotic genomes. *J. Bacteriol., 174*, 4525-4529.

Maggia, L., Nazaret, S., and Simonet, P. (1992). Molecular characterization of *Frankia* isolates from *Casuarina equisetifolia* root nodules harvested in West Africa (Senegal and Gambia). *Acta Œcol., 13*, 453-461.

Maidak, B. L., Olsen, G.J., Larsen, N., Overbeek, R., McCaughey, M. J., and Woese, C. R. (1997). The RDP (Ribosomal Database Project). *Nucl. Acids Res., 25*, 109-111.

Maréchal, J., Clement, B., Nalin, R., Gandon, C., Orso, S., Cvejic, J. H., Bruneteau, M., Berry, A. M., and Normand, P. (2000). A *rec*A gene phylogenetic analysis confirms the close proximity of *Frankia* to *Acidothermus. Int. J. Syst. Evol. Microbiol., 50*, 781-785.

Martin, B., Humbert, O., Camara, M., Guenzi, E., Walker, J., Mitchell, T., Andrew, P., Prudhomme, M., Alloing, G., Hakenbeck, R., Morrison, D. A., Boulnois, G. J., and Claverys, J.-P. (1992). A highly conserved repeated DNA element located in the chromosome of *Streptococcus pneumoniae*. *Nucl. Acids Res., 20,* 3479-3483.

Martinez-Murcia, A. J., Benlloch, S., and Collins, M. D. (1992). Phylogenetic interrelationships of members of the genera *Aeromonas* and *Plesiomonas* as determined by 16S ribosomal DNA sequencing: Lack of congruence with results of DNA-DNA hybridizations. *Int. J. Syst. Bacteriol., 42,* 412-421.

Maunuksela, L., Hahn, D., and Haahtela, K. (2000). Effect of freezing of soils on nodulation capacities of total and specific *Frankia* populations. *Symbiosis, 29,* 107-120.

Maunuksela, L., Zepp, K., Koivula, T., Zeyer, J., Haahtela, K., and Hahn, D. (1999). Analysis of *Frankia* populations in three soils devoid of actinorhizal plants. *FEMS Microbiol. Ecol., 28,* 11-21.

McEwan, N. R., Wheeler, C. T., and Milner, J. J. (1994). Strain discrimination of cultured and symbiotic *Frankia* by RFLP-PCR. *Soil Biol. Biochem., 26,* 541-545.

Mirza, M. S., Hahn, D., and Akkermans, A. D. L. (1992). Isolation and characterization of *Frankia* strains from *Coriaria nepalensis*. *Syst. Appl. Microbiol., 15,* 289-295.

Mirza, M. S., Hahn, D., Dobritsa, S. V., and Akkermans, A. D. L. (1994). Phylogenetic studies on uncultured *Frankia* populations in nodules of *Datisca cannabina*. *Can. J. Microbiol., 40,* 313-318.

Mirza, M. S., Janse, J. D., Hahn, D., and Akkermans, A. D. L. (1991). Identification of atypical *Frankia* strains by fatty acid analysis. *FEMS Microbiol. Lett., 83,* 91-97.

Mort, A., Normand, P., and Lalonde, M. (1983). 2-O-Methyl-D-mannose, a key sugar in the taxonomy of *Frankia*. *Can. J. Microbiol., 29,* 993-1002.

Mullis, K. B., and Faloona, F. (1987). Specific synthesis of DNA *in vitro* via a polymerase-catalysed chain reaction. *Method. Enzymol., 155,* 335-350.

Murray, R. G. E., Brenner, D. J., Colwell, R. R., De Vos, P., Goodfellow, M., Grimont, P. A. D., Pfennig, N., Stackebrandt, E., and Zavarzin, G.A. (1990). Report of the ad hoc committee on approaches to taxonomy within the *Proteobacteria*. *Int. J. Syst. Bacteriol., 40,* 213-215.

Murry, M. A., Konopka, A. S., Pratt, S. D., and Vandergon, T. L. (1997). The use of PCR-based typing methods to assess the diversity of *Frankia* nodule endophytes of the actinorhizal shrub *Ceanothus*. *Physiol. Plant., 99,* 714-721.

Murry, M. A., Zhang, D., Schneider, M., and de Bruijn, F. J. (1995). Use of repetitive sequences and the polymerase chain reaction (rep-PCR) to fingerprint the genomes of *Frankia* isolates. *Symbiosis, 19,* 223-240.

Mylvaganam, S., and Dennis, P. P. (1992). Sequence heterogeneity between the two genes encoding 16S rRNA from the halophilic archaeabacterium *Haloarcula marismortui*. *Genetics, 130,* 399-410.

Nalin, R., Normand, P., and Domenach, A. M. (1997). Distribution and N_2-fixing activity of *Frankia* strains in relation to soil depth. *Physiol. Plant., 99,* 732-738.

Navarro, E., Nalin, R., Gauthier, D., and Normand, P. (1997). The nodular microsymbionts of *Gymnostoma* spp. are *Elaeagnus*-infective *Frankia* strains. *Appl. Environ. Microbiol., 63,* 1610-1616.

Navarro, E., Jaffre, T., Gauthier, D., Gourbiere, F., Rinaudo, G., Simonet, P., and Normand, P. (1999). Distribution of *Gymnostoma* spp. microsymbiotic *Frankia* strains in New Caledonia is related to soil type and to host-plant species. *Mol. Ecol., 8,* 1781-1788.

Nazaret, S., Cournoyer, B., Normand, P., and Simonet, P. (1991). Phylogenetic relationships among *Frankia* genomic species determined by use of amplified 16S rRNA sequences. *J. Bacteriol., 173,* 4072-4078.

Nazaret, S., Simonet, P., Normand, P., and Bardin, R. (1989). Genetic diversity among *Frankia* isolated from *Casuarina* nodules. *Plant Soil, 118,* 241-247.

Nick, G., and Lindstrom, K. (1994). Use of repetitive sequences and the polymerase chain reaction to fingerprint the genomic DNA of *Rhizobium galegae* strains and to identify the DNA obtained by sonicating the liquid cultures and root nodules. *Syst. Appl. Microbiol., 17,* 265-273.

Nick, G., Paget, E., Simonet, P. Moiroud, A., and Normand, P. (1992). The nodular endophytes of *Coriaria* spp. form a distinct lineage within the genus *Frankia*. *Mol. Ecol., 1,* 175-181.

Nickel, A., Hahn, D., Zepp, K., and Zeyer, J. (1999). *In situ* analysis of introduced *Frankia* populations in root nodules of *Alnus glutinosa* grown under different water availability. *Can. J. Bot., 77,* 1231-1238.

Nickel, A., Pelz, O., Hahn, D., Saurer, M., Siegwolf, R., and Zeyer, J. (2001). Effect of inoculation and leaf-litter amendment on establishment of nodule-forming *Frankia* populations in soil. *Appl. Environ. Microbiol., 67*, 2603-2609.

Normand, P., and Chapelon, C. (1997). Direct characterization of *Frankia* and of close phyletic neighbors from an *Alnus viridis* rhizosphere. *Physiologia Plantarum, 99*, 722-731.

Normand, P., Gouy, M., Cournoyer, B., and Simonet, P. (1992). Nucleotide sequence of *nifD* from *Frankia alni* strain ArI3: Phylogenetic inferences. *Mol. Biol. Evol., 9*, 495-506.

Normand, P., Orso, S., Cournoyer, B., Jeannin, P., Chapelon, C., Dawson, J., Evtushenko, L., and Misra, A. K. (1996). Molecular phylogeny of the genus *Frankia* and related genera and emendation of the family *Frankiaceae*. *Int. J. Syst. Bacteriol., 46*, 1-9.

Nübel, U., Engelen, B., Felske, A., Snaidr, J., Wieshuber, A., Amann, R. I., Ludwig, W., and Backhaus, H. (1996). Sequence heterogeneities of genes encoding 16S rRNAs in *Paenibacillus polymyxa* detected by temperature gradient gel electrophoresis. *J. Bacteriol., 178*, 5636-5643.

Quispel, A., Svendsen, A. B., Schripsema, J., Baas, W. J., Erkelens, C., and Lugtenberg, J. (1989). Identification of a dipterocarpol as isolation factor for the induction of primary isolation of *Frankia* from root nodules of *Alnus glutinosa* (L.) Gaertn. *Mol. Plant-Microbe Interact., 2*, 107-112.

Ramirez-Saad, H., Janse, J., and Akkermans, A.D.L. (1998). Root nodules of *Ceanothus caerulens* contain both the N_2-fixing *Frankia* endophyte and a phylogenetically related Nod^-/Fix^- actinomycete. *Can. J. Microbiol., 44*, 140-148.

Ritchie, N. J., and Myrold, D. D. (1999). Geographic distribution and genetic diversity of *Ceanothus*-infective *Frankia* strains. *Appl. Environ. Microbiol., 65*, 1378-1383.

Rochefort, D. A., and Benson, D. R. (1990). Molecular cloning, sequencing, and expression of the glutamine synthetase II (*gln*II) gene from the Actinomycete root nodule symbiont *Frankia* sp. strain CpI1. *J. Bacteriol., 172*, 5335-5342.

Roller, C., Ludwig, W., and Schleifer, K.-H. (1992). Gram-positive bacteria with a high DNA G+C content are characterized by a common insertion within their 23S rRNA genes. *J. Gen. Microbiol., 138*, 1167-1175.

Rosbrook, P.A., Burggraaf, A. J. P., and Reddell, P. (1989). A comparison of two methods and different media for isolating *Frankia* from *Casuarina* root nodules. *Plant Soil, 120*, 187-193.

Rossello-Mora, R., and Amann, R. I. (2001). The species concept for prokaryotes. *FEMS Microbiol. Rev., 25*, 39-67.

Rouvier, C., Nazaret, S., Fernandez, M. P., Picard, B., Simonet, P., and Normand, P. (1992). *rrn* and *nif* intergenic spacers and isoenzyme patterns as tools to characterize *Casuarina*-infective *Frankia* strains. *Acta Œcol., 13*, 487-495.

Rouvier, C., Prin, Y., Reddell, P., Normand, P., and Simonet, P. (1996). Genetic diversity among *Frankia* strains nodulating members of the family *Casuarinaceae* in Australia revealed by PCR and restriction fragment length polymorphism analysis with crushed root nodules. *Appl. Environ. Microbiol., 62*, 979-985.

Saiki, R. K., Gelfand, D. H., Stoffel, S., Scharf, S. J., Higuchi, R., Horn, G. T., Mullis, K. B., and Erlich, H. A. (1988). Primer directed enzymatic amplification of DNA with a thermostable DNA polymerase. *Science, 239*, 487-491.

Schwintzer, C. R., and Tjepkema, J. D. (1990). *The biology of* Frankia *and actinorhizal plants*. San Diego, CA: Academic Press.

Sellstedt, A., Wullings, B., Nyström, U., and Gustafsson, P. (1992). Identification of *Casuarina-Frankia* strains by use of polymerase chain reaction (PCR) with arbitrary primers. *FEMS Microbiol. Lett., 93*, 1-6.

Shi, Y., and Ruan, J. (1992). DNA base composition and homology values in the classification of some *Frankia* strains. *Acta Microbiol. Sin., 32*, 133-136.

Simon, L., Jabaji-Hare, S., Bousquet, J., and Lalonde, M. (1989). Confirmation of *Frankia* species using cellular fatty acid analysis. *Syst. Appl. Microbiol., 11*, 229-235.

Simonet, P., Bosco, M., Chapelon, C., Moiroud, A., and Normand, P. (1994). Molecular characterization of *Frankia* microsymbionts from spore-positive and spore-negative nodules in a natural alder stand. *Appl. Environ. Microbiol., 60*, 1335-1341.

Simonet, P., Grosjean, M. C., Misra, A. K., Nazaret, S., Cournoyer, B., and Normand, P. (1991). *Frankia* genus-specific characterization by polymerase chain reaction. *Appl. Environ. Microbiol., 57*, 3278-3286.

Simonet, P., Nguyen-Thi, L., Teissier du Cros, E., and Bardin, R. (1988). Identification of *Frankia* strains by direct DNA hybridization of crushed nodules. *Appl. Environ. Microbiol., 54*, 2500-2503.

Simonet, P., Normand, P., Moiroud, A., and Bardin, R. (1990). Identification of *Frankia* strains in nodules by hybridization of polymerase chain reaction products with strain-specific oligonucleotide probes. *Arch. Microbiol., 153,* 235-240.

Stackebrandt, E., and Goebel, B. M. (1994). Taxonomic note: A place for DNA-DNA reassociation and 16S rRNA sequence analysis in the present species definition in bacteriology. *Int. J. Syst. Bacteriol., 44,* 846-849.

Stackebrandt, E., Ludwig, W., and Fox, G. E. (1985). 16S rRNA oligonucleotide cataloguing. *Method. Microbiol., 17,* 75-107.

Stackebrandt, E., Rainey, F. A., and Ward-Rainey, N. L. (1997). Proposal for a new hierarchical classification system, *Actinobacteria* classis nov. *Int. J. Syst. Bacteriol., 47,* 479-491.

Stern, M. J., Ames, G. F.-L., Smith, N. H., Robinson, E. C., and Higgins, C. F. (1984). Repetitive extragenic palindromic sequences: A major component of the bacterial genome. *Cell, 37,* 1015-1025.

St-Laurent, L., Bousquet, J., Simon, L., and Lalonde, M. (1987). Separation of various *Frankia* strains in the *Alnus* and *Elaeagnus* host specificity groups using sugar analysis. *Can. J. Microbiol., 33,* 764-772.

St-Laurent, L., and Lalonde, M. (1987). Isolation and characterization of *Frankia* strains isolated from *Myrica gale*. *Can. J. Bot., 65,* 1356-1363.

van Dijk, C. (1984). *Ecological aspects of spore formation in the* Frankia-Alnus *symbiosis.* Ph.D. Thesis, University of Leiden, The Netherlands.

van Dijk, C., and Sluimer, A. (1994). Resistance to an ineffective *Frankia* strain type in *Alnus glutinosa* (L.) Gaertn. *New Phytol., 128,* 497-504.

van Dijk, C., Sluimer, A., and Weber, A. (1988). Host range differentiation of spore (+) and spore (–) strain types of *Frankia* in stands of *Alnus glutinosa* and *Alnus incana* in Finland. *Physiol. Plant., 72,* 349-358.

Vandamme, P., Pot, B., Gillis, M., de Vos, P., Kersters, K., and Swings, J. (1996). Polyphasic taxonomy, a consensus approach to bacterial systematics. *Microbiol. Rev., 60,* 407-438.

Versalovic, J., Koeuth, T., and Lupski, J. R. (1991). Distribution of repetitive sequences in eubacteria and application to fingerprinting of bacterial genomes. *Nucl. Acids Res., 19,* 6823-6831.

Versalovic, J., Schneider, M., de Bruijn, F. J., and Lupski, J. R. (1994). Genomic fingerprinting of bacteria using repetitive sequence-based polymerase chain reaction. *Method. Mol. Cell. Biol., 5,* 25-40.

Wayne, L. G., Brenner, D. J., Colwell, R. R., Grimont, P. A. D., Kandler, P., Krichevsky, M. I., Moore, L. H., Moore, W. E. C., Murray, R. G. E., Stackebrandt, E., Starr, M. P., and Trueper, H. G. (1987). Report of the ad hoc committee on reconciliation of approaches to bacterial systematics. *Int. J. Syst. Bacteriol., 37,* 463-464.

Weber, A. (1986). Distribution of spore-positive and spore-negative nodules in stands of *Alnus glutinosa* and *Alnus incana* in Finland. *Plant Soil, 96,* 205-213.

Weber, A. (1990). Host specificity and efficiency of nitrogenase activity of *Frankia* strains from *Alnus incana* and *Alnus glutinosa*. *Symbiosis, 8,* 47-60.

Wheeler, C. T., Hooker, J. E., Crowe, A., and Berrie, A. M. (1986). The improvement and utilization in forestry of nitrogen fixation by actinorhizal plants with special reference to *Alnus* in Scotland. *Plant Soil, 90,* 393-406.

Winker, S., and Woese, C. R. (1991). A definition of the domains *Archaea, Bacteria* and *Eucarya* in terms of small subunit ribosomal RNA characteristics. *Syst. Appl. Microbiol., 14,* 305-310.

Woese, C. R. (1987). Bacterial evolution. *Microbiol. Rev., 51,* 221-271.

Woese, C. R., Kandler, O., and Wheelis, M. L. (1990). Towards a natural system of organisms: Proposal for the domains *Archaea, Bacteria,* and *Eucarya. Proc. Natl. Acad. Sci. USA, 87,* 4576-4579.

Wolters, D. J., Akkermans, A. D. L., and van Dijk, C. (1997a). Ineffective *Frankia* strains in wet stands of *Alnus glutinosa* L. Gaertn. in the Netherlands. *Soil Biol. Biochem., 29,* 1707-1712.

Wolters, D. J., van Dijk, C., Zoetendal, E. G., and Akkermans, A. D. L. (1997b). Phylogenetic characterization of ineffective *Frankia* in *Alnus glutinosa* (L.) Gaertn. nodules from wetland soil inoculants. *Mol. Ecol., 6,* 971-981.

Zepp, K., Hahn, D., and Zeyer, J. (1997). Evaluation of a 23S rRNA insertion as target for the analysis of uncultured *Frankia* populations in root nodules of alders by whole cell hybridization. *Syst. Appl. Microbiol., 20,* 124-132.

Chapter 3

FRANKIA ECOLOGY

M. VALDÉS
Instituto Politécnico Nacional, Escuela de Ciencias Biológicas, Departamento de Microbiología, 11340 México, DF, Mexico

1. INTRODUCTION

At the first "Conference on *Frankia* and Actinorhizal Plants" held in 1978, the term actinorhiza was adopted to refer to the root-nodule symbiosis between the N_2-fixing actinomycete *Frankia* and about 200 angiosperm species. Since then, studies on actinorhizal symbiosis have provided great insight into several aspects of the functioning of actinorhizas. However, although the research effort covers different host genera involved with this symbiosis, it represents only a small number of host plants because the host genera belong to eight plant families, some of them very divergent.

Actinorhizal plants are useful for reforestation and reclamation of marginal soils. For this reason, the actinorhizal symbiosis has a great potential for agroforestry and forestry in developing countries. Among the actinorhizal trees, only one, *Casuarina*, grows in tropical and subtropical areas, but fortunately many of these areas have tropical mountain ecosystems whose temperate environment allows the growth of the temperate actinorhizal plants. The use of actinorhizal plants can contribute to the pressing problem of deforestation. For one-half of the world's population (in developing countries), the main fuel is firewood; for example, more than 90% of the wood cut in Africa is burned as fuel and this has resulted in severe depletion of forests. Two-thirds of Latin America's original forests are gone or seriously depleted and Thailand has lost one-quarter of its forest within 10 years (NRC, 1984). Destruction of forests in the Third World will continue as long as the causes of deforestation, the lack of sufficient fuel and fodder, are not adequately addressed. The only options to stop deforestation are to minimize its causes.

One option is to plant fast-growing N_2-fixing trees, which do not require chemical fertilizers and are able to grow quickly, even in marginal soils. Thus, these

K. Pawlowski and W. E. Newton (eds.), Nitrogen-fixing Actinorhizal Symbioses, 49–71.
© 2008 *Springer*.

trees, growing in difficult sites, will colonize the eroded land and return the soils to production. *Casuarina* species have been and can be increasingly employed in subtropical and tropical areas, not only as a source of firewood but also for reforestation and reclamation of degraded soils (Figure 1). *Alnus* species have the greatest potential for use in temperate regions. Being highly beneficial to neighboring plants, they (as well as other actinorhizal species) may be used as nurse trees either in mixed plantations with valuable tree species or as a source of timber in pure or mixed stands (Dawson, 1986; Zhang-Zhongze *et al.*, 2000).

Besides the symbiotic association with *Frankia*, which results in the development of N_2-fixing nodules, the actinorhizal plants can associate through their root system with other microorganisms, for example, with mycorrhizal fungi. This association enables the plants to get phosphorus, other minerals, and water from the soil and, in addition, helps to protect them from detrimental microorganisms. Very often, a synergistic tripartite symbiosis results (Chatarpaul *et al.*, 1989; Valdés and Galicia, 1997; Valdés *et al.*, 2003).

To develop a large-scale production of actinorhizal plants implies the need for inoculation with an appropriate *Frankia* strain. Selection of an appropriate strain requires knowledge, not only of its effectiveness as a nitrogen-fixer, but also of its ability to form nodules promptly, to compete with the indigenous *Frankia*

Figure 1. Reclamation of a highly eroded site near Mexico City, utilizing Casuarina equisetifolia*. Altitude: 2300 m. Note that there is no soil in the site, just the parent rock.*

community, and to persist in the soil. Unfortunately, much of this information is still not available because of difficulties encountered in the isolation and identification of *Frankia* strains. Microbial ecology has depended on conventional microbial techniques for a long time and these techniques do not allow identification and monitoring of the inoculated strains. Recently, these nodulation- and re-isolation-dependent techniques have been replaced with molecular methods, which have the potential to provide ecological studies with meaningful and reliable results. Further, *Frankia* cultures are not necessary when utilizing these new procedures because the presence of the organism can be detected by a few specific molecules within it rather than by the whole organism itself.

This chapter reports the recent advances in knowledge of *Frankia* strains: as soil microorganisms; their population size and activity relative to various soil factors, including the presence of host and non-host plants; and their relationship to other microorganisms. Even with these advances, additional research on the physiology of *Frankia* is vital and fundamental to a better understanding of the ecology of this micro-symbiont.

2. *FRANKIA* AS A SOIL MICROORGANISM

Frankia cells occupy two distinct ecological niches, the root nodule and the soil. Although there is not much information on the physiological activities of *Frankia* in the soil, the classical plant-trap bioassays, using potential host plants inoculated with soil suspensions, have shown that *Frankia* is usually found in soils under host plants (Huss-Danell, 1997; Smolander, 1990; van Dijk, 1979), including sites where either no hosts have grown for a long time (Burleigh and Dawson, 1994; Huss-Danell and Frej, 1986; Rodríguez-Barrueco, 1968; Zitzer and Dawson, 1992) or in soils under non-host plants (Gauthier *et al.*, 2000; Huss-Danell, 1997; Huss-Danell and Frej, 1986; Jamann *et al.*, 1992; Jeoug and Myrold, 2001; Maunuksela *et al.*, 1999; Paschke and Dawson, 1992; Smolander and Sundman, 1987). In other words, the ability to enhance the *Frankia* population and the infecting capacity of a soil is not limited to sites populated with actinorhizal plants. *Frankia* has been found in very diverse ecosystems (Table 1) ranging from circumpolar soils (Huss-Danell *et al.,* 1999) to volcanic rocky soils (Burleigh and Dawson, 1994). There are also soils, such as sub-surface peat soils (Averby and Huss-Danell, 1988) and soils from the extreme northern areas, parts of Siberia and northernmost North America (Huss-Danell, 1997), without recorded infective *Frankia.*

In their native habitats, actinorhizal plants are mostly *Frankia*-nodulated. However, actinorhizal nodules have also been found on exotic actinorhizal plants. Although there are no known native actinorhizal plants in Hawaii, *Myrica faya*, an exotic plant, was found to be nodulated and *Myrica*-infecting *Frankia* was found in volcanic deposits in Hawaii (Burleigh and Dawson, 1994). In Mexico, where the

Table 1. Actinorhizal species and their natural distribution in ecosystems.

Family	*Genus*	*Ecosystem or Soils*
Betulaceae	*Alnus*	Riparian, wet or swampy soils, temperate forest, mountain valleys, circumpolar soils.
Casuarinaceae	*Allocasuarina*	Sand dunes, extremely poor soils.
	Casuarina	Subtropical forests, some species in riparian, chaparral, sand dunes.
	Gymnostoma	Secondary forests, ultramafic soils, volcanic soils, ornamental.
Coriariaceae	*Coriaria*	Temperate mountain forests, highly disturbed temperate forests, dry woods, rocky places.
Datiscaceae	*Datisca*	Temperate and cold forests, stream banks.
Elaeagnaceae	*Elaeagnus*	Sand dunes, disturbed agriculture lands, sometimes weedy.
	Hippophaë	Seashores, sand dunes, cliffs, dry river beds, valleys in mountains.
	Shepherdia	Temperate forests.
Myricaceae	*Comptonia*	Dry open forest stands.
	Myrica	The widest geographic distribution in the world.
Rhamnaceae	*Adolphia*	Temperate to subhumid climates, eroded slopes, very poor soils.
	Ceanothus	Chaparral from mid-to-low elevations.
	Colletia	Xerophitic matorral.
	Discaria	Matorral.
	Kentrothamus	Xerophitic matorral.
	Retanilla	Xerophitic matorral.
	Talguenea	Xerophitic matorral.
	Trevoa	Xerophitic matorral.
Rosaceae	*Cercocarpus*	Disturbed mountain forest, forest ecosystems of semiarid regions.
	Cowania	Xeric chaparral.
	Dryas	Tundra.
	Purshia	Forest ecosystems of semiarid lands.

local flora does not include plants in the Elaeagneaceae, *Elaeagnus angustifolia* was introduced as an arboretum specimen growing in a greenhouse in Mexico City and, after a few months, several plants became nodulated (M. Valdés, unpublished results). Furthermore, *Casuarina* trees, which are native to Australia, are commonly nodulated in the countries where they have been introduced - it is still not known if these trees were introduced as seeds or plants - even in very dissimilar sites, such as Florida (USA), Uruguay, Mexico, and China. *Frankia* appears not only to occur in certain soils devoid of host plants but to be abundant in these soils (Burleigh and Dawson, 1994; Gauthier *et al.*, 2000; Smolander, 1990; Smolander and Sunsdman, 1987). The occurrence and viability of *Frankia* in soil devoid of host plants can be explained because these microorganisms have resistant forms, such as spores, and can grow saprophytically.

There is other evidence for possible saprophytic growth by *Frankia*, such as the ability of pure cultured strains to grow *in vitro* and the influence on their numbers and nodulation capacity of soil properties, such as moisture content (Burleigh and Dawson, 1994; Dawson *et al.,* 1989; Huguet *et al.*, 2001; Kohls *et al.*, 1999; Pratt *et al.*, 1997), pH (Smolander, 1990), soil nutrients (Righetti *et al.*, 1986; Sanginga *et al.*, 1989), and organic matter content (Nickel *et al.*, 2001). Moreover, the ability of *Frankia* to produce siderophores (Arahou *et al.*, 1998; Aronson and Boyer, 1992) might also correlate with its saprophytic growth. Iron is an essential component of the enzyme nitrogenase, which is required for nitrogen fixation. Siderophore-mediated iron uptake in the soil has been reported for several N_2-fixing microorganisms. The ability of *Frankia* to synthesize antibiotics (Haansuu *et al.*, 2001) might also be involved in their saprophytic growth becausse the calcimycin antibiotic, called frankiamide, showed a strong activity against Gram-positive bacteria and several plant-pathogenic fungi. However, the physiological role of these compounds remains unknown.

Several species of actinorhizal plants form nodules on newly formed soils, such as *Dryas drummondi* on glacial soils at Glaciar Bay, Alaska (Laurence *et al.*, 1976), *Hippophaë rhamnoides* at the coastal dunes of The Netherlands (Oremus, 1980), and *Myrica cerifera* on the coastal dunes in the state of Virginia (USA). *Frankia* bacteria are also found in Hawaiian volcanic soils. However, the nodulation capacities of younger deposits are 450-to-1200 times lower than those of older deposits (Burleigh and Dawson, 1994).

The presence of *Frankia* in newly formed soils and in soils devoid of host plants indicates that the dispersal of *Frankia* occurs commonly. *Frankia* may be moved in the environment by biotic agents, as confirmed by the significant levels of *Frankia* in the nests of several bird species (Paschke and Dawson, 1993). Dispersal may also occur abiotically because hyphal and spore preparations can remain infectious when air dried (Burleigh and Torrey, 1990; Rodríguez-Barrueco, 1968; Tortosa and Cusato, 1991). Thus, water and wind may play a role in the dispersal of *Frankia* propagules.

The ability of certain actinorhizal trees to form N_2-fixing nodules is not restricted to their roots. Some species may form nodules on their stems as occurs with *Casuarina equisetifolia* and *C. glauca*, both of which exhibit aerial nodulation (Prim *et al.*, 1991; Valdés and Cruz-Cisneros, 1996; Figure 2). Aerial nodules are identical in morphology and structure to the underground nodules and are formed as adventitious roots from the stem. This type of nodule occurs in areas with prevailing high environmental humidity.

The abundance of *Frankia* cells in the soil with the ability to infect the roots of actinorhizal plants has been associated with a variety of factors, such as the presence of actinorhizal host plants, soil depth, geographic position, moisture, temperature, organic matter content, and vegetative cover. Certainly, there are other factors that have not yet been observed.

Figure 2. Aerial nodules on the trunk of Casuarina equisetifolia *growing in a tropical region where environmental humidity is very high and rainfall is 2400 mm. Altitude is 1020 m.*

2.1. Quantification of Frankia *Populations in the Soil*

The capacity of a soil to colonize and to induce the formation of N_2-fixing nodules (the so-called "infection capacity") is commonly estimated through a plant-trap bioassay. The infection capacity is not necessarily related to the number of *Frankia* bacteria in the soil because infection capacity of a soil can vary with the seasons even though the actual frankial population (in terms of genomic units) remains the same (Myrold and Huss-Danell, 1994). In Jamaican soils, *Myrica*-infective *Frankia* (both infection capacity and genomic units) vary from site to site in accordance with soil type (high natural fertility and high soil moisture) and soil history, independent of the presence of host plants (Zimpfer *et al.*, 1997). Nevertheless, such quantitative estimates in terms of infection units are needed if the ecology of the microsymbiont is to be better understood.

To evaluate the infection capacity of a soil, a combination of a host-plant bioassay plus PCR-based methods is recommended (Gauthier *et al.* 2000; Jeong and Myrold, 2001; Myrold and Huss-Danell, 1994; Myrold *et al.*, 1994). However, only large differences in infection capacities are likely to be indicative of differences in

Frankia population in the soil. A comparison of both methods, quantification of genomic units and nodulation in plant bioassays, revealed that the population of *Frankia* capable of nodulating a particular host (the physiologically active bacteria) ranged from 0.2-2940 nodulation units per gram of soil, and represented only a small fraction of the total population of *Frankia*, which ranged from 2000-92000 genomic units per gram of soil (Myrold and Huss-Danell, 1994). Now, recently developed molecular techniques have been designed to detect specific *Frankia* in nodules (nodulation-dependent methods). They are attractive for studying natural populations because they avoid the need for the isolation and culture of recalcitrant microorganisms. These PCR-based approaches, which use RPLF analysis of PCR-amplified fragments of the 16S rRNA gene and other marker genes (*glnII*, the intergenic spacers (IGS) of 16S-23S rRNA, *nifH-D* or *nifD-K*), have demonstrated significant diversity among both cultivated and symbiotic *Frankia* strains (see chapter 2 in this volume).

Most studies have focused on the frankial populations in nodules, which represent only a fraction of the physiologically active *Frankia* in the soil, rather than the total *Frankia* population. So, in the future, new approaches to study the total *Frankia* population need to be developed to better understand the ecology of this microorganism. The recently developed molecular approaches are highly relevant because they open the door to more studies on the effects of environmental parameters on the dynamics and structure of both indigenous and exotic *Frankia* communities and populations in the root of plants and in the soil.

2.2. Host and Non-host Plants

The physiological status of a specific *Frankia* population in the soil might be triggered and controlled by environmental factors, such as the presence of vegetation (host and non-host plants) that favors saprophytic growth of this population and, in some cases, increases its competitive abilities related to root-nodule formation (Nickel *et al.*, 2001). Nutrient resources might be obtained from root exudates because *Frankia* strains are able to colonize and grow on the root surface of host and non-host plants with no addition of external carbon sources (Ronkko *et al.*, 1993; Smolander *et al.*, 1990). Alternative carbon resources might be obtained from the decomposition of organic materials, such as leaf litter. Leaves of *Casuarina* contain compounds that promote the growth of *Frankia* strains that nodulate *C. cunninghamiana* (Zimpfer *et al.*, 1999). *Frankia* strains, which are inoculated and incubated in soil with no host plants but amended with leaf litter of *Alnus glutinosa*, remain infective and competitive for nodulation on *A. glutinosa* with the indigenous *Frankia* strains (Nickel *et al.*, 2001).

A variety of methods have been used to classify *Frankia* strains for taxonomical purposes; these include soluble protein patterns, isoenzyme patterns, fatty acid composition, DNA relatedness, restriction analysis of genomic and plasmid DNA, etc. They have led to the conclusion that culturable *Frankia* strains are heterogeneous and can be divided into two main groups, the *Alnus-Casuarina* infective group and the Elaeagnaceae infective group (see chapter 4 in this volume).

In the past, actinorhizal plants have been divided into promiscuous and non-promiscuous species. Non-promiscuous or specific actinorhizal genera are *Allocasuarina, Casuarina, Elaeagnus* and *Hippophaë*. The genus *Alnus* is also classified as specific, although *Alnus* spp. can to some extent be nodulated by *Frankia* strains from the Eleagnaceae infective group (Bosco *et al*., 1992). *Myrica* was considered a promiscuous actinorhizal genus for a long time (Baker, 1987) because, in the greenhouse, several species of *Myrica* are effectively nodulated by most isolated *Frankia* strains. However, 16S-rDNA restriction patterns have shown that *Myrica gale* does not exhibit such promiscuity in its natural environment and its nodulating frankiae belong to an *Alnus-Casuarina* infective phylogenetic subgroup (Huguet *et al.,* 2001). A separate study confirmed this result by yielding only a few dominant 16S-rDNA sequences from nodules from *M. gale*, again indicating either symbiont specialization or niche selection of particular ecotypes (Clawson and Benson, 1999). In contrast, 16S-rDNA sequences of *Frankia* strains inhabiting nodules of other plants in the Myricaceae family have shown that *M. pensylvanica* nodules harbor diverse *Frankia* strains and might serve as a reservoir for *Frankia* strains that infect plants from other actinorhizal families (Clawson and Benson, 1999). The genus *Gymnostoma* was once considered promiscuous but, based on both phenotypic and genotypic characteristics, nodules of *Gymnostoma* spp. were shown to contain *Frankia* strains belonging to the *Elaeagnaceae* infective group and that these strains are rather closely related to one another (Navarro *et al.,* 1997).

Actinorhizal trees differ in their frankial requirements, thus, *Frankia* specificity for both nodulation and effectiveness for nitrogen fixation is relevant for selecting suitable strains for preparing inoculants. Both nodulation properties and N_2-fixing abilities differ among *Frankia* strains isolated from plants in the Rhamnaceae family; some strains were significantly more effective than others in improving the dry weight of the host plants (Carú and Cabello, 1998; 1999). All bacterial strains isolated from Casuarinaceae were reported to produce very effective N_2-fixing nodules (Sellstedt, 1995), however, other results indicate that specific nitrogenase activity (ARA), both *in vitro* and *in planta*, of *Frankia* nodulating *C. equisetifolia* may vary significantly among strains (Vásquez *et al*., 2000). *In planta* nitrogenase activities of some Rhamnaceae and Casuarinaceae frankiae are shown in Table 2.

There are also *Frankia* strains that cannot fix N_2. These ineffective strains form a large fraction of the *Frankia* population of wet soils of *Alnus glutinosa* stands in The Netherlands. Although *Alnus glutinosa* exhibits a variable degree of incompatibility to root-nodule formation by ineffective *Frankia*, the ineffective bacteria are not strictly dependent on susceptible *A. glutinosa* for the maintenance of their population size (Wolters *et al*., 1999).

Table 2. Values of nitrogenase activity of nodules from Retanilla ephedra *(Rhamnaceae) and* C. equisetifolia *(Casuarinaceae) plants inoculated with different* Frankia *strains. Data were taken after 3 and 4 months, respectively, and are expressed as nmol of C_2H_2 h^{-1} mg^{-1} dry weight nodule (modified from Carú and Cabello, 1999, and Vásquez* et al., *1999).*

Strain	Nitrogenase activity *(nmol of C_2H_2 h^{-1} mg^{-1})*
Retanilla ephedra	
ReI4	13.5
ReI6	24.6
ChI1	09.8
TtI42	25.5
TqI15	16.3
Casuarina equisetifolia	
Ce1	51.13
Ce6	14.36
Ce15	22.57
Ce17	109.60
BR (reference)	44.62

2.2.1. Actinorhiza-like Nodules

Actinorhizal plants can produce actinorhiza-like nodules in the absence of *Frankia*; they exhibit a morphogenic response in the presence of other soil microorganisms that are similar to that induced by *Frankia*, which indicates that the plant controls nodule formation. The infection of *Elaeagnus angustifolia* by a cucomopine-strain of *Agrobacterium rhizogenes* formed calli, which produced roots carrying nodule-like structures, called pseudo-actinorhizae (Berg *et al.*, 1992). These structures were anatomically indistinguishable from the nodules induced by *Frankia* on its hosts but were formed in the absence of the bacterium.

The fungus, *Penicillium nodositatum*, is also able to induce nodule formation on the roots of the alders, *Alnus glutinosa* and *A. incana* (Capellano *et al.*, 1987; Sequerra *et al.*, 1992; Wolters *et al.*, 1999). The genus *Penicillium* (Hyphomycetes) is perhaps the most ubiquitous of all soil fungi and is found from the equator to the polar regions, although it favors the temperate and colder regions (Barron, 1969). The role of these so-called myco-nodules is unclear because they do not fix N_2 and so cannot benefit the plant, and any benefit the fungus receives is not known. There are probably other rhizospheric microorganisms which can also enter nodules and might play other still unknown roles. For example, several actinomycetes that are capable of reducing acetylene are commonly isolated from actinorhizal nodules gathered from widespread locations (Niner *et al.*, 1996; L. Wall, personal communication).

3. ABIOTIC SOIL FACTORS AND *FRANKIA* POPULATIONS

3.1. Soil Depth

Nodulation can be scarce or absent in actinorhizal *Casuarina* and *Allocasuarina* spp. (Cruz-Cisneros and Valdés, 1990; Frioni *et al.*, 1991; Vásquez *et al.*, 2000), even in their natural environment in Australia (Reddell *et al.*, 1986). Dawson *et al.* (1989), trying to identify the soil factors responsible for the inconsistent nodulation of these actinorhizal trees in Australia, concluded that the failure to find nodules in the field might be a due to the nodules of either *Casuarina* or *Allocasuarina* spp. being formed in deep soil strata that are difficult to sample. They suggested that perhaps fine-root proliferation and subsequent nodulation in these species occur primarily at depths where permanent soil moisture exists. Their results indicate that soils with the greatest nodulation capacity are those with the greatest capacity to afford adequate water, moderate temperature, and aeration to both roots and soil microorganisms. For all topographic positions studied, soil samples from depths greater than 20cm promoted 70% more nodulated *Casuarina* seedlings than soil samples from the top 20cm.

Using the restriction pattern of the *nifD-K* intergenic sequence to identify *Frankia* strains, *Elaeagnus angustifolia*-infective *Frankia* strains were found at least down to 60-cm soil depth. The density of *Frankia* decreased with depth and correlated with the decline of soil organic-matter content, but the overall diversity of the population remained similar. Individual strains showed different distributions along the depth gradient. Further, nodulation capacity increased with soil depth, whereas free-living nitrogen fixation was not affected (Nalin *et al.,* 1997).

3.2. Organic Matter Content

According to Reddell *et al.* (1986), the density of the *Frankia* population should be dependent on soil conditions. The bacteria have to survive under difficult edaphic conditions and they can form spores in order to survive under different conditions for long periods. Nitrogen-fixing activity could be a function that provides an advantage to *Frankia* strains in low nitrogen soils, but they need energy to fix N_2. Carbon sources to supply this energy might be provided by both the host rhizosphere as well from rhizospheres of non-host plants because both stimulate *Frankia* growth. This stimulation of saprophytic growth of *Frankia* by non-host plants increases the ability of *Frankia* to disperse and adapt to diverse new environments.

Nickel *et al.* (2001) studied the effect of alder leaf litter amendment on (i) the native *Alnus glutinosa*-infective *Frankia* population and (ii) an added *Frankia* inoculant in a soil devoid of host plants. They found a positive response in terms of both nodulation capacity and nitrogen acquired (^{15}N dilution method) both by the native *Frankia* population and by the added inoculant. These results show that, through either inoculation or leaf litter amendment, *Frankia* populations can be established in soils. These populations remain infective on their host plants and successfully compete for nodule formation with other indigenous or inoculated *Frankia* populations, thereby increasing plant growth performance.

Some mineral soils contain water-soluble compounds that stimulate the growth of some *Frankia* strains, whereas other strains respond negatively when exposed to extracts of peat soil (Sheppard *et al.*, 1988). Furthermore, addition of montmorillonite (a clay mineral with high adsorption capacity) to the soil doubles the nodule number on alders, a fact that can be explained by the adsorption of inhibitory compounds secreted by the tree itself (Smolander *et al.*, 1988).

3.3. Drought, Temperature and Salinity

In many soils, a population of colonizing *Frankia* bacteria may only rarely occur or may even be absent. In these cases, host plants have to be inoculated with a selected microsymbiont and, as mentioned in section 3.1, a factor that can influence the efficiency of infection is the moisture content of the soil.

Several actinorhizal trees, such as members of the Casuarinaceae, are noted for their salt and drought tolerance and have been extensively used on dry and sodic soils. These trees are used in arid or semi-arid areas: (i) for reclamation of marginal lands; (ii) in agroforestry and forestry; (iii) in reforestation of rural and urban sites; and (iv) to stabilize coastal soils or sand dunes. Casuarinas are widely planted throughout the tropical and warm regions of Egypt, India, China, Mexico, and the Pacific region. In these dry and warm areas, standard inoculation procedures used for soil reclamation, have often been ineffective (Dommergues and Bosco, 1998) and no commercial *Frankia* inoculants are available (Benoit and Berry, 1990; Valdés *et al.,* 2001). Thus, the preparation of an inoculant that contains a specifically adapted microsymbiont is crucial and it should be preferentially utilized in an optimized inoculation procedure instead of the standard treatment.

Water-retentive polymers have been developed as carriers for *Frankia* strains in inoculants (Kohls *et al.*, 1999). These polymers have been available to tree growers and are used as a dip to reduce moisture stress and to increase nutrient availability to transplanted seedlings. They can absorb up to 400-times their dry weight in moisture. Application of these polymers to roots in combination with *Frankia* inocula increased nodulation of *A. glutinosa* in the field and of *C. equisetifolia* in a controlled environment. Nodule numbers and dry weight per plant were 2-3-times higher for plants inoculated with *Frankia* in water-retentive polymers compared with plants inoculated with *Frankia* in water alone. In some cases, early growth of actinorhizal plants was enhanced by the treatment with the *Frankia*-laden polymers. An alternative inoculation technique utilized *Frankia* cells encapsulated in alginate beads, which allow the bacteria to grow within the beads (Frioni *et al.*, 1994).

Drought has been found to limit nodulation of *Ceanothus* in the southern California (USA) chaparral (Kummerow *et al.*, 1978). The actinorhizal shrub *Ceanothus* is a dominant genus in the California chaparral where, at mid-to-low elevations, it can form up to 90% of the vegetation cover. In addition to drought stress limiting nitrogen fixation in this type of ecosystem, a further deficiency of fixed nitrogen results as a consequence of the recurrent fires that burn above-ground vegetation. Not only do the fires lead to a deficiency of fixed nitrogen in the chaparral, the winter rains following the fires exacerbate the situation by leaching a significant fraction of the remaining fixed nitrogen from the bare soils.

Pratt *et al.* (1997) were able to prove that water availability is a major factor limiting *Ceanothus* seedling nodulation in this ecosystem. The seedlings lacked nodules at typical dry sites, whereas those growing at a naturally occurring well-hydrated site were often nodulated. Furthermore, irrigation induced nodulation, whereas adjacent non-irrigated control seedlings remained nodule-free. Other observations of post-fire development in the Californian chaparral showed that both decreased soil moisture and increased soil temperatures were correlated with decreases in nitrogenase activity (ARA) (Ellis and Kummerow, 1988).

When investigating the effect of temperature and sources of *Frankia* strains on the Australian tree *Casuarina cunninghamiana*, Reddell *et al.* (1986) observed that both biomass and nitrogen content of plants increased with increasing soil temperature up to 25°C, but were significantly decreased at 15°C and 30°C. The sensitivity to temperatures below 20°C appears relevant to the distribution of different *Casuarina* species, which are limited to southern Australia (warm region).

Members of the Casuarinaceae are noted for their ability to grow under saline conditions. *C. equisetifolia* and *C. glauca* are particularly tolerant to saline stress. However, the *Frankia* strains isolated from *Casuarina* spp. nodules show different sensitivities to NaCl *in vitro*. For instance, strain CcO1 is highly tolerant and grows in the presence of up to 500mM NaCl, whereas strain CcI3 is sensitive to 100mM NaCl. These results have not been confirmed under field conditions because no *Frankia* have been found in highly saline soil (Hafeez *et al.*, 1999).

Other actinorhizal plants, such as *Myrica cerifera,* are intolerant to high salinity. No nodules are formed in highly saline soil, which also prevents the germination of *M. cerifera* seeds (Young *et al.*, 1992). In contrast, *Frankia* strains isolated from plants in the Rhamnaceae family are more salt tolerant *in vitro* than strains nodulating other actinorhizal plants, however, both nitrogenase activity (ARA) and vesicle formation were affected by saline stress in all strains examined (Carrasco and Carú, 1995). More research is needed to identify the factors controlling the capabilities of both micro- and macro-symbionts, both alone and in symbiosis, to grow and fix N_2 under field conditions.

3.4. Nitrogen and Phosphorus

It is very important to identify the nutritional constraints on actinorhizal plant growth in a particular soil because they might be related to responses to *Frankia* inoculation. Growth of inoculated actinorhizal plants in soils deficient in mineral nutrients is well documented in both laboratory and field. Nutrient availability can affect nodule formation, *e.g.*, the addition of a fixed-N-free nutrient solution improved the poor nodulation of *A. glutinosa* seedlings in the field at a depth of 15–25cm (Kurdali and Domenach, 1991).

Combined-N inhibits nodulation of actinorhizal host plants (Huss-Danell, 1990; see chapter 6). Nitrate inhibition of nodulation acts at a very early point in the signal exchange between plant and bacterium, before root-hair deformation in plants that are infected intracellularly *via* root hairs (Cérémonie *et al.,* 1999). Its point of action during the intercellular-infection process is not known (Kohls and Baker, 1989; see chapter 6). However, low levels of soil-N are tolerated, *e.g.*, nodulation of the commercially valuable trees, *Alnus glutinosa* (infected *via* root hairs) and *Elaeagnus angustifolia* (infected intercellularly), was unaffected by total soil-N contents from 0.09-0.20% (Zitzer and Dawson, 1992). Further, when actinorhizal plants are associated with non-N_2-fixing plants, the resulting utilization of available fixed-N can prevent its inhibitory effect on nodulation. For example, in Italy, common oak (*Quercus robur*) growing in plantations with Italian alder utilized a considerable amount of the fixed-N from the symbiotic alder (Buresti *et al.*, 1991).

As with other N_2-fixing plant-microbe symbioses, the fixation of atmospheric N_2 in actinorhizal nodules can be severely and negatively impacted by phosphorus limitation. Low phosphorus status is a frequent limitation to nodulation of naturally occurring *Casuarina* and *Allocasuarina* species in Australia (Reddell *et al.,* 1986). A synergistic growth response to inoculation with *Frankia* occurs only when P is applied. Once the phosphorus deficiency is alleviated, there is a positive interaction between *Frankia* and phosphorus (Reddell *et al.*, 1988; Wall *et al.,* 2000). In the presence of phosphorus, *C. cunninghamiana* showed was a significant growth response both to inoculation with *Frankia* and to fixed-nitrogen fertilizer. Sanginga *et al.* (1989) demonstrated through the interaction between P and N treatments (using inoculated and N-fertilized *C. equisetifolia* plants) that there is a greater P requirement for growth under symbiotic nitrogen fixation than for plant growth with fertilizer nitrogen, when soil P is low. Growth of *C. equisetifolia* seedlings dependent on symbiotically fixed N was more sensitive to the increase of P at a low level (30 mg P kg^{-1} soil) than was growth for seedlings supplied with combined N. Arahou *et al.* (1996) found differences in the interaction of nitrogen and phosphorus with iron supply. When they cultivated nodulated *C. glauca* seedlings in different nutrient solutions in absence of KNO_3, the nodule dry weight increased more in the presence of Fe than in its absence. The interaction of P and Fe was different and it appeared that the presence of P encouraged Fe deficiency in the plants.

3.5. pH, Aluminum, Nickel, Antibiotics, Heavy Metals

Among the most important factors controlling *Frankia* soil populations is the pH of the soil solution. The growth of nodulated alders is inhibited by acidic pH

(Burggraaf and Shipton, 1982; Faure-Raynaud *et al.*, 1986). Both *A. glutinosa* and *Myrica gale* have a narrow pH optimum near neutrality at which nodulation is optimal (Smolander *et al.*, 1988; Rodríguez-Barrueco, 1968); *Frankia* cells in pure culture show optimal growth at the same pH. It appears that nodulation of *Alnus* species is strongly correlated with soil pH up to pH 8.0, but not with other common soil properties (Smolander and Sundman, 1987). For soil pH values ranging from 4.9 to 7.1, *A. glutinosa* nodule weight was found to be negatively correlated with soil pH, the highest nodule numbers being found at pH 4.9. *Elaeagnus angustifolia* trees, however, showed the highest nodule numbers at a soil pH of 6.6 (Zitzer and Dawson, 1992).

Susceptibility to antibiotics *in vitro*, using Difco antibiotic discs, is a useful characteristic for classifying *Frankia* strains. Among 39 strains tested, Dobritsa (1998) found that all of them displayed resistance to nalidixic acid and most were resistant to lincomycine, but showed low levels of susceptibility to polymixin B. Levels of susceptibility to carbenicillin, novobiocin, tetracycline, vancomycin, streptomycin, ampicillin, erythromycin, kanamycin, and penicillin G varied widely for the different strains. Among these strains, host specificity, antibiotic sensitivity, and pigment production showed some correlation to within three host-compatibility groups (*Alnus*-group, *Elaeagnus*-group, and *Casuarina*-group). A connection between sensitivity to antibiotics and symbiotic properties was also found in other studies. For example, using mutagenesis of spores from *Frankia* cells isolated from plants of the Rhamnaceae family, Carú and Cabello (1998) found some spontaneous rifampicin- or chloramphenicol-resistant mutants that differed in their symbiotic characteristics, such as infectivity and nitrogenase activity, from their parent strains. Further, in nodules formed on roots of *Retanilla ephedra*, mutant strains showed higher specific nitrogenase activities (ARA). Tisa *et al.* (1999) found that two variants of a strain that differed in their ability to grow on fatty acids and dicarboxylates, respectively, also differed in their antibiotic sensitivity.

Several actinorhizal species, *e.g.*, from the Casuarinaceae (Diem and Dommergues, 1990) and some *Elaeagnus* spp. (Fessenden, 1979), show high resistance to heavy metals, which makes them suitable for the reclamation of lands spoiled by mining and industrial waste. *Frankia* strains have differing sensitivities to heavy metals. Using a growth-inhibition assay, all of the tested strains were sensitive to low concentrations (<0.5 mM) of Ag^{+}, AsO_2^{-}, Cd^{2+}, SbO_2^{+} and Ni^{2+}; most strains were less sensitive to Pb^{2+}, CrO_4^{2-}, AsO_4^{3-} and $S_2O_2^{2-}$; and about one third were resistant to levels of Cu^{2+} as high as 20 mM (Richards *et al.*, 2002).

Surprisingly, aluminum was found to stimulate the *in vitro* growth of the microsymbiont. Both acid-sensitive and acid-tolerant strains were able to grow at a low pH (4.8) when appropriate concentrations of monomeric aluminum were present in the culture medium. Aluminum concentrations from 125-to-500 μM increased the growth of all the isolates from *C. cunninghamiana*, *Alnus viridis*, *A. rubra*, *A. jorullensis*, *M. pensilvanica*, and *Ceanothus jepsonii* (from soil beneath) (Igual and Dawson, 1999). These findings suggest that the survival and growth of *Frankia* strains in acidic soils worldwide could be greater than previously thought, provided the soils contain aluminum.

Nickel is essential for active hydrogenase functioning in free-living *Frankia*. It is well known that nitrogenase produces H_2 when fixing N_2 and H_2-uptake activity has been reported in most *Frankia* strains examined (Sellstedt and Smith, 1990; Mattsson and Sellstedt, 2002). Uptake hydrogenase could be an important factor in the efficiency of nitrogen fixation (Maier and Triplett, 1996) and so it is of biotechnological interest to search for increased efficiency of H_2 uptake in N_2-fixing bacteria. Mattsson and Sellstedt (2002) demonstrated that addition of Ni to the growth medium stimulated H_2 uptake (by [NiFe] hydrogenases) in both *Alnus*-infective and *Casuarina*-infective *Frankia* strains. Importantly, they also showed that the strategy for regulating hydrogenase in *Frankia* is different from that usd by other nitrogen-fixers.

4. BIOTIC SOIL FACTORS: MYCORRHIZAL FUNGI AND *FRANKIA*

Associations between fungi and roots are the rule in nature; any lack of association by plants is an exception that confirms the rule. Among the rhizospheric micro-organisms closely associated with plant roots, both arbuscular and ecto-mycorrhizal fungi play major roles in the productivity and health of plants. Most actinorhizal plants, in addition to forming symbiosis with the nitrogen-fixing actinomycete *Frankia,* are known to form symbioses with several species of ectomycorrhizal fungi and/or arbuscular fungi (see, *e.g.*, Provorov *et al*., 2002; Rose, 1980; Trappe, 1979; Valdés and Sánchez Francia, 1996).

The simultaneous occurrence of the mycorrhizal fungi together with the N_2-fixing bacteria generates a multipartite symbiosis able to significantly improve both nitrogen fixation by *Frankia* and the growth of the host plant. This enhancement occurs through the complementary activities developed by the microbial members of the symbiosis. Mycorrhization of plant roots improves the access of the plant to water and soil minerals, which is important because root nodules are unable to develop and to fix N_2 when the host plant is suffering from either drought stress or the lack of appropriate nutritional elements required for nitrogen fixation. These nutrients are obtained from the soil by the roots of actinorhizal plants in association with endo-mycorrhizal and/or ecto-mycorrhizal fungi; this association leads to the development of roots which are physiologically and, in case of ectomycorrhiza, also morphologically different from other roots. These roots are called mycorrhizas, which are defined as "a symbiosis in which an external mycelium of a fungus supplies soil-derived nutrients to a plant root" (Smith and Read, 1997). The associated fungal mycelium is more efficient in soil exploration and nutrient mobilization than the plant roots alone.

Ectomycorrhizal symbioses mostly involve trees and basidiomycetes, which grow between root outer cortical cells (the so-called "Hartig net") and cover the roots with a thick fungal sheath (Agerer, 2001). Endomycorrhizal symbioses occur for more than 80% of all terrestrial plants. Here, the microsymbionts are members of the order Glomales, whose hyphae can enter root cortical cells and form branched structures for nutrient exchange; these are mostly arbuscules, which led to the designation of "arbuscular mycorrhiza" (Smith and Read, 1997). Arbuscular mycorrhizae are particularly important for the phosphate nutrition of the plant. Despite the fact that much of the available literature has focused on the links between nutrient uptake by mycorrhizas and the resulting increase in plant biomass, mycorrhizas can be defined on the basis of a of broader perspective in terms of fitness in nature. D. J. Read (1998) has proposed to define mycorrhizas as "structures in which a symbiotic union between fungi and plant roots leads to increases in fitness of one or more partners". Other classification methods are based on the different mycelial exploration types with respect to their ecophysiological effectivity (Agerer, 2001). Endomycorrhizal associations can also have non-nutritional impacts on plants. These effects are related to enhanced resistance to pathogens and to environmental extremes as well as completion of the life cycle, survival in the regeneration niche, and plant seed quality and vigor (Read, 1998).

Endomycorrhizal fungi are specialized members of the huge population of microorganisms that colonize the rhizosphere of plants. They are completely dependent on the plant for organic carbon and thus independent of the soil's organic C resources, which are scarce and irregularly distributed. Thanks to their access to the plant carbon supply, these fungi are in a good position to compete with saprophytes in the mobilization of N, P, and other nutrients (Smith and Read, 1997). Endomycorrhizal fungi are present in most soils and infection occurs spontaneously. In some sites, such as plantations, trees are inoculated with mycorrhizal fungi to ensure the establishment of healthy and vigorous actinorhizal trees because it is crucial that phosphate and other micronutrients are mobilized for an effective root-nodule symbiosis.

The occurrence of different types of mycorrhizas associated with actinorhizal plants and their interactions with *Frankia* has been reviewed by Diem (1996). The importance of these tripartite symbioses for the promotion of growth of actinorhizal plants in poor soils has been documented extensively (Chatarpaul *et al.*, 1989; Diem and Dommergues, 1990; Isopi *et al.*, 1994; Sempavalan *et al.*, 1995; Valdes *et al.*, 2003). However, the degree of dependence on the mycorrhizas for the uptake of phosphate, water and other nutrients is still poorly documented, so further research is needed to provide information on the specific nutritional requirements of the nodulated actinorhizal plants.

Although host specificity is very low in arbuscular mycorrhizal associations (Smith and Read, 1997), it is higher for the associations with ectomycorrhizal fungi. The susceptibility of *Allocasuarina* and *Casuarina* to endo- and ecto-mycorrhizal colonization is different. Reddell *et al.* (1986) observed that *Allocasuarina* spp. form ectomycorrhizas more commonly than *Casuarina* spp. do; casuarinas do not form complete ectomycorrhizae when inoculated with ectomycorrhizal fungi. In contrast, more arbuscular mycorrhizal fungi are associated with *Casuarina* than with *Allocasuarina.* Field observations on the inoculation of the Casuarinaceae confirm the data obtained under controlled conditions (Theodoru and Reddell, 1991; Thoen *et al.,* 1990).

The effect of mycorrhizas on plant growth is impressive. Among the ectomycorrhizal fungi that enter a symbiosis with alder, *Alpova diplopoeus* is the only one that forms a real ectomycorrhiza with a well-developed Hartig net in the root cortex of *A. crispa* and *A. rugosa* (Godbut and Fortin, 1963). This mycorrhizal fungus plus macronutrients increased nitrogen fixation by 136% over that of control plants in *Frankia*-inoculated *A. rubra*. The infection of *Frankia*-inoculated plants with *A. diplopoeus* was more effective (by 30%) in increasing nitrogen content than fertilization with macronutrients (Rojas *et al.*, 2002). Other ectomycorrhizal symbionts of alders are *Telephora terrestris*, which enters a very effective symbiosis with *A. rubra* (Miller *et al.*, 1992), and *Pisolithus tinctorius*. *P. tinctorius* increased the growth of *A. accuminata* in the greenhouse by 192% compared to the control plants. Similar growth increases (182%) were obtained by inoculation with the arbuscular mycorrhizal fungus, *Glomus versiforme* (Valdés and Galicia, 1997). Similarly, inoculation of *Frankia*-nodulated *A. glutinosa* (Fraga-Beddiar and Le Tacon, 1990), *A. cordata* (Lumini *et al.*, 1994), and *Casuarina equisetifolia* (Gauthier and Diem, 1982) with the arbuscular mycorrhizal fungus *Glomus mosseae*, had a strong beneficial effect.

Effects of endomycorrhization on fruit production can be even stronger. Visser *et al.* (1990) reported yield increases of four- and five-times, respectively, for *Shepherdia canadensis* and *Elaeagnus communata*, after inoculation with arbuscular mycorrhizal fungi in comparison with un-inoculated plants. In contrast to these results, *Comptonia peregrina* and *Myrica gale* failed, in some sites, to form effective arbuscular mycorrhizas in the field (Crocker and Schwintzer, 1994). This result is probably due to the fact that they can improve their phosphate uptake by proteoid roots in response to phosphate deprivation and, thus, do not benefit much from arbuscular mycorrhization.

The occurrence of so-called myconodules has been reported for the Casuarinaceae family (Duhoux *et al.*, 2001). Small non-N_2-fixing modified lateral roots, which were colonized by septate filamentous fungi (presumably arbuscular mycorrhizal fungi) were reported many years ago for the families Pedocarpeaceae, Araucariaceae and Phyllocladeaceae and were designated as myconodules (Bond, 1963). Two species of *Gymnostoma, G. deplancheanun* and *G. nodiflorum*, which are endemic to New Caledonia, bear this kind of nodule and are colonized by an arbuscular mycorrhizal fungus belonging to the genus *Glomus*. Their role might be linked to the ecological conditions of the host plants, which are pioneers in exposed and rocky habitats, under ferric soils characterized for having low concentration of N, P and K and toxic levels of Ni, Co and Cr (Duhoux *et al.*, 2001).

5. CONCLUSIONS

Actinorhizal symbioses are exceptional because of their present and potential uses in forestry, agroforestry, revegetation, improvement of marginal soils, and fuel production, all of which affect the entire world, but all of which should be targeted to solve the many problems occurring in the developing countries and so help the millions of people living in marginal conditions.

Among the capabilities of the actinorhizal plants is their vital role as pioneer plants. This ability is due, in part, to the microbial associations they develop simultaneously with symbiotic microorganisms, including the N_2-fixing *Frankia* and ectomycorrhizal and endomycorrhizal fungi that provide the plants with many important nutrients and protect them against several phytopathogens. There are undoubtedly other rhizospheric microorganisms, which are also stimulated in their roots and which might play other unknown roles.

Although our knowledge concerning the actinorhizal symbiosis has increased greatly during the last three decades, there remains a need for more studies that provide an understanding of the distribution of *Frankia* and the factors affecting the functionality of *Frankia* populations in the environment. It is a hope that all the doors opened throughout the developed world and the sophisticated new techniques available to study the ecology of *Frankia* will shut the door of global forest poverty.

ACKNOWLEDGEMENTS

This chapter is dedicated to the memory of Roberto Cruz-Cisneros, an invaluable plant ecologist. The research carried out on *Frankia* and actinorhizal plants was initiated with the financial support provided by the US NAS/NRC through a grant from the Agency for the International Development. Support from the National Council for Science and Technology (CONACYT), Mexico, and the National Polytechnic Institute (IPN) is also acknowledged.

REFERENCES

Agerer, R. (2001). Exploration types of ectomycorrhizae. A proposal to classify ectomycorrhizal mycelial stems according to their patterns of differentiation and putative ecological importance. *Mycorrhiza, 11*, 107-114.

Arahou, M., Diem, H. G., and Sasson, A. (1998). Influence of iron depletion on growth and production of catechol siderophores by different *Frankia* strains. *W. J. Microbiol. Biotechnol., 14*, 31-36.

Aronson, D. B., and Boyer, G. L. (1992). *Frankia* produces a hydroxamate siderophore under iron limitation. *J. Plant Nutrition, 15*, 2993-2201.

Arveby, A. S., and Huss-Danell, K. (1988). Presence and dispersal of infective *Frankia* in peat and meadow soils in Sweden. *Biol. Fert. Soils, 6*, 39-44.

Baker, D. D. (1987). Relationship among pure-cultured strains of *Frankia* based on host specificity. *Physiol. Plant., 70*, 245-248.

Barron, G. L. (1968). *The genera of Hypomycetes from soil*. Baltimore, MD: Williams and Wilkins.

Benoit, L. F., and Berry, A. M. (1990). Methods for production and use of actinorhizal plants in forestry, low maintenance landscapes, and revegetation. In C. R. Schwintzer and J. D. Tjepkema (Eds.), *The biology of* Frankia *and actinorhizal plants* (pp. 281-194). New York, NY: Academic Press.

Berg, R. H., Liu, L., Dawson, J. O., Savka, M. A., and Farrand, S. K. (1992). Induction of pseudoactinorhizae by the plant pathogen *Agrobacterium rhizogenes. Plant Physiol., 98*, 777-779

Bond, G. (1963). The root nodules of non-legume angiosperms. In P. S. Nutman and B. Mosse (Eds.), *Symbiotic associations* (pp. 72-91). London, UK: Cambridge University Press.

Bosco, M., Fernandez, M. P., Simonet, P., Materassi, R., and Normand, P. (1992). Evidence that some *Frankia* sp strains are able to cross boundaries between *Alnus* and *Elaeagnus* host specificity groups. *Appl. Environ. Microbiol., 58*, 1569-1576.

Buresti, E., Domenach, A.-M., Bosco, M., and Moiroud, A. (1991). Comparison between *Quercus robur/Alnus cordata* mixed plantations and *Quercus robur* in monoculture. In M. Polsinelli,

R. Materassi and M. Vicenzini (Eds.), *Nitrogen fixation with non-legumes* (pp. 651-662). Dordrecht, The Netherlands: Kluwer Academic Publishers.

Burggraaf, A. J. P., and Shipton, W. A. (1983). Studies on the growth of *Frankia* isolates in relation to infectivity and nitrogen fixation (acetylene reduction). *Can. J. Bot., 61*, 2774-2782.

Burleigh, S. H., and Torrey, J. G. (1990). Effectiveness of different *Frankia* cell types as inocula for the actinorhizal plant *Casuarina*. *Appl. Environ. Microbiol., 56*, 2565-2567.

Burleigh, S. H., and Dawson, J. O. (1994). Occurrence of *Myrica*-nodulating *Frankia* in Hawaiian volcanic soils. *Plant Soil, 164*, 283-289.

Capellano, A., Dequatre, B., Valla, G., and Moiroud, A. (1987). Root-nodule formation by *Penicillium* sp. on *Alnus glutinosa* and *Alnus incana*. *Plant Soil, 104,* 45-51.

Carrasco, A., and Carú, M. (1995). Efecto de NaCl sobre el crecimiento y actividad de nitrogenasa de cepas de *Frankia* aisladas de Ramnaceas. *Acta Microbiológica, 6,* 153-161.

Carú, M., and Cabello, A. (1998). Isolation and characterization of the symbiotic phenotype of antibiotic-resistant mutants of *Frankia* from Rhamnaceae. *W. J. Microbiol. Biotechnol., 14,* 205-210.

Carú, M., and Cabello, A. (1999). Infectivity and effectivity of some *Frankia* strains from the Rhamnaceae family. *Arid Soil Res., 13,* 53-59.

Cérémonie, H., Debellé, F., and Fernandez, M. P. (1999). Structural and functional comparison of *Frankia* root hair deforming factor and rhizobia Nod factor. *Can. J. Bot., 77*, 1293-1301.

Clawson, M. L., and Benson, D. R. (1999). Natural diversity of *Frankia* in actinorhizal root nodules from promiscuous hosts in the Myricaceae. *Appl. Environ. Microbiol., 65,* 4521-4527.

Crocker, L. J., and Schwintzer, C. R. (1994). Soil conditions affect the occurrence of cluster roots in *Myrica gale* L. in the field. *Soil Biol. Biochem., 26*, 615-622.

Cruz-Cisneros, R., and Valdés, M. (1990). Ecological aspects of the actinorhizal plants growing in the basin of Mexico. *Nitrogen Fixing Tree Res. Rep., 8,* 42-47.

Chatarpaul, L., Chakravarty, P., and Subramaniam, P. (1989). Studies on tetrapartite symbiosis. 1. Role of ecto- and endomycorrhizal fungi and *Frankia* on the growth performance of *Alnus incana*. *Plant Soil, 118*, 145-150.

Dawson, J. O. (1986). Actinorhizal plants: Their use in forestry and agriculture. *Outlook Agric., 15*, 202-208.

Dawson, J. O., Kowalski, D. G., and Dart, P. J. (1989). Variation with soil depth, topographic position and host species in the capacity of soils from an Australian locale to nodulate *Casuarina* and *Allocasuarina* seedlings. *Plant Soil, 118*, 1-13.

Diem, H. G. (1996). Les mycorrhizes des plantes actinorrhizienes. *Acta Bot. Gall., 143,* 581-592.

Diem, H. G., and Dommergues, Y. R. (1990). Current and potential uses and management of Casuarinaceae in the tropics and subtropics. In C. R. Schwintzer and J. D. Tjepkema (Eds.), *The biology of* Frankia *and actinorhizal plants* (pp. 316-342). New York, NY: Academic Press.

Dobritsa, S. V. (1998). Grouping of *Frankia* strains on the basis of susceptibility to antibiotics, pigment production and host specificity. *Int. J. Syst. Bacteriol., 48*, 1265-1275.

Dommergues, Y., Duhoux, E., and Diem, H. G. (1999). *Les arbres fixateurs d'azote.* Montpellier, France: Editions Espaces 34.

Dommergues, Y. R., and Bosco, M. 2000. The contribution of N_2-fixing trees to soil productivity and rehabilitation in tropical, subtropical and Mediterranean regions. In N. S. Subba Rao and Y. R. Dommergues (Eds.), *Microbial interactions in agriculture and forestry* (pp. 65-96). Plymouth, UK: Science Publishers, Inc.

Duhoux, E., Rinaudo, G., Diem, H. G., Auguy, F., Fernandez, D., *et al.* (2001). Angiosperm *Gymnostoma* trees produce root nodules colonized by arbuscular mycorrhizal fungi related to *Glomus*. *New Phytol., 149,* 115-125.

Ellis, B. A., and Kummerow, J. (1988). N_2 (C_2H_2-C_2H_4) fixation in two species of *Ceanothus* seedlings in second year postfire chaparral. *Plant Soil, 109,* 207-213.

Faure-Raynaud, M., Bonnefoy-Poirier, M. A., and Moiroud, A. (1986). Influence de pH acides sur la viabilité d'isolats de *Frankia*. *Plant Soil, 96,* 347-358.

Fessenden, R. J. (1979). Use of actinorhizal plants for land reclamation and amenity planting in the USA and Canada. In J. C. Gordon, C. T. Wheeler and D. A. Perry (Eds.), *Symbiotic nitrogen fixation in the management of temperate forests* (pp. 403-419). Corvallis, OR: Oregon State University Press.

Fraga-Beddiar, A., and Le Tacon, F. (1990). Interactions between a VA mycorrhizal fungus and *Frankia* associated with alder (*Anus glutinosa* (L.) Gaertn.). *Symbiosis, 9*, 247-257.

Frioni, L., Soinelli, A., and Maggi, A. (1991). Nodulación y fijación de nitrógeno en especies de *Casuarina* y *Allocasuarina* cultivadas en el país. *Boletín de Investigación 30,* Universidad de la República, Montevideo, Uruguay.

Frioni, L., Le Roux, C., Dommergues, Y. R., and Diem H. G. (1994). Inoculant made of encapsulated *Frankia*: Assessment of *Frankia* growth within alginate beads. *W. J. Microbiol. Biotechnol. 10,* 118-121.

Galicia, M. C., and Valdés, M. (1997). Interaction between actinorhizal and mycorrhizal associations on *Alnus acuminata* ssp. *glabrata*. *Ciencia Forestal México, 22,* 3-14.

Gauthier, D., Diem, H. G., and Dommergues, Y. R. (1981). Infectivité et effectivité des souches de *Frankia* isolées de nodules de *Casuarina equisetifolia* et d'*Hippophaë rhamnoides*. *C. R. Acad. Sci. Paris Série III, 293,* 489-491.

Gauthier, D., Jaffré, T., and Prin, Y. (2000). Abundance of *Frankia from Gymnostoma* spp. in the rhizosphere of *Alphitonia neocaledonica,* a non-nodulated *Rhamnaceae* endemic to New Caledonia. *Eur. J. Soil Biol., 36,* 169-175.

Godbout, C., and Fortin, J. A. (1983). Morphological features of synthesized ectomycorrhizae of *Alnus crispa* and *A. rugosa*. *New Phytol., 94,* 249-262.

Haansuu, J. P., Klika, K., Söderholm, P., Ovcharenko, V., Pihlaja, K., *et al.* (2001). Isolation and biological activity of frankiamide. *J. Ind. Microbiol. Biotechnol., 27,* 62-66.

Hafeez, F. Y., Hameed, S., and Malik, K. A. (1999). *Frankia* and *Rhizobium* strains as inoculum for growing trees in a saline environment. *Pakistan J. Bot., 31,* 173-182.

Huguet, V., Batzli, J. M., Zimpfer, J. F., Normand, P., Dawson, J. O., and Fernandez, M. P. (2001). Diversity and specificity of *Frankia* strains in nodules of sympatric *Myrica gale, Alnus incana* and *Shepherdia canadensis* determined by *rrs* gene polymorphism. *Appl. Environ. Microbiol., 67,* 2116-2122.

Huss-Danell, K. (1990). The physiology of actinorhizal nodules. In C. R. Schwintzer, and J. D. Tjepkema (Eds.), *The biology of* Frankia *and actinorhizal plants* (pp. 129-156). New York: Academic Press.

Huss-Danell, K. (1997). Actinorhizal symbioses and their N_2 fixation. Tansley Review No. 93. *New Phytol., 136,* 375-405.

Huss-Danell, K., and Frej, A. (1986). Distribution of *Frankia* strains in forest and afforestation sites in Northern Sweden. *Plant Soil, 90,* 407-418.

Huss-Danell, K., Sverrisson, H., Hahlin, A-S., and Danell, K. (1999). Occurrence of *Alnus*-infective *Frankia* and *Trifolium*-infective *Rhizobium* in circumpolar soils. *Arctic Antarctic Alpine Res., 31,* 400-406.

Igual, J. M., and Dawson, J. O. (1999). Stimulatory effects of aluminum on *in vitro* growth of *Frankia*. *Can. J. Bot. 77,* 1321-1326.

Isopi, R., Lumini, E., Frattegiani, M, Puppi, G., Bosco, M., *et al.* (1994). Inoculation of *Alnus cordata* with selected microsymbionts: Effects of *Frankia* and *Glomus* spp. on seedling growth and development. *Symbiosis, 17,* 237-245.

Jamann, S., Fernandez, M. P., and Moiroud, A. (1992). Genetic diversity of Elaeagnaceae-infective *Frankia* strains isolated from various soils. *Acta Oecol., 13,* 395-405.

Jeong, S. C., and Myrold, D. D. (2001). Population size and diversity of *Frankia* in soils of *Ceanothus velutinus* and Douglas-fir stands. *Soil Biol. Biochem., 33,* 931-941.

Kohls, S. J., and Baker, D. D. (1989). Effects of substrate nitrate concentration on symbiotic nodule formation in actinorhizal plants. *Plant Soil, 118,* 171-179.

Kohls, S. J., Baker, D. D., Kremer, D. A., and Dawson, J. O. (1999). Water-retentive polymers increase nodulation of actinorhizal plants inoculated with *Frankia*. *Plant Soil, 214,* 105-115.

Kummerow, J. A., Neel, J. V., and Fishbeck, K. (1978). Symbiotic nitrogen fixation in *Ceanothus* roots. *Am. J. Bot., 65,* 63-69.

Kurdali, F., and Domenach, A. M. (1991). Evaluation of nodulation capacity in *Alnus glutinosa* seedlings grown in soil sampled from various depths. *Acta Oecol., 12,* 397-401.

Lawrence, D. B., Schoenike, A., Quispel, A., and Bond, G. (1967). The role of *Dryas drummondii* in vegetation development following ice recession at Glacier Bay, Alaska, with special reference to its nitrogen fixation by root nodules. *J. Ecol., 55,* 793-813.

Lumini, E., Bosco, M., Puppi, G., Isopi, R., Grattegiani, M., *et al.* (1994). Field performance of *Alnus cordata* Loisel (Italian alder) inoculated with *Frankia* and VA-mycorrhizal strains in mine-spoil afforestation plots. *Soil Biol. Biochem., 26,* 659-661.

Maier, R. J., and Triplett, E. W. (1996). Toward more productive, efficient, and competitive nitrogen-fixing symbiotic bacteria. *CRC Crit. Rev. Plant Sci., 15,* 191-234.

Mattsson, U., and Sellstedt, A. (2002). Nickel affects activity more than expression of hydrogenase protein in *Frankia. Curr. Microbiol., 44,* 83-93.

Maunuksela, L., Zepp, K., Koivula, T., Zeyer, J., Haahtela, K., and Hahn, D. (1999). Analysis of *Frankia* populations in three soils devoid of actinorhizal plants. *FEMS Microbiol. Ecol., 28,* 11-21.

Miller, S. L., Coo, C. D., and Molina, R. (1992). Early colonization or red alder and Douglas fir by ectomycorrhizal fungi and *Frankia* in soils from the Oregon coast range. *Mycorrhiza, 2,* 53-61.

Myrold, D. D., and Huss-Danell, K. (1994). Population dynamic of *Alnus*-infective *Frankia* in a forest soil with and without host tree. *Soil. Biol. Biochem., 26,* 533-540.

Myrold, D. D., Hilger, A. B., Huss-Danell, K., and Martin, J. K. (1994). Use of molecular methods to enumerate *Frankia* in soil. In K. Ritz, J. Dighton, and K. E. Giller (Eds.), *Beyond the biomass* (pp. 127-136). Chichester, UK: John Wiley and Sons.

Nalin, R., Normand, P., and Domenach, A-M. (1997). Distribution and N_2-fixing activity of *Frankia* strains in relation to soil depth. *Physiol. Plant., 99,* 732-738.

National Research Council (1984). *Casuarinas: Nitrogen-fixing trees for adverse sites.* Washington, D.C.: National Academy Press.

Navarro E., Nalin, R., Gauthier D., and Normand, P. (1997). The nodular microsymbionts of *Gymnostoma* spp. are *Elaeagnus*-infective *Frankia* strains. *Appl. Environ. Microbiol., 63,* 1610-1616.

Nickel, A., Pelz, O., Hahn, D., Saurer, M., Siegwolf, R., and Zeyer, J. (2001). Effect of inoculation and leaf litter amendment on establishment of nodule-forming *Frankia* populations in soil. *Appl. Environ. Microbiol., 6,* 2603-2609.

Oremus, P. A. (1980). Occurrence and infective potential of the endophyte of *Hippophae rhamnoides* L. ssp. *rhamnoides* in coastal sand-dune areas. *Plant Soil, 56,* 123-139.

Paschke, M. W., and Dawson, J. O. (1992). The occurrence of *Frankia* in tropical forest soils of Costa Rica. *Plant Soil, 142,* 63-67.

Paschke, M. W., and Dawson, J. O. (1993). Avian dispersal of *Frankia. Can. J. Bot., 71,* 1128-1131.

Pratt, S. D., Konopka, A. S., Murry, M. A., Ewers, F. W., and Davis, S. D. (1997). Influence of soil moisture on the nodulation of post fire seedlings of *Ceanothus* spp. growing in the Santa Monica Mountains of Southern California. *Physiol. Plant., 99,* 673- 679.

Prin, Y., Duhoux, E., Diem., H. G., Roederer, Y., and Dommergues, Y. R. (1991). Aerial nodules in *Casuarina cunninghamiana. Appl. Environ. Microbiol., 57,* 871-874.

Provorov, N. A., Borisov, A. Y., and Tikhonovich, I. A. (2002). Developmental genetics and evolution of symbiotic structures in nitrogen-fixing nodules and arbuscular mycorrhiza. *J. Theor. Biol. 214,* 215-232.

Read, D. J. (1998). Mycorrhiza. The state of the art. In Varma A. and Hock, B. (Eds.), *Mycorrhiza* (pp. 3-34). Berlin, Germany: Springer-Verlag.

Reddell, P., Bowen, G. D., and Robson, A. D. (1986). Nodulation of Casuarinaceae in relation to host species and soil properties. *Aust. J. Bot., 34,* 435-444.

Reddell, P., Rosbrook, P. A., Bowen, G. D. and Gwaze, D. (1988). Growth responses in *Casuarina cunninghamiana* plantings to inoculation with *Frankia. Plant Soil, 108,* 79-86.

Richards, J. W., Krumholz, G. D., Chval, M. S., and Tisa, L. S. (2002). Heavy metal resistance patterns of *Frankia* strains. *Appl. Environ. Microbiol., 68,* 923-927.

Righetti, T. L., Chard, C. H., and Backhause, R. A. (1986). Soil and environmental factors related to nodulation in *Cowania* and *Purshia. Plant Soil, 91,* 147-160.

Rodriguez-Barrueco, C. (1968). The occurrence of the root nodule endophyte of *Alnus glutinosa* and *Myrica gale* in soils. *J. Gen. Microbiol., 52,* 189-194.

Rojas, N. S., Perry, D. A., Li, C. Y., and Ganio, M. L. (2002). Interactions among soil biology, nutrition, and performance of actinorhizal plant species in the H.J. Andrews Experimental Forest in Oregon. *Appl. Soil Ecol., 19,* 13-26.

Rose, S. L. (1980). Mycorrhizal associations of some actinomycete nodulated nitrogen-fixing plants. *Can. J. Bot., 58,* 1449-1454.

Ronkko, R., Smolander, A., Nurmiaho-Lassila, E. L., and Haahtela, K. (1993). *Frankia* in the rhizosphere of nonhost plants: A comparison with root-associated nitrogen-fixing *Enterobacter, Klebsiella* and *Pseudomonas. Plant Soil, 153,* 85-95.

Sanginga, N., Danso, S. A. A., and Bowen, G. D. (1989). Nodulation and growth response of *Allocasuarina* and *Casuarina* species to phosphorous fertilization. *Plant Soil, 118*, 125-132.

Sellstedt, A., and Smith, G. D. (1990). Nickel is essential for active hydrogenase in free-living *Frankia* isolated from *Casuarina. FEMS Microbiol. Lett., 70*, 137-140.

Sellstedt, A. (1995). Specificity and effectivity in nodulation by *Frankia* on southern hemisphere actinorhiza. *FEMS Microbiol. Lett., 125,* 231-236.

Sempavalan, J., Wheeler, C. T., and Hooker, J. E. (1995). Lack of competition between *Frankia* and *Glomus* for infection and colonization of roots of *Casuarina equisetifolia* (L.). *New Phytol., 130*, 429-436.

Sequerra, J., Capellano, A., Faure-Raynard, M., Moiround, A. (1994). Root hair infection process and myconodule formation on *Alnus incana* by *Penicillium nodositatum. Can. J. Bot., 72,* 955-962.

Sheppard, L. J., Hooker, J. E., Wheeler, C. T., and Smith, R. I. (1988) Glasshouse evaluation of the growth of *Alnus rubra* and *Alnus glutinosa* on peat and acid brown earth soils when inoculated with four sources of *Frankia. Plant Soil, 110,* 187-198.

Smith, S. E., and Read D. J. (1997). *Mycorrhizal symbiosis*. London, UK: Academic Press.

Smolander, A., and Sundman, V. (1987). *Frankia* in acid soils of forest devoid of actinorhizal plants. *Physiol. Plant., 70*, 297-303.

Smolander, A., Nurmiaho-Lassila, E. L., and Sundman, V. (1988). Effect of a clay mineral (montmorillonite) on the nodulation of *Alnus* and on the nitrogenase activity of *Frankia* in pure culture. *Symbiosis, 6*, 37-52.

Smolander, A. (1990). *Frankia* populations in soil under different tree species with special emphasis on soil under *Betula pendula. Plant Soil, 121*, 1-10

Smolander, A., Ronkko, R., Nurmiaho-Lassila, E. L., and Haahtela, K. (1990). Growth of *Frankia* in the rhizosphere of *Betula pendulata*, a nonhost tree species. *Can. J. Microbiol., 36,* 649-656.

Theodoru, C., and Reddell, P. (1991). *In vitro* synthesis of ectomycorrhizas on Casuarinaceae with a range of ectomycorrhizal fungi. *New Phytol., 118*, 279-288.

Thoen, D., Sougoufara, B., and Dommergues, Y. R. (1990). *In vitro* mycorrhization of *Casuarina* and *Allocasuarina* species by *Pisolithus* isolates. *Can. J. Bot., 68*, 2537-2542.

Tisa, L. S., Chval, M. S., Krumholz, G. D., and Richards, J. (2000). Antibiotic resistance patterns of *Frankia* strains. *Can. J. Bot., 77,* 1257-1260.

Tortosa, R. D., and Cusato, M. (1991). Effective nodulation of rhamnaceous actinorhizal plants induced by air dry soils. *Plant Soil, 131*, 229-233.

Trappe, J. M. (1979). Mycorrhiza-nodule-host interrelationships in symbiotic nitrogen-fixation: A quest in need of questions. In J. C. Gordon, C. T. Wheeler and D. A. Perry (Eds.), *Symbiotic nitrogen fixation in the management of temperature forests* (pp. 84-94). Corvallis, OR: Oregon State University Press.

Valdés, M., and Cruz-Cisneros, R. (1996). Root and stem nodulation of *Casuarina* in Mexico. *Forest Farm Com. Tree Res. Rep., I*, 61-65.

Valdés, M., and Sánchez Francia, D. (1996). Response of *Alnus* and *Casuarina* to endomycorrhizal inoculation. *Rev. Mex. Mic., 12,* 65-77.

Valdés, M., and Galicia, M. C. (1997). Interaction between actinorhizal and mycorrhizal associations on *Alnus accuminata* ssp. *glabrata. Cienc. For. Mex., 22*, 3-14.

Valdés, M., Pérez, N. O., and Vásquez, L. (2001). La bacteria filamentosa *Frankia.* In E. Martínez-Romero and J. Martínez-Romero (Eds.), *Microbios en línea* (pp. 237-257). UNAM, Mexico: DGSCA Press.

Valdés, M., Cayetano, A., Leyva, M. A., and Camacho, A. D. (2004). Promoción del crecimiento de *Casuarina equisetifolia* (L.) por diferentes microorganismos simbiontes. *Terra, 22,* 207-215.

Vásquez, L., Pérez, N-O., and Valdés, M. (2000). Isolation and symbiotic characteristics of Mexican *Frankia* strains associated with *Casuarina. Appl. Soil Ecol., 14,* 249-255.

Vásquez, L., Cruz-Cisneros, R., and Valdés, M. (2000). Presencia de nódulos fijadores de nitrógeno en raíces de *Casuarina* spp. en México. *Cienc. For. Mex., 25,* 93-102.

van Dijk, C. (1979). Endophyte distribution in the soil. In J. C. Gordon, C. T. Wheeler and D. A. Perry (Eds.), *Symbiotic nitrogen fixation in the management of temperature forests* (pp. 84-94). Corvallis, OR: Oregon State University Press.

Visser, S., Danielson, R. M., and Parkinson, D. (1990). Field performance of *Elaeagnus communata* and *Shepherdia canadensis* (Elaeagnaceae) inoculated with soil containing *Frankia* and vesicular-arbuscular mycorrhizal fungi. *Can. J. Bot., 69*, 1321-1328.

Wall, L. G., Hellsten, A., and Huss-Danell, K. (2000). Nitrogen, phosphorus, and the ratio between them affect nodulation in *Alnus incana* and *Trifolium pratense*. *Symbiosis, 29,* 91-105.

Wolters, D. J., Van Dijk, C., Zoetendal, E. G., and Akkermans, A. D. L. (1997). Phylogenetic characterization of inffective *Frankia* in *Alnus glutinosa* (L). Gaertn. nodules from wetland soil inoculants. *Mol. Ecol., 6,* 971-981.

Wolters, D. J., van Dijk, C., Akkermans, A., and Woldendorp, J. W. (1999). Ineffective *Frankia* and host resistance in natural populations of *Alnus glutinosa* (L.) Gaertn. *Acta Oecol., 20,* 71-79.

Young, D. R., Sande, E., and Peters, G. A. (1992). Spatial relationships of *Frankia* and *Myrica cerifera* on a Virginia, USA, barrier island. *Symbiosis, 12,* 209-220.

Zhang-Zhongze, Z., Chenggang, Z., Xingyuan, H., and Yuying, W. (2000). Effect of inoculation on the *Frankia-Hipphophaë* symbiosis and influence of *Hippophaë* on *Populus* growth in mixed plantations. In N. S. Subba Rao and Y. R. Dommergues (Eds.), *Microbial interactions in agriculture and forestry* (pp.155-162). Plymouth, UK: Science Publishers Inc.

Zimpfer, J. F., Smyth, C. A., and Dawson, J. O. (1997). The capacity of Jamaican mine spoils, agricultural and forest soils to nodulate *Myrica cerifera, Leucaena leucocephala* and *Casuarina cunninghamiana*. *Physiol. Plant., 99,* 664-672.

Zimpfer, J. F., Kennedy, G. J., Smyth, C. A., and Dawson, J. O. (1999). Localization of *Casuarina* infective *Frankia* near *Casuarina equisetifolia* trees in Jamaica. *Can. J. Bot., 77,* 1248-1256.

Zitzer, S. F., and Dawson, J. O. (1992). Soil properties and actinorhizal vegetation influence nodulation of *Alnus glutinosa* and *Elaeagnus angustifolia* by *Frankia*. *Plant Soil, 140,* 197-204.

Chapter 4

EVOLUTION OF ACTINORHIZAL HOST PLANTS AND *FRANKIA* ENDOSYMBIONTS

S. M. SWENSEN[1] AND D. R. BENSON[2]
[1]*Department of Biology, Ithaca College, Ithaca, NY 14850, U.S.A.*
[2]*Department of Molecular and Cell Biology, University of Connecticut, Storrs, CT 06269-3044, U.S.A.*

1. INTRODUCTION

Actinorhizal symbioses are found in a remarkable variety of flowering plants. Although the host plants are quite diverse, the diversity of their actinomycete symbionts, *Frankia* sp. strains, is just beginning to be understood. In this chapter, we review the significant advances in our understanding of the evolutionary relationships (phylogeny) of both host plants and *Frankia* endosymbionts. We consider different lines of evidence from molecular, morphological, and fossil sources. We also compare host and endosymbiont phylogeny in an effort to address the question of reciprocal evolution (co-evolution) between hosts and endosymbionts. Finally, we consider the future prospects and directions for further study on the evolution of actinorhizal symbioses.

The recent advances in the area of phylogenetic analysis are the result of new molecular techniques that have revolutionized the way that plants and bacteria can be studied genetically and evolutionarily. Most significant among these techniques is the polymerase chain reaction (PCR), which amplifies DNA fragments from target organisms through the use of taxon- and sequence-specific DNA primers. The development of this technique has two great advantages. First, it has removed the necessity of obtaining pure cultures of *Frankia*, which is often difficult or impossible from many species of plant hosts and from soil samples, for DNA isolation. Second, for both host plants and bacteria, the ability to amplify target genes by PCR and then directly sequence them has also eliminated the laborious DNA-cloning steps that were once necessary components of sequencing studies.

K. Pawlowski and W. E. Newton (eds.), Nitrogen-fixing Actinorhizal Symbioses, 73–104.
© 2008 *Springer*.

In addition to the inclusion of molecular data, another shift in evolutionary studies has been toward the use of cladistics as a methodological approach to estimating phylogeny (Hennig 1950; 1960). Phylogenetic systematics (or cladistics) utilizes only synapomorphies (or shared derived characters) in defining taxon groupings which is different from methods that define groups based on overall similarity among all characters (phenetics). Cladistics, therefore, avoids the use of sympleiosmorphies (or primitively shared characters) that would be the result of shared (but ancient) ancestry and focuses on the use of derived characters that are unique to specific taxon groups. Actinorhizal host phylogeny has been well documented due to the ease of obtaining plant material and DNA, however, the picture of *Frankia* phylogeny has only recently emerged. Direct amplification of *Frankia* DNA from host nodules has accelerated the rate at which strain DNA can be sequenced and, as a result, has provided a broader and more inclusive view of *Frankia* phylogeny. An understanding of both host and endosymbiont phylogeny is critical to unraveling the evolutionary history of this association.

2. HOST-PLANT SYSTEMATICS AND PHYLOGENY

2.1. Traditional Classifications of Host Plants

Actinorhizal nodules were first described on *Alnus glutinosa* (European Alder; Meyen, 1829) although, at that time, the nodule microsymbionts were considered to be parasitic plants. In 1866, Woronin carefully described the anatomy of *Alnus glutinosa* nodules and concluded that the microsymbionts were fungal parasites. It was not until 1898 that Hiltner correctly identified the microsymbionts as N_2-fixing bacteria by comparison with bacteria in the N_2-fixing nodules of legumes (Hiltner, 1898; Woronin, 1866).

Since the initial description of the actinorhizal symbiosis in plant roots, *ca.* 194 plant hosts have been identified among the dicotyledonous angiosperms (Table 1). The hosts are mostly woody plants, classified among eight different families and four subclasses of flowering plants according to the classification system of Cronquist (1981). This classification system is essentially the same as other prominent morphologically-based systems (Dahlgren, 1980; Takhtajan, 1980; Thorne, 1992) with respect to the taxonomic placement of actinorhizal plants. These schemes differ mainly in their placement of the family Coriariaceae. Cronquist places this family in the order Ranunculales, whereas others place it in an order called Rutales, also known as Sapindales. The families Coriariaceae and Datiscaceae are considered to be the most divergent of the actinorhizal host families. They appear to have no close relatives and their placements in these taxonomic systems are the most uncertain. For the purposes of this chapter, we will follow the taxonomic systems of Cronquist (1981) and APG (1998).

Table 1. Classification of actinorhizal species and frequency of nodulation within families.

Subclass[a]	*Family*	*# Nodulated /Total Genera*	*Genus*	*No. of Species*
Hamamelidae	Betulaceae	1/6	*Alnus*	47
	Casuarinaceae	4/4	*Allocasuarina*	54
			Casuarina	16
			Ceuthostoma	2
			Gymnostoma	18
	Myricaceae	2/3	*Comptonia*	1
			Myrica	28
Rosidae	Elaeagnaceae	3/3	*Elaeagnus*	38
			Hippophae	2
			Shepherdia	2
	Rhamnaceae	8/55	*Adolphia*[b]	2
			Ceanothus	31
			Colletia	4
			Discaria	5
			Kentrothamnus	1
			Retanilla	2
			Talguenea	1
			Trevoa	2
	Rosaceae	5/100	*Cercocarpus*	4
			Chamaebatia	1
			Cowania	1
			Dryas	1
			Purshia	2
Magnoliidae	Coriariaceae	1/1	*Coriaria*	5
Dilleniidae	Datiscaceae	1/3	*Datisca*	2

[a]Classification according to Cronquist (1981).
[b]*Adolphia infesta* (Kunth) Meisn. and *Colubrina infesta* (Kunth) Schltdl. are synonymous with basionym *Ceanothus infestus* Kunth.

2.2. Molecular Systematics of Nodulating Host Plants

In the plant world, the first gene broadly utilized for molecular systematics was *rbcL*, an evolutionarily conserved chloroplast gene that encodes the large subunit of the carbon-fixing enzyme, ribulose 1,5-bisphosphate carboxylase/oxygenase (EC 4.1.1.39), also known as "rubisco". The first published major molecular systematic study of plants included 499 flowering plants (Chase *et al.*, 1993). Among these were three subfamilies of legumes, the family Ulmaceae that contains the non-legume *Rhizobium*-host *Parasponia*, and seven families (Betulaceae, Casuarinaceae, Coriariaceae, Datiscaceae, Myricaceae, Rhamnaceae, and Rosaceae) that contain actinorhizal hosts. In contrast to the morphologically-centered classification system of Cronquist (1981), which indicated considerable taxonomic distance among host plants, the *rbcL*-based phylogeny placed the families with N_2-fixing root-nodule symbioses together as part of a single clade. The clade consisted of a portion of Cronquist's subclass Rosidae as well as members of the subclass Hamamelidae and was called "rosid I" or "eurosid I" *sensu* (Angiosperm Phylogeny Group, 1998). The eurosid I group contains both actinorhizal and rhizobial host plants.

Due to the large size of the data set of Chase *et al.* (1993) and the computational difficulty associated with analyzing data sets of this size, the resulting phylogeny could only be regarded as preliminary. Thus, further studies of the phylogeny of N_2-fixing root-nodule symbioses were undertaken. Soltis *et al.* (1995) analysed *rbcL* sequences from 99 plant species belonging to the rosid I clade of Chase *et al.* (1993), and included many non-nodulating species as well as additional actinorhizal species that were not part of the Chase *et al.* (1993) study. The results of this analysis were congruent with those of Chase *et al.* (1993) in that all families with N_2-fixing root-nodule symbioses grouped together as part of a single "nitrogen-fixing" clade, along with several plant families that lack these symbioses. This clade was moderately well supported by the data, as determined by a decay value = 2 in a decay analysis (Bremer, 1988; Donoghue *et al.* 1992). The conclusion from this study was that only a single lineage of flowering plants achieved the necessary genetic background to engage in nodular N_2-fixing symbioses.

The initial broad-scope molecular-systematic studies in plants were based solely on *rbcL*. However, it is widely recognized that a single gene reflects only the phylogeny of that gene itself, and may not necessarily reflect organismal phylogeny (Doyle, 1992; Page and Charleston, 1997). Thus, in 1997, Soltis *et al.* published a molecular phylogeny of 223 flowering plants based on the nuclear 18S-ribosomal gene that encodes a portion of the RNA that makes up cytosolic ribosomes. The 18S analyses provided additional evidence for the existence of a N_2-fixing clade, but did not yield exactly the same taxonomic groupings as was found with *rbcL* and, in some cases, no recognizable nitrogen-fixing group was apparent. For example, in the 18S analyses where a N_2-fixing clade was present, the clade did not include the families Rosaceae or Fabaceae, which were part of the *rbcL*-based N_2-fixing clade. Also, in some analyses, the 18S N_2-fixing clade included the families Malvaceae, Tiliaceae, and Bombaceae that were not part of the *rbcL*-based grouping. Two families (Coriariaceae, Myricaceae) sampled in previous *rbcL* studies were not included in the 18S analysis.

A third broad-based angiosperm molecular-systematic study was conducted and incorporated the data from three separate genes, *rbcL*, 18S rDNA, and the chloroplast-encoded *atpB* gene (Soltis *et al.*, 2000). The *atpB* gene encodes the B subunit of the chloroplast enzyme, ATP synthase. This enzyme couples proton translocation across the thylakoid membrane of the chloroplast with ATP synthesis. Like *rbcL* and 18S rDNA, *atpB* is highly conserved evolutionarily and is, therefore, useful for inferring the phylogeny at the level of genus or family and above (Hoot *et al.*, 1995). This study included 560 species and a total of 4733 base pairs of sequence data. The computational difficulty of analyzing this large data set was, ironically, alleviated by increasing taxon-sampling density and by increasing the number of characters analyzed, *e.g.*, total base pairs of sequence data (Graybeal,

1998; Hillis, 1996). This study represents one of the best estimates thus far of angiosperm phylogeny and a summary tree is shown in Figure 1.

This analysis provides the strongest evidence to date for a single N_2-fixing clade within the eurosid I group. The N_2-fixing clade includes all families known to

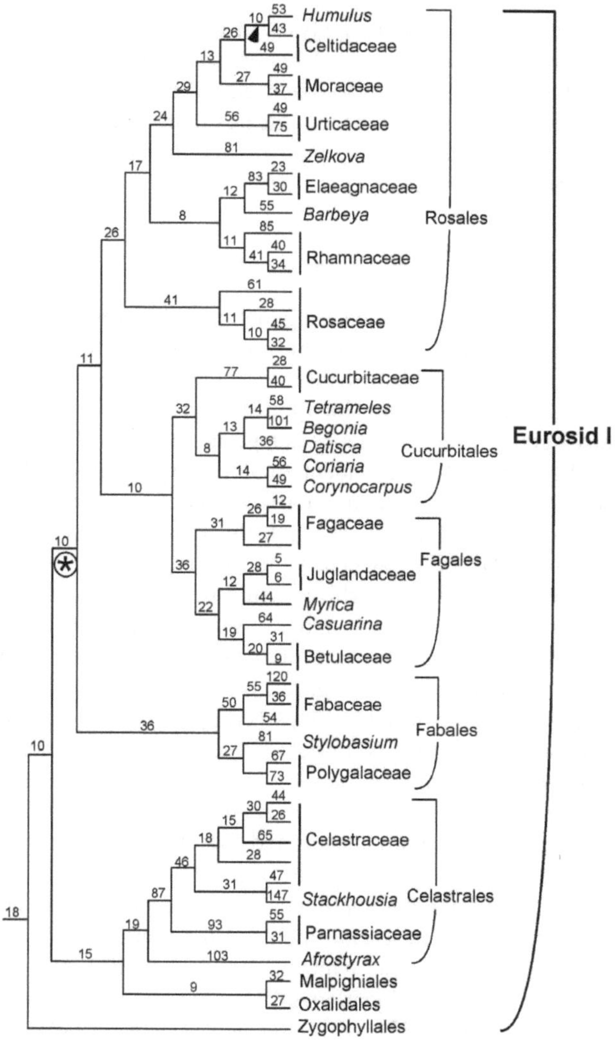

Figure 1. One of 8000 most parsimonious trees obtained via phylogenetic analysis of a three-gene data set (modified from Soltis et al., *2000). Asterisk indicates position of the N_2-fixing clade. Numbers are branch lengths (ACCTRAN optimization). Arrow denotes the single group not found in all most parsimonious trees.*

contain hosts of nodulating, N_2-fixing bacteria and is supported by a bootstrap value of 68% (Felsenstein, 1985), which represents a higher level of support than found in previous analyses. Within the N_2-fixing clade, the legumes occupied a basal (ancestral) position, although this result was not supported by a parsimony jackknife analysis (Soltis *et al.,* 2000). The basal position of legumes was also detected in an *rbcL*-only analysis of Pawlowski *et al.* (2003).

These broad-scope phylogenetic analyses of angiosperms have provided a very different view of relationships among plants engaging in N_2-fixing symbioses than was implied by their morphologically-based classification. The presence of a N_2-fixing clade within the eurosid I lineage indicates that nodular N_2-fixing symbioses do not occur sporadically among the angiosperms but rather are the property of a single lineage of plants.

2.3. *Evolution of Nodulation in Actinorhizal Plants*

If members of the N_2-fixing clade share a genetic predisposition toward nodulation, one might expect recurrent evolution of nodulation in this group. Differences between rhizobial and actinorhizal symbioses include different bacteria, different nodule anatomies, and the separate location of legumes and actinorhizal hosts in angiosperm phylogeny. Thus, evidence suggests that rhizobial and actinorhizal symbioses are the result of independent evolutionary events and several studies have concluded that the different actinorhizal symbioses are also the result of multiple evolutionary events (Jeong *et al.,* 1999; Pawlowski *et al.,* 2003; Roy and Bousquet, 1996; Swensen, 1996; Swensen and Mullin, 1997).

Studies of the evolution of actinorhizal symbioses have all utilized the *rbcL* gene and, for the most part, reveal four major groups of actinorhizal hosts that are interspersed with non-actinorhizal species in a pattern suggestive of multiple origins or multiple losses. A summary tree (Figure 2), based on the analysis of Pawlowski *et al.* (2003), shows the actinorhizal groups (groups B-E) plus the legumes (group A). The four actinorhizal groups comprise: (i) actinorhizal members of the family Rosaceae (D); (ii) actinorhizal members of family Rhamnaceae and Elaeagnaceae (C); (iii) *Coriaria* and *Datisca* (E); and (iv) higher Hamamelidae that includes actinorhizal Betulaceae, Casuarinaceae, and Myricaceae (B). Swensen (1996) addressed the question of homology of the symbioses in the four different actinorhizal groups by mapping anatomical and morphological features of root nodules onto the *rbcL* tree. These data included the number and shape of *Frankia* symbiotic vesicles (Baker and Schwintzer, 1990; Lalonde, 1979; Murray *et al.,* 1985; Zhang *et al.,* 1984), the location of *Frankia* vesicles within infected cells (Akkermans and van Dijk, 1981; Silvester and Harris, 1989), the location of infected cells within the nodule (Akkermans *et al.,* 1983; Calvert *et al.,* 1979; Newcomb and Pankhurst, 1982), evidence for gaseous diffusion barriers in the nodule and hemoglobin concentration (Tjepkema 1978; 1979), and the mode of infection of the host plants (Berry *et al.,* 1986; Berry and Sunell 1990; Callaham *et al.,*

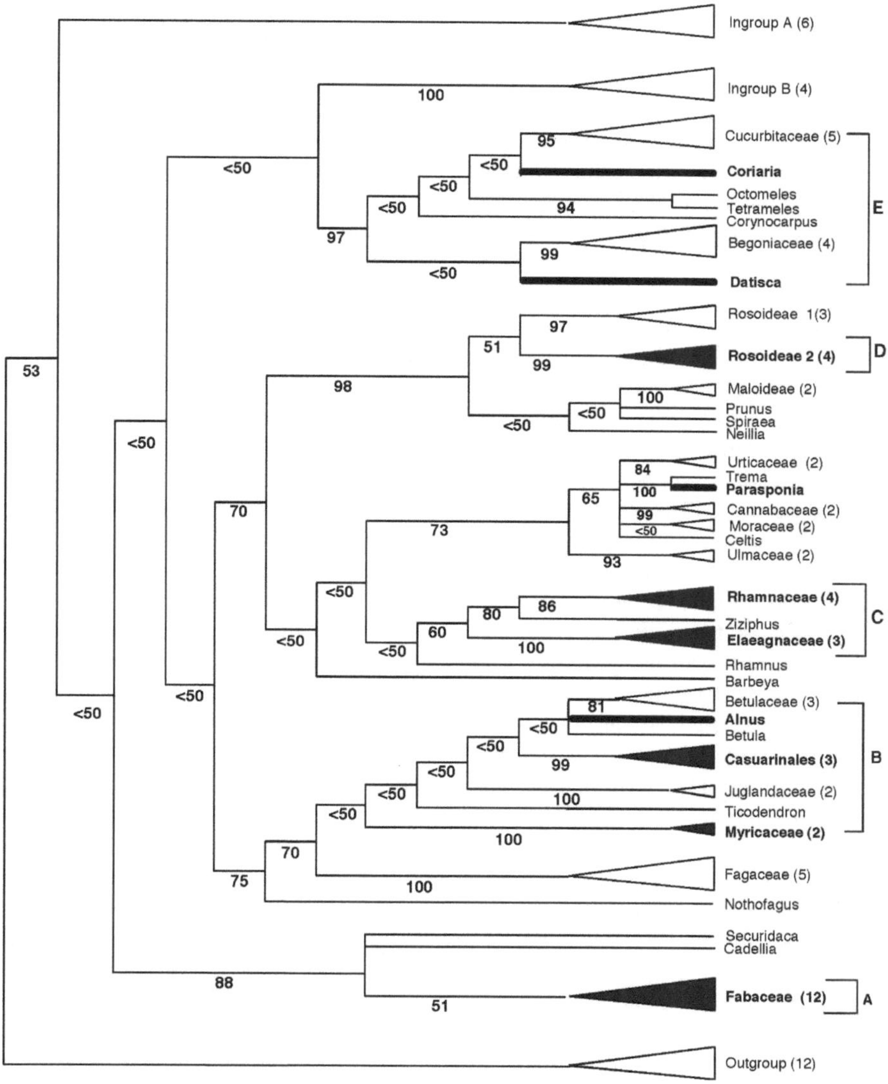

Figure 2. Summary tree based on 72 equally parsimonious trees resulting from analysis of rbcL *genes (taken from Pawlowski* et al., *2003).*
Bold lines or triangles represent groups containing species with root nodules.
Numbers below the branches are support values obtained from parsimony jackknife analysis (branches with values < 50% were not found in the parsimony jackknife analysis).
Closely grouped species are represented by group, family, or subfamily names followed by the number of analyzed species in parentheses.

1979; Kohls *et al.,* 1994; Liu and Berry 1991; Miller and Baker, 1985; Racette and Torrey 1989a). There were clear differences in these characters in actinorhizal species belonging to each of the four groups, suggesting that groups B-E were the result of different symbiotic origins and that actinorhizal symbioses originated at least four times in the evolution of flowering plants.

The analysis of Roy and Bousquet (1996) was conducted using Neighbor-Joining on (i) actinorhizal species alone and (ii) on actinorhizal species plus their non-actinorhizal relatives. The first analysis revealed only three major groupings; a group consisting of members of families Myricaeae, Betulaceae, and Casuarinaceae, a group consisting of *Coriaria* and *Datisca*, and a group consisting of members of families Elaeagnaceae, Rhamnaceae, and Rosaceae. The second analysis revealed the same four groupings as found by Swensen (1996) and underscores the importance of inclusive taxon sampling in phylogenetic analyses.

In addition to phylogenetic analysis of plant hosts, Jeong *et al.* (1999) calculated a divergence time for actinorhizal groups based on previously estimated rates of nucleotide change in *rbcL*. According to these calculations, the actinorhizal clades diverged between 170-429 million years ago, before the divergence of their *Frankia* symbionts as estimated by 16S rDNA divergence; however, this date also precedes the appearance of the first unambiguous angiosperm fossils (Crane *et al.,* 1995; see discussion of molecular clocks below).

Because of the separate groupings of different actinorhizal hosts, the plant phylogenies suggest that symbioses evolved more than once in evolutionary history, which supports the notion that a genetic predisposition in members of the N_2-fixing clade would result in recurrent evolution of symbiotic associations. However, the phylogenetic placement of actinorhizal species among non-actinorhizal species, often belonging to the same family, also supports the idea that loss of the symbiosis may be a common feature of this evolutionary process. In contrast to gaining a symbiotic association, the loss of a symbiosis might easily result from a single mutation in either the host or endosymbiont. Ecological pressures or local adaptations by either partner may preclude the formation of the symbiosis, if the energy costs are too high. In habitats either with abundant soil nitrogen or under high stress, host plants generally do not engage in symbioses (Benoit and Berry, 1990; Dixon and Wheeler, 1983; Dommergues *et al.,* 1984). Factors that decrease photosynthetic activity correspondingly decrease N_2-fixing activity (Dawson and Gordon, 1979; Gordon and Wheeler, 1978). Symbiotic species are likely to expend more energy than non-symbiotic species in the same environment and, in addition, provide fixed nitrogen to their competitors, which may over-shade them. Such limitations might explain why many actinorhizal species do not persist beyond early successional habitats (Côte *et al.,* 1988; Dawson, 1990; Neave *et al.,* 1989).

2.4. Evolutionary Origins of Actinorhizal Symbioses

Phylogenetic reconstructions of actinorhizal plants have suggested multiple evolutionary origins for the actinorhizal symbioses, however, the relative timing of these origins is unclear. Current molecular phylogenies do not indicate which of the actinorhizal groups was the first one to diverge and so which might have been the first symbiotic hosts. Indeed, the first actinorhizal host probably left no record of its existence. Fossil evidence exists for many actinorhizal species and these data provide minimum age estimates for the symbioses. Combined with the information from molecular phylogenetic analyses, a somewhat clearer picture of actinorhizal symbiotic evolution has emerged, although many questions still remain.

2.4.1. Angiosperm Fossil Dates and Molecular Dates

The eurosid I clade, where all nodular N_2-fixing symbioses are found, makes up a portion of the eudicot angiosperms. The eudicots are the most common angiosperms, comprising approximately 75% of all angiosperm species. The eudicots are recognized as a monophyletic group (a group containing all the descendants of a common ancestor) on the basis of both molecular and morphological analyses. Triaperturate pollen grains are diagnostic for the eudicots and first appear in the fossil record approximately 125 Myr at the Barremian-Aptian boundary of the lower Cretaceous (Crane *et al.,* 1995). In contrast to this fossil evidence, estimates of angiosperm age based on rates of molecular evolution of the genes, *gapC* and *rbcL* (Martin *et al.,* 1989; Martin *et al.,* 1993), suggest a much earlier appearance for angiosperms and eudicots (in the Triassic or Jurassic). It is now recognized that the substitution rates of *rbcL* vary, even among closely related groups, and depart from a constant molecular clock (Bousquet *et al.,* 1992; Clegg *et al.,* 1993; Gaut *et al.,* 1992; Savard *et al.,* 1994). Even relatively small rate variations may have large effects on the calculation of specific dates and so dates calculated under the assumption of a molecular clock need to be supported by evidence that the genes evolve in a clock-like manner (*e.g.*, by the relative rate test; Sarich and Wilson, 1967). In addition, there is no undisputed fossil evidence that supports an earlier appearance of eudicots (Crane *et al.,* 1995). If we accept the 125 Myr origin for eudicots, their appearance comprises part of a rapid diversification of angiosperms that occurred during the Mid-Cretaceous. This rapid and complex pattern of diversification has made phylogenetic reconstruction difficult because there are few intermediate taxa or characters that enable resolution of all the branching events of the radiation.

Paleobotanical studies indicate that the earliest members of the eudicot assemblage include ranunculids (subclass Magnoliidae), such as the family Chloranthaceae of the order Piperales, and the "lower" hamamelids (subclass Hamamelidae), such as the Platanaceae and Trochodendraceae families (Crane, 1989a). Radiating from within this basal assemblage are the more abundant eudicot groups, including the "higher" hamamelids (Betulaceae, Casuarinaceae, Juglandaceae, Myricaceae, and Fagaceae), and the subclasses Rosidae, Caryophyllidae, Dilleniidae, and Asteridae. Molecular phylogenies (*e.g.*, Chase *et al.,* 1993) largely agree with this paleobotanical view of eudicot evolution, but are not necessarily consistent with the taxonomic boundaries defined by Cronquist. For example, the rosid I clade consists of plants that belong to the higher hamamelids, Rosidae, and Dilleniidae. Although Coriariaceae are placed in the Magnoliidae by Cronquist, the molecular analyses clearly dispute this placement because the bulk of magnoliids appear in more basal positions in the phylogenies. Thus, all actinorhizal host plants appear to have evolved after the 125-Myr origin of the first eudicots.

The oldest recognizable eudicot families with extant actinorhizal species include the "higher" hamamelid families Betulaceae, Casuarinaceae, and Myricaceae. Pollen from fossil species related to these families (the Normapolles group) has been described from the Cenomanian of the Cretaceous (95 Myr; Sims *et al.,* 1999). Fossil pollen from Myricaceae has been dated to the Santonian (87.5-84 Myr; Muller, 1981) and the Cenomanian (97.5-91 Myr; Gladkova, 1962). Fossil pollen of Betulaceae (similar to *Alnus*) also appears in the Upper Cretaceous (dated to 87.5-66 Myr), whereas fossils of reproductive structures from *Alnus* are described from the Paleocene (Crane, 1989b). Fossil pollen evidence for Casuarinaceae also appears in the Paleocene (Muller, 1981).

The remaining actinorhizal groups, for which pollen fossils are known, appeared much later in the geologic record. Pollen from the families Elaeagnaceae, Rhamnaceae, and Rosaceae dates to the Oligocene (39-22.5 Myr; Muller, 1981). Pollen from Coriariaceae is dated to the upper Miocene (11-5 Myr; Muller, 1981) although palynological records from New Zealand indicate *Coriaria* pollen from possibly as early as the lower Miocene (Mildenhall, 1980). A phylogenetic study of *Coriaria* species calculated a divergence time of 63-59 Myr years between species groups based on *mat*K and *rbcL* molecular clocks (Yokoyama *et al.,* 2000). Although no fossil record is known for members of the genus *Datisca*, a record exists for the fossil form, *Tetramelioxylon prenudiflora*, comparable to the extant species, *Tetrameles nudiflora*, which is a member of the segregate family Tetramelaceae, once considered part of Datiscaceae (Lakhanpal, 1970; Swensen *et al.,* 1998). This megafossil dates to the early Eocene (55 Myr).

2.4.2. Actinorhizal Nodule Fossils

The fossil record primarily dates the evolutionary origin of the plant hosts rather than the existence of the symbiosis, and there is much less direct evidence to show that these plants were engaged in symbioses at or near the time of their diversification. The first nodule fossil was described by Karavaev (1959) from the Tertiary, purportedly from *Alnus*, although with little support for this taxonomic identification. An earlier putative nodule fossil is described from mesofossil assemblages from the Santonian of the Upper Cretaceous (84 Myr; Herendeen *et al.,* 1999). Fossilized nodules have been excavated from more recent deposits in lake sediments from Vermont dating to 11.5 Myr (Baker and Miller, 1980). These fossil nodules were identified as belonging to Elaeagnaceae (*Elaeagnus* or *Shepherdia*) based on comparisons to living nodules. Other than these direct reports, the evidence for nodulation is circumstantial. The location of actinorhizal plant remains is typically an early successional habitat either following the retreat of glaciers or in disturbed areas along stream banks, and in poor soils. The ancient habitat of actinorhizal plants appears to be similar to that of extant nodulated species and, combined with direct reports of nodule fossils, supports the idea that actinorhizal plants may have been nodulated for much of their evolutionary history.

Fossil evidence points to the higher hamamelid genera of the Betulaceae, Casuarinaceae, and Myricaceae as the earliest diverging actinorhizal species, however to date, molecular phylogenies can provide no indication as to which of the four clades of actinorhizal species was the first to diverge. It is possible that these plants were nodulated near the time of their divergence because fossil records indicate pollen as well as nodules appearing in the Upper Cretaceous, not long after the major diversification of the angiosperms. These early actinorhizal groups may have served as reservoirs of infective *Frankia* that were then able nodulate other more recently diverged actinorhizal hosts. Indeed, studies of *Frankia* diversity in older actinorhizal hosts (below) appear to support this idea.

3. *FRANKIA* PHYLOGENY AND EVOLUTION

3.1 Frankia *Taxonomy*

A discussion of *Frankia* taxonomy can be found elsewhere in this volume, so only a brief outline is given here. The genus name *Frankia* was created by Brunchorst to honor his mentor, A. B. Frank (1839-1900), a Swiss microbiologist (Brunchorst, 1886-1888; 1887). At that time, the microbial symbiont was considered to be a fungus because of its filamentous morphology within plant tissue. Opinions on whether the endophyte was fungal or bacterial varied for many years (reviewed in Lechevalier and Lechevalier, 1979; Lechevalier, 1984; Lechevalier and Lechevalier, 1989; Lechevalier, 1994; Quispel, 1990). The issue was unambiguously resolved only when Silver (1964) and Becking, DeBoer, and Houwink (1964) studied the endophytes in nodules of *Myrica cerifera* and *Alnus glutinosa,* respectively, by electron microscopy and revealed their prokaryotic nature.

In 1970, Becking (1970) affirmed the genus name; he also created six new species to supplement the four previously named by others according to the host plants in which they resided. Subspecies of *F. alni* were added after isolates became available (Lalonde *et al.*, 1988). Unfortunately, cross-inoculation groups do not always correlate with plants of origin, so a nomenclature based on that character can be misleading (Baker, 1987; Torrey, 1990; Lechevalier, 1994). Other than *F. alni,* the species and subspecies that have been proposed have not found common usage in the literature. *Frankia* nomenclature at the species level is ripe for revision or revival given the amount of phylogenetic information currently available about many strains both in culture and in root nodules.

In practice, most *Frankia* strains are still referred to by acronyms based either on plant of origin or by strain identification numbers that use a coding system devised in 1983 (Lechevalier, 1983). The acronymic system (Callaham *et al.*, 1978) incorporates the first letters of the host plant genus and species followed by a strain number. The first confirmed isolate, *Frankia alni* strain CpI1, is accordingly referred to as CpI1 (*Comptonia peregrina* Isolate No. 1) (Callaham *et al.,* 1978) or

HFPCpI1 (HFP = Harvard Forest Petersham), strain ArI4 is from *Alnus rubra*, EAN1 is from *Elaeagnus angustifolia*, etc. This approach is widely used even though it can be misleading. For example, strain PtI1 was isolated from *Purshia tridentate*, although it does not infect *Purshia* or other plants tested (Baker, 1987). Several *Casuarina* isolates, *e.g.*, CcI2, R43 (Zhang *et al.,* 1984), CeI5 (Torrey, 1990), G2, and D11 (Diem *et al.,* 1982), and isolates from *Purshia* and *Ceanothus* (Torrey, 1990), do not nodulate their original hosts but will nodulate members of the Elaeagnaceae, Myricaceae and/or *Gymnostoma* from the Casuarinaceae, *i.e.*, their host range is similar to isolates from the Elaeagnaceae (Group 3 below). Thus, the system for strain designation based on host plant origin, although correctly descriptive in most cases, can also lead to confusion.

A cataloguing system was introduced in 1983 to standardize strain designations as isolates proliferated in the literature. It incorporates a code based on host plant and lab of origin and lab number. Thus, CpI1 is designated HFP070101, where 07 = *Comptonia*, 01 = *peregrina*, 01 = either first isolate or lab number. The system eliminates the problem of different labs using the same acronym but is still subject to the problem of defining strains based on host plant of origin. The system is not commonly encountered in the literature, although culture collections, such as the ARS-NRRL (http://nrrl.ncaur.usda.gov), use the designation.

In some ways, nomenclature systems were swamped soon after their derivation by the success of investigators in isolating *Frankia* strains. Once considered difficult, it became clear in the 1980s that strains from five of eight actinorhizal plant families were readily isolated using simple microbiological techniques. The main ingredients missing from earlier attempts were time and separation from faster-growing contaminants (Benson and Silvester, 1993). The value of maintaining a large collection of numbered strains became less obvious because many isolates were quite similar or identical to others. The exception to the isolate bonanza was, and remains, strains in nodules of the Rosaceae, Coriariaceae, Datiscaceae, and *Ceanothus* from the Rhamnaceae, which remain uncultivated.

3.2. Molecular Systematic Studies

3.2.1. Phylogenetic Position among Bacteria

The first attempt to place *Frankia* among other actinobacteria used oligonucleotide cataloguing of the 16S-rRNA genes of the alder strains AirI1 and Ag45/mut15 (Hahn *et al.* 1989). A close relationship between *Frankia*, *"Blastococcus",* and *Geodermatophilus* was noted and the family *Frankiaceae* was defined as consisting of these three genera.

Later, more inclusive studies, using nearly complete 16S-rDNA sequences and partial RecA sequence comparisons, indicated that the nearest known relative to *Frankia* strains in phylogenetic terms is *Acidothermus cellulolyticus* and that *Frankia* is further removed from *Geodermatophilus* than previously thought (Marechal *et al.,* 2000; Normand *et al.,* 1996). *Acidothermus* is a cellulolytic bacterium isolated from hot springs at Yellowstone National Park and shares with *Frankia* an abundance of hopanoid lipids in its cell envelope (Marechal *et al.,*

2000). It grows optimally at around pH 5.2 and 55°C. Its relationship with *Frankia* is an unusual one that is perhaps understandable only in light of a shared ability to hydrolyze cellulose (Safo-Sampah and Torrey, 1988). All *Frankia* sp. strains are now classified in a separate lineage among the high Mol % G+C Gram-positive actinomycetes with the following position in the bacterial hierarchy (from http://www.cme.msu.edu/Bergeys/taxonomyinfo.html):

> Phylum "*Actinobacteria*"; Class "*Actinobacteria*"; Subclass *Actinobacteridae*; Suborder *Frankineae*; Family *Frankiaceae*; Genus *Frankia*

3.2.2. Phenetic Groupings of Strains within the Genus Frankia

Many studies performed in the 1980s and early 1990s focused on grouping strains in culture using biochemical markers, physiological characteristics, and broad genome comparisons, including DNA-DNA hybridization (Akimov and Dobritsa, 1992; An *et al.,* 1985; Bloom *et al.,* 1989; Fernandez *et al.,* 1989) and Low Frequency Restriction Fragment (LFRF) analysis (Beyazova and Lechevalier, 1992). Much of this work has been reviewed (Benson and Silvester, 1993; Lechevalier, 1994; Schwencke and Caru, 2001; Wall, 2000). The general conclusion from these studies is that clear groups of *Frankia,* which have particular host ranges, can be identified with some accuracy. Two primary groups, whose members were given the trivial names "elaeagnus" and "alder" strains, were firmly established and their host ranges clarified by large-scale inoculation experiments (Torrey, 1990). Strains isolated from *Casuarina* were recognized as being similar to alder strains, but not infective on *Alnus* sp., and became known as "casuarina strains". The entire group of strains that infect members of the Rosaceae, Datiscaceae, Coriariaceae and *Ceanothus* of the Rhamnaceae was missing in early studies because no isolates that could infect or reinfect these hosts were available.

DNA-DNA reassociation studies delineated at least eleven (and probably many more) genomic species of *Frankia* (Akimov and Dobritsa, 1992; An *et al.,* 1985; Fernandez *et al.,* 1989). An *et al.* (1985) described "genogroup 1" as a common group of isolates that infect alders and "genogroup 2" as consisting of *Casuarina* isolates G2 and D11 (shown later to infect *Elaeagnus* but not *Casuarina*). Genogroup 1 contains many alder strains that were subsequently used for physiological and biochemical work and include CpI1, ACoN24d, AvcI1, ArI3 and ArI4 (An *et al.,* 1985).

Another eight genomic species were added by Fernandez *et al.* (1989) with three additional *Alnus*-compatible, five *Elaeagnus*-compatible, and one *Casuarina*-compatible genomic species, along with several unclassified strains. Two more genomic species were found in a study of broad host-range elaeagnus strains that infected both *Elaeagnus sp.* and *Alnus sp.* (Lumini *et al.,* 1996). Of the nine genospecies identified by Akimov *et al.* (1991), only one can be equated with previously identified groups because a different set of strains was used for analysis. Subsequent work employing RFLP of PCR products and DNA sequences has confirmed the existence of several subgroups within the family *Frankiaceae.* Targets in these analyses included the intergenic regions between the 16S-23S rRNA genes and *nif* genes, including *nifD-nifK* (Jamann *et al.,* 1993; Lumini *et al.,* 1996; Rouvier *et al.,* 1996) and *nifH-nifD* (Cournoyer *et al.,* 1993).

3.2.3. Molecular Phylogenetics within the Genus Frankia

Dozens of full and partial 16S-rDNA sequences of *Frankia* strains have been used to reveal the broad phylogenetic structure of the genus and to place the various genospecies into a phylogenetic context. Because the 16S-rRNA gene has a variability of about 5–6% among the major groups, its resolution in molecular-phylogeny studies places strains in major clades with less resolution at the tips of phylogenetic trees. Increased resolution is provided by protein-encoding genes and even more is afforded by variable intergenic spacer regions. Because of their variability, intergenic spacer regions are frequently used for studying variants in ecological contexts.

Different regions of the 16S-rRNA gene have been used in phylogenetic, ecological and diversity studies of *Frankia* strains. Depending on the task at hand, fragment sizes have been used that incorporate much of the variability to be found among *Frankia* while retaining a reasonable fragment size (Clawson and Benson, 1999b). For phylogenetic studies, it has become apparent that more sequence information is better than less, because major clades of *Frankia* tend to have different sister relationships depending on the region of the molecule used. An alignment of 17 *Frankia* 16S-rDNA sequences (1380 bp), representing the major phylogenetic groups, reveals the clustering of variable regions (Figure 3). A total of 78 (5.6%) variable sites can be identified among these 17 strains with 33 in the first 400 bp and 45 in the remaining 980 bp.

In an attempt to provide a phylogenetic framework within the genus, Nazaret *et al.* (1991) used a 274-bp variable region (bp 861-1128 in Figure 3) from members of the genomic species identified previously (Fernandez *et al.,* 1989). Despite the short sequence used, two major groups were identified; one group was infective on members of the Elaeagnaceae and Rhamnaceae (a *Colletia* isolate represented the Rhamnaceae) and a second group was infective on the actinorhizal hamamelids (*Myrica, Alnus* and *Casuarina*). However, some *Frankia* strains isolated from *Casuarina* nodules fell into the elaeagnus group. The problematic isolate PtI1, a Nod^-/Fix^- actinomycete isolated from the roots of *Purshia tridentata* of the Rosaceae, was on a separate branch removed from the other strains.

Other studies using the 16S-rDNA gene have provided shape to the unfolding *Frankia* phylogeny. Small subunit rDNA sequences have been obtained from nodules or isolates that infect nearly all known genera of actinorhizal plants, including the actinorhizal members of the Cucurbitales (*Coriaria* and *Datisca*) (Benson *et al.,* 1996; Mirza *et al.,* 1994; Nick *et al.,* 1992), the Rosaceae (Bosco *et al.,* 1994), and Rhamnaceae (Benson *et al.,* 1996; Clawson *et al.,* 1998; Murry *et al.,* 1997; Ramirez-Saad *et al.,* 1998; Ritchie and Myrold 1999). Depending on the variable region used, alder strains form a sister group either with unisolated strains from the Rosaceae and

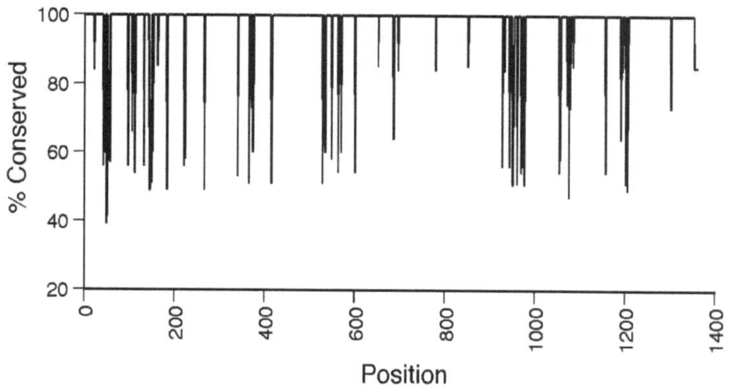

Figure 3. Variability in conservation of nucleotides in 16S rDNA from 16 Frankia *strains. Information generated from the alignment of 16S rDNAs shown in Figure 4, using ClustalX and graphing the resulting column scores.*

Coriariaceae (Bosco *et al.,* 1994; Clawson *et al.,* 1998; Nick *et al.,* 1992) or with elaeagnus strains, albeit with low statistical support (Wolters *et al.,* 1997).

A detailed study by Normand *et al.* (1996), using an expanded 16S database of 17 nearly-full-length sequences, provided the basic structure of the genus. It indicated a relatively close relationship between *Frankia* and *Acidothermus* but only a distant relationship with *Geodermatophilus*. Four main clusters within the genus were recognized:-

(i). Cluster 1 contained two subclusters of alder strains plus other alder, myrica and casuarina strains;

(ii). Cluster 2 included sequences amplified from *Coriaria, Datisa,* and *Dryas* nodules;

(iii). Cluster 3 contained elaeagnus strains; and

(iv). Cluster 4 was broader in scope and contained Nod$^-$/Fix$^-$ strains from *Coriaria, Datisca, Purshia,* and *Alnus* nodules. It branched close to elaeagnus strains, but was unstable in bootstrap analyses and did not appear in parsimony. These latter organisms generally appear on poorly supported long branches somewhere within or outside the frankia clade (Niner *et al.,* 1996; Ramirez-Saad *et al.,* 1998).

Although the branching order of the terminal taxa was unstable, depending on both the region of the molecule analyzed and the method of analysis, three of the four major clusters remained robust. The stage was thereby set for ensuing work that has provided a fairly comprehensive view of the major patterns, and exceptions to the patterns, among *Frankia* in root nodules. From these and other studies, three phylogenetic groups, designated Groups or Clusters 1, 2 and 3 have emerged with reasonable clarity (Benson *et al.,* 1996; Benson and Clawson, 2000; Clawson *et al.,*

1998; Hönerlage *et al.,* 1994; Nick *et al.,* 1992; Normand *et al.,* 1996; Wolters *et al.,* 1997). The basic shape of the *Frankia* phylogeny is shown in Figure 4.

The tree (Figure 4) was constructed using Neighbor-Joining with *Acidothermus* as the outgroup. Essentially the same tree is found using UPGMA, Maximum Likelihood, Tree-PUZZLE and Maximum Parsimony. It is important to note that phylogenetic trees constructed from highly conserved genes need few corrections

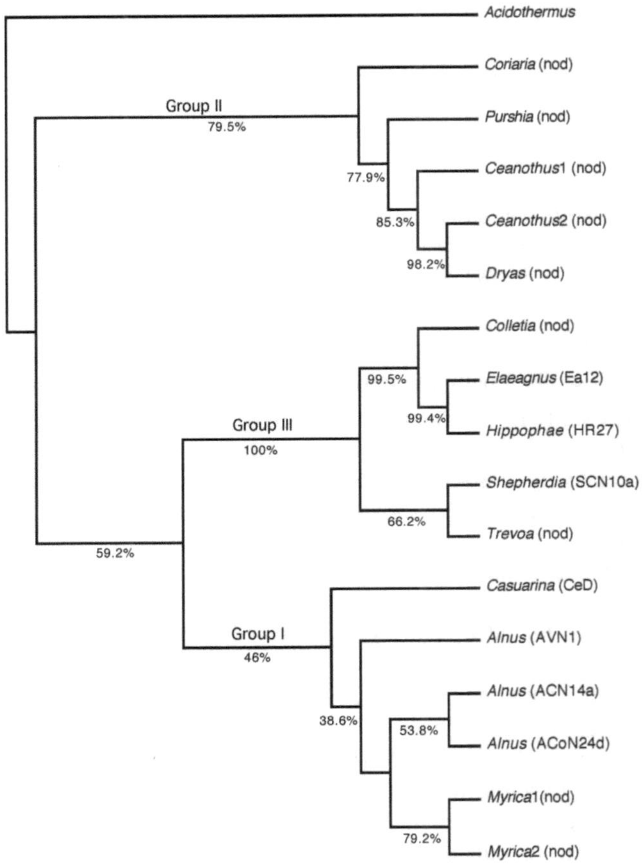

Figure 4. Frankia *phylogeny based on Neighbor-Joining analysis of complete 16S rDNA sequences. Numbers below branches are confidence values obtained from bootstrap analysis. Groups or Cluster numbers of* Frankia *are shown above branches.*
Accession numbers are: Acidothermus, *X70635;* Coriaria *(nod), AF063641;* Purshia *(nod), AF034776;* Ceanothus*1 (nod), AF063639; Ceanothus2 (nod), U69265; Dryas (nod), L40616;* Colletia *(nod), AF063640;* Elaeagnus *(Ea12), L40618;* Hippophae *(HR27), L40617;* Shepherdia *(SCN10a), L40619;* Trevoa *(nod), AF063642;* Casuarina *(CeD), M55343;* Alnus *(AVN1), L40613;* Alnus *(ACN14a), M88466;* Alnus *(AcoN24d), L40610;* Myrica*1 (nod), L40622;* Myrica*2 (nod), AF158687. The designation "nod" refers to sequences amplified directly from root nodules.*

for rate variation or multiple substitution events. Thus, simple distance methods, such as the UPGMA, provide the same trees as other methods. Features of the trees, confirmed by additional analyses using different molecules (Clawson, 1999), are: (i) the basal position of Group 2; (ii) the sister relationship between Groups 1 and 3; and (iii) the relatively high bootstrap support for each of the major groups. The protein-encoding genes, which have been used to supplement phylogenetic work based on rRNA genes, are *nif* (Cournoyer *et al.,* 1993; Jeong *et al.,* 1999; Normand and Bousquet 1989), *glnII* (Cournoyer and Lavire, 1999), *glnA* (Clawson, 1999), and *recA* (Marechal *et al.,* 2000). Most of these targets have been used to place *Frankia* among other actinomycetes but the same Clusters or Groups are consistently found.

3.2.4. Timing of Divergences

Placing a time of divergence of the various *Frankia* lineages is imprecise, as is dating bacterial divergences in general. The rate of change in 16S rRNA has been estimated at either nearly 1% per 50 Myr, using ecological events (Ochman and Wilson, 1987; Ochman *et al.,* 1999), or between 1-2% per 50 Myr, using maternally inherited insect symbionts (Moran *et al.,* 1993). The latter estimate is based on *Buchnera* strains that apparently have an accelerated rate of evolution (Moran, 1996) and using it to calibrate *Frankia* divergences might be expected to yield artificially recent time estimates. In addition, strains from at least two of the three major *Frankia* groups have a significant free-living existence and, thus, may exhibit different rates of evolution. Nevertheless, using one or both calibrations, the divergence of the alder group from the elaeagnus group has been estimated at 46-91 Myr (Jeong *et al.,* 1999), 25-50 Myr (Normand *et al.,* 1996), and at 131 +/– 56 Myr (Cournoyer *et al.,* 1993). The earlier divergence of the Group 2 strains from the other groups has been estimated at 62-130 Myr (Jeong *et al.,* 1999) and 100-200 Myr (Normand *et al.,* 1996). Because of the broad span of time represented by the various estimates, it is difficult to discern if the divergences correspond to plant divergences, to geographic separation over time, or to speciation pressures independent of the plant, however, it can be concluded that they occurred a long time ago. What is clear is that Group 2 strains occupy a basal position in the phylogeny and that *Frankia* strains probably associated with plant roots at least from the earliest radiation of angiosperms in the Cretaceous.

3.3. Characteristics of Frankia *Groups (Endosymbiont-controlled Characters)*

Infective and effective strains from each natural clade exhibit characteristic host ranges, morphology, and modes of infection (Benson and Silvester, 1993). Consistent with their sister relationship, Groups 1 and 3 (equivalent to Clusters 1 and 3 of Normand *et al.* (1996)) have strains that are more similar to each other than either are to strains belonging to Group 2.

3.3.1. Symbiont Compatibility

The term symbiont compatibility includes both the "host range" of individual *Frankia* sp. strains and the range of *Frankia* sp. strains that can infect a single plant species. The host range of *Frankia* strains can be defined only in a general sense because each strain may have a gradient of specificity for different plant species depending on the ecological context (Clawson and Benson, 1999a; Clawson and Benson, 1999b; Huguet *et al.,* 2001; Lumini *et al.,* 1996). Likewise, it is difficult to circumscribe a range of *Frankia* strains that can infect a plant species without sufficient sampling to assess the natural diversity of strains within root nodules. Such sampling has, thus far, has been done on very few plant species.

An illustration of this approach is the case of Myricaceae species that are generally recognized as promiscuous hosts. A different pattern of diversity was found for each of three species native to the Northeast United States when assessed by 16S-rDNA sequences amplified from nodules (Clawson and Benson, 1999b). *Myrica gale*, described as promiscuous in greenhouse trials, harbors only a few dominant Group 1 strains in the field (Clawson and Benson, 1999b; Huguet *et al.,* 2001). *Myrica pennsylvanica* houses many strains from Groups 1 and 3, with none dominating, whereas *Comptonia peregrina* houses diverse strains but only from Group 1 with some indication that a few of the strains are more common than others. Thus, even within a plant family or genus, the range of strains that actually inhabit root nodules may be either broad or narrow depending on the context. In the case of the Group 1 *M. gale* strains, it is unclear whether the environment (wet areas) selects for the dominant strain or if the dominant strain is selected on the basis of its ability to best infect the specific host. With the above difficulties in mind, host compatibility can be defined in a broader sense at the group rather than strain or species levels. A general outline of symbiont compatibility is shown in Figure 5.

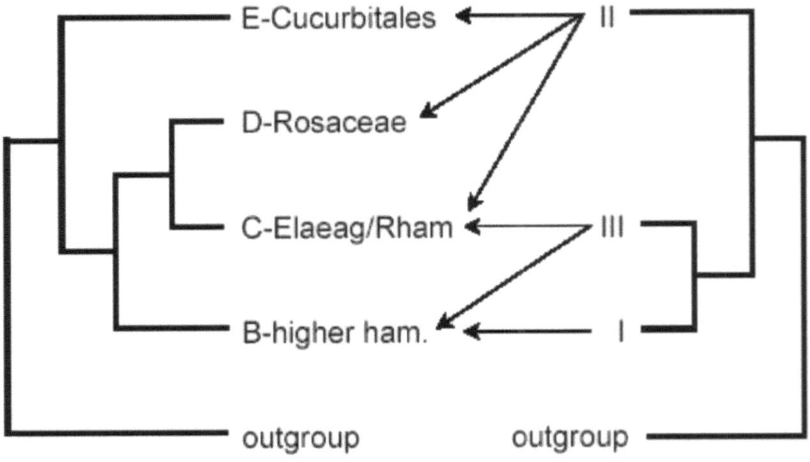

Figure 5. General outline of host-symbiont compatibility.

Group 1 strains have been found only in nodules from the "higher" hamamelid families, Betulaceae, Casuarinaceae and Myricaceae (Benson and Clawson, 2000), and may have evolved in concert with the hamamelids. Although some strains tested from this group can nodulate members of the Myricaceae, some are the primary symbionts of the Casuarinaceae, and some of *Alnus* sp., but no strains are known that infect both *Alnus* and *Casuarina* spp. (Rouvier *et al.,* 1996; Simonet *et al.,* 1999; Torrey, 1990). Thus, either subgroups within Group 1 specialize in certain plants or specific plants in the hamamelids have evolved to specialize in certain strains.

Group 3 strains are symbionts of the hamamelid genera *Myrica* and *Gymnostoma* (Navarro *et al.,* 1997; Racette and Torrey, 1989b), plus all genera of the Elaeagnaceae and the nodulated Rhamnaceae, except for *Ceanothus* (Clawson *et al.,* 1998). A few members of this group can infect alders (Lumini and Bosco 1996) and some have been isolated from root nodules of *Casuarina sp.* (Gauthier *et al.,* 1999; Navarro *et al.,* 1997; Torrey, 1990), of members of the Rosaceae (D.R. Benson, unpublished data), and of *Ceanothus* (Murry *et al.,* 1997). Strains from the enigmatic Group 2 (Cluster 2 of Normand *et al.* (1996)) reside in nodules from *Ceanothus* sp. of the Rhamnaceae, the Coriariaceae, Datiscaceae, and Rosaceae. They have not yet been isolated in pure culture. Consequently, little is known concerning the host range of individual members of Group 2, although cross-inoculation studies using crushed nodules suggest that symbionts from *Dryas, Ceanothus*, *Datisca*, and *Coriaria* are in the same cross-inoculation group (Kohls *et al.,* 1994; Mirza *et al.,* 1994; Torrey, 1990). Group 2 is characterized by low variability among the reported 16S-rRNA gene sequences, a situation that may indicate either a bottleneck in their evolutionary development or simply insufficient sampling compared to other actinorhizal species. This group of strains, on the other hand, infects the broadest range of hosts, including plants from four families in three of the major plant lineages.

3.3.2. Vesicle Morphology and Ontogeny

Frankia "vesicles" are spherical, lipid-enveloped, cellular structures that contain nitrogenase (Benson and Silvester, 1993). They are functionally analogous to heterocysts in cyanobacteria. The laminated lipid envelope contains an abundance of pentacyclic bacteriohopanes that presumably serve as a barrier to O_2 diffusion (Berry *et al.,* 1993). Vesicles are induced by exposure to low fixed-nitrogen levels plus O_2 with the vesicle envelope thickening as O_2 levels increase (Parsons *et al.,* 1987).

In culture, vesicle morphology is essentially the same for both alder and elaeagnus strains (Groups 1 and 3) and generally can be induced by nitrogen starvation (Benson and Silvester, 1993). In symbiosis, vesicle morphology varies among plant genera sometimes even within plant families. For example, Group 1 isolates form no vesicles in *Casuarina* root nodules, spherical vesicles in alder nodules, and elongated club-shaped vesicles in *Myrica* and *Comptonia.* Group 3 isolates that infect another member of the Casuarinaceae, *Gymnostoma*, form typical spherical vesicles in symbiosis (Racette and Torrey, 1989a). The spherical shape of vesicles is also exhibited by Group 3 strains in the Rhamnaceae (except for *Ceanothus*) and members of the Elaeagnaceae. In contrast, the Group 3 strains in *Myrica* nodules form elongated, club-shaped vesicles.

Thus, vesicle morphology in symbiosis is a poor character for defining evolutionary relationships among Group 1 and 3 *Frankia* or even among plants. A single lineage, such as the "higher" hamamelids, has symbionts with spherical vesicles, club-shaped vesicles, or no vesicles depending on plant genus, not on the *Frankia* strain provided. Vesicle morphology in symbiosis more likely depends on the physical or physiological barriers to O_2 diffusion provided by the plant and exploited by the actinomycete. These barriers include modified cell walls, hemoglobins, or elevated respiration (Berg, 1983; Berg and McDowell, 1988; Berg *et al.,* 1999; Jacobsen-Lyon *et al.,* 1995; Laplaze *et al.,* 2000; Mirza *et al.,* 1994; Silvester *et al.,* 1990; Tjepkema, 1983; Tjepkema *et al.,* 2002).

In nodules from the Datiscaceae, Coriariaceae, Rosaceae, and *Ceanothus*, vesicles are clearly different from those produced by Groups 1 or 3 *Frankia*, but no isolates are available for comparison to those made in culture (Newcomb and Wood, 1987). Nitrogen-fixing *Frankia* cells in *Coriaria* and *Datisca* nodules are oriented perpendicular to a large central vacuole and are interdigitated with mitochondria that are proposed to scavenge any O_2 in the vicinity (Mirza *et al.,* 1994). *Frankia* within the nodules from the Rosaceae and *Ceanothus* of the Rhamnaceae form vesicles but they are non-septate oval structures that are markedly different from typical vesicles produced by members of Groups 1 or 3 (Newcomb *et al.,* 1979; Newcomb, 1981). The unique vesicles seen in these root nodules provide support for the idea that Group 2 *Frankia* are more distantly related to both Groups 1 and 3.

3.3.3. Mode of Infection

The mode of plant-root infection may be particularly informative about the course of actinorhizal plant evolution because it is a clear case where direct cell-cell interaction must have co-evolved between the symbionts. *Frankia* enter the roots of compatible plants either by root hair infection (RHI) or by direct intercellular penetration (IP; see chapter by Berry and Wall in this volume). Group 1 strains infect *via* RHI (Berry and Torrey, 1983; Berry *et al.,* 1986; Callaham *et al.,* 1979; Torrey, 1976); Group 3 strains infect either by IP or RHI depending on the plant being infected (Bosco *et al.,* 1992; Cournoyer *et al.,* 1993; Miller and Baker, 1986; Racette and Torrey, 1989b); and Group 2 strains infect hosts by IP (Berry and Sunell, 1990; Kohls *et al.,* 1994; Liu and Berry, 1991; Miller and Baker, 1985; Valverde and Wall, 1999).

Discussions of the evolutionary importance of modes of infection must be tempered with the understanding that the infection process has been visualized in relatively few plant species using only a few *Frankia* strains. Nevertheless, the fact that both Group 3 and the basal Group 2 strains penetrate *via* IP suggests that this is the ancestral route of infection. This route then became unavailable, perhaps by mutation in the plant or bacteria, to the progenitor to Group 1. Because only the hamamelids are known to be infected by RHI and not by IP, *Frankia* in Groups 1 and 3 most likely developed the ability to infect root hairs in concert with the hamamelids at some point after they diverged from Group 2 but before they diverged from each other (see Figure 4).

4. HOST-*FRANKIA* INTERACTION AND EVOLUTION

4.1. Promiscuity

Actinorhizal symbioses may proceed evolutionarily towards more specialization, which leads to a greater probability of losing the symbiosis in more recently derived lineages (Maggia and Bousquet, 1994). In support of this hypothesis, more ancient symbioses, such as those involving *Myrica* or *Gymnostoma* sp., appear to involve a broader range of strains, at least when tested in the greenhouse. A narrower range of strains infects more recently derived genera, such as *Casuarina, Allocasuarina* or *Alnus*. Although this hypothesis may prove to be correct, promiscuity has not been examined in the majority of actinorhizal plant species and it is not as commonly observed in the field as it is in the greenhouse (Clawson and Benson, 1999b; Huang *et al.,* 1985; Navarro *et al.,* 1999). Furthermore, because *Alnus* sp. can be infected by a subset of strains from Group 3 (Lumini *et al.,* 1996) and because *Gymnostoma* species show no indication of promiscuity when examined in the field (Navarro *et al.,* 1999), plus the observed variation in the levels of natural diversity among the Myricaceae endosymbionts (Clawson and Benson, 1999b), the issue of promiscuity as related to evolutionary history needs to be reexamined.

It is important to recognize that *Frankia* strains have evolved along with the plants. They too may be viewed as more promiscuous or more specialized depending on how the terms are applied. For example, elaeagnus strains (Group 3) have been detected on or in nodules from the greatest number of families (six of eight) but are only infective and effective in plants from four families in two of four major clades (Hamamelidae and Urticales). The rosaceous strains (Group 2) infect plants from four families but from three major clades (Cucurbitales, Rosaceae and Urticales). Group 1 strains are the most limited and infect members of three families but only in one major clade, the "higher" hamamelids. Thus, at deeper phylogenetic levels, the basal Group 2 strains are evolutionarily the most promiscuous, whereas Group 1 strains are the most specialized.

4.2. Co-evolution between Host Plants and Frankia

Molecular phylogenetic analyses have provided estimates of the evolutionary history of both host plants and *Frankia*. It is, therefore, reasonable to ask whether these phylogenies provide any evidence of co-evolution. The term co-evolution was not explicitly defined in the original paper that described the phenomenon (Ehrlich and Raven, 1964), and since then, the term has been widely used and variously defined. Co-evolution has been used to refer to a variety of evolutionary circumstances that range from reciprocal genetic changes in interacting species to general evolutionary changes in interacting species (Thompson, 1989). Fortunately, the use of the term co-evolution was redirected by Janzen (1980), who emphasized the original component of *reciprocity* of evolutionary change in interacting species. Strictly co-evolving species would be expected to produce congruent branching patterns in phylogenetic trees.

4.2.1. Congruence among Phylogenetic Trees

Jeong *et al.* (1999) produced and compared phylogenetic trees for both *Frankia* (16S) and hosts (*rbcL*) in an effort to detect whether co-evolutionary changes have occurred between the symbionts. Despite the fact that branching patterns did not strictly correspond, analysis using the COMPONENT 2.0 program (Page, 1994) indicated that the match between the phylogenies was significantly better than if a random plant phylogeny was being compared to the *Frankia* phylogeny. Using the tree mapping option of COMPONENT, eight co-divergent branching events were identified between host and symbiont trees, which was taken to indicate that co-evolution between the host and symbiont had occurred. However, duplication events were also identified that indicate some divergences in the symbiont that were not mirrored in the host. Likelihood-ratio tests did not provide significant support for congruence between the two phylogenies, and molecular-clock tests indicated that *Frankia* 16S genes and host *rbcL* genes do not exhibit proportional amounts of differentiation, which would be expected under a strict co-evolutionary model.

In the above analysis, Group 3 strains were assumed to be associated only with members of the Elaeagnaceae and Rhamnaceae, whereas many also infect members of the Betulaceae, Myricaceae, and Casuarinaceae. Moreover, most of the co-evolutionary events detected were within strain Groups that would be expected to cross-infect other plant species within the same symbiont compatibility group (*e.g.*, alder strains also infect *Myrica* and *Comptonia*). A more general portrayal compares branching orders at deeper phylogenetic levels, such as is shown in Figure 5 where major groups of *Frankia* are aligned with the major plant clades.

Inspection of Figure 5 reveals that *Frankia* lineages and host lineages have not co-diverged. The only probable co-evolutionary path observed is between Group 1 *Frankia* and the higher hamamelids. Group 1 strains only infect hosts *via* RHI. If there are both plant and bacterial characters associated with RHI, then this adaptation may have limited Group 1 strains to infecting only higher hamamelid hosts. This restriction may be a mechanism for co-evolution among the Group 1 strains and hamamelids, but evidence must be presented that RHI by frankiae evolved in response to association with Group B hosts, and that those hosts also responded with some adaptation to allow this. At present, not enough information besides correlation is available to support this hypothesis.

Additional co-evolution may be present at the genus level between the *Alnus* and *Casuarina*-infective strains. However, the fact that *Myrica* species can be nodulated by both sets of strains suggests that alders and casuarinas have lost the ability to be nodulated by each other's strains, perhaps through geographical isolation rather than reciprocal co-evolution involving increasing specialization of the microorganism and plant.

Not enough information is available about compatibility groups in other actinorhizal genera to draw further conclusions about co-evolution at the genus level but it may occur. The fact that individual strains come to dominance in certain ecological situations suggests that specific plant/microbe/niche combinations may yet be discovered.

One contributing factor to the patterns observed may lie in the correspondence between geography and the Group of strains found in root nodules. For example, most nodulated Rhamnaceae are found in South America and New Zealand and are nodulated by Group 3 strains. The exception is *Ceanothus* species that are found exclusively in North America, especially in the west. These plants are nodulated by strains from Group 2. Strains from this Group are also the primary symbionts of nodulated members of the Rosaceae that are likewise limited to western North America. One can imagine a scenario wherein actinorhizal Rhamnaceae were at one time infected by progenitors of both groups and then became focused on one or the other lineage in different geographic contexts. Similarly, *Gymnostoma* species are nodulated by Group 3 strains in their native range, whereas other members of the Casuarinaceae are nodulated by relatively few strains from Clade 1. Navarro *et al.* (1999) propose that the difference in symbionts may be due to an early allopatric distribution of the plants and presumably symbionts (*Gymnostoma* in the north and east of Gondwana islands; *Casuarina/Allocasuarina* in the drier Australia). It is possible that further studies will identify additional correlations between geography and strain specialization and may provide a rationale for present-day patterns of symbiont compatibilities.

4.2.2. Complicating Factors

Lack of congruence among host and endosymbiont trees may be the result of many different events that can decouple co-evolution. Duplication, sorting, and horizontal gene transfer are three categories of changes that would cause phylogenetic trees to disagree (Thompson, 1998). Duplications in a host-symbiont interaction refer to within-host speciation events that result in multiple strains infecting the same host lineage. Sorting events are apparent absences of a symbiont in a host that previously had that symbiont. This may be caused either by extinction of the endosymbiont or by the endosymbiont simply not occupying all of its potential hosts (a patchy distribution). Finally, horizontal transfer results in a host shift, possibly by a gene transfer that enables infection of different or additional host species. The exchange of genes that affect host range among rhizobia is known (Laguerre *et al.,* 1992; Sullivan *et al.,* 1995), however, the extent to which symbiotic gene transfer occurs among *Frankia* is not known.

The aforementioned events refer to changes in the endosymbiont, but changes in the host may similarly decouple co-evolution. For example, a plant species may acquire the ability to be nodulated by *Frankia* strains that already nodulate unrelated hosts. In this case, the evolutionary paths of the endosymbiont and the plant, up until that point, have been completely disengaged and one would not expect congruence. Nodulated plants may expand their geographic distributions and either encounter other strains capable of infecting them or, as a result of the geographic shift, lose older associations. Permanent losses of symbiotic associations in certain lineages would result in greater taxonomic separation between nodulating groups than would otherwise be apparent.

Although strict co-evolution is not apparent among plant and frankia phylogenies, one pattern is particularly striking. The basal Group 2 of the *Frankia* phylogeny infects the broadest taxonomic group of plants, including the Coriariaceae/Datiscaceae, Rhamnaceae, and Rosaceae lineages. This pattern may mean either that Group 2 *Frankia* are particularly good at establishing new symbioses with aspiring actinorhizal plants or that strains from the basal group are the remnant of a more ancient symbiosis, which was present before the divergence of present-day lineages. Extinction of the symbiosis is apparently common and is evident from the patchy distribution of the nodulation phenotype within the nitrogen-fixing clade, sometimes even at the species level (Figure 2). Indeed, if there is any average probability of losing the symbiosis as speciation proceeds through evolution, the result would be a pattern such as seen in Figure 2, where a sporadic distribution of the nodulation phenotype is dispersed among non-nodulated taxa. Group 2 *Frankia* may also be the only strains that require the presence of the host, *i.e.*, they are obligate symbionts. Such dependence might be expected if the symbiosis were established early on during evolution.

Most *Frankia* and their hosts are not exclusively symbiotic and, thus, the evolutionary changes that can occur either during free-living stages or among non-infective populations should not be ignored. Free-living bacteria are likely to exchange DNA with infective strains and, therefore, contribute to overall genetic diversity and plasticity of the group. Without sampling the diversity of free-living *Frankia* strains in the soil, only a subset of the genus is represented in the phylogeny. Likewise, by including only symbiotic plants in phylogenetic analyses, a partial view of host phylogeny will result.

Finally, our ability to accurately estimate the evolutionary history of both the plants and the bacteria rests on the density of taxon sampling, the choice of genes that provide appropriate information on organismal history, the total amount of data available, the accuracy of sequence alignments, and the robustness of phylogenetic reconstruction methods. Even with all factors optimized, there is no guarantee of recovering actual evolutionary history. Thus, the confidence levels associated with branching patterns also provide limits on the confidence we place in apparent congruencies in tree comparisons.

4.3. Conclusions

The lack of evidence for strict co-evolution between *Frankia* and their hosts based on comparisons of phylogenies perhaps should not be surprising given the many factors that allow the endosymbionts and hosts to follow independent evolutionary trajectories. Nonetheless, the ability to detect specific co-evolutionary events may be beyond the capability of these phylogenetic analyses. Co-evolutionary interactions may be occurring at the level of populations and cannot be reflected in phylogenies based on highly conserved plant and bacterial genes. Perhaps better examples of co-evolution between hosts and bacteria will come from phylogenetic analysis at lower taxonomic levels among closely related host species and their endosymbionts.

5. FUTURE DIRECTIONS

5.1. Studies of Hosts

Recent phylogenetic work has set the stage for new studies that can focus more specifically on the evolution of actinorhizal symbioses. The host phylogenies provide insights into target groups of plants, where patterns of symbiotic gain and loss might best be explored and possibly correlated with environmental factors. Additional phylogenetic work, focusing on other genes in combination with *rbcL*, may provide a better idea of the relative diversification pattern of the four actinorhizal clades, which at present is unclear. Finally, assessing the homology of different actinorhizal symbioses may also be accomplished by looking at expression patterns of genes important to the symbiosis (nodulins) and by comparing these patterns and the gene themselves to other symbiotic systems.

5.2. Studies of Frankia

Studies on the ecology and diversity of *Frankia* strains are just beginning to shed light on the evolution of the symbiosis. As more information becomes available concerning the richness and evenness of strain diversity in nodules, evidence is emerging that the biogeography of *Frankia* sp. strains depends to a variable extent on the presence of the plant and also on edaphic factors that both select for particular sets of strains and permit strain persistence in the soil. More studies of this type are needed to develop hypotheses about factors that affect strain and actinorhizal plant distribution. Because of the difficulties involved, few studies have addressed the populations of groups of *Frankia* strains in different kinds of soil. Such information would aid in placing *Frankia* symbionts among the general population of soil bacteria and among closely related and slow-growing actinomycetes that often emerge from isolation trials.

Finally, comparisons based on 16S-rDNA sequencing have provided a wealth of information about the major clades of *Frankia* and the overall phylogenetic structure of the genus. The time is ripe for expanding the repertoire of molecules used for phylogenetic inference and so begin to extend beyond the alignment of broad groups of strains and plants and to ask questions about whether specific strains have adapted to particular soil conditions, climatic zones, or subpopulations of plants. Some results from this type of work are already available (Clawson and Benson, 1999b; Navarro *et al.*, 1999; Ritchie and Myrold, 1999).

REFERENCES

Akimov, V. N., and Dobritza, S. V. (1992). Grouping of *Frankia* strains on the basis of DNA relatedness. *Syst. Appl. Microbiol., 15,* 327-379.

Akimov, V. N., Dobritsa, S. V., and Stupar, O. S. (1991). Grouping of *Frankia* strains by DNA-DNA homology: How many genospecies are in the genus *Frankia*? In M. Polsinelli, R. Materassi and M. Vincenzini (Eds.), *Nitrogen fixation* (pp. 635-636). Dordrecht, The Netherlands: Kluwer Academic Publishers.

Akkermans, A. D. L., Hafeez, F., Roelofsen, W., Chaudhary, A. H., and Baas, R. (1983). Ultrastructure and nitrogenase activity of *Frankia* grown in pure culture and in actinorrhizae of *Alnus, Colletia* and *Datisca.* In C. Veeger and W. E. Newton (Eds.), *Advances in nitrogen fixation research* (pp. 311-319). The Hague, The Netherlands: Nijhoff/Junk.

Akkermans, A. D. L., and van Dijk, C. (1981). Non-legumious root-nodule symbioses with actinomycetes and *Rhizobium.* In W. Broughton (Ed.), *Nitrogen fixation* (vol. 1, pp. 57-103). London, UK: Oxford University Press.

An, C. S., Riggsby, W. S., and Mullin, B. C. (1985). Relationships of *Frankia* isolates based on deoxyribonucleic acid homology studies. *Int. J. Syst. Bacteriol., 35,* 140-146.

Angiosperm Phylogeny Group (1998). An ordinal classification for the families of flowering plants. *Ann. Missouri Bot. Gard., 85,* 531-553.

Baker, D. D. (1987). Relationships among pure cultured strains of *Frankia* based on host specificity. *Physiol. Plant., 70,* 245-248.

Baker, D. D., and Miller, N. G. (1980). Ultrastructural evidence for the existence of actinorhizal species in the late Pleistocene. *Can. J. Bot., 58,* 1612-1620.

Baker, D. D., and Schwintzer, C. R. (1990). Introduction. In C. R. Schwintzer and J. D. Tjepkema (Eds.), *The biology of* Frankia *and actinorhizal plants* (pp. 1-13). New York, NY: Academic Press.

Becking, J. H. (1970). *Frankiaceae* fam. nov. (Actinomycetales) with one new combination and six new species of the genus *Frankia* Brunchorst 1886, 174. *Inter. J. System. Bacteriol., 20,* 201-220.

Becking, J. H., De Boer, W. E., and Houwink, A. L. (1964). Electron microscopy of the endophyte of *Alnus glutinosa. Antonie Leeuwenhoek J. Microbiol. Serol., 30,* 343-376.

Benoit, L. F., and Berry, A. M. (1990). Methods for the production and use of actinorhizal plants in forestry, low maintenance landscapes, and revegetation. In C. R. Schwintzer and J. D. Tjepkema (Eds.), *The biology of* Frankia *and actinorhizal plants* (pp. 281-297). New York, NY: Academic Press.

Benson, D. R., and Clawson, M. L. (2000). Evolution of the actinorhizal plant symbioses. In E. W. Triplett (Ed.) *Prokaryotic nitrogen fixation: A model system for analysis of biological processes* (pp. 207-224). Wymondham, UK: Horizon Scientific Press.

Benson, D. R., and Silvester, W. B. (1993). Biology of *Frankia* strains, actinomycete symbionts of actinorhizal plants. *Microbiol. Rev., 57,* 293-319.

Benson, D. R., Stephens, D. W., Clawson, M. L., and Silvester, W. B. (1996). Amplification of 16S rRNA genes from *Frankia* strains in root nodules of *Ceanothus griseus, Coriaria arborea, Coriaria plumosa, Discaria toumatou,* and *Purshia tridentata. Appl. Environ. Microbiol., 62,* 2904-2909.

Berg, R. H. (1983). Preliminary evidence for the involvement of suberization in infection of *Casuarina. Can. J. Bot., 61,* 2910-2918.

Berg, R. H., Langenstein, B., and Silvester, W. B. (1999). Development in the *Datisca-Coriaria* nodule type. *Can. J. Bot., 77,* 1334-1350.

Berg, R. H., and McDowell, L. (1988). Cytochemistry of the wall of infected *Casuarina* actinorhizae. *Can. J. Bot., 66,* 2038-2047.

Berry, A. M., Harriott, O. T., Moreau, R. A., Osman, S. F., Benson, D. R., and Jones, A. D. (1993). Hopanoid lipids compose the *Frankia* vesicle envelope, presumptive barrier of oxygen diffusion to nitrogenase. *Proc. Natl. Acad. Sci. USA, 90,* 6091-6094.

Berry, A. M., McIntyre, L., and. McCully, M. E. (1986). Fine structure of root hair infection leading to nodulation in the *Frankia-Alnus* symbiosis. *Can. J. Bot., 64,* 292-305.

Berry, A. M., and Sunell, L. A. (1990). The infection process and nodule development. In C. R. Schwintzer and J. D. Tjepkema (Eds.), *The biology of* Frankia *and actinorhizal plants* (pp. 61-81). New York, NY: Academic Press.

Berry, A. M., and Torrey, J. G. (1983). Root hair deformation in the infection process of *Alnus rubra. Can. J. Microbiol., 61,* 2863-2976.

Beyazova, M., and Lechevalier, M. P. (1992). Low-frequency restriction fragment analysis of *Frankia* strains (*Actinomycetales*). *Int. J. Syst. Bacteriol., 42,* 422-433.

Bloom, R. A., Mullin, B. C., and Tate, III, R. L. (1989). DNA restriction patterns and DNA-DNA solution hybridization studies of *Frankia* isolates from *Myrica pennsylvanica* (bayberry). *Appl. Environ. Microbiol., 55,* 2155-2160.

Bosco, M., Fernandez, M. P., Simonet, P., Materassi, R., and Normand, P. (1992). Evidence that some *Frankia* sp. strains are able to cross boundaries between *Alnus* and *Elaeagnus* host specificity groups. *Appl. Environ. Microbiol., 58,* 1569-1576.

Bosco, M. S., Jamann, S., Chapelon, C., Simonet, P., and Normand, P. (1994). *Frankia* microsymbiont in *Dryas drummondii* nodules is closely related to the microsymbiont of *Coriaria* and genetically distinct from other characterized *Frankia* strains. In H. A. Hegazi, M. Fayez, and M. Monib (Eds.), *Nitrogen fixation with non-legumes* (pp. 173-183). Cairo, Egypt: The American University in Cairo Press.

Brunchorst, J. (1886-1888). Über einige Wurzelanschwellungen, besonders diejenigen von *Alnus* und den *Elaeagnaceen. Unters. Bot. Inst. Tübingen, 2,* 150-177.

Brunchorst, J. (1887). Die Struktur der Inhaltskörper in den Zellen einiger Wurzelanschwellungen. *Bergens Mus. Aarsb.,* 235.

Bousquet, J., Strauss, S. H., Doerksen, A. H., and Price, R. A. (1992). Extensive variation in the rate of *rbcL* gene sequences among seed plants. *Proc. Natl. Acad. Sci. U.S.A., 89,* 7844-7848.

Bremer, K. (1988). The limits of amino acid sequence data in angiosperm phylogenetic reconstruction. *Evolution, 42,* 795-803.

Callaham, D., Del Tredici, P., and Torrey, J. G. (1978). Isolation and cultivation in vitro of the actinomycete causing root nodulation in *Comptonia. Science, 199,* 899-902.

Callaham, D., Newcomb, W., Torrey, J. G., and Peterson, R. L. (1979). Root hair infection in actinomycete-induced root nodule initiation in *Casuarina, Myrica,* and *Comptonia. Bot. Gaz., 140 (Suppl.),* S1-S9.

Calvert, H. E., Chaudhary, A. H., and Lalonde, M. (1979). Structure of an unusual nodule root symbiosis in a non-leguminous herbaceous dicotyledon. In J. C. Gordon, C. T. Wheeler, and D. A. Perry (Eds.), *Symbiotic nitrogen fixation in the management of temperate forests* (p. 474). Corvallis, OR: Forest Research Laboratory, Oregon State University Press.

Chase, M. W., Soltis, D. E., Olmstead, R. G., Morgan, D., Les, D. H., *et al.* (1993). Phylogenetics of seed plants: An analysis of nucleotide sequences from the plastid gene *rbc*L. *Ann. Mo. Bot. Gard., 80,* 528-580.

Clawson, M. L. (1999). *The diversity, ecology and phylogeny of* Frankia *in actinorhizal plants.* Ph.D. Thesis, University of Connecticut, Storrs CT.

Clawson, M. L., and Benson, D. R. (1999a). Dominance of *Frankia* strains in stands of *Alnus incana* subsp. *rugosa* and *Myrica pennsylvanica. Can. J. Bot., 77,* 1203-1207.

Clawson, M. L., and Benson, D. R. (1999b). Natural diversity of *Frankia* strains in actinorhizal root nodules from promiscuous hosts in the family Myricaceae. *Appl. Environ. Microbiol., 65,* 4521-4527.

Clawson, M. L., Carú, M., and Benson, D. R. (1998). Diversity of *Frankia* strains in root nodules of plants from the families Elaeagnaceae and Rhamnaceae. *Appl. Environ. Microbiol., 64, 3539*-3543.

Clegg, M. T. (1993). Chloroplast gene sequences and the study of plant evolution. *Proc. Natl. Acad. Sci. U.S.A., 90,* 363-367.

Côte, B, Carlson, R. W., and Dawson, J. O. (1988). Leaf photosynthetic characteristics of seedlings of actinorhizal *Alnus* spp. and *Elaeagnus* spp. *Photosynth. Res., 16,* 211-218.

Cournoyer, B., Gouy, M., and Normand, P. (1993). Molecular phylogeny of the symbiotic actinomycetes of the genus *Frankia* matches host-plant infection processes. *Mol. Biol. Evol., 10,* 1303-1316.

Cournoyer, B., and Lavire, C. (1999). Analysis of *Frankia* evolutionary radiation using *glnII* sequences. *FEMS Microbiol. Lett., 177,* 29-34.

Crane, P. R. (1989a). Paleobotanical evidence on the early radiation of nonmagnoliid dicotyledons. *Plant Syst. Evol., 162,* 165-191.

Crane, P. R. (1989b). Early fossil history and evolution of the Betulaceae. In P. R. Crane and S. Blackmore (Eds.), *Evolution, systematics, and fossil history of the Hamamelidae* (pp. 87-116). Oxford, UK: Clarendon Press.

Crane, P. R., Friis, E. M., and Pedersen, K. R. (1995). The origin and early diversification of angiosperms. *Nature, 374,* 27-33.

Cronquist, A. (1981). An integrated system of classification of flowering plants. New York, NY: Columbia University Press.

Dahlgren, M. R. T. (1980). A revised system of classification of the angiosperms. *Bot. J. Linn. Soc., 80,* 91-124.

Dawson, J. O. (1990). Interactions among actinorhizal and associated species. In C. R. Schwintzer and J. D. Tjepkema (Eds.), *The biology of* Frankia *and actinorhizal plants* (pp. 299-316). New York, NY: Academic Press.

Dawson, J. O., and Gordon, J. C. (1979). Nitrogen fixation in relation to photosynthesis in *Alnus glutinosa. Bot. Gaz., 140 (Suppl.),* S70-S75.

Dixon, R. O. D., and Wheeler, C. T. (1983). Biochemical, physiological, and environmental aspects of symbiotic nitrogen fixation. In J. C. Gordon and C. T. Wheeler (Eds.), *Biological nitrogen fixation in forest ecosystems: Foundations and applications* (pp. 107-171). The Hague, The Netherlands: Nijhoff/ Junk.

Dommergues, Y. R., Diem, H. G., Gauthier, B. L., Dreyfus, B. L., and Cornet, F. (1984). Nitrogen-fixing trees in the tropics: Potentialities and limitations. In C. Veeger and W. E. Newton (Eds.), *Recent advances in nitrogen fixation research* (pp. 7-13). The Hague, The Netherlands: Nijhoff/ Junk.

Donoghue, M. J., Olmstead, R. G., Smith, J. F., and Palmer, J. D. (1992). Phylogenetic relationships of Dipsacales based on *rbcL* sequences. *Ann. Mo. Bot. Gard., 79,* 333-345.

Doyle, J. J. (1992). Gene trees and species trees: molecular systematics as one-character taxonomy. *Syst. Bot., 17,* 144-163.

Erlich, P. R., and Raven, P. H. (1964). Butterflies and plants: A study in coevolution. *Evolution, 18,* 586–608.

Felsenstein, J. (1985). Confidence limits on phylogenies: An approach using the bootstrap. *Evolution, 39,* 783-791.

Fernandez, M. P., Meugnier, H., Grimont, P. A. D., and Bardin, R. (1989). Deoxyribonucleic acid relatedness among members of the genus *Frankia. Int. J. Syst. Bacteriol., 39,* 424-429.

Gaut, B. S., Musa, S. V., Clark, W. D., and Clegg, M. T. (1992). Relative rates of nucleotide substitution on the *rbcL* locus of monocotyledonous plants. *J. Mol. Evol., 35,* 292-303.

Gauthier, D., Jaffre, T., and Prin, Y. (1999). Occurrence of both *Casuarina*-infective and *Elaeagnus*-infective *Frankia* strains within actinorhizae of *Casuarina collina*, endemic to New Caledonia. *Eur. J. Soil. Biol., 35,* 9-15.

Gladkova, A. N. (1962). Fragments of the history of the Myricaceae family. *Pollen Spores, 4,* 345.

Gordon, J. C., and Wheeler, C. T. (1978). Whole plant studies on photosynthesis and acetylene reduction in *Alnus glutinosa. New Phytol., 80,* 179-186.

Graybeal, A. (1998). Is it better to add taxa or characters to a difficult phylogenetic problem? *Syst. Biol., 47,* 9-17.

Hahn, D., Lechevalier, M. P., Fischer, A., and Stackebrandt, E. (1989). Evidence for a close phylogenetic relationship between members of the genera *Frankia, Geodermatophilus,* and *"Blastococcus"* and emendation of the family *Frankiaceae. System. Appl. Microbiol., 11,* 236-242.

Hennig, W. (1950). *Grundzüge einer theorie der phylogenetischen systematik.* Berlin, Germany: Deutscher Zentralverlag.

Hennig, W. (1966). *Phylogenetic systematics.* Urbana, IL: University of Illinois Press.

Herendeen, P. S., Magallon-Puebla, S., Lupia, R., Crane, P. R., and Kobylinksa, J. (1999). A preliminary conspectus of the Allon flora from the late Cretaceous (late Santonian) of central Georgia, U.S.A. *Ann. Missouri Bot. Gard., 86, 407*-471.

Hillis, D. M. (1996). Inferring complex phylogenies. *Nature, 383,* 130-131.

Hiltner, L. (1898). Über Entstehung und physiologische Bedeutung der Wurzelknöllchen. *Forst. Naturwiss. Z., 7,* 415-423.

Hönerlage, W., Hahn, D., Zepp, K., Zeyer, J., and Normand, P. (1994). A hypervariable 23S rRNA region provides a discriminating target for specific characterization of uncultured and cultured *Frankia*. *Syst. Appl. Microbiol., 17,* 433-443.

Hoot, S. B., Culham, A., and Crane, P. (1995). The utility of *atpB* gene sequences in resolving phylogenetic relationships: Comparison with *rbcL* and 18S ribosomal DNA sequences in the Lardizabalaceae. *Ann. Missouri Bot. Gard., 82,* 194-207.

Huang, J.-B., Zhao, Z.-Y., Chen, G.-X., and Liu, H.-C. (1985). Host range of *Frankia* endophytes. *Plant Soil, 87,* 61-65.

Huguet, V., McCray Batzli, J., Zimpfer, J. F., Normand, P., Dawson, J. O., and Fernandez, M. P. (2001). Diversity and specificity of *Frankia* strains in nodules of sympatric *Myrica gale, Alnus incana,* and *Shepherdia canadensis* determined by *rrs* gene polymorphism. *Appl. Environ. Microbiol., 67,* 2116-2122.

Jacobsen-Lyon, K., Jensen, E. O., Jorgensen, J. E., Marcker, K. A., Peacock, W. J., and Dennis, E. S. (1995). Symbiotic and nonsymbiotic hemoglobin genes of *Casuarina glauca*. *Plant Cell, 7,* 213-23.

Jamann, S., Fernandez, M. P., and Normand, P. (1993). Typing method for N_2-fixing bacteria based on PCR-RFLP - application to the characterization of *Frankia* strains. *Mol. Ecol., 2,* 17-26.

Janzen, D. H. (1980). When is it co-evolution? *Evolution, 34,* 611-612.

Jeong, S. C., Ritchie, N. J., and Myrold, D. D. (1999). Molecular phylogenies of plants and *Frankia* support multiple origins of actinorhizal symbioses. *Mol. Phylogen. Evol., 13,* 493-503.

Karavaev, M. N. (1959). Nitrogen-fixing nodules on the roots of fossil alder (*Alnus* sp.). (In Russian.) *Bot. Zh. (Leningrad), 44,* 1000-1001.

Kohls, S. J., Thimmapuram, J., Buschena, C. A., Paschke, M. W., and Dawson, J. O. (1994). Nodulation patterns of actinorhizal plants in the family Rosaceae. *Plant Soil, 162,* 229-239.

Laguerre, G., Mazurier, S. I., and Amarger, N. (1992). Plasmid profiles and restriction fragment length polymorphism of *Rhizobium leguminosarum* bv. *viciae* in field populations. *FEMS Microbiol. Ecol., 101,* 17-26

Lalonde, M. (1979). Immunological and ultrastructural demonstration of nodulation of the European *Alnus glutinosa* (L.) Gaertn. host plant by an actinomycetal isolate from North American *Comptonia peregrina* (L.) Coult. root nodule. *Bot. Gaz., 140 (Suppl.),* S35-S43.

Lalonde, M., Simon, L., Bousquet, J., and Seguin, A. (1988). Advances in the taxonomy of *Frankia:* Recognition of species *alni* and *elaeagni* and novel subspecies *pommerii* and *vandijkii*. In H. Bothe, F. J. De Bruijn, and W. E. Newton (Eds.), *Nitrogen fixation: Hundred years after* (pp. 671-680). Stuttgart, Germany: Gustav Fischer.

Lankhanpal, R. N. (1970). Tertiary floras of India and their bearing on the historical geology of the region. *Taxon, 19,* 675-694.

Laplaze, L., Duhoux, E., Franche, C., Frutz, T., Svistoonoff, S., *et al.* (2000). *Casuarina glauca* prenodule cells display the same differentiation as the corresponding nodule cells. *Mol. Plant-Microbe Interact., 13,* 107-112.

Lechevalier, M. P. (1983). Cataloging *Frankia* strains. *Can. J. Microbiol., 61,* 2964-2967.

Lechevalier, M. P. (1984). The taxonomy of the genus *Frankia*. *Plant Soil, 78,* 1-6.

Lechevalier, M. P. (1994). Taxonomy of the genus *Frankia* (*Actinomycetales*). *Int. J. Syst. Bacteriol., 44,* 1-8.

Lechevalier, M. P., and Lechevalier, H. A. (1979). The taxonomic position of the actinomycetic endophytes. In J. C. Gordon, C. T. Wheeler and D. A. Perry (Eds.) *Symbiotic nitrogen fixation in the management of temperate forests* (pp. 111-122). Corvallis, OR: Forest Research Laboratory, Oregon State University.

Lechevalier, M. P., and Lechevalier, H. A. (1989). Genus *Frankia* Brunchorst 1886, 174^{AL}. In S. T. Williams, M. E. Sharpe and J. G. Holt (Eds.), *Bergey's manual of systematic bacteriology* (pp. 2410-2417). Baltimore, MD: Williams and Wilkins.

Liu, Q., and Berry, A. M. (1991). The infection process and nodule initiation in the *Frankia-Ceanothus* root nodule symbiosis. *Protoplasma, 163,* 82-92.

Lumini, E., and Bosco, M. (1996). PCR-restriction fragment length polymorphism identification and host range of single-spore isolates of the flexible *Frankia* sp. strain UFI 132715. *Appl. Environ. Microbiol., 62,* 3026-3029.

Lumini, E., Bosco, M., and Fernandez, M. P. (1996). PCR-RFLP and total DNA homology revealed three related genomic species among broad-host-range *Frankia* strains. *FEMS Microbiol. Ecol., 21,* 303-311.

Maggia, L., and Bousquet, J. (1994). Molecular phylogeny of the actinorhizal Hamamelidae and relationships with host promiscuity toward *Frankia*. *Mol. Ecol., 3,* 459-467.

Marechal, J., Clement, B., Nalin, R., Gandon, C., Orso, S., *et al.* (2000). A *recA* gene phylogenetic analysis confirms the close proximity of *Frankia* to *Acidothermus*. *Int. J. Syst. Evol. Microbiol., 50,* 781-785.

Martin, W., Gierl, A., and Saedler, H. (1989). Molecular evidence for pre-Cretaceous angiosperm origins. *Nature, 339,* 46-48.

Martin, W., Lydiate, D., Brinkmann, H., Forkmann, G., Saedler, H., and Cerff, R. (1993). Molecular phylogenies in angiosperm evolution. *Mol. Biol. Evol., 10,* 140-162.

Meyen, J. (1829). Über das hervorwachsen parasitischer gebilde aus den wurzeln anderer pflanzen. *Flora (Jena), 12,* 49-64.

Mildenhall, D. C. (1980). New Zealand late Cretaceous and Cenozoic plant biogeography: A contribution. *Palaeogeogr. Palaecl., 31,* 197-233.

Miller, I. M., and Baker, D. D. (1985). The initiation, development and structure of root nodules in *Elaeagnus angustifolia* L. (Elaeagnaceae). *Protoplasma, 128,* 107-119.

Miller, I. M., and Baker, D. D. (1986). Nodulation of actinorhizal plants by *Frankia* strains capable of both root hair infection and intercellular penetration. *Protoplasma, 131,* 82-91.

Mirza, M. S., Hahn, D., Dobritsa, S. V., and Akkermans, A. D. L. (1994). Phylogenetic studies on uncultured *Frankia* populations in nodules of *Datisca cannabina*. *Can. J. Microbiol., 40,* 313-318.

Mirza, M. S., Hameed, S., and Akkermans, A. D. L. (1994). Genetic diversity of *Datisca cannabina*-compatible *Frankia* strains as determined by sequence analysis of the PCR-amplified 16S rRNA gene. *Appl. Environ. Microbiol., 60,* 2371-6.

Mirza, M. S., Pawlowski, K., Hafeez, F. Y., Chaudhary, A. H., and Akkermans, A. D. L. (1994). Ultrastructure of the endophyte and localization of *nifH* transcripts in root nodules of *Coriaria nepalensis* Wall. by *in situ* hybridization. *New Phytol., 126,* 131-136.

Mirza, S. M., Akkermans, W. M., and Akkermans, A. D. L. (1994). PCR-amplified 16S rRNA sequence analysis to confirm nodulation of *Datisca cannabina* L. by the endophyte of *Coriaria nepalensis* Wall. *Plant Soil, 160,* 147-152.

Moran, N. A. (1996). Accelerated evolution and Muller's rachet in endosymbiotic bacteria. *Proc. Natl. Acad. Sci. U.S.A., 93,* 2873-2878.

Moran, N. A., Munson, M. A., Baumann, P., and Ishikawa, H. (1993). A molecular clock in endosymbiotic bacteria is calibrated using the insect hosts. *Proc. R. Soc. Lond. B Biol. Sci., 253,* 167-171.

Muller, J. (1981). Fossil pollen records of extant angiosperms. *Bot. Rev., 47,* 1-142.

Murry, M. A., Konopka, A. S., Pratt, S. D., and Vandergon, T. L. (1997). The use of PCR-based typing methods to assess the diversity of *Frankia* nodule endophytes of the actinorhizal shrub *Ceanothus*. *Physiol. Plant., 99,* 714-721.

Murry, M. A., Zhang, Z., and Torrey, J. G. (1985). Effect of oxygen on vesicle formation, acetylene reduction and oxygen-uptake kinetics in *Frankia* sp. HFPCcI3 isolated from *Casuarina cunninghamiana*. *Can. J. Microbiol., 31,* 804-809.

Navarro, E., Jaffre, T., Gauthier, D., Gourbiere, F., Rinaudo, G., *et al.* (1999). Distribution of *Gymnostoma* spp. microsymbiotic *Frankia* strains in New Caledonia is related to soil type and to host-plant species. *Mol. Ecol., 8,* 1781-1788.

Navarro, E., Nalin, R., Gauthier, D., and Normand, P. (1997). The nodular microsymbionts of *Gymnostoma* spp. are *Elaeagnus*-infective *Frankia* Strains. *Appl. Environ. Microbiol., 63,* 1610-1616.

Nazaret, S., Cournoyer, B., Normand, P., and Simonet, P. (1991). Phylogenetic relationships among *Frankia* genomic species determined by use of amplified 16S rDNA sequences. *J. Bacteriol., 173,* 4074-4078.

Neave, I. A., Dawson, J. O., and DeLucia, E. H. (1989). Autumnal photosynthesis is extended in nitrogen-fixing European black alder when compared with white basswood: Possible adaptive significance. *Can. J. Forest Res., 19,* 12-17.

Newcomb, W. (1981). Fine structure of the root nodules of *Dryas drummondii* Richards (Rosaceae). *Can. J. Bot., 29,* 2500-2514.

Newcomb, W., Callaham, D., Torrey, J. G., and Peterson, R. L. (1979). Morphogenesis and fine structure of the actinomycetous endophyte of nitrogen-fixing root nodules of *Comptonia peregrina*. *Bot. Gaz., 140 (Suppl.),* S22-S34.

Newcomb, W., and Pankhurst, C. E. (1982). Fine structure of actinorhizal nodules of *Coriaria arborea* (Coriariaceae). *New Zeal. J. Bot., 20,* 93-103.

Newcomb, W., and Wood, S. M. (1987). Morphogenesis and fine structure of *Frankia* (Actinomycetales): The microsymbiont of nitrogen-fixing actinorhizal root nodules. *Int. Rev. Cytol., 109,* 1-88.

Nick, G., Paget, E., Simonet, P., Moiroud, A., and Normand, P. (1992). The nodular endophytes of *Coriaria* sp. form a distinct lineage within the genus *Frankia*. *Mol. Ecol., 1,* 175-181.

Niner, B. M., Brandt, J. P., Villegas, M., Marshall, C. R., Hirsch A. M., and Valdes, M. (1996). Analysis of partial sequences of genes coding for 16S rRNA of actinomycetes isolated from *Casuarina equisetifolia* nodules in Mexico. *Appl. Environ. Microbiol., 62,* 3034-3036.

Normand, P., and Bousquet, J. (1989). Phylogeny of nitrogenase sequences in *Frankia* and other nitrogen-fixing microorganisms. *J. Mol. Evol., 29,* 436-447.

Normand, P., Orso, S., Cournoyer, B., Jeannin, P., Chapelon, C., *et al.* (1996). Molecular phylogeny of the genus *Frankia* and related genera and emendation of the family Frankiaceae. *Int. J. Syst. Bacteriol., 46,* 1-9.

Ochman, H., Elwyn, S., and Moran, N. A. (1999). Calibrating bacterial evolution. *Proc. Natl. Acad. Sci. U.S.A., 96,* 12638-12643.

Ochman, H., and Wilson, A. C. (1987). Evolution in bacteria: Evidence for a universal substitution rate in cellular genomes. *J. Mol. Evol., 26,* 74-86.

Page, R. D. M. (1994). Maps between trees and cladistic analysis of historical associations among genes, organisms, and areas. *Syst. Biol., 43,* 58-77.

Page, R. D. M., and Charleston, M. A. (1997). From gene to organismal phylogeny: Reconciled trees and the gene tree/species tree problem. *Mol. Phylogenet. Evol., 7,* 231-240.

Parsons, R., Silvester, W. B., Harris, S., Gruijters, W. T. M., and Bullivant, S. (1987). *Frankia* vesicles provide inducible and absolute oxygen protection for nitrogenase. *Plant Physiol., 83,* 728-731.

Quispel, A. (1990). Discoveries, discussions, and trends in research on actinorhizal root nodule symbioses before 1978. In C. R. Schwintzer and J. D. Tjepkema (Eds.), *The biology of* Frankia *and actinorhizal plants* (pp. 15-33). San Diego, CA: Academic Press.

Racette, S., and Torrey, J. G. (1989a). The isolation, culture and infectivity of a *Frankia* strain from *Gymnostoma papuanum* (Casuarinaceae). *Plant Soil, 118,* 165-170.

Racette, S., and Torrey, J. G. (1989b). Root nodule initiation in *Gymnostoma* (Casuarinaceae) and *Shepherdia* (Elaeagnaceae) induced by *Frankia* strain HFPGpI1. *Can. J. Bot., 67,* 2873-2879.

Ramirez-Saad, H., Janse, J. D., and Akkermans, A. D. L. (1998). Root nodules of *Ceanothus caeruleus* contain both the N_2-fixing *Frankia* endophyte and a phylogenetically related Nod^-/Fix^- actinomycete. *Can. J. Microbiol., 44,* 140-148.

Ritchie, N. J., and Myrold, D. D. (1999). Geographic distribution and genetic diversity of *Ceanothus*-infective *Frankia* strains. *Appl. Environ. Microbiol., 65,* 1378-1383.

Rouvier, C., Prin, Y., Reddell, P., Normand, P., and Simonet, P. (1996). Genetic diversity among *Frankia* strains nodulating members of the family Casuarinaceae in Australia revealed by PCR and restriction fragment length polymorphism analysis with crushed root nodules. *Appl. Environ. Microbiol., 62,* 979-985.

Roy, A., and Bousquet, J. (1996). The evolution of the actinorhizal symbiosis through phylogenetic analysis of host plants. *Acta Bot. Gall., 143,* 635-650.

Safo-Sampah, S., and Torrey, J. G. (1988). Polysaccharide-hydrolyzing enzymes of *Frankia* (Actinomycetales). *Plant Soil, 112,* 89-97.

Sarich, V., and Wison, A. C. (1967). Rates of albumin evolution in primates. *Proc. Natl. Acad. Sci. U.S.A., 58,* 142-148.

Savard, L., Li, P., Strauss, S. H., Chase, M. W., Michaud, M., and Bousquet, J. (1994). Chloroplast and nuclear gene sequences indicate late Pennsylvanian time for the last common ancestor of extant seed plants. *Proc. Natl. Acad. Sci. U.S.A., 91,* 5163-5167.

Schwencke, J., and Caru, M. (2001). Advances in actinorhizal symbiosis: Host plant-*Frankia* interactions, biology, and application in arid land reclamation. *Arid Land Res. Manag. 15,* 285-327.

Silver, W. S. (1964). Root nodule symbiosis. I. Endophyte of *Myrica cerifera*. *J. Bacteriol., 87,* 416-421.

Silvester, W., Harris, S. L., and Tjepkema, J. D. (1990). Oxygen regulation and hemoglobin. In C. R. Schwintzer and J. D. Tjepkema (Eds.), *The biology of* Frankia *and actinorhizal plants* (pp. 157-176). New York, NY: Academic Press.

Simonet, P., Navarro, E., Rouvier, C., Reddell, P., Zimpfer, J., *et al.* (1999). Co-evolution between *Frankia* populations and host plants in the family Casuarinaceae and consequent patterns of global dispersal. *Environ. Microbiol., 1,* 525-533.

Sims, H. J., Herendeen, P. S., Lupia, R., Christopher, R. A., and Crane, P. R. (1999). Fossil flowers with Normapolles pollen from the Upper Cretaceous of southeastern North America. *Rev. Palaeobot. Palynol., 106,* 131-151

Soltis, D. E., Soltis, P. S., Chase, M.W., Mort, M. E., Albach, D. C., *et al.* (2000). Angiosperm phylogeny inferred from 18S rDNA, *rbcL,* and *atpB* sequences. *Bot. J. Linn. Soc., 133,* 381-461.

Soltis, D. E., Soltis, P. S., Nickrent, D. L., Johnson, L. A., Hahn, W. J., *et al.* (1997). Angiosperm phylogeny inferred from 18S ribosomal DNA sequences. *Ann. Missouri Bot. Gard., 84,* 1-49.

Soltis, D. E., Soltis, P. S., Morgan, D. R., Swensen, S. M., Mullin, B. C., *et al.*, (1995). Chloroplast gene sequence data suggest a single origin of the predisposition for symbiotic nitrogen fixation in angiosperms. *Proc. Natl Acad. Sci. U.S.A., 92,* 2647-2651.

Sullivan, J. T., Patrick, H. N., Lowther, W. L., Scott, D. B., and Ronson, C. W. (1995). Nodulating strains of *Rhizobium loti* arise through chromosomal symbiotic gene transfer in the environment. *Proc. Natl. Acad. Sci. U.S.A., 92,* 8985-8989.

Swensen, S. M. (1996). The evolution of actinorhizal symbioses: Evidence for multiple origins of the symbiotic association. *Amer. J. Bot., 83,* 1503-1512.

Swensen, S. M, Luthi, J. N., and Rieseberg, L. H. (1998). Datiscaceae revisited: Monophyly and the sequence of breeding system evolution. *Syst. Bot., 23,* 157-169.

Swensen, S. M., and Mullin, B.C. (1997). Phylogenetic relationships among actinorhizal plants. The impact of molecular systematics and implications for the evolution of actinorhizal symbioses. *Physiol. Plant., 99,* 565-573.

Takhtajan, A. (1980). Outline of the classification of flowering plants (Magnoliophyta). *Bot. Rev., 46,* 225-348.

Thompson, J. N. (1989). Concepts of coevolution. *Trends Ecol. Evol., 4,* 179-183.

Thorne, R. T. (1992). Classification and geography of the flowering plants. *Bot. Rev., 58,* 225-359.

Tjepkema, J. D. (1978). The role of oxygen diffusion from the shoots and nodule roots in nitrogen fixation by root nodules of *Myrica gale* L. *Can. J. Bot., 61,* 2898-2909.

Tjepkema, J. D. (1979). Oxygen relations in leguminous and actinorhizal nodules. In J. C. Gordon, C. T. Wheeler, and D. A. Perry (Eds.), *Symbiotic nitrogen fixation in the management of temperate forests* (pp. 175-186). Corvallis, OR: Forest Research Laboratory, Oregon State University.

Tjepkema, J. D. (1983). Hemoglobins in the nitrogen-fixing root nodules of actinorhizal plants. *Can. J. Bot., 61,* 2924-2929.

Tjepkema, J. D., Cashon, R. E., Beckwith, J., and Schwintzer, C. R. (2002). Hemoglobin in *Frankia*, a nitrogen-fixing actinomycete. *Appl. Environ. Microbiol., 68,* 2629-26231.

Torrey, J. G. (1976). Initiation and development of root nodules of *Casuarina* (Casuarinaceae). *Amer. J. Bot., 63,* 335-344.

Torrey, J. G. (1990). Cross-inoculation groups within *Frankia*. In C. R. Schwintzer and J. D. Tjepkema (Eds.), *The biology of* Frankia *and actinorhizal plants* (pp. 83-106). New York: Academic Press.

Valverde, C., and Wall, L. G. (1999). Regulation of nodulation in *Discaria trinervis* (Rhamnaceae)-*Frankia* symbiosis. *Can. J. Bot., 77,* 1302-1310.

Wall, L. G. (2000). The actinorhizal symbiosis. *J. Plant Growth Regul., 19,* 167-182.

Wolters, D. J., van Dijk, C., Zoetendal, E. G., and Akkermans, A. D. L. (1997). Phylogenetic characterization of ineffective *Frankia* in *Alnus glutinosa* (L.) Gaertn. nodules from wetland soil inoculants. *Mol. Ecol., 6,* 971-981.

Woronin, M. (1866). Über die bei der schwarzerle (*Alnus glutinosa*) und bei der gewöhnlichen gartenlupine (*Lupinus mutabilis*) auftretenden wurzelanschwellungen. *Mem. Acad. Imp. Sci. St. Petersburg, Ser. 7 T., 10,* 1-13.

Yokoyama, J., Suzuki, M., Iwatsuki, K., and Hasebe, M. (2000). Molecular phylogeny of *Coriaria*, with special emphasis on the disjunct distribution. *Mol. Phylogenet. Evol., 14,* 11-19.

Zhang, Z., Lopez, M., and Torrey, J. G. (1984). A comparison of the cultural characteristics and infectivity of *Frankia* isolates from root nodules of *Casuarina* species. *Plant Soil, 78,* 79-90.

Chapter 5

OXYGEN RESPONSES, HEMOGLOBIN, AND THE STRUCTURE AND FUNCTION OF VESICLES

W. B. SILVESTER[1], R. H. BERG[2], C. R. SCHWINTZER[3] AND J. D. TJEPKEMA[3]

[1]*Department of Biological Sciences, University of Waikato, Private Bag 3105, Hamilton, New Zealand*

[2]*Donald Danforth Plant Science Center, 975 N. Warson Road, St Louis, MO 63132, U.S.A.*

[3]*Department of Biological Sciences, University of Maine, Orono, ME 04469-5722, U.S.A.*

1. INTRODUCTION

Frankia nitrogenase is, like all other nitrogenases, highly sensitive to O_2 and the chemistry of its activity means it can only function in an O_2-free environment. This O_2 sensitivity is central to many aspects of the ecology and physiology of all nitrogenase-containing organisms, and many of the symbiotic states and ecological niches of nitrogenase-expressing microbes are determined by that sensitivity. For anaerobes, the problem is simply solved, but for those organisms dependent on atmospheric O_2 to generate their energy supply, the separation of nitrogenase from respiration represents a fascinating physiological dilemma. The dilemma centers on the need both to provide a high O_2 flux to drive the energy demand of nitrogenase and to protect nitrogenase from denaturation by free O_2. For all aerobic N_2-fixing organisms, this situation requires a high O_2 flux adjacent to a vanishingly low O_2 tension at the site of N_2 fixation.

A variety of strategies for solving the so-called O_2 dilemma have been described (Gallon, 1981) and they range from total avoidance, *i.e.*, the anaerobic strategy, to living with it, *i.e.*, the strategy of balancing nitrogenase synthesis with inactivation. The majority of organisms, both free-living and in symbiosis, adopt intermediate strategies that rely on external O_2 barriers coupled with high O_2 utilization at the nitrogenase site to maintain a steep O_2 gradient. Among the many physiological

K. Pawlowski and W. E. Newton (eds.), Nitrogen-fixing Actinorhizal Symbioses, 105–146.
© 2008 *Springer*.

conditions that provide O_2 protection, two main families of strategies can be identified.

First, the majority of N_2-fixing bacteria rely on an external environmental source of O_2 regulation. Thus, many free-living bacteria live in microaerophilic environments in water, soil, or decaying organic matter and make use of the diffusion resistance of the medium and respiratory O_2 uptake of associated microbes to control the ambient pO_2. Similarly, the symbiotic state appears to be specifically adapted to provide the necessary pO_2 conditions for efficient nitrogenase activity. The legume nodule is the most sophisticated and physiologically intriguing example of this phenomenon and is dealt with in volume 7, *Nitrogen-fixing Leguminous Symbioses*, of this series.

Second, a few bacteria maintain their own physiological or structural O_2 diffusion barrier. These organisms can express functional nitrogenase in air and have a unique adaptability to a wide range of pO_2. This group includes only two types, the heterocystous cyanobacteria and the filamentous actinomycete *Frankia.* Both produce specialized thick-walled cells, called heterocysts and vesicles, respectively, which provide the requisite diffusion resistance to sustain nitrogenase at atmospheric pO_2. The history and verification of this phenomenon in *Frankia* vesicles forms much of the substance of this chapter.

Actinorhizal nodules then, like cyanobacterial symbioses, have two potential mechanisms of O_2 control, one at the level of the vesicle and another at the level of the nodule structure and its physiology. Because the vesicle is common to both free-living *Frankia* and symbiotic *Frankia*, knowledge of the unique structural and physiological properties of the vesicle must be central to this review. Aspects of this subject have been reviewed previously (Benson and Silvester, 1993; Huss-Danell, 1997; Silvester *et al.,* 1990). Here, we summarize and update the information on the role of O_2 in the structure and physiology of *Frankia* and its nodule systems. We also review the occurrence of hemoglobin in nodules and *Frankia* and its function in controlling O_2.

2. VESICLE STRUCTURE AND OXYGEN RESPONSES OF *FRANKIA IN VITRO*

2.1. Vesicle Structure

The *Frankia* vesicle is worthy of developmental study from a basic biology viewpoint because it is a rare example of a prokaryotic structure that is highly specialized in form and function. It holds information on how a prokaryote can accomplish formation of a sphere that contains many cells, is housed in a microaerophilic environment, and is specialized to accomplish the complex function of nitrogen fixation. An extensive review of the structural aspects of *Frankia* in plants and *in vitro* is available (Newcomb and Woods, 1987) and here we summarize the structure of the *Frankia* vesicle with respect to its development and O_2 protection of nitrogenase.

Evidence that *Frankia* vesicles are the site of nitrogen fixation comes from studies showing that: (i) *Frankia* cultures develop vesicles at tips of growing vegetative hyphae when induced to fix N_2 (Fontaine *et al.,* 1984; Torrey and Callaham, 1982); (ii) nitrogenase activity fractionates with these vesicles (Noridge and Benson, 1986); and (iii) the enzyme immunolocalizes specifically in the vesicle (Meesters, 1987).

Vesicle structure is illustrated in Figure 1A-E. The vesicle is always attached to a stalk cell (S) and, in phase-contrast images (Figure 1A), these two structures have in common a phase-bright "halo" relative to vegetative cells (V). This halo is due to birefringence from layers of lipid material that accumulate on these cells, which are stained red with the fluorescent dye Nile Red (Lamont *et al.,* 1988). These built-up layers comprise the vesicle envelope, which is the proposed O_2-diffusion barrier discussed in Section 2.2. Vesicles in cultures of *Frankia* are about 2.5μm in diameter and septate.

Because the component layers of the envelope are in the range of 35-47Å (Berg, 1994; Torrey and Callaham, 1982), the resolving power of transmission electron microscopy is required to analyze their organization in the vesicle envelope (Figure 1B-F). In thin sections of chemically fixed cells, the envelope (Figure 1B, VE) surrounds the vesicle and stalk, and component layers of the envelope can be seen as electron-translucent layers within the envelope (Figure 1D). Chemically fixed specimens show an electron-translucent zone surrounding the vesicle, the "void space" (Figure 1B, VS). This feature has been interpreted either as due to extraction of envelope lipids by dehydration solvents or as a shrinkage artifact resulting from dehydration-induced shrinkage of hydrophilic vesicle cytoplasm relative to rigid lipid-rich envelope. That the void space is due to the latter was shown by Lancelle *et al.* (1985), using thin sections of freeze-substituted *Frankia* vesicles (Figure 1B-E). In parallel samples, the void space was present in chemically fixed sections (Figure 1B) and absent in freeze-substituted *Frankia* vesicles (Figure 1C). The void space forms between the vesicle cell membrane and the envelope, evidently by shrinkage of the cytoplasm of vesicle cells. Lancelle *et al.* (1985) also show that ultra-rapid freezing preserves a number of unusual cytoplasmic inclusions (fibers and tubules) in *Frankia* vesicles that are not preserved by chemical fixation. This is the case for symbiotic vesicles as well and is discussed in Section 3.4, along with their possible role in vesicle development. Of particular interest are bundles of filaments (Figure1D-E, small arrowheads), found only in vesicles and stalks, that associate with granular bodies at the top of the stalk (Figure 1D, large arrowhead), extend longitudinally into the stalk (Figure 1D-E), and terminate at the basal septum (Figure 1E, large arrowhead). In septate vesicles, in culture or in symbiosis, the envelope is significantly thicker in the stalk cell (Figure 1C-E, electron-translucent region of the wall).

The chemical composition of the envelope has been identified as a mixture of hopanoid lipids (Berry *et al.,* 1993), which are pentacyclic triterpenoids. Vesicle envelopes have a specific class of hopanoid, bacteriohopanetetrol phenylacetate (Berry *et al.,* 1993) that reflects the difference of the vesicle envelope structure from

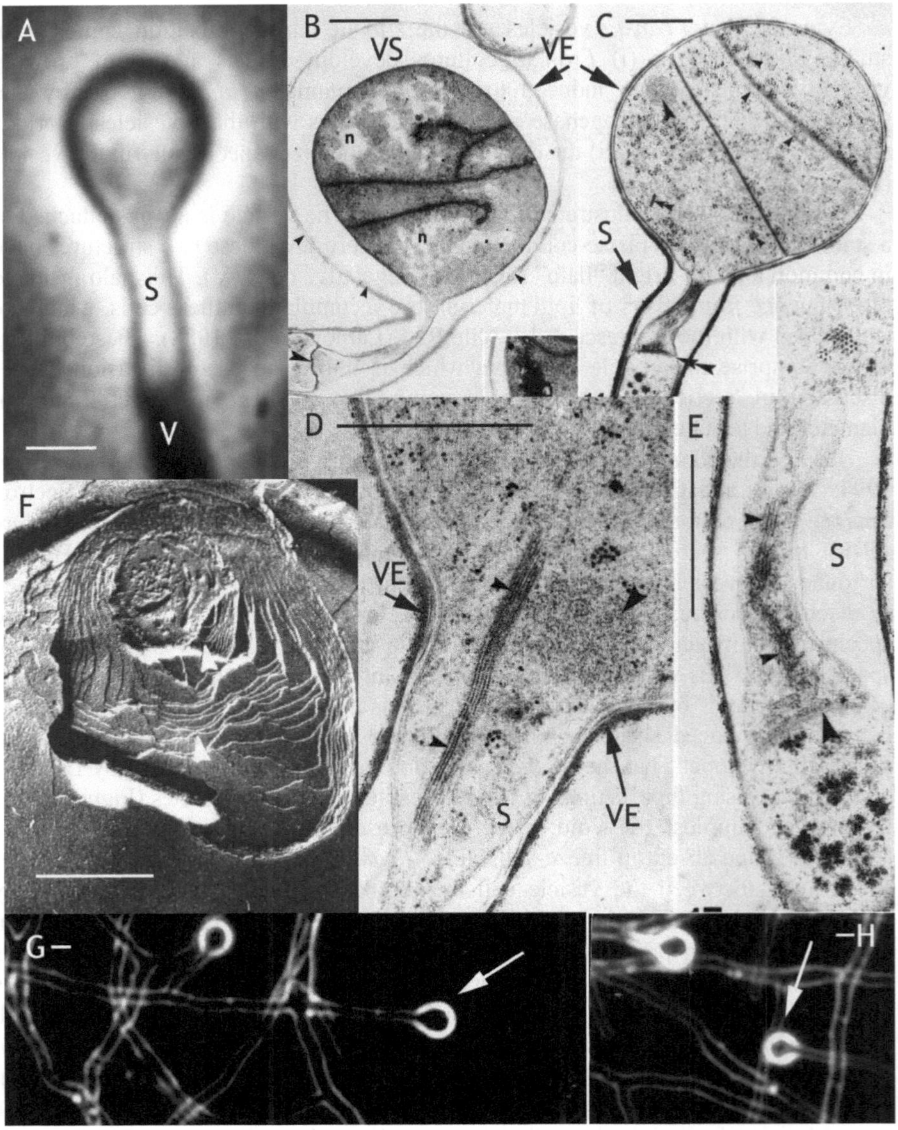

Figure 1. Frankia *vesicle structure.* Frankia *strains HFPCcI3 (A-F) and HFPArI3 (G,H). Marker = 1 μm in A,G,H and 0.5 μm in B-F.*
(A) Phase contrast of the birefringent halo of the envelope surrounding vesicle and stalk (s) (v = vegetative cell). Vesicle ultrastructure, chemically fixed (B) compared to freeze-substituted (C). (D,E) Fiber bundle in freeze-substituted stalk cell. (B-E) reproduced from Lancelle et al. (1985), with permission. (F) envelope lipid layers shown in freeze fracture. (G,H) Dark field envelope birefringence, which is much dimmer at 2kPa O_2 (G) than at 21kPa O_2 (H). (Reproduced from Harris and Silvester (1992), with permission).

the *Frankia* hyphal wall, which has relatively few, loosely organized lipid layers (Berg and McDowell, 1987). The uniform thickness of these layers suggests that they are crystals (of hopanoid), which perhaps results in a sufficient density of molecules to provide diffusive resistance to O_2 diffusion.

In their initial analysis of *Frankia* vesicle ultrastructure, Torrey and Callaham (1982) note the resistance of the envelope to conventional EM stains and report that more oxidative stains, such as provided by fixation in permanganate, improve the imaging of the envelope. Indeed, *en bloc* staining with chromic acid enables analysis of the envelope in thin sections of symbiotic vesicles (see section 3.4.2). Torrey and Callaham (1982) also show that freeze fracture provides an effective means for imaging the envelope, albeit a problematic approach for quantifying numbers of lipid layers because of the need for cross fractures.

2.2. Responses to O_2

2.2.1. Physiological Responses

The re-isolation of *Frankia* (Callaham *et al.,* 1978) made available to the scientific community an organism with many unique and fascinating properties. In total contrast to rhizobia, which are able to express nitrogenase only at extremely low (<0.1kPa) pO_2 (Tjepkema and Evans, 1975), *Frankia* shows maximum nitrogenase activity at atmospheric pO_2 (Gauthier *et al.,* 1981; Tjepkema *et al.,* 1980; 1981). Further, nitrogenase expression in cultured *Frankia* occurs coincidentally with the formation of vesicles (Tjepkema *et al.,* 1980), eliciting an interesting parallel with the heterocyst in cyanobacteria.

The normal site of nitrogenase within the vesicle has been confirmed by a number of approaches. First, when vesicles are separated from their subtending hyphae, nitrogenase activity is found only in the vesicle fraction (Noridge and Benson, 1986; Tisa and Ensign, 1987). Second, nitrogenase has been directly identified in vesicles by immunolocalization studies (Meesters, 1987; Meesters *et al.,* 1987). Third, the correlative work on vesicle induction and nitrogenase activity (Murry *et al.,* 1984a; Tjepkema *et al.,* 1980) all identify the vesicle as the site of nitrogenase and the site of O_2 protection. The above results for air-grown *Frankia* cultures contrast with those for *Frankia* grown at very low pO_2 of 0.1-0.3kPa. At these O_2-limiting conditions, vesicles do not form, but nitrogenase is expressed in the hyphae (Murry *et al.,* 1985).

The hypothesis that the vesicle, as well as being the site of nitrogenase, is also the site of O_2-diffusion limitation and control is further supported by work on the uptake kinetics of O_2 and C_2H_2 by *Frankia* in liquid culture. Isolated nitrogenase has a $K_m(C_2H_2)$ of ca. 0.6kPa, which is the same as for most intact N_2-fixing cells. However, the $K_m(C_2H_2)$ for *Frankia* is ca. 2kPa and does not change when V_{max} is altered by changing the assay pO_2 (Murry *et al.,* 1984a). This result was interpreted as a diffusion-limited system in which the apparent $K_m(C_2H_2)$ is the result of the enzyme kinetics modified by diffusion limitation. In addition, the same authors showed that in ammonia-grown cells, which lack vesicles, the uptake rate for O_2 showed a simple kinetic curve that saturated at 18μM O_2 (K_m = 1μM), whereas nitroogenase-derepressed cultures (grown on N_2) gave two $K_m(O_2)$ values of 1μM and 170μM. This result adds further support to the conclusion that the vesicle represents a compartment with a significantly greater diffusion resistance than the hyphae filaments (Murry *et al.*, 1984a; Murry *et al.*, 1985).

2.2.2. Adaptation to varying pO_2

Perhaps the greatest insights into O_2 responses have been achieved by studying the structural and physiological responses of *Frankia* cultures to varying pO_2 in the growth media. When *Frankia* HFPCcI$_3$ vesicles were initiated in culture at pO_2 levels of 0.1, 0.3, 0.5, 5 and 20kPa, two important results were observed (Murry *et al.*, 1985; Parsons *et al.*, 1987). First, cultures at low pO_2 (< 0.3kPa) failed to produce vesicles (see above). Second, and most significantly, the pO_2 optimum for nitrogenase activity shifted with the growth pO_2. Thus, cultures grown at 20kPa O_2 had a nitrogenase-activity optimum at near 20kPa O_2, whereas those grown at 5kPa O_2 had a nitrogenase-activity optimum at near 5kPa O_2 (Figure 2). This result demonstrated for the first time that not only is *Frankia* able to express nitrogenase at atmospheric pO_2, it also is able to adapt to a wide range of pO_2 and "....that vesicle formation and the development of the O_2-protection mechanism of nitrogenase are regulated by ambient pO_2..." (Murry *et al.*, 1985).

Although *Frankia* is able to adapt to a wide variety of pO_2, nitrogenase is still sensitive to O_2 outside the ambient ranges and pulses of O_2 have been used to destroy nitrogenase in order to determine recovery rates (Baker and Huss-Danell, 1986; Harris and Silvester, 1992).

The thick walls of the vesicles were first postulated as the site of O_2-diffusion resistance by Torrey and Callaham (1982) and they also showed that the envelope

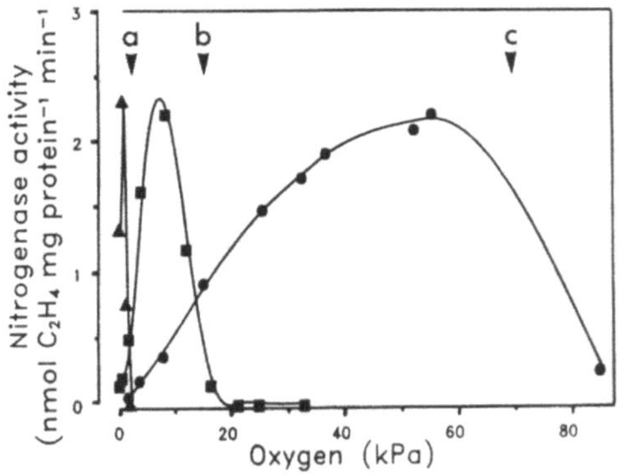

Figure 2. Effect of pO_2 on nitrogenase activity in Frankia *grown at three O_2 levels.* Frankia *was grown at 2, 16 and 70kPa O_2 and assayed for nitrogenase at various pO_2 starting at 0.4kPa. Response curves are for cultures grown at 2kPa, 16kPa and 70kPa. The arrows a, b and c indicate the level of O_2 at which the culture was grown (reproduced from Parsons* et al. *(1987), with permission).*

outside the wall is a multi-laminated structure made up of many layers of lipid. The vesicle envelope shows up as a crystalline array, which responds to polarized light and, under TEM, is shown to be made up of many lipid monolayers.

Assuming that the vesicle envelope is the site of diffusion resistance, Parsons *et al.* (1987) postulated that any adaptation to varying pO_2 must be evident in the properties of that envelope. They showed that the vesicle envelope displayed remarkable bright properties when viewed under dark-field microscopy, where the apparent envelope thickness could be readily viewed and measured (Figure 1G-H). Cultures of $HFPCcI_3$ grown at 2, 16, or 70kPa O_2 grew well and adapted their nitrogenase activity to those levels of O_2 (Figure 2). The vesicle envelopes at 2kPa O_2 were almost transparent under bright field, whereas those at 70kPa O_2 were bright glowing spheres. Apparent thickness varied from 1.38 to 3.17 units over the O_2 range. Most significantly, however, was the change in lipid layers, which ranged from an average of 17 at 4kPa O_2 to an average of 40 at 70kPa O_2.

The adaptive nature of the vesicle envelope has been confirmed by the use of Nile Red fluorescent stain, which is specific for lipid (Lamont *et al.,* 1988) and which showed the fluorescent staining to be coincident with changes in dark-field images of pO_2-adapted cultures.

O_2 protection within a thick-walled cell, such as a vesicle, can only be effective if it is accompanied by an O_2 sink. Support for the dual role of diffusion resistance and respiratory O_2 consumption is provided by the interaction between C substrate and optimum pO_2 for nitrogenase activity. Early work on requirements for heterotrophic growth of *Frankia* showed an absolute requirement for an organic acid, such as malate, fumarate or succinate (Tjepkema *et al.,* 1981). However, Lopez *et al.* (1986) showed that other substrates were useable and that the type of substrate affected the pO_2 optimum of nitrogenase activity. Cells using organic acids, *e.g*., propionate, had maximum nitrogenase activity at 15-20kPa O_2, whereas cells able to assimilate glucose or trehalose showed lower nitrogenase activity at a pO_2 optimum of 10kPa. The conclusion from these results is that the higher respiration rate associated with organic-acid consumption provides a lower pO_2 and better O_2 control.

Growing *Frankia* in continuous culture has allowed the dynamics of adaptation to be followed (Harris and Silvester, 1992). They grew HFPCcI3 at various pO_2 at dilution rates of 0.100-0.125 day^{-1} and used an O_2 shock to destroy nitrogenase. Although the induction of vesicle formation was shown to be independent of pO_2, both the thickness of the vesicle envelope and the timing of nitrogenase recovery were specifically controlled by the growth pO_2. Thus, a culture, which was grown at 2kPa O_2, shocked with air at 21kPa O_2, and then maintained at 21kPa O_2, took 36h for nitrogenase activity to reappear, whereas a culture returned to 2kPa O_2 after the O_2 shock regained activity within a few hours. The lag in activity was measured as the time required for both existing and new vesicles to achieve sufficient thickness in the envelope.

Prior to the discovery of the vesicle and its properties, the heterocystous cyanobacteria were considered the only genuinely aerobic N_2 fixers capable of growth in the dark under N_2-fixing conditions (Jensen and Cox, 1983). Now, the full range of *Frankia* species has been added to this class of organisms.

3. RESPONSES IN SYMBIOSIS

3.1. Nodule Structure

Root nodules of legumes and actinorhizal plants show some external similarities but they differ enormously in many of the details both of structure and function. The legume nodule is remarkably similar across many species studied (Bergersen, 1982), having a central zone of infected tissue that is bounded by a tightly packed diffusion-limiting inner cortex (Figure 3). A similar structure occurs in *Parasponia*, the only non-legume genus to have a rhizobial endophyte. In both cases, the infected cells contain high concentrations of hemoglobin and the vascular bundles are outside the infected zone.

In keeping with the diverse taxonomic status of actinorhizal species, the actinorhizal nodule is diverse in both morphology and anatomy and is quite distinct in many respects from the legume nodule (Figure 3). Nodules are bounded by a periderm, which may have either lenticels or, if not, nodule roots attached to the nodule apex as in the genera *Casuarina*, *Datisca*, *Myrica*, *Comptonia* and *Gymnostoma*. These roots have large air spaces and likely function in O_2 transport to the nodules (Bond, 1952; Tjepkema, 1978). The infected cells are variously disposed in actinorhizal nodules (Figure 3) and the disposition of lenticels and nodule roots and the situation of vascular bundles either amongst or within the infected cortex provides preliminary evidence of well ventilated nodules.

3.2. Early Physiological Studies

Early physiological studies on actinorhizal nodules were pioneered by George Bond and colleagues at Glasgow University, mostly using detached nodules and a ^{15}N tracer as an assay for nitrogenase. Using detached *Alnus glutinosa* nodules, which were exposed to 1, 5, 12 or 21kPa O_2 in the presence of $^{15}N_2$, Bond and MacConnell (1955) showed a proportionate reduction in $^{15}N_2$ uptake at the lower pO_2 and, with parallel respiration studies, suggested that respiration and N_2 fixation were strongly linked. Similar experiments with *Myrica gale* nodules (Bond, 1957) showed little reduction in $^{15}N_2$ uptake down to 5kPa O_2, but complete loss of N_2-fixation activity at 1kPa O_2.

This work was extended to four actinorhizal species (Bond, 1961) and showed that optimal N_2 fixation occurs at 12-25kPa O_2, depending on species, and that there is a steep decline in activity at higher pO_2. Bond concluded that the relationship between pO_2 and nitrogenase activity at the lower values of the pO_2 curve was due to the requirement for respiration to drive nitrogenase. On the other hand, with considerable prescience, he concluded that the inhibition of nitrogenase activity at higher pO_2 was a direct effect of O_2 on nitrogenase. By the 1970s, a number of other experiments, with many actinorhizal species, had confirmed the optimal range

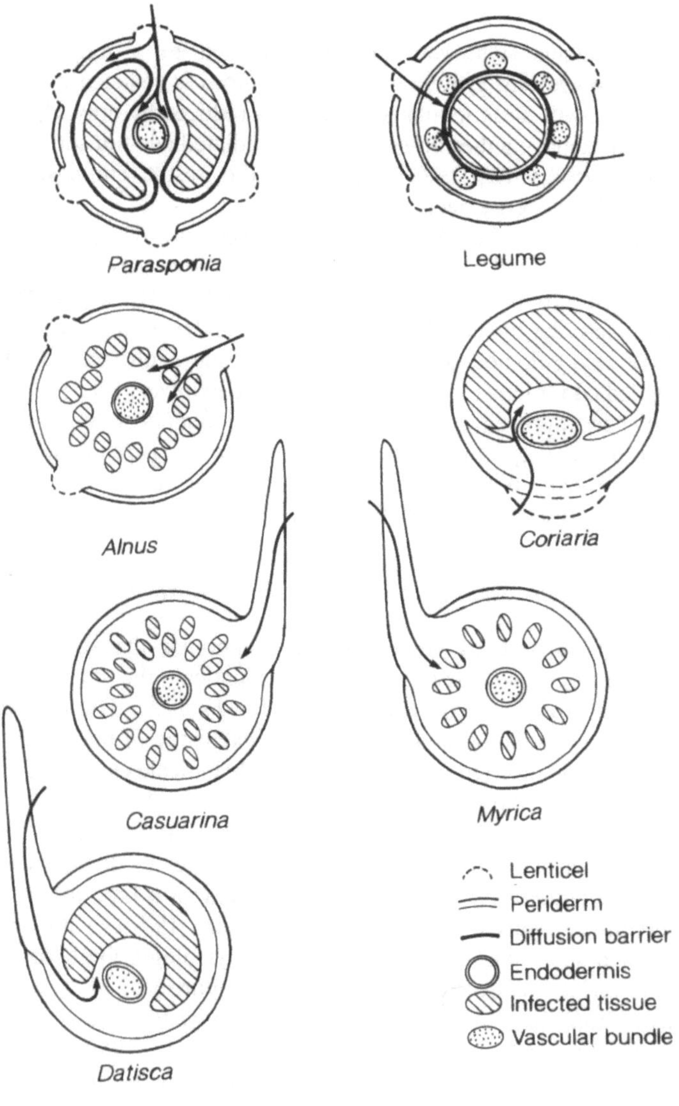

Figure 3. Diagrammatic representation of nodule structure demonstrating the main types. Arrows represent putative major O_2-diffusion pathways (reproduced from Silvester et al. *(1990), with permission).*

for nitrogenase activity as 12-25kPa O_2 (Fessenden *et al.,* 1973; Waughman, 1972; Wheeler *et al.,* 1979) and also concluded, despite the limitations of the available techniques, that the sensitivity to O_2 could not be entirely explained by O_2 needed for respiration and that there was an apparent nitrogenase-specific O_2 inhibition in actinorhizal nodules.

Growth of whole plants with root systems exposed to a range of pO_2 (MacConnell, 1959) showed that growth of *Alnus glutinosa* is reduced by 50% at 5kPa O_2, whereas the growth of unnodulated plants was reduced by only 15%. This species has been shown to have a thermo-osmotically mediated gas-transport system from shoots to roots (Schröder, 1989) and thus, the result emphasizes the importance of O_2 to fuel the energy needs of nitrogenase.

An unusual method of supplying O_2 to the nodule is seen in *Myrica gale*, which like *Alnus glutinosa*, is adapted to wet soils with very low pO_2. *Myrica gale* produces upward–growing roots at the tips of its nodule lobes. These nodule roots enhance nitrogenase activity by enhancing O_2 uptake under flooded conditions (Tjepkema, 1978). *Myrica gale* nodules are also able to resume nitrogenase activity in spring under flooded conditions but at reduced rates (Schwintzer, 1985).

Two major advances allowed rapid progress in the understanding of many aspects of *Frankia* physiology. First, the advent of the C_2H_2-reduction assay, initially used on actinorhizal nodules by Stewart *et al.* (1967), and second, the re-isolation of *Frankia* as a pure culture (Callaham *et al.*, 1978).

3.3. Comparisons with Legume Nodules and the Internal O_2 Environment

Comparisons with legume nodules emphasize that, although the structure and biochemistry of the nitrogenase enzyme system is identical, the physiology of nitrogenase and nodule structure are very different for *Frankia*. These essential differences center on the ventilation and, thus, the gas exchange of the two nodule types. An O_2 microelectrode inserted into a legume nodule indicates a dramatic drop in pO_2 in the inner cortex and shows the center bacteroid zone to be at very low pO_2 (Tjepkema, 1979; Tjepkema and Yocum, 1974). Similarly, vacuum infiltration of India ink into nodules confirms that there is a distinct barrier to diffusion in the inner cortex of legume nodules, which has been identified as a layer of compact cells lacking intercellular spaces (Tjepkema and Yocum, 1974). Although all legume nodules studied show the above structure and physiology, the actinorhizal nodule shows a wide variety of structures and responses (Figure 3).

O_2-microelectrode analysis and India-ink infiltration show that the actinorhizal nodules of *Alnus*, *Datisca* and *Coriaria* (Silvester and Harris, 1989; Tjepkema, 1979; Tjepkema *et al.*, 1988a) are well aerated with a continuous air spaces leading from lenticels in the surrounding periderm to the surfaces of infected cells. *Myrica* nodules are somewhat intermediate in having pockets of infected cells, which show low pO_2 are not penetrated by India ink, and are embedded in a well-aerated cortex (Tjepkema, 1979; Tjepkema, 1983a). In strong contrast, the nodules of *Casuarina* possess a strong discontinuity and the groups of infected cells are in a low O_2 environment. Dye infiltration (Zeng *et al.*, 1989) and electron micrograph studies (Berg and McDowell, 1987) confirm that the infected cells of *Casuarina* are surrounded by a dense layer of cortical cells and are not in contact with air.

Further weight to the distinction between legume and actinorhizal nodules is provided by their contrasting temperature/O_2 interactions. Legume nodules show a low Q_{10} for respiration and for nitrogenase activity (Tjepkema and Yocum, 1973; Waughman, 1977), which has been interpreted as the center of the nodule being essentially diffusion limited due to the inner cortical diffusion barrier. In contrast, nitrogenase activity in *Alnus* nodules is strongly temperature limited (Waughman, 1972) and O_2 consumption at 5 and 20kPa O_2 shows a Q_{10} of approximately 2.0 (Tjepkema, 1979).

So, in contrast to legume nodules, actinorhizal nodules, with the exception of *Casuarina,* are relatively well ventilated, apparently lack significant diffusion boundaries, show a high Q_{10} for respiration and nitrogenase activity, and in keeping with a well-ventilated system, show an inhibition of nitrogenase activity at pO_2 values above ambient.

3.4. The Vesicle in Symbiosis

In symbiosis with plants, *Frankia* in most cases fix N_2 in vesicle structures equivalent to those formed *in vitro*. The story becomes even more interesting because vesicle structure is different in different hosts, the apparent result of the environment unique to each host's symbiotic cell. For example, the same *Frankia* strain forms typical alder-type vesicles in alder and typical *Myrica*-type club-shaped vesicles in the *Myrica* relative, *Comptonia* (Lalonde, 1979). Here, we summarize vesicle structure in different plant hosts as represented by five different classes of N_2-fixing *Frankia* structures (Torrey, 1985). In all cases, these are in symbiotic host cells, which are cortical cells of modified lateral roots. For more details, the reader is referred to an excellent review (Newcomb and Wood, 1987). We adopt their terminology in that vesicles formed during symbiosis are termed *symbiotic vesicles*, whereas those formed *in vitro* are termed *Frankia vesicles*.

3.4.1. Symbiotic Vesicle Structure and Development - Alder Symbiotic Vesicles

Considered the classic example of the symbiotic vesicle, the alder-type has a form that is the closest to that of the *Frankia* vesicle, which is a septate (multi-cellular) sphere with subtending stalk (Figure 4). This vesicle type is found in a variety of actinorhizal plants (Newcomb and Wood, 1987), including members of *Alnus*, the Elaeagnaceae, and some members of the Rhamnaceae. This vesicle is about 4-5μm in diameter compared to a diameter of about 2.5 μm for vesicles in culture. As is universal with *Frankia*, the vesicle forms at the tips of hyphae. The *Frankia* hyphae enter a plant cell *via* infection threads (Berg *et al.,* 1999) and, in alder, these invasive hyphae branch and proliferate in the cell center in the form of vegetative hyphae (Figure 4A; v). Vesicles differentiate in the cell cortex (Figure 4A; sv) to produce a large number in the periphery of a cortical cell of the nodule. Interspersed with the infected cells are uninfected cells (Figure 4A; uc) that contain amyloplasts.

The alder-type vesicle is highly septate (Figure 4B-C), with some septa forming complete cross walls (Figure 4C; double asterisk) and others terminating in the

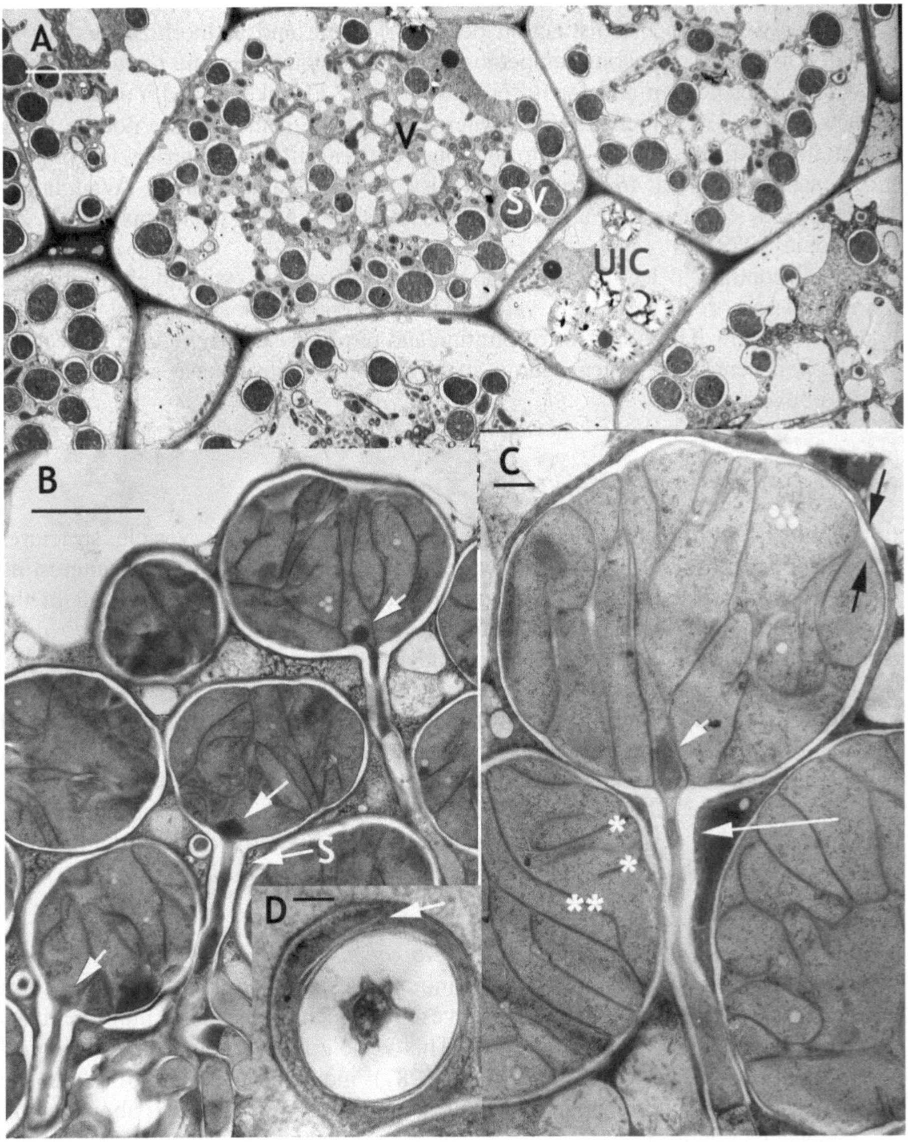

Figure 4. Ultrastructure of the alder-type vesicle, in Alnus serrulata *(A, chemically fixed, B-D freeze substituted; marker = 10 μm in A, 2 μm in B, 0.5 μm in C, and 0.1 μm in D). (A) Vegetative hyphae (V) proliferate at cell center, symbiotic vesicles (SV) differentiate at cell periphery, UIC = uninfected cell. (B) Stalk cell (S) joins vesicle near granular bodies (arrows), envelope is electron-translucent and thickest on stalk. (C) Complete (double-) and incomplete (single-asterisks) septa; envelope is uneven at septa-surface junctions (opposing arrows). (D) Stalk envelope in cross section. (From Berg* et al. *(1999) [A]; 1994 [B,D]; and Berry (1994) [C], with permission).*

cytoplasm (Figure 4C; single asterisk). The former make the vesicle multi-cellular and the latter, in effect, increase the surface area of vesicle cell membranes in perhaps an adaptation that increases available sites for membrane electron transport and energy production for nitrogen fixation. Immunolabeling shows that nitrogenase appears in alder-type vesicles only after septa are formed (Huss-Danell and Bergman, 1990), with the highest immunolabeling density only after full development of vesicle septation. By this stage in vesicle development, the vesicle envelope is fully developed, providing a diffusion barrier for protecting nitrogenase from O_2 inactivation.

The structure of the envelope is discussed below, where it is explained that the envelope can be imaged as an electron-translucent halo when tissue is prepared by freeze substitution (see also Section 2.1.2 for the structure of the vesicle *in vitro*). In Figure 4B-C, high pressure-frozen and freeze-substituted tissue alder vesicles are medially sectioned to show the envelope structure on both vesicle and subtending stalk (S). The envelope halo sometimes varies in thickness around the vesicle (Figure 4C), with the greatest number of lipid layers at septa junctions with the vesicle surface (Figure 4C, opposing arrows), which may result from accumulation of envelope lipid diffusing from septal membrane sites of synthesis (Berg, 1994).

Stalk cells are remarkable for the amount of lipid layers in their envelope (Figure 4B-D), especially at junctions with the vesicle (Figure 4C, arrow). This accumulation constricts stalk-cell cytoplasm to a small channel (Figure 4D, arrow indicates loosely associated surface lipid layer). This accumulation might act to reduce O_2 diffus ion from vegetative cells, which have very few lipid layers and therefore likely a higher pO_2.

Another interesting feature of stalk cells is the inclusions seen in freeze-substituted cells. Bundles of fibers, terminating at the basal septum of the stalk cell (Figure 5A, s and arrow; Figure 5B, asterisk) extend longitudinally along the stalk, terminating where stalk cytoplasm joins vesicle cell cytoplasm. Always adjacent to this terminus in the vesicle is a granular body (Figure 5A, asterisk; see also Figure 1D-E). Because the fibers always associate with the granular body (Figure 4B-C, arrowheads), it is likely that together they have an important purpose in either vesicle development or function. As is the case for vesicles *in vitro* (Lancelle *et al.*, 1985), fibers in symbiosis are only found in stalks or vesicles and the cytoplasm of the stalk is continuous with vesicle-cell cytoplasm. In the early stage of vesicle differentiation before septation, a fibrous body appears in the center of the vesicle (Figure 5C; Berg, 1994) and is oriented longitudinally to the stalk axis. In some cells of nearly mature vesicles, bundles of fibers traverse the vesicle-cell cytoplasm (Figure 6A) and some of these clearly bind the vesicle-cell septal membrane (Figure 6B, arrow). The possibility exists, then, that these fibers are involved in vesicle morphogenesis and bind developing septal walls and stalk septum or other anchors. Perhaps related to this situation, it is not unusual to find vesicle septa terminating at the neck of the stalk, associated with the stalk fiber bundle (Figure 4B-C; Figure 5A, arrowhead). These fibers might function as contractile proteins in vesicle-cell division and morphogenesis, and perhaps hold the key to understanding how *Frankia* accomplishes the feat of building a spherical, multi-cellular prokaryotic structure.

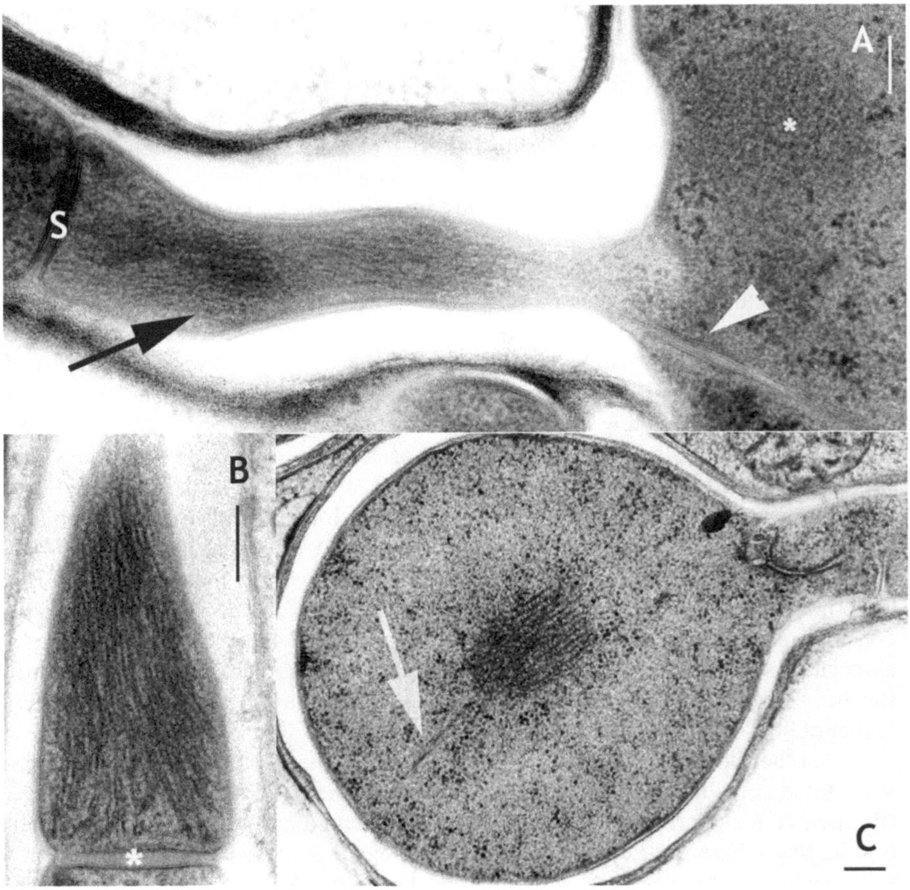

Figure 5. Stalk ultrastructure in high pressure-frozen/freeze-substituted Alnus serrulata. *Markers = 0.2 µm. (A) Stalk cell longitudinal fiber bundle (arrow) terminates at the basal end near the stalk cell septum (S) and at the apical end where stalk cell cytoplasm merges with that of vesicle cells, always associating with a granular body (asterisk, also see arrows in Figure 5B-C). Vesicle septa commonly terminate near apex of stalk cell (arrowhead). Stalk cell envelope (electron-translucent surface) is thickest at vesicle juncture, tapering towards stalk basal septum. (B) Fibers terminate at septum and at lateral walls of stalk cell (reproduced from Berg, 1994, with permission). (C) Early in vesicle differentiation a fibrous body in vesicle center orients longitudinally with the stalk axis, in this case with fibers extending toward the vesicle cell wall (arrow).*

When molecular approaches to understanding *Frankia* biology become available, it will be worthwhile to investigate whether these fibers are homologous to the bacterial protein FtsZ, a tubulin homolog that forms fiber arrays that contract to carry out bacterial cell division (Bramhill, 1997).

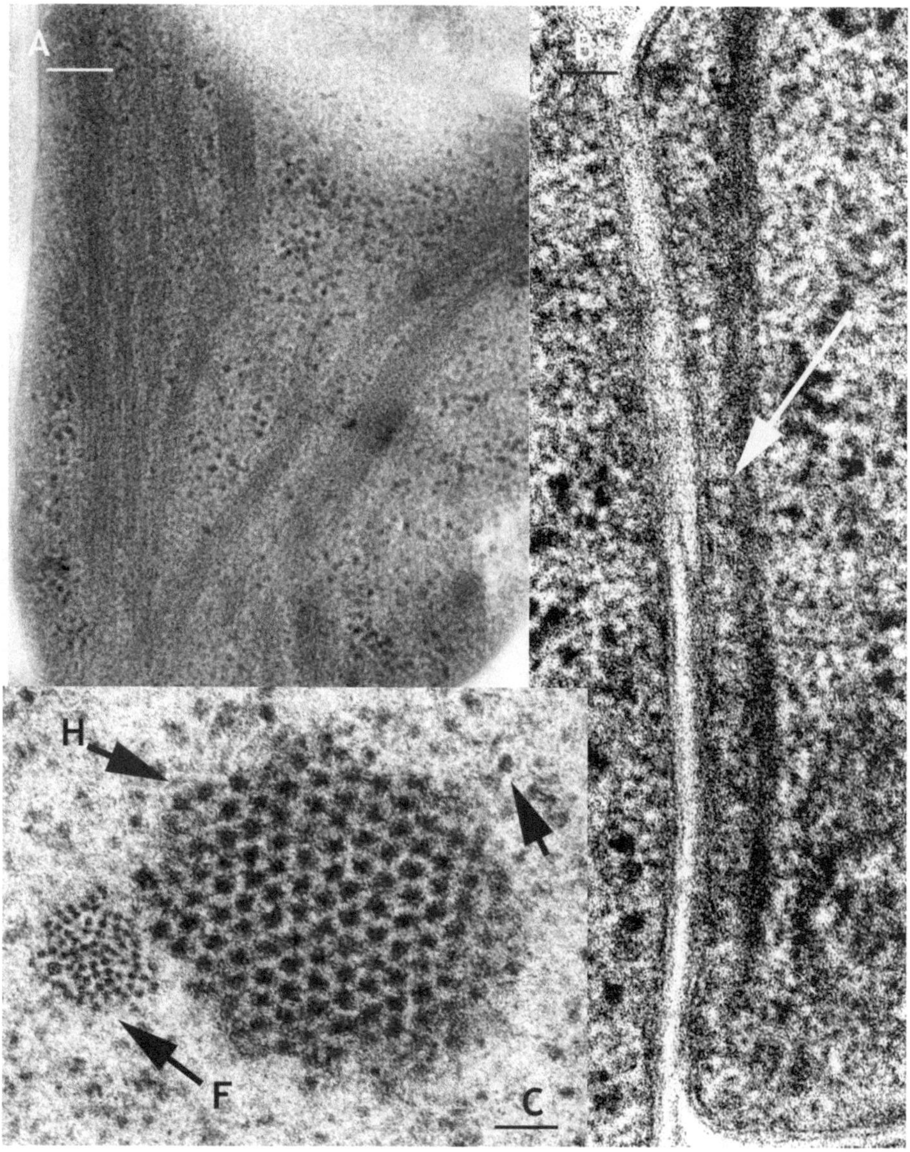

Figure 6. Cytoplasmic inclusions in high pressure-frozen/freeze-substituted alder-type vesicles (Alnus serrulata). *Markers = 100 nm.*
(A), 50 nm (B,C). A number of fiber bundles traverse this vesicle cell cytoplasm. (Reproduced from Berg (1994), with permission).
(B) A single fiber bundle with several cross-links (e.g., arrow) to a septum membrane.
(C) Hexagonal inclusion (H) of particles which also are present free in vesicle cytoplasm (arrow). Fiber bundle cross section (F).

An abundant inclusion within vesicles is a hexagonal crystal (Figure 6C, h) that is formed from particles, which are also found free in the vesicle-cell cytoplasm (Figure 6C, unlabeled arrow). These associate with, and are sometimes bound to, septa (Berg, 1994). Their function is unknown, but they may be an excess of the free particles, which could be some sort of macromolecular complex. The smaller inclusion in Figure 6C is a cross section of a fiber bundle (also present in vesicles *in vitro*; Lancelle *et al.,* 1985).

3.4.2. Envelope Ultrastructure - Alder Symbiotic Vesicles

The symbiotic vesicle envelope, like the envelope of vesicles *in vitro*, does not stain with conventional stains for electron microscopy and instead produces an electron-translucent halo, the "void space", that is both shrinkage artifact and unstained envelope. Using chromic-acid fixation, it is possible to contrast lipid layers of the envelope in symbiotic vesicles *in vivo* (Berg, 1994), so providing a means both for quantitative analysis of envelope ultrastructure and for evaluating the void space artifact. The vesicle in Figure 7A tells the story. The envelope is stained with chromic acid (arrows in Figure 7A-B point to the same region), showing it is retained after processing in dehydration solvents. A prominent void space (Figure 7A, vs) is an artifact due to shrinkage of vesicle cytoplasm away from the envelope during solvent dehydration, which occurs in relatively big increments in chemical-fixation protocols. As is the case for vesicles *in vitro* (Lancelle *et al.,* 1985, see Section 2.1.2), freeze substitution eliminates the shrinkage artifact and allows ultrastructural analysis of the envelope as an electron-translucent halo. This effect is clearly shown by a comparison of the alder symbiotic-vesicle envelope stained by chromic-acid fixation (Figure 7C) with that imaged by high pressure-freezing/freeze-substitution (Figure 7D). Occasionally, individual lipid layers of the envelope can be discerned in freeze-substituted material, as shown for the envelope of *Myrica cerifera* in Figure 7E.

Quantitative analysis by either counting the number of chromic acid-stained layers or by measuring envelope thickness in freeze-substituted samples gave similar results (Berg, 1994). For *Alnus serrulata*, the vesicle envelope averages about 65nm thick or about 14 lipid layers. For *Elaeagnus umbellata* (alder-type vesicle), the envelope is about 39nm thick or about nine layers. These are statistically significant differences in thickness and likely reflect differences in nodule aeration. The morphology of the envelope is similar in these two hosts, with the stalk envelope having the greatest number of layers (about 40 in *A. serrulata* and 30 in *E. umbellata*).

3.4.3. Symbiotic Vesicle Structure and Development - Ceanothus *Symbiotic Vesicles*

The symbiotic vesicle in *Ceanothus* and in members of the Rosaceae (the *Ceanothus*-type) is pear-shaped, non-septate, and lacks a stalk cell (Figure 8B). In *Ceanothus*, the vesicle diameter is about 2μm and, in the other symbioses, this diameter varies up to 4μm (Newcomb and Wood, 1987). In *Ceanothus*, as in the alders, vesicles form at the periphery of the host cell (Figure 8A, sv, v = vegetative

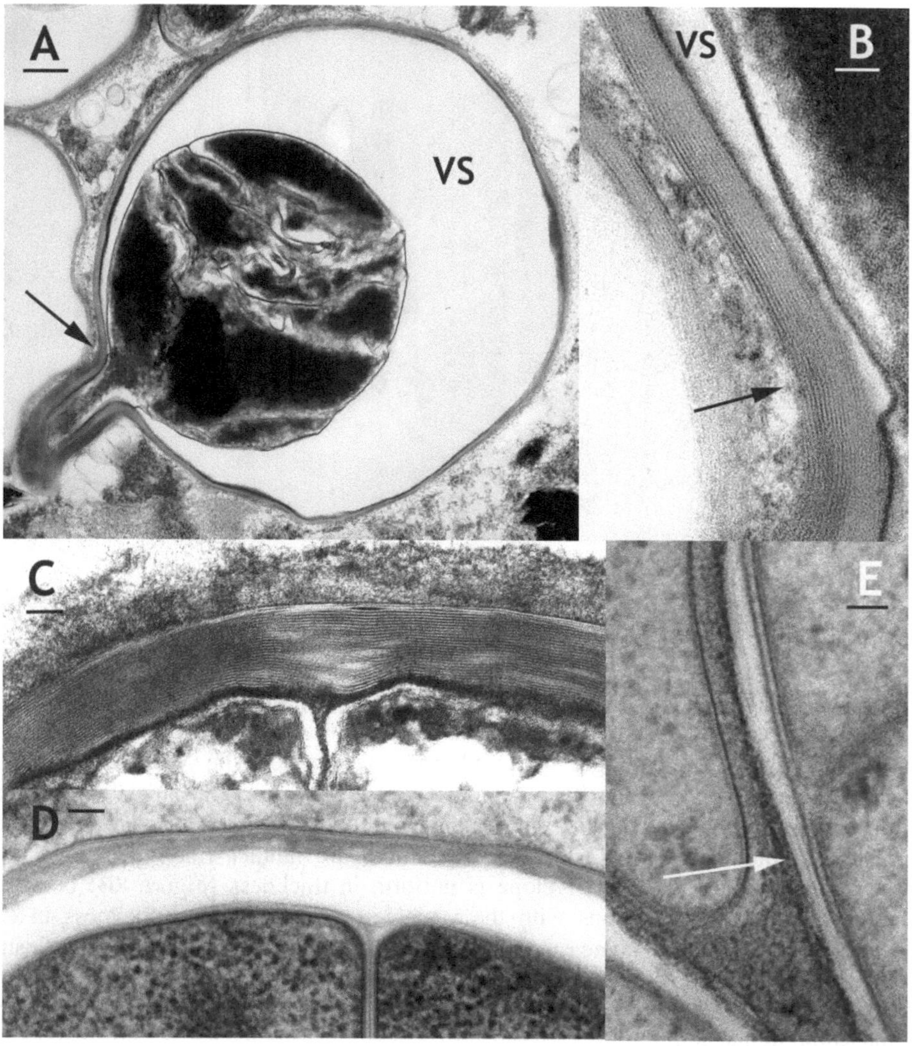

Figure 7. Cytochemistry of the vesicle envelope.
Markers = 0.5 µm (A), 50 nm (B-E).
(A,B) Chromic acid-fixed symbiotic vesicle in Elaeagnus umbellata, *showing the void space artifact (VS) relative to the envelope, which is stained (e.g., arrow, pointing to the same in both micrographs).*
(C,D) comparison of envelope structure stained with chromic acid (C) or imaged as an electron-translucent region free of the void space artifact, in freeze-substituted material (D); Alnus serrulata, *reproduced from Berg, 1994, with permission.*
(E) Vesicle envelope in freeze-substituted Myrica cerifera*—verification that individual lipid layers survive dehydration solvents (arrow).*

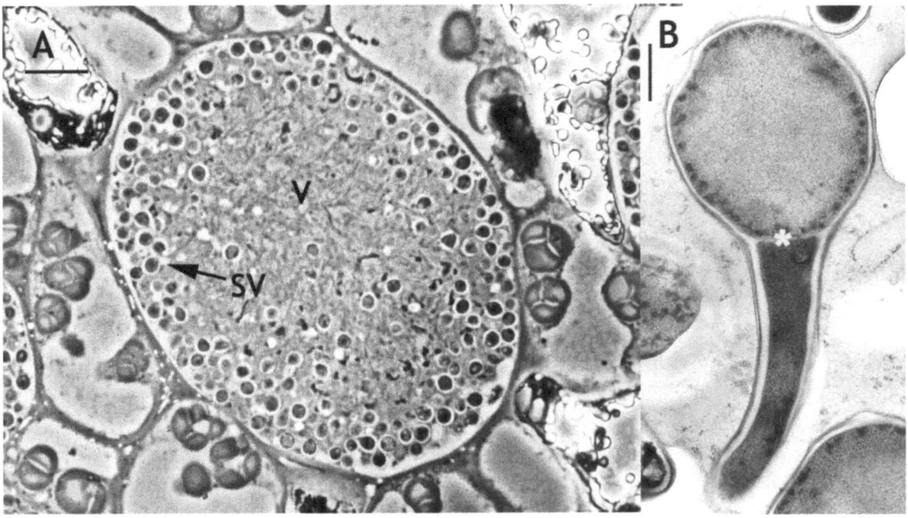

Figure 8. The Ceanothus-*type vesicle, in* Ceanothus americanus.
Markers = 10 µm (A), 0.5 µm (B).
(A) In the large infected cells of Ceanothus, *vegetative hyphae (V) proliferate in cell center and symbiotic vesicles (SV) differentiate at cell periphery.*
(B) The Ceanothus-*type vesicle is pear-shaped and unicellular, being separated from vegetative hypha by a septum (asterisk) (reproduced from Berg (1994) with permission).*

hyphae), which is much larger than symbiotic cells of alder. The vesicle is separated from the vegetative hypha by a septum (Figure 8B, asterisk). This vesicle type differs from the alder-type in being smaller and single-celled. Correlated with the lack of septum formation and stalk cell, there are neither fiber inclusions nor hexagonal bodies, and the envelope is uniform in thickness (Figure 8B) because there are no septa junctions with the vesicle surface. Envelope thickness in *C. americanus* (Berg, 1994) measures 43nm (about 8 lipid layers) with no significant increase in thickness in the adjacent hyphal cell, unlike a stalk cell. This value is not statistically different from the vesicle envelope thickness in *Elaeagnus umbellata* measured in that study.

3.4.4. Symbiotic Vesicle Structure and Development - Myrica *Symbiotic Vesicles*

The *Myrica*-type vesicle (in *Comptonia* and *Myrica*) departs from the spherical motif by being club-shaped and elongate (Figure 9). They can extend over a third of the length of the host cell (Figure 9A, arrows), differentiating peripherally (towards one side of the cell) from vegetative hyphae at cell center (Figure 9A, v). They gradually increase in diameter towards their tip, which is about 2µm in diameter (Figure 9D). There are disorganized septa (Figure 9D, arrowhead), some that join with the vesicle surface and, as in the alder-type vesicle, in these regions the envelope is usually thicker (Figure 9D, arrow). There are cross-wall septa (Figure 9D, asterisk), but there is no stalk cell. There is an inclusion, the "striated

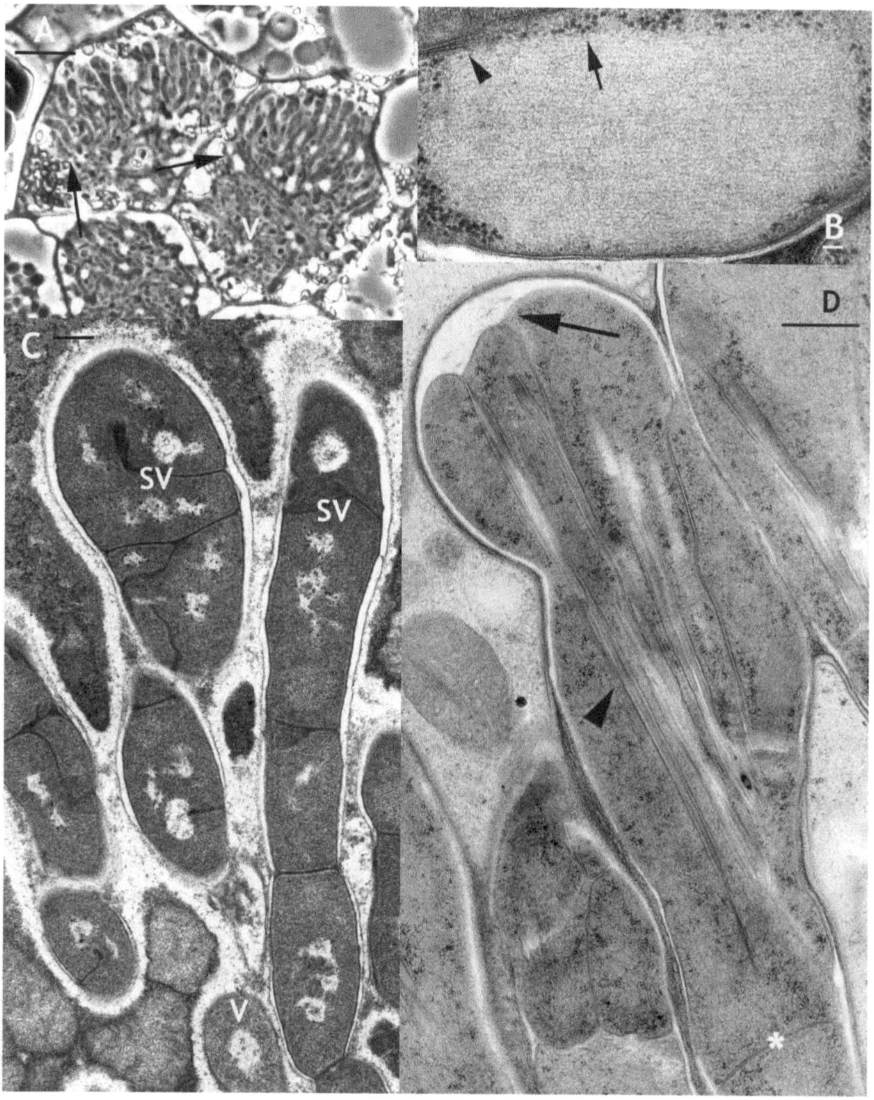

Figure 9. The Myrica*-type vesicle, in* Myrica cerifera.
Markers = 10 μm (A), 100 nm (B), and 0.5 μm (C,D).
(B,D) freeze-substituted, (C) permanganate-fixed.
(A) Club-shaped vesicles (arrows) differentiate on one side of the cell relative to vegetative hyphae (V). (B) A striated body inclusion in the vesicle associates with septum (arrowhead) and particles (arrow) in vesicle cytoplasm. (C) Vesicle envelope thickness increases from very little on vegetative hyphae (V) to its greatest thickness at vesicle apex (SV). (D) Complexes of septa align axially in the vesicle (arrowhead), which also contains a cross septum (asterisk). Envelope is thickest where septa junction its surface (arrow).
(D) Reproduced from Berg (1994) with permission.

body" (Figure 9B), that may be homologous to the fibers or hexagonal crystals of the alder-type vesicle. The striated body sometimes binds septa (Figure 9B, arrowhead) and particles, which are similar in structure to those that bind hexagonal bodies in alder (Figure 7H), bind its surface (Figure 9B, arrow). Based on the likely involvement of alder-type inclusions in septum formation, one would expect similar inclusions in this septa-forming vesicle. There might be a causal relationship between formation of longitudinal arrays of septa and the club shape. The envelope is about 64nm thick in non-thickened (*i.e.*, not at septa) regions, corresponding to about 6 lipid layers (Berg, 1994); both values are not statistically different from the envelope in alder-type vesicles but significantly higher than the envelope thickness in *Elaeagnus*. When nodules are fixed in permanganate, the envelope is an electron-translucent halo in contrast to the heavily oxidized tissue components and shows a gradual envelope thickening from little on the vegetative hyphae (Figure 9C, v) to maximum thickness in the tip region of the vesicles (Figure 9C, sv).

3.4.5. Symbiotic Vesicle Structure and Development – Casuarina, a Special Case

Symbiotic vesicles do not form in *Casuarina* and *Frankia*, in this case, fixes N_2 while in the hyphal form. Figure 10B is an image of a 1µm-thick section (using a high voltage EM) clearly showing that only hyphal filaments exist in mature infected cells of *Casuarina* and are distributed evenly throughout the cell (Berg and McDowell 1987; 1988). A second remarkable feature is that these N_2-fixing hyphae are non-septate (Berg and McDowell, 1987) and are separated from invasive hyphae by a septum. A third unusual feature is that walls of infected cells in *Casuarina* become lignified in response to infection (Berg and McDowell, 1988) and can be stained with the lignin stain phloroglucinol (Figure 10A, longitudinal section, files of infected cells marked with asterisks, x = xylem). As discussed below (Section 3.5.), by modifying the physical characteristics of the wall, this lignification may have a role in suppressing vesicle formation during nitrogen fixation.

3.4.5. Symbiotic Vesicle Structure and Development – Datisca/Coriaria Vesicle

Two features distinguish this vesicle type; its location and structure. In contrast to other actinorhizal symbioses, in *Datisca* and *Coriaria*, the vegetative hyphae proliferate at the cell periphery (Figure 10C, v) and symbiotic vesicles differentiate near the cell center (Figure 10C, sv) and "point" at the intact vacuole in the cell center (C, vac) to form a hollow sphere of tightly packed vesicles (Berg *et al.*, 1999; Newcomb and Pankhurst, 1982). The vesicle is essentially a single-cell rod-shaped hypha with the diameter of vegetative hyphae (about 1µm) and 3-4µm in length. The envelope is no different in thickness than that of vegetative hyphae (Silvester *et al.*, 1999). An unusual feature of the sphere of symbiotic vesicles is that a layer of mitochondria covers them (Hafeez *et al.*, 1984a; Mirza *et al.*, 1994; Silvester *et al.*, 1999). The implications of this arrangement in O_2 protection are discussed below. The confocal micrograph in Figure 10D is of living *Coriaria* nodule cells stained with rhodamine 123, which loads into mitochondria, and illustrates the ring of mitochondria (arrow) surrounding the bases of the sphere of symbiotic vesicles.

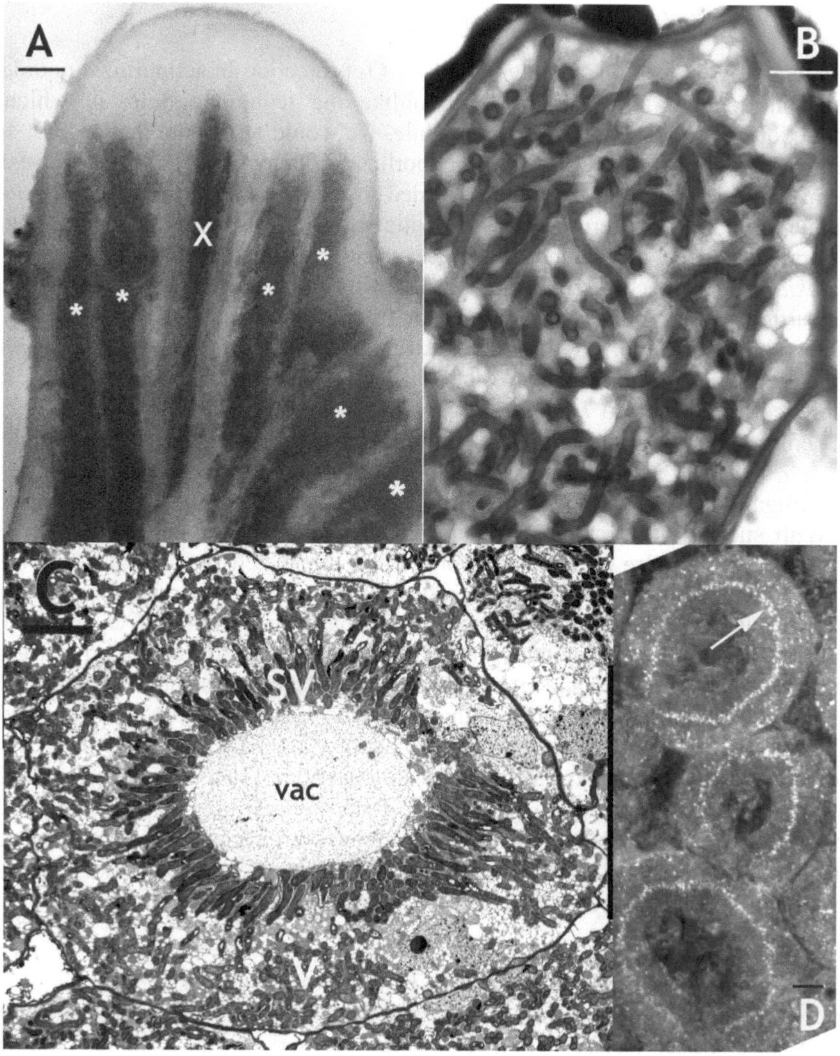

Figure 10. Special adaptations in Casuarina *(A,B) and* Datisca/Coriaria *nodule types (C,D). Marker = 100 µm (A), 5 µm (B-D).*
(A) Walls of infected cells in Casuarina *become lignified, shown here by staining with phloroglucinol (asterisks mark longitudinal files of infected cells, x = xylem). (B) Hyphae fix nitrogen in symbiotic cells of* Casuarina *(high voltage EM micrograph of 1 µm thick section). (C) Symbiotic cell in* Datisca *showing "reversed" anatomy: vegetative hyphae (V) at cell cortex and tightly-packed sphere of rod-shaped vesicles (SV) pointing to the central vacuole (vac). (D) Blanket of mitochondria (arrow) covers the sphere of vesicles; confocal micrograph of live tissue stained with rhodamine 123* (Coriaria).
Reproduced (C) from Berg et al. *(1999), and (D) from Silvester* et al. *(1999), with permission.*

3.5. Responses to O_2

It is tempting to seek one simple model for O_2 protection in actinorhizal nodules as found in legume nodules. However, unlike the legume nodule, in which O_2 protection is provided solely by the nodule, we would anticipate that at least some part of the O_2-protection system in actinorhizal nodules is in the *Frankia* vesicles themselves as well as in the nodule. Added to this complication is the variation in nodule structure as determined by the wide variety of host families and genera of actinorhizas. Thus, at one end of a spectrum, in the classical *Alnus* nodule, normal *Frankia* vesicles occur and are organized within infected cells as clusters that face outwards. In this nodule type, we might predict that much of the O_2 protection resides in the vesicle. At the other extreme is the *Casuarina* nodule in which *Frankia*, though forming good vesicles in culture, lacks vesicles in symbioses. By inference from work on *Frankia in vitro*, the lack of vesicles indicates a very low pO_2 environment within the nodule. Intermediate conditions are seen in nodules of *Myrica*, *Coriaria* and *Datisca* in which club-shaped vesicles are produced, either centrifugally in *Myrica* or centripetally in *Coriaria* and *Datisca*.

With such a wide range of nodule anatomy and *Frankia* differentiation, it is difficult to generalize about O_2 responses and associated physiologies and so the different types of nodules are considered separately. Given the range of physiology, it is surprising that the energy cost of nitrogenase, measured as CO_2 evolved per unit of C_2H_4 production, neither varies significantly across five actinorhizal genera nor between these genera and eight legume genera (Tjepkema and Winship, 1980).

The O_2 responses are of three distinct varieties. First, they may be the immediate effect of different pO_2 levels applied to unadapted (or air-grown) nodules. Second, they may involve short-term adaptation (minutes-hours) to a changed pO_2 in which case physiological adaptation is implied. Finally, they may involve longer-term application of altered pO_2, which brings about both structural and physiological adaptation to the altered environment. The short-term effects are considered prior to describing the major types of response by the plants.

3.5.1. Short-term Effects of Acetylene and O_2 on Nitrogenase Activity

Root nodules of actinorhizal plants, whether attached or excised, show two important short-term effects (transients) that are fundamental to understanding O_2 effects. First is the so-called C_2H_2-induced decline, which was first observed in legume nodules (Minchin *et al.*, 1983) and is also apparent to varying degrees in actinorhizal nodules (Rosendahl and Huss-Danell, 1988; Silvester and Harris, 1989; Tjepkema *et al.*, 1988b; see section *3.5.2.*). This effect arises with some plants when C_2H_2 reduction is used to measure nitrogenase activity. The second effect is the transient decrease in nitrogenase activity in response to rapid small shifts in pO_2 during nitrogenase assays (Silvester and Winship, 1990; see section *3.5.3.*).

3.5.2. Acetylene-induced Decline in Nitrogenase Activity

The effect of adding C_2H_2 or removing N_2 (replaced by Ar) on legume nodules is to set in motion a chain of activities that causes a rapid decline in electron transfer through nitrogenase. This phenomenon, termed the C_2H_2-induced decline, is well documented for many, but not all, legumes (Minchin *et al.,* 1983; Witty *et al.,* 1984). In some cases, there may be a partial recovery of activity (Schuller *et al.,* 1988), but full recovery has not been recorded in legumes. Both C_2H_2- and Ar-induced declines occur only in those legume nodules that show a tolerance to O_2 levels above ambient; nodules that are not well protected against high pO_2 show no decline (Witty *et al.,* 1984). The decline phenomenon, which is also initiated by root disturbance, moisture stress, and defoliation (Durand *et al.,* 1987; Hartwig *et al.,* 1987; Minchin *et al.,* 1986) has been explained as a sequence of events that starts with the cessation of ammonia production and triggers a rapid increase in the diffusion resistance of the nodule (Witty *et al.,* 1984). Those nodules, which are already highly limited for O_2, then show a rapid decline in nitrogenase activity, which can be reversed by increasing the pO_2 of the assay.

Actinorhizal nodules show a similar decline but, in keeping with the wide range of nodule structures and *Frankia* morphologies, there is a wide variety of resulting effects. The decline does not occur with *Frankia in vitro* as shown by fast-flow C_2H_2-reduction assays in liquid culture on two species of *Frankia*, which were suspended on either nylon mesh or filter paper, where no decrease in C_2H_4 production occurred over a 12-minute assay period (Silvester and Winship, 1990). Thus, the conclusion must be that the decline is a nodule effect and the mechanism sought, as in the legume nodule, is in structures outside the symbiotic vesicle.

A variety of actinorhizal species have been investigated for C_2H_2-induced decline but, in contrast to legumes, the decline is often followed by either a partial or full recovery (Silvester and Winship, 1990; Tjepkema *et al.,* 1988b). The magnitude of the decline and its subsequent recovery is dependent on the host species, growth conditions, and plant age.

The extent of the problem is illustrated by the work on C_2H_2 decline in several species of *Alnus*. *Alnus incana* spp. *rugosa*, when grown aeroponically in the laboratory, showed a marked decline and normally a complete recovery within 20 minutes (Silvester and Winship, 1990). The same species taken from the field showed little or no recovery (Schwintzer and Tjepkema, 1997), however when grown hydroponically, the plants recovered slowly but only partially. The related European species *Alnus incana* spp. *incana* showed no recovery (Rosendahl and Huss-Danell, 1988), whereas *Alnus rubra* showed full recovery (Tjepkema *et al.,* 1988b). *Elaeagnus*, a genus with similar nodule anatomy to *Alnus*, exhibited a decline with partial recovery (Johnson *et al.,* 1997). Furthermore, in *Casuarina cunninghamiana*, both a large decline and full recovery (Tjepkema and Murry, 1989) and a very small decline with minimum recovery (Silvester and Winship, 1990; Figure 11) have been reported.

In contrast to the above observations, the performance of *Coriaria* and *Datisca* is unequivocal. These genera, from separate but related families, have very similar nodule anatomy and display large C_2H_2- and Ar-induced declines with no recovery (Harris and Silvester, 1994; Tjepkema, 1997; Tjepkema *et al.,* 1988b).

Without doubt, the most intensively studied genus is *Myrica*. In unstressed laboratory-grown plants, nitrogenase in *Myrica* typically shows a marked C_2H_2-induced decline with a recovery to 80% or more of the original rate (Schwintzer and

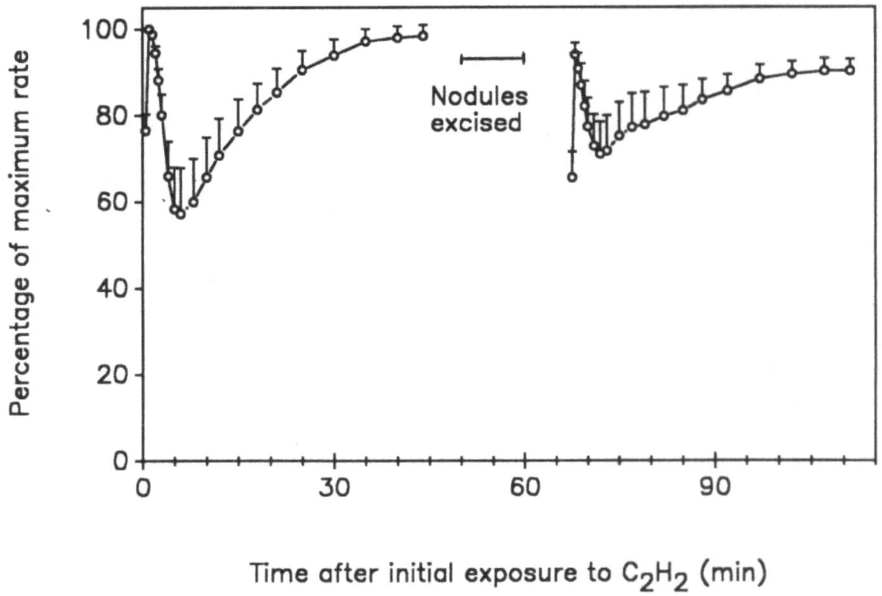

Figure 11. Acetylene-induced decline in nodules of Casuarina cunninghamiana *before and after excision.*
(Reproduced from Tjepkema and Murry, 1989, with permission).

Tjepkema, 1994; Silvester and Winship, 1990; Tjepkema and Schwintzer, 1992; Tjepkema *et al.,* 1988a). In contrast, laboratory-grown plants, which are stressed by extended darkness or water deficit prior to the acetylene-reduction assay (Schwintzer and Tjepkema, 1994), and field-collected nodules (Schwintzer and Tjepkema, 1997) have substantially deeper declines and much less recovery. As with legumes, the decline is initiated by cessation of ammonia production, which in turn may lead to depletion of either an amino acid or other substance involved in the generation of reductant for nitrogenase (Tjepkema and Schwintzer, 1992).

Although it may appear that the presence or absence of an C_2H_2-induced decline is fundamental to our understanding of gas exchange in nodules, other explanations are possible. For example, plant (nodule) age is known to be important in defining the presence of an C_2H_2-induced decline. In *Coriaria* nodules, the equilibrium rate changes from *ca.* 60% of the maximum rate in three-month-old nodules to 83% in six-month-old plants, with only a negligible decline in both nine- and 12-month-old plants (Harris and Silvester, 1994). This result was attributed to increases in the time required for gas diffusion in older nodules rather than a lack of an C_2H_2-induced decline, which was simply being masked by the extra tissue in the gas pathway of older nodules (Harris and Silvester, 1994). Thus, the presence or absence of a measured C_2H_2-induced decline may be due mainly to the age or response time of the nodule, but the spontaneous recovery from a decline is of fundamental significance in understanding the O_2-protection mechanism.

3.5.3. O_2-induced Transients

When root nodules are exposed to higher pO_2 in a flow-through continuous assay system, they often show a transient decrease in nitrogenase activity. These transients were first observed in legume nodules (Hunt *et al.,* 1987) and were later shown, by using spectroscopic monitoring of leghemoglobin during H_2 evolution by nitrogenase, to be accompanied by a sharp rise in internal pO_2 of the nodule.

The same effect has been observed in *Alnus* (Rosendahl and Huss-Danell, 1988; Silvester *et al.,* 1988b), *Casuarina* (Silvester and Winship, 1990), *Myrica* (Silvester *et al.,* 1988a) and *Coriaria* (Silvester and Harris, 1989). The phenomenon manifests itself as a massive drop in nitrogenase activity following a small increase in pO_2 in the flow-through cuvette; nitrogenase activity then spontaneously recovers often to a rate higher than the initial nitrogenase-activity rate. The event can only be witnessed when continuous steady-state monitoring of nitrogenase activity is conducted and, in many cases, nitrogenase activity may decrease to zero for a short period of time. The event is very short lived (1-5 minutes) and is not associated with a short-term change in diffusion resistance (Silvester and Winship, 1990).

O_2-induced transients in legumes were considered to be an effect either close to or within the bacterium (King *et al.,* 1988) and related to O_2 concentrations close to nitrogenase. Although this suggestion cannot be tested for rhizobia because of their intense sensitivity to O_2, it has been tested for *Frankia in vitro* (Silvester and Winship, 1990). In order to overcome the problem of gas diffusion through liquid media, *Frankia* were suspended on nylon mesh in a cuvette and exposed to changes in pO_2. Transient decreases in nitrogenase activity were observed that are identical to those seen with nodules, in that the transient response is rapid, is proportional to the magnitude of the applied O_2 pulse, and is spontaneously recoverable (Silvester and Winship, 1990).

It now appears likely that the step-up in pO_2, which initiates the transient decrease in nitrogenase activity, produces a transient increase in pO_2 at the nitrogenase site, which directly or indirectly affects either nitrogenase itself or its electron-transport chain. The spontaneous recovery of nitrogenase activity is explained as an increase in O_2-limited respiration (Winship and Tjepkema, 1985) within or close to the bacterium, which then restores the low internal equilibrium pO_2 level. The rapid switch-off of nitrogenase activity that is required to support this mechanism was first described in *Azotobacter* by Dalton and Postgate (1969) and given the name of conformational protection. This mechanism involves an FeS protein, which binds to nitrogenase making it inactive, but tolerant to O_2, for short periods (Robson and Postgate, 1980).

O_2-induced transients therefore appear to be a feature common to N_2-fixing bacteria, whether in culture or in symbioses, and they have features in common across many different species and systems. In many ways, the recoverable C_2H_2-induced declines that are seen in *Myrica, Alnus*, and *Casuarina* resemble these transients, although the fact that an C_2H_2-induced decline is not observed with *Frankia in vitro* argues for a somewhat different mechanism.

3.5.4. Adaptation to Varying pO_2 and O_2 Protection

Four major nodule anatomies are displayed by actinorhizas and these give rise to differences in *Frankia* anatomy and the physiology of nitrogenase. Each type will be described separately prior to an attempt to distill an overall general hypothesis of O_2 protection. As both *Frankia* and the nodule display possible gas-diffusion properties and responses to O_2, it is the balance between these two responses that is liable to change with different nodule types. Any consideration of nodule physiology will need to explain such phenomena as C_2H_2-induced decline, O_2-induced transients, and both long- and short-term adaptation to pO_2 as well as nodule structure.

3.5.5. O_2 Adaptation and O_2 Protection in Alnus

Alnus nodules are exceptionally well-ventilated, they have many lenticels (Bond, 1974), and they contain an interlacing network of air spaces (Wheeler *et al.,* 1979). When infiltrated under vacuum with India ink, the spaces are shown to be connected to the outside (Tjepkema, 1979). Nitrogenase activity in *Alnus rubra* shows a broad optimum response to O_2 being largely independent of pO_2 over the range from 15-30kPa O_2, but is severely limited by pO_2 both above and below that range (Wheeler *et al.,* 1979; Winship and Tjepkema, 1985). The infected cells form a broad band of clumps of cells in the mid cortex surrounding the central stele and *Frankia* shows well formed vesicles towards the outside of the infected cells (Figure 3).

Protection of nitrogenase from the effects of O_2 in *Alnus* nodules is the result of a combination of diffusion limitation coupled with a high rate of O_2-sensitive respiration (Winship and Tjepkema, 1983). The permeability of the O_2 pathway in nodules is extremely temperature sensitive and the combined effects of temperature and pO_2 on nodules are consistent with a nodule in which nitrogenase is located within a series of compartments. Some compartments of high permeability are active at low pO_2, whereas others of low permeability are only active at high pO_2 (Winship and Tjepkema, 1985). This model helps explain the physiological responses but does not identify the actual sites of diffusion resistance.

Alnus plants grown at different pO_2 levels showed very little difference in growth between 10 and 30kPa O_2 (MacConnell, 1959), thus implying that plants can adapt to varying pO_2. Such adaptation must be accompanied by physiological and/or anatomical changes that can be used to identify the sites of O_2 protection and diffusion limitation. Using this hypothesis, plants were grown at 5, 21 and 40kPa O_2 and shown to adapt nitrogenase activity to each of the growth pO_2 levels (Silvester *et al.,* 1988b). Thus, plants grown at 5kPa O_2 had a broad nitrogenase optimum at 5-9kPa O_2, whereas those grown at 40kPa O_2 optimized nitrogenase activity at 15-30kPa O_2. Significant changes in nodule anatomy were also observed, including lenticel size, intercellular spaces, and most significantly the dark-field appearance of vesicles. Thus, at high pO_2, nodules had restricted lenticel development, smaller intercellular spaces (with very small ones between infected cells), and relatively thick-walled vesicles (Silvester *et al.,* 1988b).

Furthermore, the vesicles from this experiment were examined by freeze-fracture electron microscopy (Abeysekera *et al.,* 1990). The distal parts of vesicles showed significant variation in the number of lipid laminae; a two-fold increase in the number of laminae was observed with the eight-fold increase in pO_2 from 5-40kPa. These results indicate that, although vesicle-envelope thickening is certainly a component of the O_2-protection mechanism, it is not the only part.

The emerging picture of O_2 protection and adaptation in *Alnus* nodules is a combination of changes in vesicle envelope and nodule structure, backed up by respiratory O_2 uptake. The results of physiological studies (Winship and Tjepkema, 1983, 1985), structural adaptation (Silvester *et al.,* 1988b), anatomical study (Abeysekera *et al.,* 1990), and theoretical network simulation (Winship and Silvester, 1989) all support the model in which nitrogenase exists in several compartments in a nodule. The O_2-diffusion pathway is to some extent controlled by diffusion through the nodule, but this is a minor part of the resistance. A second source of resistance is through the packing of cells together in clusters, with some cells or parts of cells exposed to intercellular spaces and some not. Finally, the major control component of the O_2-diffusion pathway is the vesicle, which in all cases contains many lipid layers and adapts significantly to changing ambient pO_2. The situation in the *Alnus* nodule is probably the easiest to explain because it implicates the symbiotic-vesicle envelope as a major diffusion barrier.

3.5.6. O_2 Adaptation and O_2 Protection in Casuarina

Frankia in *Casuarina* nodules do not form vesicles (Berg, 1983; Berg and McDowell, 1987) despite the fact that these frankiae are capable of forming vesicles when grown in culture. Work on vesicle formation and function (Murry *et al.,* 1985) concluded that infected cells are at a very low pO_2 and operate in an environment similar to legume nodules. This situation is consistent with the high level of hemoglobin found in *Casuarina* nodules (see section *4.1.2.*). *Casuarina* nodules are surrounded by a solid periderm that lacks lenticels; the connection to the outside is *via* the nodule root (Figure 3). Dye infiltration (Zeng *et al.,* 1989) and EM studies (Berg and McDowell, 1987) show that the internal air spaces, although connected to air *via* the nodule root, become attenuated near the infected cell and are absent amongst infected cells.

Casuarina nodules show responses in both nitrogenase activity and respiration to changes in pO_2 that are remarkably similar to those of legume nodules (Tjepkema and Murry, 1989). Both nodule types show a linear increase in both respiration and nitrogenase activity over the full range of sub-optimal pO_2 that is consistent with a diffusion barrier within the nodule. *Casuarina* nodules also show a respiration component that is saturated at *ca.* 5kPa O_2, suggesting the presence of a compartment that is presumably outside the diffusion barrier (Tjepkema and Murry, 1989).

These results led to the conclusion that *Casuarina* is unique among actinorhizal nodules in that the nodule provides a major component of the diffusion barrier and

O_2 protection to *Frankia*. This barrier could be represented either by the close packing of cells in the infected area so presenting a cytoplasmic diffusion requirement for O_2 or by the lignified/suberized cell walls of the infected cells (Berg, 1983; Berg and McDowell, 1988). Moreover, both these components could act in concert to determine an inner low pO_2 area in which *Frankia* can express nitrogenase in hyphae rather than in vesicles.

3.5.7. O_2 Adaptation and O_2 Protection in Myrica

Isolates from nodules containing *Myrica*-type vesicles form typical *Frankia* vesicles *in vitro* and alder-type vesicles in alder (Lalonde, 1979; Torrey and Callaham, 1982). This illustrates the potential for host-plant influence on vesicle structure. As shown in a study of *Myrica gale* (Tjepkema, 1983a), the O_2 concentration is low in the region of infected cells due, at least in part, to smaller intercellular air spaces near infected cells and the lack of these spaces in groupings of infected cells. The walls of infected cells in *Myrica* become lignified during infection (Tjepkema and Asa, 1987) and this wall modification might be an important component in slowing the diffusion of O_2 into infected cells (Zeng and Tjepkema, 1994).

The *Myrica*-type vesicle could be considered as an intermediate between the alder-type and the lack of vesicles in *Casuarina*. Like *Casuarina,* hemoglobin concentration is high in *Myrica* nodules (Tjepkema and Asa, 1987; see section *4.1.2.*). However, in contrast to *Casuarina*, there is an envelope on the *Myrica*-type vesicle (Figure 9) that is comparable in thickness to that in the alder-type vesicle (Berg, 1994), indicating a microsymbiont response to O_2.

Myrica gale nodules, when subjected to various O_2 levels, show a response in nodule anatomy but not in vesicle envelope (Silvester *et al.,* 1988a). At 40kPa O_2, the nodules formed substantially smaller nodule roots (a source for nodule aeration) and decreased intercellular air spaces in both the nodule cortex and the region of infected cells. Changes in the walls of infected cells were not evident by light microscopy. Overall, these nodules adapt structurally to increased pO_2 by altering the entry of O_2 into nodule tissue through nodule roots.

The sites of O_2 resistance within nodules have also been studied by using silicone grease to eliminate gas entry into nodules. These experiments indicate that the nodule root is a major entry point for O_2 into nodules (Zeng and Tjepkema, 1994) and they have been likened to O_2-gathering antennae (Silvester *et al.*, 1988a). However, when nodules are sliced (0.1-mm thick), nitrogenase activity is hardly affected, being still O_2 limited up to 18kPa O_2 (Zeng and Tjepkema, 1994). Further proof that the resistance to O_2 diffusion is at the host-cell surface is provided by the observation that 2kPa CO added to the gas phase had little effect on nodule respiration (Zeng and Tjepkema, 1994).

All of these results strongly indicate that the site of O_2-diffusion resistance lies within the nodule and most likely at the host-cell walls. Because the diffusion resistance of *Myrica* nodules does not respond to pO_2 changes (Tjepkema and Schwintzer, 1992; Silvester *et al.,* 1988a) but is very sensitive to temperature (Zeng and Tjepkema, 1995), it may involve the lignified wall. The lignified wall would have a fixed resistance at constant temperature but may decrease in permeability at lower temperature, analogous to polyethylene, which has a Q_{10} of 1.8 over the temperature range 0-30°C (Zeng and Tjepkema, 1995).

O_2 protection in *Myrica* appears to be a combination of cell-wall resistance coupled with a contribution from the vesicle envelope that exists in *Myrica* symbiotic vesicles.

3.5.8. O_2 Adaptation and O_2 Protection in Ceanothus

There are no experimental data on the response to varying pO_2 in *Ceanothus* nodules. The nodules appear well aerated because very little nodule hemoglobin is present (Silvester *et al.,* 1990, see also section *4.1.2.*). The thickness of the vesicle envelope in the *Ceanothus*-type vesicle is not as thick as in the alder-type nodule (Berg, 1994, and see discussion above), but this does not mean it is inadequate to function as an O_2-diffusion barrier. The difference in vesicle morphology relative to the alder-type may be genetically determined. Based on 16S-rRNA gene sequences, the *Ceanothus* microsymbiont is placed with the microsymbiont of *Dryas* (Ritchie and Myrold, 1999), another *Frankia* that forms *Ceanothus*-type vesicles. It would be interesting to see if nitrogen-fixing cultures of these microsymbionts form *Ceanothus*-type vesicles *in vitro.*

3.5.9. O_2 Adaptation and O_2 Protection in Coriaria *and* Datisca

These two genera, belonging to two distinct angiosperm families, show remarkably similar nodule anatomy and physiology that is quite distinct from all other actinorhizal plants. The nodules develop asymmetrically with a mass of infected cells displacing the stele to one side (Calvert, 1979; Mirza *et al.,* 1994). Unlike other actinorhizal nodules, all cells in the infected cortex contain *Frankia* (Figure 3). *Frankia* proliferates in the perimeter of the cell and develops tightly packed symbiotic vesicles facing towards a central vacuole (Berg *et al.,* 1999). Anatomical studies and India-ink infiltration shows that all the infected cells of both genera are surrounded by a network of air spaces that are connected to the outside (Silvester and Harris, 1989; Tjepkema *et al.,* 1988a) *via* a nodule root in *Datisca* and *via* a single lateral lenticel in *Coriaria.* The relatively unrestricted gas-diffusion path in nodules of both species is further illustrated by the short time required to reach peak ethylene concentration in fast-flow C_2H_2-reduction assays. When either *Coriaria* (Silvester and Harris, 1989) or *Datisca* (Tjepkema, 1997) are exposed to C_2H_2 in a fast-flow system, which is capable of resolving reactions at 5-s intervals, maximum ethylene output is achieved in approximately 30s. The speed of these responses is greatly in excess of any other actinorhizal species and confirms ready access of gas to the interior of the nodule.

Both genera show significant C_2H_2-induced declines with either no recovery in *Coriaria* (Silvester and Harris, 1989) or a small recovery in *Datisca* (Tjepkema, 1997). This decline in nitrogenase activity is accompanied by a parallel decline in respiration, results that are similar in many ways to those of legume nodules.

When *Coriaria* plants are grown with root systems at pO_2 levels from 5-40kPa, there is no significant change in either plant performance or level of nodulation, which indicates that these plants, like *Alnus*, can adapt their root and nodule activities to a wide pO_2 range (Silvester and Harris, 1989). The effects on root structure include a minor increase in periderm thickness at high pO_2 and an increase in lenticel size at low pO_2. Most notably, however, was the increase in suberization of the periderm around the whole circumference and the increased thickening of the internal periderm that wraps around the infected cell mass. Long-term morphological responses to a wide range of pO_2 are minor in *Coriaria* compared to those seen in *Alnus* and *Myrica*. What are important however, are the short-term adaptations.

When *Coriaria* nodules of an air-grown plant are exposed to quite rapid changes in pO_2 from 10-60kPa over a period of 4.5h, the plant quickly adapts nitrogenase activity to the new pO_2. Similarly, a plant grown with roots at 5kPa O_2, when exposed to pO_2 levels from 1.7-30kPa, adjusted the nitrogenase rate to within 20% of the initial control rate (Silvester and Harris, 1989). Each change in pO_2 is accompanied by an O_2-induced transient decrease in nitrogenase activity, which is then followed by a return to the initial control rate. These results indicate probable changes in diffusion resistance in the tissue in response to changing pO_2. Lag-phase measurements of C_2H_4 production at various pO_2 levels also indicate significant changes in nodule-diffusion resistance. The time constant (half the time required to return to the equilibrium C_2H_4 concentration) changed from *ca.* 13s at 21kPa O_2 to *ca.* 28s at 40kPa O_2 (Silvester and Harris, 1989). The most likely site of these changes is the 2-3 cell layers between the internal periderm and the endodermis. Although this layer allows India ink to penetrate into the nodule interior, it does represent a significant choke, which could osmotically change its configuration.

The situation in *Datisca* has been interpreted quite differently (Tjepkema *et al.*, 1999). Like *Coriaria* nodules, *Datisca* nodules respond very rapidly to C_2H_2 exposure and are extremely well ventilated, but in contrast to legume nodules, *Datisca* nodules show only very small changes in both respiration and nitrogenase activity when exposed to 16-30kPa pO_2 during C_2H_2 exposure. There is also little difference in nitrogenase activity when measurements are made in He/O_2 mixtures compared with those made in Ar/O_2 mixtures. These results are consistent with a system lacking significant gas-phase limitation on diffusion.

The interpretation by Tjepkema *et al.* (1999) of O_2 protection in *Datisca* is similar to that proposed for *Alnus*; nitrogenase exists in a number of compartments, maybe a continuum, of differing accessibility to O_2. They conclude that the major resistance to O_2 lies in the infected cells. The vesicles, which lie tightly packed towards the center of the cell, present only a small proportion of their surface area to the surrounding cell cytoplasm and this orientation is considered sufficient to offer significant resistance to O_2 and to explain the effects.

The inward facing vesicles of *Coriaria* and *Datisca* are highly reducing sites, as shown by tetrazolium reduction (Hafeez *et al.*, 1984a, 1984b; Silvester *et al.*, 1999), and nitrogenase has been located in these sites by *in situ* hybridization of *nif*H transcripts (Mirza *et al.*, 1994). These structures, although rod-shaped, are comparable to symbiotic vesicles, but they do not have a thickened envelope (Silvester *et al.*, 1999) and, therefore, do not function in the same way as *Alnus* vesicles. There is no doubt that the vesicles are operating at a very low pO_2 as defined by intense cyanotetrazolium-chloride (CTC) fluorescence in vesicles of normal *Coriaria* nodules.

An unusual concentration of mitochondria at the base of symbiotic vesicles occurs both in *Datisca* (Hafeez *et al.,* 1984b) and in *Coriaria* (Mirza *et al.,* 1994), indicative of a possible role in O_2 regulation. A detailed study of the location and possible role of these mitochondria (Silvester *et al.,* 1999) showed that the dense blanket of mitochondria at the base of the vesicle area was absolutely defined by the reduced-CTC volume of the symbiotic vesicles. This mitochondrial blanket also forms a discontinuous blanket around the intercellular spaces. This mitochondrial blanket was assigned a major role in mopping up O_2 (Silvester *et al.,* 1999), but this proposal has been questioned (Tjepkema *et al.*, 1999) because the respiration rate required for these mitochondria to protect the symbiotic vesicles is many times greater than the actual observed respiration rate of the nodules. Despite this uncertainty of any role in the O_2-protection catena, their very specific location, concentrated as a blanket, is indicative of a significant role for these mitochondria in nodule physiology.

O_2 protection and control in the *Coriaria/Datisca* nodule is enigmatic and unresolved. The evidence suggests that the two nodule types behave differently. In *Datisca*, it is proposed that there is no variable diffusion barrier and the very small response to external pO_2 is explained by the presence of a series of compartments with varying degrees of O_2 protection. These compartments are sequentially activated and deactivated during changes in pO_2, thus keeping the mean activity relatively unchanged. The major resistance to O_2 is proposed to be the tightly packed mass of symbiotic vesicles (Tjepkema *et al.,* 1999).

In *Coriaria*, although the evidence is interpreted as a variable O_2 barrier being present, it is also possible that the lag-phase results represent various compartments along a continuum of O_2 accessibility in the same manner as *Datisca*. However, additional O_2 protection is provided in the infected cell. Between the gas-filled intercellular space and the frankiae, there are (in order) a plant-cell wall, cell membrane, vegetative hyphae, mitochondria, and the vesicle wall. Each of these components likely contributes to O_2 reduction by either its permeability or uptake capacity. These resistances in series, a catena, act in concert to provide the reducing environment in which nitrogenase can operate (Silvester *et al.,* 1999).

4. HEMOGLOBINS IN ACTINORHIZAL NODULES AND *FRANKIA*

4.1. Hemoglobins in Legume Nodules

Legume nodules contain a hemoglobin that is produced by the host plant. This hemoglobin, termed leghemoglobin, is localized in the infected (*Rhizobium*-containing) cells that are the site of nitrogen fixation. These cells occur at the center of the nodule and are surrounded by a layer of tightly packed cortical cells that forms a diffusion barrier (Tjepkema and Yocum, 1973). Leghemoglobin is thought to function by facilitating the transport of O_2 within the infected cells, such that the O_2 concentration is sufficient in the interior of the cell to support respiration while also being low enough at the cell periphery to avoid damage to nitrogenase (Appleby, 1992).

4.2. Hemoglobins in Actinorhizal Nodules

4.2.1. Presence and Amounts of Hemoglobin

There has long been evidence for the presence of hemoglobin in actinorhizal nodules (Davenport, 1960) and now it has been established that actinorhizal nodules contain one or more hemoglobins (Christensen *et al.,* 1991; Fleming *et al.,* 1987; Tjepkema and Asa, 1987). The amounts of hemoglobin present vary widely among actinorhizal genera (Silvester *et al.,* 1990) with the highest concentrations being found in the nodules of *Casuarina* and *Myrica*, both of which contain hemoglobin concentrations approaching those found in legume nodules (Fleming *et al.,* 1987; Tjepkema and Asa, 1987). In contrast, *Alnus* nodules contain only *ca.* 25% of the hemoglobin in *Casuarina* and *Myrica* and other genera, including *Ceanothus*, *Elaeagnus*, *Coriaria*, and *Datisca*, contain even less (Silvester *et al.,* 1990).

4.2.2. Plant-produced Hemoglobins

Casuarina glauca nodules contain a soluble hemoglobin that has a typical optical absorption spectrum. Moreover, the spectrum shows the expected shifts in the optical absorption peaks between the ferrous (Hb), carboxyferrous (HbCO), and oxyferrous (HbO_2) forms of hemoglobin (Fleming *et al.,* 1987). This hemoglobin is a monomer with a molecular mass estimated by gel filtration of 17.5kDa. The kinetics of its reactions with O_2 and CO are similar to those of soybean leghemoglobin and *Parasponia* hemoglobin. The O_2-association rate is very rapid ($41\mu M^{-1}s^{-1}$) and the dissociation rate is moderate ($5.5s^{-1}$) (Fleming *et al.,* 1987). This hemoglobin has three components, termed Hb I, Hb II, and Hb III, identified by isoelectric focusing. Of these, Hb I is the most abundant (Fleming *et al.*, 1987) and has a 44% amino-acid sequence identity with the leghemoglobin of soybean nodules (Kortt *et al.*, 1988).

C. glauca has two types of hemoglobin genes, symbiotic and non-symbiotic. Three closely related symbiotic genes have been identified and designated as *symA*, *symB*, and *symC* (Jacobsen-Lyon *et al.,* 1995). These genes are only expressed in the nodule. The deduced amino-acid sequence of *symA* is almost identical with the protein sequence for Hb I. The sequences for *symB* and *symC* are at least 97% identical with *symA* (Jacobsen-Lyon *et al.,* 1995). A non-symbiotic gene, which encodes Hb II (Christensen *et al.,* 1991), has also been identified. This gene is expressed in roots and shoots and is only barely detectable in nodules (Jacobsen-Lyon *et al.,* 1995). The deduced amino-acid sequence for Hb II is 80% homologous with the hemoglobins of *Trema* (a non-symbiotic genus closely related to *Parasponia*) and *Parasponia*, but only 53% homologous with *Casuarina* Hb I (Christensen *et al.,* 1991). Thus, this gene is very different from the family of symbiotic genes.

Within *C. glauca* nodules, hemoglobin has been localized using immunocytochemistry and is found in the N_2-fixing cells that contain *Frankia* (the infected cells), but not in the uninfected cells. Within infected cells, hemoglobin is found in the host cytoplasm and nucleus, but not in the *Frankia* hyphae (Goodchild and Miller, 1997). Localization of hemoglobin-gene (*hb*) transcripts in *C. glauca* also shows that hemoglobin is restricted to the infected cells (Gherbi *et al.,* 1997). Developmentally, *hb* transcripts are first detected in the youngest infected cells near the nodule lobe apex and reach maximum concentration in the mature infected cells just below this region. Comparable to the developmental sequence observed in legumes, transcripts of *nif*H mRNA, a *Frankia* nitrogenase structural gene, are not detectable in the youngest infected cells but are found in the mature infected cells (Gherbi *et al.,* 1997). Thus, production of hemoglobin precedes production of nitrogenase in the development of *Casuarina* nodules as it does in legume nodules.

Hemoglobin has also been extracted from root nodules of *Casuarina cunninghamiana*, *Casuarina equisetifolia*, and *C. glauca* that were inoculated with four different *Frankia* sources each (Sellstedt *et al.,* 1991). For each plant species, there is a strong linear correlation between concentration of hemoglobin in the nodules and dry mass of the whole plants. There is also a moderate linear correlation with nitrogen content of the whole plant, but no correlation with nitrogen concentration within the nodules (Sellstedt *et al.,* 1991). Varying amounts of hemoglobin also occur in root nodules of these same *Casuarina* species, when inoculated with two different *Frankia* sources. Moreover, the nodules contain lignin-like compounds in the cell walls which may serve as a diffusion barrier (Sellstedt *et al.,* 1991). These results suggest that hemoglobin plays an important role in nitrogen fixation by these plants.

Hemoglobin from *Myrica gale* nodules has been less extensively studied. *M. gale* produces a typical plant hemoglobin with optical absorption peaks for Hb, HbCO, and HbO_2 similar to those of hemoglobins from *Casuarina*, *Parasponia*, and various legume nodules. This hemoglobin has a molecular mass estimated by gel filtration of 38.5kDa and is probably a dimer (Pathirana and Tjepkema, 1995). In addition, *M. gale* produces a second hemoglobin-like compound with a molecular mass of 16.7kDa that is consistent with a monomeric hemoglobin. The absorption spectrum of the CO form of this compound is also consistent with it being a hemoglobin. This compound may be a separate hemoglobin and not just a monomeric form of the dimer because its absorption maximum is at 419nm in the Soret region instead of 416.2nm characteristic of the dimer (Pathirana and Tjepkema, 1995).

The function of the plant-produced *Casuarina* and *Myrica* hemoglobins is probably facilitated O_2 diffusion similar to that of hemoglobins in legume nodules. Evidence for this function is suggested both by the presence of relatively high concentrations of hemoglobin (Pathirana and Tjepkema, 1995) and by the likely diffusion barrier in the walls of the infected cells that probably restricts O_2 diffusion into the cells (Berg and McDowell, 1988; Zeng, Tjepkema, and Berg, 1989).

4.2.3. Alder Hemoglobin: A Possible Frankia *Hemoglobin*

A hemoglobin has also been purified from the nodules of *Alnus glutinosa* (Suharjo and Tjepkema, 1995). This hemoglobin is a monomer, has typical optical absorption spectra for the Hb, HbCO, and HbO_2 forms, and a molecular mass estimated by gel filtration of 15.3kDa. There is no evidence for a second hemoglobin in these nodules. In nodule homogenates, much of this hemoglobin is associated with *Frankia* vesicle clusters. The relatively low concentrations of hemoglobin (Suharjo and Tjepkema, 1995; Tjepkema and Asa, 1987) found in these nodules and its association with *Frankia* vesicle clusters suggests that at least part of this hemoglobin may be produced by *Frankia* rather than the host plant. This suggestion is further supported by the recent discovery that several strains of *Frankia,* including a strain isolated from *Alnus,* produce a hemoglobin when grown in culture (Beckwith *et al.,* 2002; Tjepkema *et al.,* 2002).

4.3. Hemoglobins Produced by Frankia *in vitro*

4.3.1. Characteristics of Frankia *Hemoglobins*

Cultures of *Frankia* strain CcI3, originally isolated from root nodules of *C. cunninghamiana*, produce a hemoglobin with optical absorption spectra for Hb, HbCO, and HbO_2 typical of a hemoglobin, and a molecular mass determined by gel filtration of 14.1kDa. This hemoglobin has extremely rapid O_2-binding kinetics with an association rate constant of $206\mu M^{-1}\ s^{-1}$ and an O_2-dissociation rate constant of $56s^{-1}$; the equilibrium O_2-binding constant is 274nM (Tjepkema *et al.,* 2002). A similar hemoglobin is produced by cultures of *Frankia* strain EAN1pec, originally isolated from *Elaeagnus angustifolia.* This hemoglobin has typical optical absorption spectra, a molecular mass of 13.4kDa, and an O_2-dissociation rate constant of $131s^{-1}$ (Beckwith *et al.,* 2002).

4.3.2. Production of Hemoglobin is Widespread among Frankia *Strains*

All five strains of *Frankia* examined produce hemoglobin (Beckwith *et al.,* 2002). These are, in addition to strains CcI3 and EAN1pec (previously discussed), strain ArI3 that was originally isolated from *Alnus rubra*, strain EUN1f that was originally isolated from *Elaeagnus umbellata*, and strain Cc.1.17 originally isolated from *Colletia cruciata*. These five strains represent a wide range of the genetic diversity that was previously identified in the genus *Frankia* and include four genomic species (Fernandez *et al.,* 1989) and three genospecies (Akimov and Dobritsa, 1992). The presence of hemoglobin in all of these genetically diverse strains suggests that hemoglobin may be produced throughout the genus *Frankia.* It is also probable that the hemoglobin, which is produced by *Frankia* in culture, is also produced in the nodules of actinorhizal plants. As discussed earlier, the hemoglobin extracted from *Alnus glutinosa* nodules is likely to be produced at least in part by *Frankia.* Moreover, *M. gale* produces low concentrations of a hemoglobin-like monomer in addition to the plant-produced dimmer, which is found in relatively high concentrations (Pathirana and Tjepkema, 1995). This monomer may be a *Frankia* hemoglobin.

4.3.3. Conditions that Affect Production of Frankia *Hemoglobins*

All five *Frankia* strains (see above) produce hemoglobin in N_2-fixing cultures, *i.e.*, without supplied fixed-N, as well as in cultures supplied with ammonium, *i.e.*, where nitrogen fixation is absent or minimal (Beckwith *et al.,* 2002). Hemoglobin production in 14-day-old cultures and 28-day-old cultures of these five strains varies with strain, culture age, and nitrogen nutrition. In the N_2-fixing cultures, the hemoglobin concentration in four of the five strains increases with age, but remains unchanged in one strain. In contrast, in cultures supplied with ammonium, the hemoglobin concentration decreases with age in two strains and remains unchanged in three strains. The effect of the amount of O_2 present in the culture has only been examined with strain EAN1pec. In this strain, O_2 strongly influences hemoglobin concentration in N_2-fixing cultures, but has little effect in cultures supplied with ammonium (Beckwith *et al.,* 2002).

4.3.4. Relationship of Frankia *Hemoglobin to Other Hemoglobins*

Hemoglobins exist in a wide variety of non-vertebrate organisms, including microbes, plants, and invertebrate animals. Their molecular adaptations and functions have been reviewed by Weber and Vinogradov (2001). Many bacteria produce a flavohemoglobin that has two domains; one domain contains heme that binds O_2 and is homologous to classical hemoglobins, whereas the second domain has reductase activity. These flavohemoglobins have several possible functions, including removal of excess NO (Poole and Hughes, 2000). A single domain hemoglobin that is related to the flavohemoglobins is produced by *Vitreoscilla.* Various lines of evidence suggest that this hemoglobin enhances O_2 uptake under conditions of low O_2 concentration (Ramandeep *et al.,* 2001).

Many bacteria also produce truncated hemoglobins (trHb). These hemoglobins share little amino-acid sequence similarity with classical hemoglobins and may be of very ancient origin (Wittenberg *et al.,* 2002). Besides having a shorter amino-acid sequence than the classical hemoglobins, their tertiary structure has a novel two-over-two α-helical sandwich fold rather than the three-over-three fold of the classical hemoglobins (Pesce *et al.,* 2000). Three major groups, Groups I, II, and III, of the truncated hemoglobins have been proposed and some bacterial species have genes for two or even all three of these groups (Wittenberg *et al.,* 2002). *Mycobacterium bovis* produces two hemoglobins, trHbN and trHbO, and these have different functions; trHbN provides protection from NO (Ouellet *et al.,* 2002), whereas trHbO enhances O_2 uptake under hypoxic conditions (Pathania *et al.,* 2002). At present, generalizations cannot be made about the extent of occurrence of truncated hemoglobins in bacteria and uncertainties about their functions remain.

Nostoc commune, a N_2-fixing cyanobacterium, produces a truncated hemoglobin that is expressed under conditions of low O_2 and fixed-N starvation. Its gene is located between two genes needed for nitrogen fixation, suggesting a function in this process (Potts *et al.,* 1992). However, in a survey of sixteen genera representing all five sections of the cyanobacteria, the gene was found only in five strains of *Nostoc* (Hill *et al.,* 1996). The *Nostoc* hemoglobin may function to scavenge O_2 to support respiration under conditions when O_2 is limiting (Hill *et al.,* 1996), and thus may not play a direct role in nitrogen fixation.

Evidence has accumulated to conclude that *Frankia* hemoglobin is a truncated hemoglobin. This evidence includes both a molecular mass and O_2-binding properties that are similar to those of the truncated hemoglobin from *Nostoc commune*. Moreover, a 202-bp gene sequence from four *Frankia* strains shares about 80% homology on average with the sequence of the truncated hemoglobin of *Streptomyces coelicolor.* Genetic analysis, which included the two strains, EAN1pec and EUN1f, from which hemoglobin was previously extracted (Beckwith *et al.*, 2002), indicates that all four sequences fit into the trHbO subgroup of Group II of the truncated hemoglobins (J. Niemann and L. S. Tisa, unpublished data).

4.3.5. Possible Functions of Frankia *Hemoglobins*

The function of *Frankia* hemoglobin is probably not directly related to nitrogen fixation based on the following two lines of evidence. First, hemoglobin is produced in *Frankia* cultures in the absence of nitrogen fixation (where fixed-N is present) and in concentrations that are roughly comparable to those produced under N_2-fixing conditions. Second, O_2 concentration in the medium does not affect the amount of hemoglobin produced in the absence of nitrogen fixation (Beckwith *et al.,* 2002).

One possible function is facilitated O_2 diffusion over short distances within *Frankia* cells because this hemoglobin has extremely rapid O_2-binding kinetics that would be well suited for this function. However, hemoglobin concentrations within *Frankia* cells may not be high enough for this function unless the hemoglobin is concentrated in a specific region of the cell, such as the area near the plasma membrane (Tjepkema *et al.,* 2002). Another possible function is in protection against excess NO as found in *Mycobacterium bovis* (Ouellet *et al.,* 2002). Clearly, much uncertainty remains about the function(s) of *Frankia* hemoglobin.

5. CONCLUDING DISCUSSION

The *Frankia* vesicle represents one of the most fascinating models of physiological response in the prokaryote world. Its role appears to be solely related to nitrogenase because cultures grown on combined-N fail to produce vesicles. Further, its response to varying pO_2 indicates its adaptive significance to the functioning of *Frankia,* both in the free-living and the symbiotic state. Thus, in the free-living state, *Frankia,* like the cyanobacteria, is able to function at high efficiency in air and by altering the vesicle envelope is able to adapt nitrogenase to a wide variety of ambient pO_2. Whether it actually does this in nature can only be conjecture because these responses have only been observed *in vitro.* The role of *Frankia* hemoglobin remains a mystery, it does not respond as if it were a fundamental part of the *in vitro* *Frankia* O_2-protection mechanism and appears to be a constitutive property.

The legume nodule has always been the model by which other nitrogenase-containing nodules are compared but, because a primary function of the legume nodule is that of both short- and long-term O_2 control and regulation, the comparison of legume nodules with those of actinorhizal and cyanobacterial nodules is tenuous. There is no doubt that no single physiological mechanism operates across the wide spectrum of actinorhizal nodules, where both the vesicle and the nodule contribute to O_2 protection. The variety of mechanisms constitutes very elegant combinations of structure and physiology which, acting in concert, bring about equal energetic efficiency across a wide diversity of structures. These structures, which range from the *Alnus* nodule where the vesicle represents the major O_2 barrier through to *Casuarina* where significant low-pO_2 sites exist within the nodule, represent a continuum of physiology and structure. The role of host hemoglobin, in contrast to that of *Frankia* hemoglobin, may be significant in *Myrica* and *Casuarina.* Both these symbioses possess high levels of hemoglobin and both have infected cells and cell masses with significant wall thickenings that support the notion of cellular O_2 transport by hemoglobin in these nodules.

It is apparent from this review and from work on energetics (Tjepkema and Winship, 1980) that no nodule system, whether legume or actinorhizal, is more efficient than any other. As respiratory O_2 uptake is an essential part of the O_2-protection mechanism in all systems, it appears that all the mechanisms, which have evolved, have done so with equal efficiency or perhaps that equally efficient systems have been selected for.

REFERENCES

Abeysekera, R. M., Newcomb, W., Silvester, W. B., and Torrey, J. G. (1990). A freeze-fracture electron microscopic study of *Frankia* in root nodules of *Alnus incana* grown at three oxygen tensions. *Can. J. Microbiol., 36,* 97-108.

Akimov, V. N., and Drobitsa, S. V. (1992). Grouping of *Frankia* strains on the basis of DNA relatedness. *Syst. Appl. Microbiol., 15,* 372-379.

Appleby, C. A. (1992). The origin and functions of haemoglobin in plants. *Sci. Prog. Oxf., 76,* 365-398.

Baker, D. D., and Huss-Danell, K. (1986). Effects of oxygen and chloramphenicol on *Frankia* nitrogenase activity. *Arch. Microbiol., 144,* 233-236.

Beckwith, J., Tjepkema, J. D., Cashon, R. E., and Schwintzer, C. R. (2002). Hemoglobins in five genetically diverse *Frankia* strains. *Can. J. Microbiol., 48,* 1048-1055.

Benson, D. R., and Silvester, W. B. (1993). Biology of *Frankia* strains, actinomycete symbionts of actinorhizal plants. *Microbiol. Rev., 57,* 293-319.

Berg, R. H. (1983). Preliminary evidence for the involvement of suberization in infection of *Casuarina. Can. J. Bot., 61,* 2910-2918.

Berg, R. H. (1994). Symbiotic vesicle ultrastructure in high pressure-frozen, freeze-substituted actinorhizae. *Protoplasma, 183,* 37-48.

Berg, R. H. (1999). *Frankia* forms infection threads. *Can. J. Bot., 77,* 1327-1333.

Berg, R. H., and McDowell, L. (1987). Endophyte differentiation in *Casuarina* actinorhizae. *Protoplasma, 136,* 104-117.

Berg, R. H., and McDowell, L. (1988). Cytochemistry of the wall of infected *Casuarina* actinorhizae. *Can. J. Bot., 66,* 2038-2047.

Berg, R. H., Langenstein, B., and Silvester, W. B. (1999). Development in the *Datisca-Coriaria* nodule type. *Can. J. Bot., 77,* 1334-1350.

Bergersen, F. J. (1982). *Root nodules of legumes: Structure and function.* Chichester, UK: Res. Stud. Press.

Berry, A. M., Harriott, O. T., Moreau, R. A., Osman, S. F., Benson, D. R., and Jones, A. D. (1993). Hopanoid lipids compose the *Frankia* vesicle envelope, presumptive barrier of oxygen diffusion to nitrogenase. *Proc. Natl. Acad. Sci. U.S.A., 90,* 6091-6094.

Bond, G. (1952). Some features of the root growth in nodulated plants of *Myrica gale* L. *Ann. Bot., 16,* 467-475.

Bond, G. (1957). Isotopic studies of nitrogen fixation in non-legume root nodules. *Ann. Bot., 21,* 513-521.

Bond, G. (1961). The oxygen relation of nitrogen fixation in root nodules. *Zeit. Alleg. Mikrobiol., 1,* 93-99.

Bond, G. (1974). Root nodule symbiosis with actinomycete-like organisms. In A. Quispel (Ed.), *The biology of nitrogen fixation* (pp. 342-378). Amsterdam, The Netherlands: North Holland Publications/American Elsevier.

Bond, G., and MacConell, J. T. (1955). Nitrogen fixation in detached non-legume root nodules. *Nature, 176,* 606.

Bramhill, D. (1997). Bacterial cell division. *Ann. Rev. Cell Develop. Biol., 13,* 395-424.

Callaham, D., Del Tredici, P., and Torrey, J. G. (1978). Isolation and cultivation *in vitro* of the actinomycete causing root nodulation in *Comptonia. Science, 199,* 899-902.

Calvert, H. E., Chaudhary, A. H., and Lalonde, M. (1979). Structure of an unusual root nodule symbiosis in a non-leguminous herbaceous dicotyledon. In J. C. Gordon, C. T. Wheeler, and D. A. Perry (Eds.), *Symbiotic nitrogen fixation in the management of temperate forests* (pp. 474-475). Corvallis, OR: Oregon State University Press.

Christensen, E. S., Dennis, J. W., Peacock, W. J., Landsmann, J., and Marcker, K. A. (1991). Hemoglobin genes in non-legumes: Cloning and characterization of a *Casuarina glauca* hemoglobin gene. *Plant Mol. Biol., 16,* 339-344.

Dalton, H., and Postgate, J. R. (1969). Growth and physiology of *Azotobacter chroococcum* in continuous culture. *J. Gen. Microbiol., 56,* 307-319.

Davenport, H. E. (1960). Haemoglobin in the root nodules of *Casuarina cunninghamiana. Nature (London), 186,* 653-654.

Durand, J.-L., Sheehy, J. E., and Minchin, F. R. (1987). Nitrogenase activity, photosynthesis and nodule water potential in soybean plants experiencing water deprivation. *J. Exp. Bot., 38,* 311-321.

Fernandez, M. P., Meugnier, H., Grimont, P. A. D., and Bardin, R. (1989). Deoxyribonucleic acid relatedness among members of the genus *Frankia. Int. J. Syst. Bacteriol., 39,* 424-429.

Fessenden, R. J., Knowles, R., and Brouzes, R. (1973). Acetylene-ethylene assay studies on excised root nodules of *Myrica asplenifolia.* L. *Soil Sci. Soc. Am. Proc., 37,* 893-898.

Fleming, A. I., Wittenberg, J. B., Wittenberg, B. A., Dudman, W. F., and Appleby, C. A. (1987). The purification, characterization and ligand-binding kinetics of hemoglobins from root nodules of the non-leguminous *Casuarina glauca - Frankia* symbiosis. *Biochim. Biophys. Acta, 911,* 209-220.

Fontaine, M. S., Lancelle, S. A., and Torrey, J. G. (1984). Initiation and ontogeny of vesicles in cultured *Frankia* sp. strain HFPArI3. *J. Bacteriol., 160,* 921-927.

Gallon, J. R. (1981). The oxygen sensitivity of nitrogenase: A problem for biochemists and micro-organisms. *Trends Biochem. Sci., 6,* 19-23.

Gauthier, D., Diem, H. G., and Dommergues, Y. (1981). *In vitro* nitrogen fixation by two actinomycete strains isolated from *Casuarina* nodules. *Appl. Environ. Microbiol., 41,* 306-308.

Gherbi, H., Duhoux, E., Franche, C., Pawlowski, K., Nassar, A., Berry, A. M., *et al.* (1997). Cloning of a full-length symbiotic hemoglobin cDNA and *in situ* localization of the corresponding mRNA in *Casuarina glauca* root nodule. *Physiol. Plant., 99,* 608-616.

Goodchild, D. J., and Miller, C. (1997). Immunogold localisation of hemoglobin in *Casuarina* root nodules. *Protoplasma, 198,* 130-134.

Hafeez, F., Akkermans, A. D. L., and Chaudhary, A. H. (1984a). Observations on the ultrastructure of *Frankia* sp. in root nodules of *Datisca cannabina* L. *Plant Soil, 79,* 383-402.

Hafeez, F., Chaudhary, A. H., and Akkermans, A. D. L. (1984b). Physiological studies on N_2-fixing root nodules of *Datisca canabina* L. and *Alnus nitida* Endl. from Himalaya region in Pakistan. *Plant Soil, 78,* 129-146.

Harris, S., and Silvester, W. B. (1994). Acetylene and argon-induced declines in nitrogenase activity in *Coriaria arborea. Soil Biol. Biochem., 26,* 641-648.

Harris, S. L., and Silvester, W. B. (1992). Oxygen controls the development of *Frankia* vesicles in continuous culture. *New Phytol., 121,* 43-48.

Hartwig, U., Boller, D., and Nosberger, J. (1987). Oxygen supply limits nitrogenase activity of clover nodules after defoliation. *Ann. Bot., 59,* 285-291.

Hill, D. R., Belbin, T. J., Thorsteinsson, M. V., Bassam, D., Brass, S., Ernst, A., *et al.* (1996). GlbN (Cyanoglobin) is a peripheral membrane protein that is restricted to certain *Nostoc* spp. *J. Bacteriol., 178,* 6587-6598.

Hunt, S., King, B. J., Canvin, D. T., and Layzell, D. L. (1987). Steady and non-steady state gas exchange characteristics of soybean nodules in relation to the oxygen diffusion barrier. *Plant Physiol., 84,* 164-172.

Huss-Danell, K. (1997). Actinorhizal symbioses and their N_2 fixation. *New Phytol., 136,* 375-405.

Huss-Danell, K., and Bergman, B. (1990). Nitrogenase in *Frankia* from root nodules of *Alnus incana* (L.) Moench: Immunolocalization of the Fe- and MoFe-proteins during vesicle differentiation. *New Phytol., 116,* 443-455.

Jacobsen-Lyon, K., Jensen, E. O., Jorgensen, J.-E., Marcker, K. A., Peacock, W. J., and Dennis, E. S. (1995). Symbiotic and nonsymbiotic hemoglobin genes of *Casuarina glauca. Plant Cell, 7,* 213-223.

Jensen, B. B., and Cox, R. P. (1983). Effect of oxygen concentration on dark nitrogen fixation and respiration in cyanobacteria. *Arch. Microbiol., 135,* 287-292.

Johnson, G., Tjepkema, J. D., and Schwintzer, C. R. (1997). The acetylene-induced decline in nitrogenase activity in *Elaeagnus umbellata. Plant Soil, 191,* 157-161.

King, B. J., Hunt, S., Weagle, G., Walsh, K.B., Pottier, R. H., Canvin, D. T., *et al.* (1988). Regulation of oxygen concentration in soybean nodules observed by *in situ* spectroscopic measurement of leghemoglobin concentration. *Plant Physiol., 87,* 296-299.

Kortt, A. A., Inglis, A. S., Fleming, A. I., and Appleby, C. A. (1988). Amino acid sequence of hemoglobin I from root nodules of the non-leguminous *Casuarina-Frankia* symbiosis. *FEBS Lett., 231,* 341-346.

Lalonde, M. (1979). Immunological and ultrastructural demonstration of nodulation of the European *Alnus glutinosa* (L.) Gaertn. host plant by an actinomycetal isolate from the North American *Comptonia peregrina* (L.) Coult. root nodule. *Bot. Gaz., 140 (Suppl.),* S35-S43.

Lamont, H. C., Silvester, W. B., and Torrey, J. G. (1988). Nile red fluorescence demonstrates lipid in the envelope of vesicles from N_2-fixing cultures of *Frankia. Can. J. Microbiol., 34,* 656-660.

Lancelle, S. A., Torrey, J. G., Helper, P. K., and Callaham, D. A. (1985). Ultrastructure of freeze-substituted *Frankia* strain HFPCcI3, the actinomycete isolated from root nodules of *Casuarina cunninghamiana. Protoplasma, 127,* 64-72.

Lopez, M. F., Young, P., and Torrey, J. G. (1986). A comparison of carbon source utilization for growth and nitrogenase activity in two *Frankia* isolates. *Can. J. Microbiol., 32,* 353-358.

MacConnell, J. T. (1959). The oxygen factor in the development and function of the root nodules of alder. *Ann. Bot., 23,* 261-268.

Meesters, T. M. (1987). Localization of nitrogenase in vesicles of *Frankia* sp. CcI7 by immunogold labelling on ultrathin sections. *Arch. Microbiol., 146,* 327-331.

Meesters, T. M., van Vliet, W. M., and Akkermans, A. D. L. (1987). Nitrogenase is restricted to the vesicles in *Frankia* strain EAN1pec. *Physiol. Plant., 70,* 267-271.

Minchin, F. R., Sheehy, J. E., and Witty, J. F. (1986). Further errors in the acetylene reduction assay: effects of plant disturbance. *J. Exp. Bot., 37,* 1581-1591.

Minchin, F. R., Witty, J. F., Sheehy, J. E., and Muller, M. (1983). A major error in the acetylene reduction assay: Decrease in nodular nitrogenase activity under assay conditions. *J. Exp. Bot., 34,* 641-649.

Mirza, M. S., Pawlowski, K., Hafeez, F. Y., Chaudhary, A. H., and Akkermans, A. D. L. (1994). Ultrastructure of the endophyte and localization of *nifH* transcripts in root nodules of *Coriaria nepalensis* wall by *in situ* hybridization. *New Phytol., 126,* 131-136.

Murry, M. A., Fontaine, M. S., and Tjepkema, J. D. (1984a). Oxygen protection of nitrogenase in *Frankia* sp. HFPArI3. *Arch. Microbiol., 139,* 162-166.

Murry, M. A., Fontaine, M. S., and Torrey, J. G. (1984b). Growth kinetics and nitrogenase induction in *Frankia* sp. HFPArI3 grown in batch culture. *Plant Soil, 78,* 61-78.

Murry, M. A., Zhang, Z., and Torrey, J. G. (1985). Effect of O_2 on vesicle formation, acetylene reduction, and O_2-uptake kinetics in *Frankia* sp. HFPCcI3 isolated from *Casuarina cunninghamiana. Can. J. Microbiol., 31,* 804-809.

Newcomb, W., and Pankhurst, C. E. (1982). Fine structure of actinorhizal root nodules of *Coriaria arborea* (Coriariaceae). *N. Z. J. Bot., 20,* 93-103.

Newcomb, W., and Wood, S. M. (1987). Morphogenesis and fine structure of *Frankia* (Actinomycetales): the microsymbiont of nitrogen-fixing actinorhizal root nodules. *Int. Rev. Cytol., 109,* 1-88.

Noridge, N. A., and Benson, D. R. (1986). Isolation and nitrogen-fixing activity of *Frankia* sp. strain CpI1 vesicles. *J. Bacteriol., 166,* 301-305.

Ouellet, H., Oellet, Y., Richard, M., Labarre, M., Wittenberg, B., Wittenberg, J., *et al.* (2002). Truncated hemoglobin HbN protects *Mycobacterium bovis* from nitric oxide. *Proc. Natl. Acad. Sci. U.S.A., 99,* 5902-5907.

Parsons, R., Silvester, W. B., Harris, S., Gruijters, W. T. M., and Bullivant, S. (1987). *Frankia* vesicles provide inducible and absolute oxygen protection for nitrogenase. *Plant Physiol., 83,* 728-73.

Pathania, R., Navani, N. K., Rajamohan, G., and Dikshit, K. L. (2002). *Mycobacterium tuberculosis* hemoglobin associates with membranes and stimulates cellular respiration of recombinant *Escherichia coli*. *J. Biol. Chem., 277,* 15293-15302.

Pathirana, S. M., and Tjepkema, J. D. (1995). Purification of hemoglobin from the actinorhizal root nodules of *Myrica gale* L. *Plant Physiol., 107,* 827-831.

Pesce, A., Couture, M., Dewilde, S., Guertin, M., Yamauchi, K., Ascenzi, P., *et al.* (2000). A novel two-over-two-helical sandwich fold is characteristic of the truncated hemoglobin family. *EMBO J., 19,* 2424-2434.

Poole, R. K., and Hughes, M. N. (2000). New functions for the ancient globin family: Bacterial responses to nitric oxide and nitrosative stress. *Mol. Microbiol., 36,* 775-783.

Potts, M., Angeloni, S. V., Ebel, R. E., and Bassam, D. (1992). Myoglobin in a cyanobacterium. *Science, 256,* 1690-1692.

Ramandeep, Hwang, K. W., Raje, M., Kim, K.-J., Stark, B. C., Dikshit, K. L., and Webster, D. A. (2001). *Vitreoscilla* hemoglobin. Intracellular localization and binding to membranes. *J. Biol. Chem., 276,* 24781-24789.

Ritchie, N. J., and Myrold, D. D. (1999). Phylogenetic placement of uncultured *Ceanothus* microsymbionts using 16S rRNA gene sequences. *Can. J. Bot., 77,* 1208-1213.

Robson, R. L., and Postgate, J. R. (1980). Oxygen and hydrogen in biological nitrogen fixation. *Annu. Rev. Microbiol., 34,* 183-207.

Rosendahl, L., and Huss-Danell, K. (1988). Effects of elevated oxygen tensions on acetylene reduction in *Alnus incana-Frankia* symbioses. *Physiol. Plant.*, 74, 89-94.

Schröder, P. (1989). Aeration of the root system in *Alnus glutinosa* L. Gaertn. *Ann. Sci. For., 46,* 310-314.

Schuller, K. A., Minchin, F. R., and Gresshoff, P. M. (1988). Nitrogenase activity and oxygen diffusion in nodules of soybean cultivar Bragg and a supernodulating mutant: Effects of nitrate. *J. Exp. Bot., 204,* 865-878.

Schwintzer, C. R. (1985). Effect of spring flooding on endophyte differentiation, nitrogenase activity, root growth and shoot growth in *Myrica gale*. *Plant Soil, 87,* 109-124.

Schwintzer, C. R., and Tjepkema, J. D. (1994). Factors affecting the acetylene to $^{15}N_2$ conversion ratio in root nodules of *Myrica gale* L. *Plant Physiol., 106,* 1041-1047.

Schwintzer, C. R., and Tjepkema, J. D. (1997). Field nodules of *Alnus incana* ssp. *rugosa* and *Myrica gale* exhibit pronounced acetylene- induced declines in nitrogenase activity. *Can. J. Bot., 75,* 1415-1423.

Sellstedt, A., Reddell, P., Rosbrook, P. A., and Ziehr, A. (1991). The relations of haemoglobin and lignin-like compounds to acetylene reduction in symbiotic *Casuarina*. *J. Exp. Bot., 42,* 1331-1337.

Sellstedt, A., Redell, P., and Rosbrook, P. A. (1991). The occurrence of haemoglobin and hydrogenase in nodules of twelve *Casuarina-Frankia* symbiotic associations. *Physiol. Plant., 82,* 458-464.

Silvester, W. B., Silvester, J. K., and Torrey, J. G. (1988b). Adaptation of nitrogenase to varying oxygen tension and the role of the vesicle in root nodules of *Alnus incana* ssp. *rugosa*. *Can. J. Bot., 66,* 1772-1779.

Silvester, W. B., Whitbeck, J., Silvester, J. K., and Torrey, J. G. (1988a). Growth, nodule morphology and nitrogenase activity of *Myrica gale* grown with roots at various oxygen levels. *Can. J. Bot., 66,* 1762-1771.

Silvester, W., Langenstein, B., and Berg, R. H. (1999). Do mitochondria provide the oxygen diffusion barrier in root nodules of *Coriaria* and *Datisca*? *Can. J. Bot., 77,* 1358-1366.

Silvester, W. B., and Harris, S. L. (1989). Nodule structure and nitrogenase activity of *Coriaria arborea* in response to varying pO_2. *Plant Soil, 118,* 97-109.

Silvester, W. B., and Winship, L. J. (1990). Transient responses of nitrogenase to acetylene and oxygen by actinorhizal nodules and cultured *Frankia*. *Plant Physiol., 92,* 480-486.

Silvester, W. B., Harris, S. L., and Tjepkema, J. D. (1990). Oxygen regulation and hemoglobin. In C. R. Schwintzer and J. D. Tjepkema (Eds.), *The biology of* Frankia *and actinorhizal plants* (pp. 157-176). San Diego, CA: Academic Press.

Stewart, W. D. P., Fitzgerald, G. P., and Burris, R. H. (1967). *In situ* studies on N_2 fixation using the acetylene reduction technique. *Proc. Natl. Acad. Sci., 68,* 2071-2078.

Suharjo, U. K. J., and Tjepkema, J. D. (1995). Occurrence of hemoglobin in the nitrogen-fixing root nodules of *Alnus glutinosa* L. *Physiol. Plant., 95,* 247-252.

Tisa, L. S., and Ensign. J. C. (1987). Isolation and nitrogenase activity of vesicles from *Frankia* sp. strain EAN1pec. *J. Bacteriol., 169,* 5054-5059.

Tjepkema, J. D. (1978). The role of oxygen diffusion from the shoots and the nodule roots in nitrogen fixation by root nodules of *Myrica gale. Can. J. Bot., 56,* 1365-1371.

Tjepkema, J. D. (1979). Oxygen relations in leguminous and actinorhizal nodules. In J. C. Gordon, C. T. Wheeler, and D. A. Perry (Eds.), *Symbiotic nitrogen fixation in the management of temperate forests* (pp. 175-186). Corvallis, OR: Oregon State University Press.

Tjepkema, J. D. (1983a). Oxygen concentration within the nitrogen fixing root nodules of *Myrica gale* L. *Am. J. Bot., 70,* 59-63.

Tjepkema, J. D. (1983b). Hemoglobins in the nitrogen-fixing root nodules of actinorhizal plants. *Can. J. Bot., 61,* 2924-2929.

Tjepkema, J. D. (1997). C_2H_2 and Ar induce rapid declines in nitrogenase activity and CO_2 evolution in nodules of *Datisca glomerata. Physiol. Plant., 99,* 648-652.

Tjepkema, J. D., and Asa, D. J. (1987). Total and CO-reactive heme content of actinorhizal nodules and the roots of some non-nodulated plants. *Plant Soil, 100,* 225-236.

Tjepkema, J. D., and Murry, M. A. (1989). Respiration and nitrogenase activity in nodules of *Casuarina cunninghamiana* and culture of *Frankia* HFP020203: Effects of temperature and partial pressure of O_2. *Plant Soil, 118,* 111-118.

Tjepkema, J. D., and Schwintzer, C. R. (1992). Factors affecting the acetylene-induced decline during nitrogenase assays in root nodules of *Myrica gale* L. *Plant Physiol., 98,* 1451-1459.

Tjepkema, J. D., and Winship, L. J. (1980). Energy requirement for nitrogen fixation in actinorhizal and legume root nodules. *Science, 209,* 279-281.

Tjepkema, J. D., and Yocum, C. S. (1973). Respiration and oxygen transport in soybean nodules. *Planta, 115,* 59-72.

Tjepkema, J. D., and Yocum, C. S. (1974). Measurement of oxygen partial pressure within soybean nodules by oxygen microelectrodes. *Planta, 119,* 351-360.

Tjepkema, J. D., Cashon, R. E., Beckwith, J., and Schwintzer, C. R. (2002). Hemoglobin in *Frankia*, a nitrogen-fixing actinomycete. *Appl. Environ. Microbiol., 68,* 2629-2631.

Tjepkema, J. D., Du, G., and Schwintzer, C. R. (1999). Response of respiration and nitrogenase activity in *Datisca glomerata* (Presl.) Baill. to changes in pO_2. *Can. J. Bot., 77,* 1367-1372.

Tjepkema, J. D., Ormerod, W., and Torrey J. G. (1980). Vesicle formation and acetylene reduction in *Frankia* sp. CpI1 cultured in defined nutrient media. *Nature, 287,* 633-635.

Tjepkema, J. D., Ormerod, W., and Torrey, J. G. (1981). Factors affecting vesicle formation and acetylene reduction (nitrogenase activity) in *Frankia* sp. CpI1. *Can. J. Microbiol., 27,* 815-823.

Tjepkema, J. D., Pathirana, M. S., and Zeng, S. (1988a). The gas diffusion pathway and hemoglobin content in actinorhizal nodules. In H. Bothe, F.J. de Bruijn, and W.E. Newton (Eds.), *Nitrogen fixation: Hundred years after* (pp. 701). Stuttgart, Germany: Gustaf Fischer.

Tjepkema, J. D., Schwintzer, C. R., and Monz, C. A. (1988b). Time course of acetylene reduction in nodules of five actinorhizal plants. *Plant Physiol., 86,* 581-583.

Tjpekema, J. D., and Evans, H. J. (1975). Nitrogen fixation by free-living *Rhizobium* in a defined liquid medium. *Biochem. Biophys. Res. Commun., 65,* 625-628.

Torrey, J. G. (1985). The site of nitrogenase in *Frankia* in free-living culture and in symbiosis. In H. J. Evans, P. J. Bottomley, and W. E. Newton (Eds.), *Nitrogen fixation research progress* (pp. 293-299). Dordrecht, The Netherlands: Kluwer Academic Publishers.

Torrey, J. G., and Callaham, D. (1982). Structural features of the vesicle of *Frankia* sp. CpI1 in culture. *Can. J. Microbiol., 28,* 749-757.

Waughman, G. J. (1972). The effect of varying oxygen tension, temperature and sample size on acetylene reduction by nodules of *Alnus* and *Hippophae. Plant Soil, 37,* 521-528.

Waughman, G. J. (1977). The effect of temperature on nitrogenase activity. *J. Exp. Bot., 28,* 949-960.

Weber, R. E., and Vinogradov, S.N. (2001). Nonvertebrate hemoglobins: Functions and adaptations. *Physiol. Rev., 81,* 569-628.

Wheeler, C. T., Gordon, E. M., and Ching, T. M. (1979). Oxygen relations of the root nodules of *Alnus rubra* Bong. *New Phytol., 82,* 449-457.

Winship, L. J., and Silvester, W. B. (1989). Modelling gas exchange by actionrhizal root nodules using network simulation analysis. In J. G. Torrey and L. J. Winship (Eds.), *Applications of continuous and steady-state methods to root biology* (pp. 121-146). Dordrecht, The Netherlands: Kluwer Academic Publishers.

Winship, L. J., and Tjepkema, J. D. (1983). The role of diffusion in oxygen protection of nitrogenase in nodules of *Alnus rubra*. *Can. J. Bot., 61*, 2930-2936.

Winship, L. J., and Tjepkema., J. D. (1985). Nitrogen fixation and respiration by root nodules of *Alnus rubra* Bong.: Effects of temperature and oxygen concentration. *Plant Soil, 87,* 91-107.

Wittenberg, J. B., Bolognesi, M., Wittenberg, B. A., and Guertin, M. (2002). Truncated hemoglobins: A new family of hemoglobin widely distributed in bacteria, unicellular eukaryotes, and plants. *J. Biol. Chem., 277,* 871-874.

Witty, J. F., Minchin, F. R., Sheehy, J. E., and Mingues, M. I. (1984). Acetylene-induced changes in the oxygen diffusion resistance and nitrogenase activity of legume root nodules. *Ann. Bot., 53,* 13-20.

Zeng, S., and Tjepkema, J. D. (1995). The resistance of the diffusion barrier in nodules of *Myrica gale* L. Changes in response to temperature but not to partial presence of O_2. *Plant Physiol., 107,* 1269-1275.

Zeng, S., and Tjepkema, J. D. (1994). The wall of the infected cell may be the major diffusion barrier in nodules of *Myrica gale* L. *Soil Biol. Biochem., 26,* 633-639.

Zeng, S., Tjepkema, J. D., and Berg, R. H. (1989). Gas diffusion pathway in nodules of *Casuarina cunninghamiana. Plant Soil, 118*, 119-123.

Chapter 6

EARLY INTERACTIONS, INFECTION AND NODULATION IN ACTINORHIZAL SYMBIOSIS

L. G. WALL[1] AND A. M. BERRY[2]

[1]*Universidad Nacional de Quilmes, Programa de Investigación en Interacciones Biológicas, R. Sáenz Peña 180, B1876BXD Bernal, Argentina*

[2]*University of California, Department of Environmental Horticulture, One Shields Avenue, Davis CA 95616, U.S.A.*

1. INTRODUCTION

Actinorhizal symbioses are N_2-fixing root nodules that are formed as an association between the actinomycete *Frankia* and a diverse group of host plants in eight angiosperm families. The interaction starts in the host-plant rhizosphere, where *Frankia* filaments colonize root surfaces and trigger changes in the host cells that, in turn, lead to successful *Frankia* entry and internal colonization of the developing root-nodule tissue. The common phylogenetic origin of both the actinorhizal plant taxa and the plants nodulated by rhizobia in a single Rosid clade (Soltis *et al.*, 1995) strongly suggests that the nodulation process is conserved in its major features between frankial and rhizobial symbioses. *Parasponia* (Ulmaceae) is the only non-leguminous host genus that is nodulated by rhizobia and, based on *rbcL* sequence analysis (Swensen, 1996), is phylogenetically more closely related to actinorhizal plants than to legumes. Comparative features of actinorhizal *vs.* legume symbioses have been reviewed (Swensen, 1996; Pawlowski and Bisseling, 1996; Pawlowski *et al.*, 2003), with more recent perspectives presented in chapter 10 of this volume.

A common feature of actinorhizal and legume symbioses is that either of two basic patterns of early nodule initiation may be found. The first involves initial entry by the microsymbiont as in *intra*-cellular penetration through a curled root hair, whereas the second pattern involves an initial phase of *inter*-cellular colonization of the root tissue, with neither root-hair involvement nor intracellular penetration until

K. Pawlowski and W. E. Newton (eds.), Nitrogen-fixing Actinorhizal Symbioses, 147–166.
© 2008 *Springer*.

the nodule primordium stage. In the actinorhizal symbioses, three of the eight host families are nodulated *via* the root-hair infection pathway, whereas the other five families (with only indirect evidence for three of these families) initiate nodules *via* intercellular colonization (Swensen and Mullin 1997). In this chapter, we present common features of both infection pathways, with emphasis on recent characterizations of the intercellular-infection pathway.

Specificity of nodulation for *Frankia* strains in relation to particular host taxa was originally demonstrated through cross-inoculation experiments (Torrey, 1990; Wall *et al.,* 2000a). Through phylogenetic comparisons of *Frankia* genomic species, in combination with molecular-genetic typing of *Frankia* strains in field nodules, host-range sets for the different classes of *Frankia* genomic species have been delineated (Benson *et al.,* 1996; Benson and Clawson, 2000; Jeong *et al.,* 1999; Normand *et al.,* 1996). These specificity groupings have been further elucidated by phylogenetic comparisons of host taxa (Benson and Clawson, 2000; Jeong *et al.,* 1999; Soltis *et al.,* 1995; Swensen and Mullin, 1997). Some exceptional phenomena remain unexplained by the current specificity groupings (Benson and Clawson, 2000), for example, some nodules contain *Frankia* strains from outside the compatible specificity grouping (Ramirez-Saad *et al.,* 1998). Moreover, differences are observed among species within a plant-host genus in both nodulation rate and timing in response to inoculation with different *Frankia* strains from a single specificity group (Chaia *et al.,* 2003; see Section 3).

Before root colonization, signaling interactions between the plant and the bacteria necessarily start in the rhizosphere and continue during the infection process and nodule development, so determining compatibility at different levels of interaction. Most probably, before both external surfaces of the organisms come into contact, signal exchanges between partners and their transduction mechanisms cause both genomes to interact through sequential patterns of gene expression and regulation. Such patterns have been extensively characterized in the legume-rhizobia symbioses (Spaink, 2000), but we still know very little about the early interactions between actinorhizal plants and *Frankia*. This chapter summarizes what is known about the different steps in the pathways of plant-*Frankia* interactions that culminate in a functional N_2-fixing root nodule.

2. EARLY INTERACTIONS

What do we understand about *early interactions* in the *Frankia* symbioses? Events that set the stage for the development of the root nodule as a symbiotic structure include molecular signaling between partners, some mechanism of specificity in the recognition process, and additional levels of interaction, both in the rhizosphere and as host-cell responses that facilitate the infection process. Evolution has selected exquisite molecular mechanisms for recognition between partners. In the legume-rhizobial symbioses, the products of the rhizobial nodulation genes (*nod* genes), the so-called Nod factors, are lipochitinoligosaccharides, which trigger a series of host responses that include deformation of root hairs, cell divisions initiated in the root cortex, and secretion of flavonoid compounds that amplify *nod*-gene expression

(Dénarié *et al.,* 1996). Based on the close phylogenetic relationship among the actinorhizal hosts and the rhizobially-nodulated taxa including *Parasponia* (Ulmaceae), the existence of a signaling mechanism that is homologous to legume-rhizobial interactions has long been postulated for the *Frankia* symbioses. However, details of this interaction remain elusive.

Root-hair deformation has been used as the bioassay to detect the putative *Frankia* signals involved in actinorhizal recognition because a molecular-genetic reporter system is lacking in *Frankia.* A root-hair deformation factor, termed Had factor, is produced in pure *Frankia* cultures, in some strains at comparable levels with or without incubation with the host root or root exudates (Cérémonie *et al.,* 1999; van Ghelue *et al.,* 1997). The *Frankia* Had factor seems to be structurally different from the rhizobial Nod factors because it is neither amphiphilic nor sensitive to chitinase. But *N*-acetyl-glucosamine, the subunit of the Nod factor backbone, has been detected in Had-factor fractions by GC-MS, leaving open the possible existence of a Nod factor-related *Frankia* compound. However, neither broad-host-range rhizobial cells nor purified Nod factors induced root-hair deformation in *Alnus glutinosa,* the actinorhizal host used in the bioassay. Moreover, although root-hair deformation induced by Had factor was inhibited by nitrate and ammonium, as occurs in legume symbioses (Heidstra *et al.,* 1994), no cell divisions occurred in the root cortex, suggesting that frankial Had factor is not functionally comparable to rhizobial Nod factors in the nodulation process (Cérémonie *et al.,* 1999).

From the plant side, flavonoid compounds are involved in the induction of microbial genes in several plant-microbe symbioses, including the rhizobial-legume symbioses (Phillips, 2000), as well as in spore germination and hyphae development of abuscular mycorrhizae (Tsai and Phillips, 1991). Flavonoid substances could be involved in the actinorhizal nodulation process, both in stimulating nodulation and in inhibiting it. Different flavonoid-like compounds from *Alnus* seed rinses either enhanced or inhibited nodulation (Benoit and Berry, 1997). Preincubation of *Frankia* in kaempferol, which was identified in *Alnus* root exudates, inhibited nodulation in this host; and a 70-fold increase in *Alnus* root kaempferol concentration was found to correlate with root exposure to light, a condition that inhibits nodulation (Hughes *et al.,* 1999). These two reports suggest that flavonoid compounds in root exudates and/or root tissue participate in positive and negative control of early actinorhizal interactions. However, both the precise identity of flavonoids that are active in nodulation and whether they affect the host or *Frankia,* remain to be determined. Another example of early interaction between actinorhizal symbionts was suggested by the pre-incubation of *Frankia* sp. strain ACN14a-tsr with root exudates of its symbiotic host, *Alnus glutinosa,* which induced modification of protein expression as revealed by 2-D gel electrophoresis. In this case, the N-terminal sequence analysis of the five most prominent proteins that were induced by root exudates showed high similarity with stress proteins (Hammad *et al.*, 2001).

3. RHIZOSPHERE COLONIZATION

Very little is known about root colonization by *Frankia*. Localized aggregations of *Frankia* filaments were seen at certain points along growing roots of *Discaria trinervis*, a few days after inoculation with a dense homogenized *Frankia* suspension and these later become sites of nodule formation (C. Valverde and L. Wall, unpublished data). Similarly, accumulations of *Frankia* and other bacteria are especially pronounced in folded regions of infected roots hairs of *Alnus* (Figure 1B; A.M. Berry, unpublished data). These aggregations may indicate the location of host exudates and microbial responses and may be of several possible types, *e.g.*, they may be chemotactic gradients of small organic molecules, adhesive sites, or foci of nodulation-related signaling compounds secreted by the host.

These accumulations of bacteria in the rhizosphere may, in turn, trigger bacterial density-dependent changes in genome expression through quorum sensing, an emerging paradigm for microbial-cell signaling, that underlies processes ranging from differentiation to population behaviors, such as biofilms (Miller and Bassler, 2001). Bacterial quorum sensing determines rhizosphere colonization in different pathogenic or symbiotic plant-microbe interactions (Lugtenberg *et al.,* 2001). Where quorum-dependent gene expression patterns have been characterized in legume symbioses, the expression of quorum-sensing signals is negatively correlated with nodulation signaling. The *rhi* genes of rhizobia are expressed in the rhizosphere and can, in some taxa, play a role in their interaction with legumes. For example, flavonoid induction of *nod* genes represses the expression of the *rhiABC* operon in *R. leguminosarum* bv. *viciae* (Rodelas *et al.,* 1999). In soybean nodulation, expression of *nod* genes is lowest at high *Bradyrhizobium* cell densities, and this quorum-based repression is mediated by the regulatory gene, *nwsB* (Loh *et al.*, 2002a). Mutants defective in *nwsB* have a competitive advantage in nodulation over wild type (Loh *et al.,* 2002b). Recently, it has been demonstrated that legume plant roots differentially detect rhizobial *vs.* plant pathogenic quorum-sensing signals (homoserinelactone derivatives), resulting in extensively altered root proteome expression (Mathesius *et al.,* 2003). Thus, there may be multiple levels of quorum-related interactions, with both positive and negative impacts on nodulation, during the establishment of root-nodule symbioses.

Nodulation in axenic seedling cultures of *Alnus rubra* was increased when rhizosphere bacteria were co-inoculated with the infective *Frankia* strain (Knowlton *et al.,* 1980). These “helper” bacteria, which included Gram negatives such as *Sinorhizobium meliloti* and *Burkholderia cepacia*, caused substantial host root-hair deformation, even in the absence of *Frankia*, suggesting that their main effect on nodulation was through preconditioning of host cellular processes and was independent of direct interaction with *Frankia* (Berry and Torrey, 1983).

In other cases, rhizosphere bacteria apparently affect *Frankia* directly. For example, in *Discaria trinervis*, the nodulation capacity of a low *Frankia*-inoculum concentration could be enhanced by co-inoculation with either infective or non-infective strains of *Frankia* (and even by co-inoculation with another actinomycete) in a dose-response manner. In this case, co-inoculation with several other typical rhizosphere bacteria, including *Rhizobium*, *Azospirillum*, *Bacillus,* or *Pseudomonas*, did not enhance nodulation, which suggests that a class of signal molecules effective in enhancing nodulation by *Frankia* is perhaps specific to actinomycetes (Gabbarini and Wall, 2002; Gabbarini and Wall, unpublished data). This enhancement of nodulation rate could also be obtained by pre-treating a low-cell-density *Frankia* inoculum with a dialysate of a bacterial suspension at high concentration; either infective or non-infective *Frankia*, or even of *Streptomyces*, was effective, but not killed *Frankia* cells (Figure 1; L. Gabbarini and L. G. Wall, unpublished data). These results open the possibility that a new type of signaling occurs in the actinorhizal symbioses, which may be cell density dependent.

Compatibility groupings in actinorhizal symbioses have been generally studied using *Frankia* isolates and actinorhizal genera in cross-inoculation experiments (Baker, 1987; Torrey, 1990; Wall *et al.*, 2000) and have revealed symbiotic specificity at the actinorhizal-plant clade/*Frankia*-clade level (Benson and Clawson, 2000). In this context, positive nodulation could be explained as recognition either of the same molecules or of the basic structural elements of different signaling molecules. Broad host-range *Frankia* strains may produce a family of different nodulation signals, like the rhizobial strain NGR234 (Pueppke and Broughton,

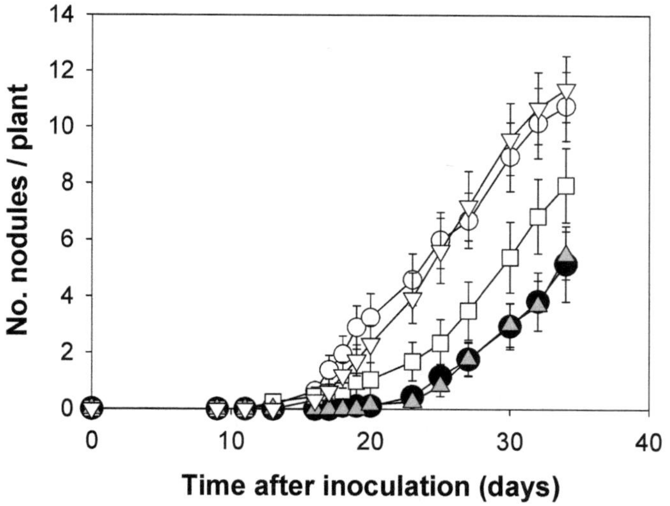

Figure 1. Influence of bacterial factors on nodulation rate.
A dilute inoculum of Frankia *BCU110501 (0.01 μL/mL) was pretreated for 36h at 28°C with dialysates of bacterial suspensions (12kD mw cutoff). Black circles, control (mineral solution); open circles,* Frankia *BCU110501 (1μL/mL); open squares,* Frankia *ArI3 (10 μL/mL); open triangles,* Streptomyces coelicolor *(10 μL/mL); grey triangles, dead (γ-radiated)* Frankia *BCU110501 (10 μL/mL).*
Inoculation treatments were applied to D. trinervis *grown in pouches (20 plants per treatment), and nodulation was recorded at intervals.*

1999). Fine-tuning in both signaling and recognition operates at an additional level within a single cross-inoculation group. For example, different nodulation rates were observed in three different *Discaria* species, *D. trinervis*, *D. chacaye* and *D. articulata*, in a cross-inoculation experiment using a set of *Frankia* strains that had been isolated originally from each of the three host species (Chaia *et al.*, 2003).

Specificity and recognition are not related to a particular infection pathway because the same *Frankia* isolate can infect different actinorhizal hosts by either root-hair infection or inter-cellular penetration (Miller and Baker, 1986). Moreover, physiological complementation occurs across host-compatibility groups. In a co-cultivation experiment, involving both *Discaria trinervi/Frankia* BCU110501 and *Alnus acuminate/Frankia* ArI3, it was found that, when both pairs were grown together, the nodulation rate of each was stimulated, whereas neither the bacteria (at low inoculum concentration) nor the plant alone affected nodulation in the reciprocal pair (L. Gabbarini and L.G. Wall, unpublished data). These results indicate the existence of signaling interactions that function broadly in both inter-cellular and intra-cellular pathways of nodulation.

Actinorhizal nodulation signals can even be generated by non-*Frankia* organisms. The fungus *Penicillium nodositatum* infects roots of *Alnus* species through root-hair penetration, induces nodule primordia, and colonizes these primordia intra-cellularly. These myconodules do not fix N_2 and the primordia do not expand beyond the single-lobe stage (Sequerra *et al.*, 1993).

Physical attachment of the microsymbiont to host root surfaces is another aspect of rhizosphere interactions. *Frankia* filaments are embedded in a mucilage layer of plant origin at either the root hair or epidermal surface of actinorhizal roots (Berry *et al.*, 1983; Prin and Rougier, 1986; Valverde and Wall, 1999a). Symbiosis-specific components of this matrix, which might contribute to attachment or recognition, have not been definitively identified. Adhesive molecules with an electron-dense patterning, which is suggestive of glycoproteins, bind bacteria at *Alnus* root-hair surfaces, but not *Frankia* specifically (Berry, 1984). *Frankia* cells can discriminate *in vitro* between different sugar-specific lectins (Chaboud and Lalonde, 1982), suggesting that *Frankia* would be able to use sugar-lectin specific binding in recognition phenomena. Nonetheless, there is no information on the role of lectins in *Frankia* infection or attachment to the root surface.

Several studies have shown that the presence of *Frankia* in a particular soil does not necessary correlate with the presence or historical occurrence of actinorhizal plant species (Huss-Danell, 1997). *Frankia* has been described as an important component of the rhizosphere community in non-actinorhizal plants (Maunuksela *et al.*, 1999; Smolander *et al.*, 1987). Rhizosphere interactions are complex. There is a growing body of experimental evidence about different mechanisms of suppression and enhancement that influence plant growth and development (Glick, 1999). Experiments with wheat and canola have shows that these crops can influence *Frankia* activity in soils (Cusato *et al.*, 2000), which raises the question of whether *Frankia* can improve crop production in certain cases, if appropriate rhizosphere conditions are established.

4. ROOT INFECTION BY *FRANKIA*

4.1. Intracellular Penetration Through Root Hairs

Actinorhizal host species, which belong to seven genera distributed among the families Myricaceae (2), Betulaceae (1), and Casuarinaceae (4), are: (i) phylogenetically closely-related in the same subclade on the basis of *rbcL* gene sequence analysis (Soltis *et al.,* 1995); (ii) share some characteristics of nodule anatomy (Swensen and Mullin, 1997); and (iii) appear among the oldest fossil records for actinorhizal plant species (Benson and Clawson, 2000). All of these genera are infected by *Frankia* initially through intracellular penetration of deformed root hairs (Berry and Sunell, 1990). Root-hair deformation is the first visible expression of the interaction between *Frankia* and actinorhizal plants infected intracellularly (Figure 2A). It is similar to the root-hair curling that accompanies infection in legume-rhizobial symbioses. However, in the actinorhizal symbioses, all of the root hairs elongating on the primary root axis after inoculation with *Frankia* undergo some degree of deformation (Berry and Torrey, 1983), whereas in legumes, only infected root hairs exhibit the curling response.

The infected root hair exhibits extensive root-hair deformation (Callaham *et al.,* 1979; Figure 2B). In a deeply-folded region of this root hair, *Frankia* transits from the outer root-hair surface across the host-cell primary wall, accompanied by synthesis of host plasma membrane and deposition of new primary wall material around the invading *Frankia* filament (Berry *et al.,* 1986). Some alteration of the preexisting cell wall near the site of infection must occur, whether by direct enzymatic degradation, by other mechanisms of wall-loosening that accompany primary wall synthesis, or by modification of the cell-wall composition prior to infection. The host cell wall that encapsulates *Frankia* forms a structure within the root hair, which is analogous to the infection thread of legume-rhizobial symbioses (Berg, 1999a), except that the actinorhizal encapsulation lacks a thread matrix.

Frankia penetration of the root hair triggers cell divisions in the root cortex subjacent to the infected root hair, in a manner somewhat similar to the initial stages of legume-nodule formation. Some of these newly-divided cells expand and subsequently become infected by *Frankia,* forming a zone called the prenodule (Figure 2C). In the mature prenodule, *Frankia* vesicles differentiate and *nif* genes are expressed (Laplaze *et al.,* 2000a). The prenodule, which occurs in the three actinorhizal families nodulated *via* root-hair penetration, would be the closest analog to the early cortical cell-division centers in legumes.

At the same time, a nodule-lobe primordium is initiated in the root pericycle near the prenodule (Figure 2C). The number of nodule primordia initiated per prenodule can vary from one to several (Callaham and Torrey, 1977). As the nodule lobe primordium expands, the nodule cortex becomes infected intracellularly by *Frankia* filaments transiting through the prenodule into the base of the primordium. Finally, the mature nodule develops from the apical meristem, as *Frankia* vesicles differentiate within the cortical cells, and nitrogen fixation is expressed (Figure 2D).

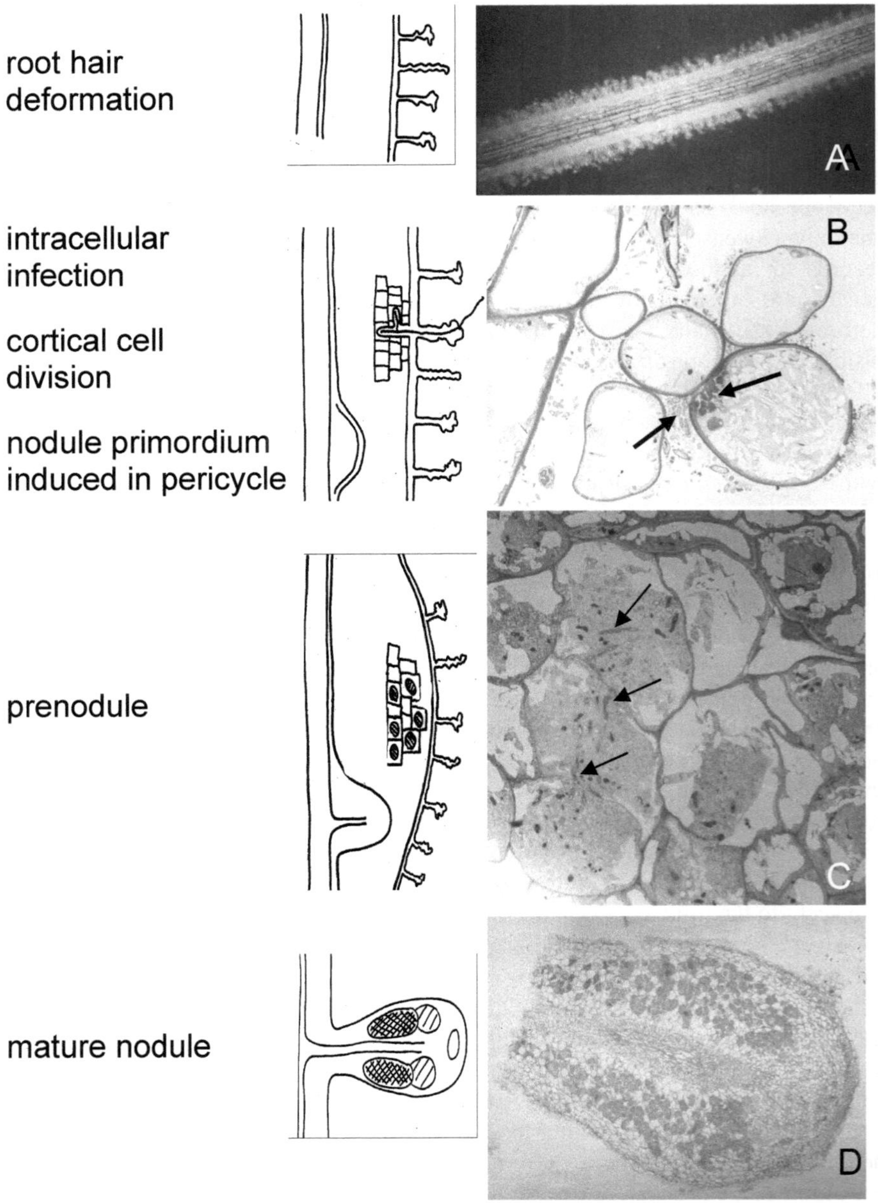

Figure 2. Stages of infection and nodulation in the intracellular (root hair) infection pathway (A,D) Alnus incana*; (B, C)* Alnus rubra *(L. A. Sunell and A. M. Berry, unpublished data). See text for arrow indications.*

The observation that multiple root hairs can be invaded by *Frankia*, but that extra root-hair infections can be aborted very early, accompanied by localized deposition of callose and hypersensitive responses (Berry *et al.,* 1986; Berry and McCully, 1990), suggests that one level of nodulation control operates at the transition from root-hair infection to prenodule expansion. Moreover, it is clear (Gentili, 2003; Gentili and Huss-Danell, 2002; 2003) that not all cortical centers of cell division give rise to prenodules and that not all prenodules are associated with the formation of true nodule lobes. These results point to the existence of separate levels of regulation of nodule development.

4.2. Intercellular Colonization

In seventeen genera, which are distributed in the families Elaeagnaceae (3), Rhamnaceae (7), Rosaseae (5), Datiscaceae (1) and Coriariaceae (1), represent over 70% of the total actinorhizal species, and are phylogenetically grouped in three different subclades on the basis of *rbcL* gene sequence analysis (Soltis *et al.,* 1995), early stages of nodulation take place (only indirect evidence for the families Rosaseae, Datiscaceae and Coriariaceae) *via* intercellular colonization of the young host root rather than through root-hair infection (Swensen and Mullin, 1997). In this infection pathway, *Frankia* filaments invade the root cortex by growing between adjacent epidermal cells and cortical cells, through the middle lamella (Figure 3B; Miller and Baker, 1985; Liu and Berry, 1991a; Valverde and Wall, 1999a). The host cells secrete pectin-rich wall material that also stains intensely with Coomassie blue, creating an expanded intercellular zone (Figure 3B; Liu and Berry, 1991a; Valverde and Wall, 1999a). Intracellular penetration by *Frankia* and infection-thread formation is initiated acropetally in developing cortical cells of the nodule-lobe primordium, following a pattern similar to that described for root-hair invaded plant species (Berg, 1999b; Berry and Sunell, 1990; Valverde and Wall, 1999a; also see above).

The time course of initiation and development of root nodules *via* the intercellular initiation pathway was investigated in *Discaria trinervis*, which is a South American actinorhizal shrub in the Rhamnaceae (Valverde and Wall, 1999a). Inoculated seedlings were examined at intervals under the light microscope after clearing with aqueous NaClO. In parallel, both semi-thin and ultra-thin sections were analyzed by light and electron microscopy. Root hairs were not deformed during the early events of nodule formation (Figure 3A). Nodule primordia were detected from 6d after inoculation (Figure 3C), while bacteria progressed through intercellular zones of the outer layers of cortical cells (Figure 3B). Seven-to-nine days after inoculation, intracellular invasion of host cells by the symbiont occurred in the nodule lobe primordium; hypertrophy of the primordium cells was associated with *Frankia* penetration (Figures 3D-E). By 16d after inoculation, septate *Frankia* vesicles had differentiated within the host cells in the mature nodule (Figure 3F). Understanding the timing of events involved in the intercellular infection pathway provides information crucial to a better understanding of the regulation of nodulation in actinorhizal plants.

Once *Frankia* hyphae are within the host cells, the filaments are retained within a layer of host primary-wall material during all stages of nodulation and vesicle maturation, whether the early infection pathway is intracellular or intercellular (Berg, 1999a ; Berry and Sunell, 1990). The host intercellular wall matrix is rich in

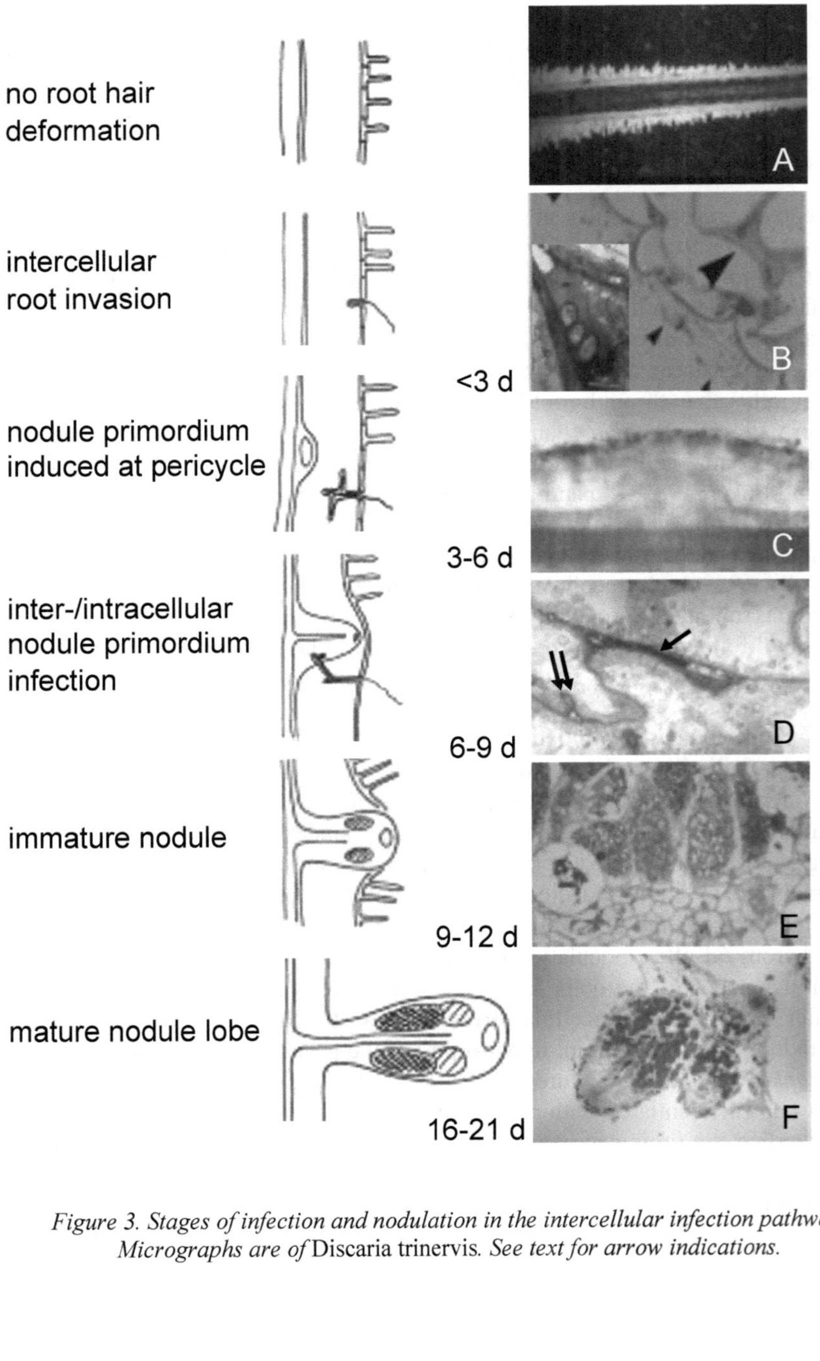

Figure 3. Stages of infection and nodulation in the intercellular infection pathway. Micrographs are of Discaria trinervis. *See text for arrow indications.*

pectins, with a higher proportion of methyl-esterified pectins in the zone of *Frankia* colonization, which could result in a low level of calcium crosslinking, less adhesiveness toward *Frankia*, or other factors of importance at the symbiotic interface (Liu and Berry, 1991b). High levels of calcium (>0.1mM), which can cross-link pectins, were found to inhibit intercellular infection by *Frankia* in *D. trinerva* (Valverde, 2000). At lower and moderate calcium concentrations, nodulation was stimulated, but only as a function of enhanced overall plant growth.

In addition to pectins, a host-derived arabinogalactan protein (AGP) epitope was observed specifically at the symbiotic interface during *Frankia* proliferation in *Alnus* nodules (Berry *et al.,* 2002), another indication of modifications in the composition of the host extracellular matrix that accompany and probably facilitate the intracellular *Frankia* stage. The role of the AGPs may be to stabilize the pectin-rich wall material, to bind the *Frankia* hyphae, and/or to serve as signaling molecules in the infection process (Berry *et al.*, 2002). *ENOD5* genes, which encode AGP homologs, are most highly expressed in legume-nodule cells at the infection-thread stage, when pectic wall material is deposited (Scheres *et al.,* 1990).

Boron is another element known to be involved in plant cell-wall stabilization and to be necessary for normal infection and nodule development in legume-rhizobial symbiosis (Bolaños *et al.,* 1994; Redondo Nieto *et al.,* 2001). Boron is essential in cyanobacteria for normal development of heterocysts, the specialized cells where nitrogen fixation take place (Bonilla *et al.*, 1990). Cyanobacterial heterocysts are equivalent to vesicles in *Frankia.* A study of boron deficiency in the establishment of the *Discaria* symbiosis suggested that boron is necessary for both partners and permits normal *Frankia* development, host cell-wall development, root infection, and nitrogen fixation (Bolaños *et al.,* 2002).

5. NODULE DEVELOPMENT

Actinorhizal nodules are perennial structures consisting of multiple nodule lobes that form a coralloid structure (Berry and Sunell, 1990). Although all nodules lobes have an apical meristem, the growth of individual lobes is limited; additional branch lobes are formed as lateral primordia in the pericycle of the preceding nodule lobe. In the *Casuarinaceae* and the *Myricaceae*, the nodule lobe meristem continues to grow as an indeterminate structure called the nodule root. Nodule roots are negatively geotropic and appear involved in gas diffusion to the nodule tissue (Tjepkema, 1978; see chapter 5 this volume).

The ontogeny of nodule lobes is characteristic of both actinorhizal symbioses and *Parasponia* and marks a clear difference from legume nodules (Hirsch and LaRue, 1997). Actinorhizal nodule primordia originate as lateral roots in the pericycle (Berry and Sunell, 1990), with a central vascular cylinder, cortex and periderm. The presence of *Frankia* in the cortex is associated with cortical hypertrophy, which results in a modified lateral root (Berry and Sunell, 1990). Multiple nodule lobes can arise at the site of a single *Frankia* infection, indicating that *Frankia* induces the formation of additional lateral roots (Angulo Carmona, 1974; Callaham and Torrey, 1977). Multiple nodule primordia may also be induced at sites of *Frankia* colonization in the intercellular infection pathway (Figure 3F). *Frankia* secretes at least two auxins (Berry *et al.,* 1989; Hammad *et al.,* 2003), a class of plant growth regulator known to induce lateral-root primordium formation.

Many genes are specifically expressed or enhanced in actinorhizal nodules compared with expression in roots, as occurs for legume nodules (Pawlowski, 1997; see chapter 9 this volume). At least one of those genes, *dg93*, is a homolog of a legume early nodulin (Okubara *et al.,* 2000). This result suggests that, despite the origin of actinorhizal nodule lobes (and *Parasponia* nodules) as lateral roots, which contrasts with the cortical origin of legume nodules, the commonalities of the nodulation process doubtless outweigh the differences.

6. REGULATION OF NODULATION

Plant symbioses are compatible interactions where the plant obtains benefits for its growth and development through interaction with the microorganism, which generally facilitates the acquisition of a particular nutrient (Werner, 1992). One important level of plant control in the actinorhizal symbioses may be by regulation of the proportion of symbiotic tissue in the plant in relation to plant biomass allocation. This balance can be achieved by controlling either the number of new infections or the development of existing infections/nodules (Wall, 2000). Different control points for the regulation of nodulation have been suggested (see Figure 4).

Two levels of feedback regulation of nodulation have been described for legumes (Caetano-Anollés and Gresshoff, 1991) and actinorhizal plants (Wall, 2000). One level is a regulatory event that occurs early during nodule formation and controls the number of infections in the growing root and also suppresses early nodule development (Valverde and Wall, 1999b; Wall and Huss-Danell, 1997). This process operates very early in the plant-microbe interaction, after intracellular penetration but definitely before nitrogen fixation takes place in mature nodules, and is called *autoregulation* of nodulation. In legumes, the molecular basis for autoregulation appears to be complex and a fundamental part of root developmental processes (Wopereis *et al.,* 2000). A second level of regulation involves fixed-N regulation of the susceptibility of roots to infection and nodulation (Parsons *et al.,* 1993). Some common mechanisms may exist among autoregulation and fixed-N regulation, but the details of the integrated picture are still lacking in both legumes and actinorhizal plants.

Nitrogen-fixing nodules induced by *Frankia* form as discrete clusters of nodules, both in experimental conditions (Valverde and Wall, 1999b; Wall and Huss-Danell, 1997) and in nature (Chaia and Raffaele, 2000). This clustering of nodules can be explained as an initial burst of nodulation following *Frankia* inoculation. Initiation of new nodules is then regulated subsequently by a feedback inhibitory mechanism that changes the susceptibility of the growing root once a certain degree of nodule induction has occurred. The young region of the growing root, and particularly the elongation region behind the root tip, is the most susceptible for infection and nodulation regardless of whether the infection pathway occurs intracellularly *via* root hairs as in *Alnus glutinosa* (Burgraaf *et al.,* 1983) or *Alnus incana* (Wall and

Huss-Danell, 1997; Wall *et al.*, 2003), or *via* intercellular colonization in *Discaria trinervis* (Valverde and Wall, 1999b; Valverde and Wall, 2002; Wall *et al.*, 2003),

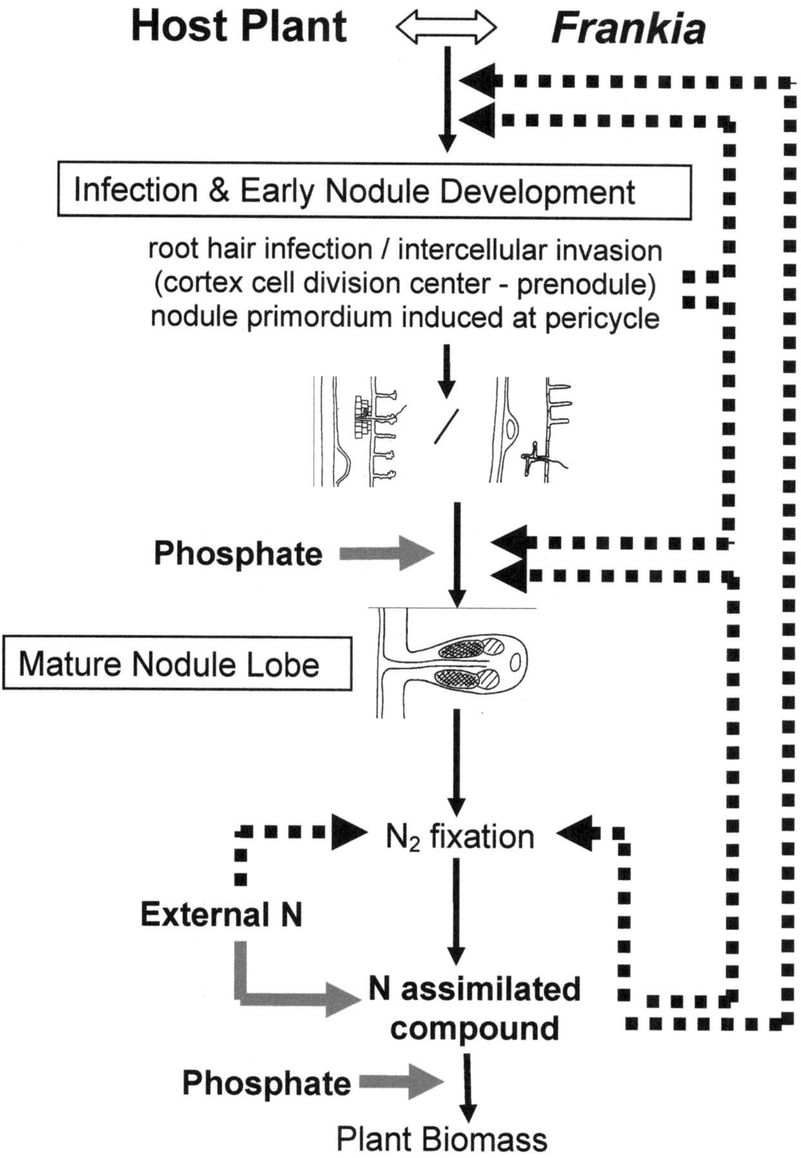

Figure 4. Control points for regulation of infection and nodulation in actinorhizal plants. Model after studies on Alnus incana *nodulation (Wall and Huss-Danell, 1997; Wall* et al., *2000, Wall* et al., *2003), and* Discaria trinervis *nodulation (Valverde and Wall, 1999b; 2002, 2003; Valverde* et al., *2002; Wall* et al., *2003).*
Dashed line with arrow indicates inhibition. Grey full line with arrow indicates stimulation.

or *Hippophae rhamnoides* (Dobritsa and Novik, 1992). This very early feedback inhibition of infection and nodule development in actinorhizal plants is analogous to autoregulation in legume-rhizobial symbioses (Caetano and Gresshoff, 1991; Wopereis *et al.,* 2000). In *D. trinervis*, autoregulation of nodulation starts locally within 3-6d of inoculation, when new infections are suppressed in the more distal part of a nodulating root. By about 9d after inoculation, autoregulatory inhibition becomes systemic, at a stage before the nodule lobe matures and nitrogen fixation commences (Valverde and Wall, 1999b; Valverde *et al.,* 2000).

The loss of susceptibility in the zone of root elongation is likely due to a change in either expression or activity of plant receptors for *Frankia* signaling, although suppression of a host-signaling compound, such as a flavonoid, cannot be ruled out. In legumes, receptor-like kinases mediate the shoot regulation of nodulation (Krusell *et al.,* 2002; Nishimura *et al.,* 2002) by an as-yet unknown mechanism. The major effects of the *har1* autoregulatory gene in *Lotus japonicus* are to limit the extent of cortical-cell division centers and to control lateral-root formation (Wopereis *et al.,* 2000). Comparisons with legume symbioses will provide an interesting area for further investigation of the control of actinorhizal nodulation, particularly because actinorhizal nodules develop from lateral-root primordia, whereas legume nodules form from cortical-cell division centers.

In double-inoculation experiments with actinorhizal plants, the development of additional nodule primordia was arrested at a stage before *Frankia* entered these primordia. When existing nodule lobes were removed, suppressed nodule primordia developed into mature nodules, regardless of whether the intra- or inter-cellular infection pathway was used (Valverde and Wall, 1999b; Wall and Huss-Danell, 1997). It is likely that, in autoregulation of the actinorhizal taxa nodulated by root-hair infection, *Frankia* is present in nascent intracellular infections, at least in some root hairs and perhaps also in small subjacent cortical division centers, as shown during autoregulation in *L. japonicum* (Wopereis *et al.,* 2000). In the host taxa nodulated by intercellular colonization, intracellular infections are confined to the nodule lobe, although *Frankia* filaments continue to be present intercellularly in the root.

Once nitrogen fixation has commenced, fixed-N inhibition controls either nodule number or total nodule biomass in both legumes (MacConnell and Bond, 1957) and actinorhizal plants (Arnone *et al.,* 1994; Kohls and Baker, 1989; Thomas and Berry, 1989). Such inhibition can result from either exogenous fixed-N application or endogenous N_2 fixation (Baker and Parsons, 1997a; 1997b). Detailed studies with split-root systems have demonstrated that fixed-N inhibition is both localized and systemic in actinorhizal symbioses (Gentili and Huss-Danell, 2002; Gentili and Huss-Danell, 2003; Pizelle, 1965; 1966).

Fixed-N inhibition of nodulation depends on the external N/P ratio that is sensed by the plant. Thus, inhibition by high inorganic-N can be counteracted if P levels are high. This P effect has been demonstrated both in root hair-infected plants (Gentili and Huss-Danell, 2003; Wall *et al.,* 2000b; Yang, 1995; Yang *et al.,* 1997) and in intercellularly-infected plants (Gentili and Huss-Danell, 2002; Valverde *et al.,* 2002). As P increases, the ratio of nodule biomass to whole-plant biomass increases. The strategy to achieve this result differs among actinorhizal plants, probably on the basis of the infection pathway. Although nodule number per plant increased in *A. incana* with increasing P (Wall *et al.,* 2000b), in *D. trinervis* the size of individual nodules increased, not the number of nodules (Valverde *et al.,* 2002). The interaction of P and N in nodulation seems to occur at the level of systemic control (Gentili, 2003). The effect of P is directly upon nodulation and can this effect can be distinguished from a nonspecific effect on plant growth (Gentili, 2003; Gentili and Huss-Danell, 2002; Israel, 1993; Reddell *et al.,* 1997; Valverde *et al.,* 2002). Furthermore, the leaf N/P ratio is negatively correlated with the proportion of nodule biomass in *D. trinervis* (Valverde *et al.,* 2002). This relationship seems to hold for many actinorhizal plants and legumes, as calculated by a meta-analysis of values currently in the literature (C. Valverde and L. Wall, unpublished data).

As discussed above, active nodules can serve as a source of inhibition of further nodulation. A clear difference exists in the mechanism for regulating nodulation between root-hair infected plants, such as *Alnus* (Wall and Huss-Danell, 1997) and intercellularly colonized taxa, such as *Discaria* (Valverde and Wall, 1999b). Interestingly, if the fixed-N feedback mechanism, which operates in nodulated plants, temporarily disappears, *i.e.*, by growing nodulated roots under an Ar atmosphere without N_2, the subsequent nodulation pattern parallels that observed for P enrichment in each infection type; thus, new nodules can develop in intracellularly-infected hosts (*Alnus*), but only increases in biomass of already-existing nodules (with no additional nodulation) occurs in intercellularly-colonized plants (*Discaria*) (Wall *et al.,* 2003). In *D. trinervis,* a similar effect results if nitrogen fixation is limited by a 6-day dark treatment (Valverde and Wall, 2003), again fixed-N feedback inhibition of new nodulation is not released. However, if active nodules are completely removed from the roots, then *de novo* infections and new nodule primordia develop both in *A. incana* (Wall and Huss-Danell, 1997) and in *D. trinervis* (Valverde and Wall, 1999b). Thus, the N/P effects on new nodulation seem to depend on some crucial differential role of intercellular or intracellular *Frankia* early in the infection pathway.

Because many plant developmental features are regulated by plant-growth regulators, plant hormones are almost certainly involved in regulation of nodulation. *Frankia* secretes several auxins (Berry *et al.,* 1989), including phenylacetic acid (Hammad *et al.,* 2003), which could influence cell division and nodule development. Auxin accumulates in lateral-root initials and in legume-nodule primordia in the early stages of organ differentiation (Mathesius *et al.,* 2000).

In seedlings of *D. trinervis*, the extent of nodulation in the mature region of the root was increased by inhibitors of ethylene synthesis (AVG) or perception (Ag^+), and reduced by ethylene inducers (CEPA and ACC; C. Valverde and L. Wall, unpublished data). The effect of the ethylene inhibitors is similar to the hypernodulation phenotype of the *sickle* mutant in *Medicago truncatula*, which is ethylene-insensitive (Penmetsa and Cook, 1997). Neither AVG nor Ag^+ affected autoregulation of nodulation in the growing root region, however, suggesting that autoregulation is not ethylene-mediated in *D. trinervis* as for the *har1* mutant of *L. japonicus* (Wopereis *et al.,* 2000). The Har1 protein has recently been characterized as a receptor-like kinase with homology to CLAVATA1, a regulator of meristem function through peptide communication (Nishimura *et al.,* 2002; Searle *et al.,* 2003).

Taken altogether, the recent information on physiological regulation supports a model of homeostatic regulation of actinorhizal symbioses that involves more than one controlling factor or signal (see also Huss-Danell, 1997; Wall, 2000). In the near future, the use of transgenic actinorhizal plants (Franche *et al.,* 1998) and proteomic and genomic comparisons between the root-hair infection and intercellular nodulation pathways will provide further insight into the mechanisms and signals involved in actinorhizal symbioses. Finally, comparisons of actinorhizal nodulation with legume-rhizobia systems will continue to highlight the many common mechanisms between these two N_2-fixing symbioses as well as the unique aspects of actinorhizal nodulation.

ACKNOWLEDGEMENTS

L. G. Wall acknowledges the Universidad Nacional de Quilmes (Argentina), CONICET (Argentina), ANPCyT (Argentina), and STINT (Sweden) for support through grants. This work was also supported by California Agricultural Experiment Station project 6258-H (A. M. Berry).

REFERENCES

Angulo Carmona, A. F. (1974). La formation des nodules fixateurs d'azote chez *Alnus glutinosa* (L.) Vill. *Acta Bot. Neerl., 23,* 257-303.

Arnone, J. A., Kohls, S. J., and Baker, D. D. (1994). Nitrate effects on nodulation and nitrogenase activity of actinorhizal *Casuarina* studied in split root systems. *Soil Biol. Biochem., 26,* 599-606.

Baker, A., Hill, G. F., and Parsons, R. (1997). Alteration of N nutrition in *Myrica gale* induces changes in nodule growth, nodule activity and amino acid composition. *Physiol. Plant., 99*, 632-639.

Baker, A., and Parsons, R. (1997). Evidence for N feedback regulation of N_2 fixation in *Alnus glutinosa* L. *J. Exp. Bot., 48*, 67-73.

Baker, D. D. (1987). Relationships among pure culture strains of *Frankia* based on host specificity. *Physiol. Plant., 70*, 245-248.

Benoit, L. F., and Berry, A. M. (1997). Flavonoid-like compounds from seeds of red alder (*Alnus rubra*) influence host nodulation by *Frankia* (Actinomycetales). *Physiol. Plant., 99*, 588-593.

Benson, D. R., and Clawson, M. L. (2000). Evolution of the actinorhizal plant symbiosis. In E. W. Triplett (Ed.), *Prokaryotic nitrogen fixation: A model system for analysis of a biological process* (pp. 207-224). Wymondham, UK: Horizon Scientific Press.

Benson, D. R., Stephens, D. W., Clawson, M. L., and Silvester, W. B. (1996). Amplification of 16S rRNA genes from *Frankia* strains in root nodules of *Ceanothus griseus*, *Coriaria arborea*, *Coriaria plumosa*, *Discaria toumatou*, and *Purshia tridentata*. *Appl. Environ. Microbiol., 62*, 2904-2909.

Berg, R. H. (1999a). Frankia forms infection threads. *Can. J. Bot., 77*, 1327-1333.

Berg, R. H. (1999b). Cytoplasmic bridge formation in the nodule apex of actinorhizal root nodules. *Can. J. Bot., 77*, 1351-1357.

Berry, A. M. (1984). The actinorhizal infection process: Review of recent research. In M. J. Klug and C. A. Reddy (Eds.), *Current perspectives in microbial ecology* (pp. 222-229). Washington, DC: ASM.

Berry, A. M., and Torrey, J. G. (1983). Root hair deformation in the infection process of *Alnus rubra. Can. J. Bot., 66*, 2863-2876.

Berry, A. M., Torrey J. G., and McCully, M. E. (1983). The fine structure of root hair wall and surface mucilage in the actinorhizal host *Alnus rubra.* In R. Goldberg (Ed.), *Plant molecular biology* (pp. 319-327). New York: Liss.

Berry, A. M., McIntyre, M., and McCully M. E. (1986). Fine structure of root hair infection leading to nodulation in the *Frankia-Alnus* symbiosis. *Can. J. Bot.*, *64*, 292-305.
Berry, A. M., and McCully, M. E. (1990). Callose-containing deposits in relation to root hair infections of *Alnus rubra* Bong. by *Frankia*. *Can. J. Bot.*, *68*, 798-802
Berry, A. M., Rasmussen, U., Bateman, K., Huss-Danell, K., Lindwall, S., and Bergman, B. (2002). Arabinogalactan proteins are expressed at the symbiotic interface in root nodules of *Alnus* spp. *New Phytol.*, *155*, 469-479.
Berry, A. M., Kahn, R. K. S., and Booth, M. C. (1989). Identification of indole compounds secreted by Frankia HFPArI3 in defined culture medium *Plant Soil*, *118*, 205-209.
Berry, A. M., and Sunell, L. A. (1990). The infection process and nodule development. In C. R. Schwintzer and J. D. Tjepkma (Eds.), *The biology of* Frankia *and actinorhizal plants* (pp. 61-81). San Diego, CA: Academic Press.
Bolaños, L., Brewin, N. J., and Bonilla, I. (1996). Effects of boron on *Rhizobium*-legume cell-surface interactions and nodule development. *Plant Physiol.*, *110*, 1249-1256
Bolaños, L., Redondo-Nieto, M., Bonilla, I., and Wall, L. G. (2002). Boron requirement for growth, nitrogen fixation and nodulation of *Frankia* BCU110501. *Physiol. Plant*, *115*, 563-570.
Bonilla, I., García-González, M., and Mateo, P. (1990). Boron requirement in cyanobacteria. Its possible role in the early evolution of photosynthetic organisms. *Plant Physiol.*, *94*, 1554-1560.
Burggraaf, A. J. P., van der Linden, J., and Tak, T. (1983). Studies on the localization of infectible cells on *Alnus glutinosa* roots. *Plant Soil*, *74*, 175-188.
Caetano-Anollés, G., and Gresshoff, P.M. (1991). Plant genetic control of nodulation. *Ann. Rev. Microbiol.*, *45*, 345-382.
Callaham, D., and Torrey, J. G. (1977). Prenodule formation and primary nodule development in roots of *Comptonia* (Myricaceae). *Can. J. Bot.*, *55*, 2306-2318.
Callaham, D., Newcomb, W., Torrey, J. G., and Peterson, R. L. (1979). Root hair infection in actinomycete-induced root nodule initiation in *Casuarina*, *Myrica* and Comptonia. *Bot. Gaz. (Chicago)*, *140 (Suppl.)*, S1-S9.
Cérémonie, H., Debellé, F., and Fernandez, M. P. (1999). Structural and functional comparison of *Frankia* root hair deforming factor and rhizobia Nod factor. *Can. J. Bot.*, *77*, 1293-1301.
Chaboud, A., and Lalonde, M. (1982). Lectin binding on surfaces of *Frankia* strains. *Can. J. Bot.*, *61*, 2889-2897.
Chaia, E., Valverde, C., and Wall, L. G. (2006). Local adaptation of *Frankia* to different *Discaria* (Rhamnaceae) host species growing in Patagonia. *Curr. Microbiol.*, *53*, 523-528.
Chaia, E., and Raffaele, E. (2000). Spatial patterns of root branching and actinorhizal nodulation in *Discaria trinervis* seedlings. *Symbiosis*, *29*, 329-341.
Cusato, M. S., and Tortosa, R. D. (2000). Interactions between *Frankia* and crops. *Phyton*, *68*, 47-53.
Dénarié, J., Debellé, F., and Promé, J. C. (1996). *Rhizobium* lipochitooligosaccharide nodulation factors: Signaling molecules mediating recognition and morphogenesis. *Annu. Rev. Biochem.*, *65*, 503-535.
Dobritsa, S. V., and Novik, S. N. (1992). Feedback regulation of nodule formation in *Hippophaë rhamnoides*. *Plant Soil*, *144*, 45-50.
Franche, C., Laplaze, L., Duhoux, E., and Bogusz, D. (1998). Actinorhizal symbioses: Recent advances in plant molecular and genetic transformation studies. *Crit. Rev. Plant Sci.*, *17*, 1-28.
Gabbarini, L., and Wall, L. G. (2002). Novel interactions in actinorhizal symbiosis. 8th New Phytologist Symposium, *Soil microbes and plant productivity*, 9-14 June 2002, Helsinki, Finland. Book of Abstracts.
Gentili, F. (2003). *Nutrient effects on nodulation and N_2 fixation in actinorhizal symbioses.* Ph.D. thesis. Swedish University of Agricultural Sciences, Agragria 392, Umeå, Sweden.
Gentili, F., and Huss-Danell, K. (2002). Phosphorus modifies the effects of nitrogen on nodulation in split-root systems of *Hippophaë rhamnoides*. *New Phytol*, *153*, 53-61.
Gentili, F., and Huss-Danell, K. (2003). Local and systemic effects of phosphorous and nitrogen on nodulation and nodule function in *Alnus incana*. *J. Exp. Bot.*, *54*, 2757-2767.
Glick, B. R. (1999). *Biochemical and genetic mechanisms used by plant growth promoting bacteria.* New York, NY: Imperial College Press.
Hammad, Y., Maréchal, J., Cournoyer, B., Norman, P., and Domenach, A.-M. (2001). Modification of the protein expression pattern induced in the nitrogen-fixing actinomycete *Frankia* sp. strain ACN14a-tsr by root exudates of its symbiotic host *Alnus glutinosa* and cloning of the *sodF* gene. *Can. J. Microbiol.*, *47*, 541-547.

Hammad, Y., Nalin, R., Maréchal, J., Fiasson, K., Pepin, R., *et al.* (2003). A possible role for phenyl acetic acid (PAA) on *Alnus glutinosa* nodulation by *Frankia*. *Plant Soil, 254,* 193-205.
Heidstra, R., Geurts, R., Franssen, H., Spaink, H. P., van Kammen, A., and Bisseling, T. (1994). Root hair deformation activity of nodulation factors and their fate on *Vicia sativa*. *Plant Physiol., 105,* 787-797.
Hirsch, A. M., and La Rue, T. A. (1997). Is the legume nodule a modified root or stem or an organ *sui generis*? *Crit. Rev. Plant. Sci., 16,* 361-392.
Hughes, M., Donnelly, C., Crozier, A., and Wheeler, C. T. (1999). Effects of the exposure of roots of *Alnus glutinosa* to light on flavonoids and nodulation. *Can. J. Bot., 77,* 1-5.
Huss-Danell, K. (1997). Actinorhizal symbioses and their N_2 fixation. *New Phytol., 136,* 375-405.
Israel, D. W. (1993). Symbiotic dinitrogen fixation and host-plant growth during development of and recovery from phosphorus deficiency. *Physiol. Plant., 88,* 294-300.
Jeong, S. C., Ritchie, N. J., and Myrold, D. D. (1999). Molecular phylogenies of plant and *Frankia* support multiple origins of actinorhizal symbioses. *Mol. Phylogen. Evol., 13,* 493-503.
Knowlton, S., Berry, A. M., and Torrey, J.G. (1980). Evidence that associated soil bacteria may influence root hair infection of actinorhizal plants by *Frankia*. *Can. J. Microbiol., 26,* 971-7.
Kohls, S. J., and Baker, D. D. (1989). Effects of substrate nitrate concentration on symbiotic nodule formation in actinorhizal plants. *Plant Soil, 118,* 171-179.
Krusell, L., Madsen, L. H., Sato, S., Aubert, G., Genua, A., *et al.* (2002). Shoot control of root development and nodulation is mediated by a receptor like kinase. *Nature, 420,* 422-426
Laplaze, L., Duhoux, E., Franche, C., Frutz, T., Svistoonoff, S., *et al.* (2000). *Casuarina glauca* prenodule cells display the same differentiation as the corresponding nodule cells. *Mol. Plant-Microbe Interact., 13,* 107-112.
Liu, Q., and Berry, A. M. (1991a). The infection process and nodule initiation in the *Frankia-Ceanothus* root nodule symbiosis. *Protoplasma, 163,* 82-92.
Liu, Q., and Berry, A. M. (1991b). Localization and characterization of pectic polysacharides in roots and root nodules of *Ceanothus* spp. during intercellular infection by *Frankia*. *Protoplasma, 163,* 93-101.
Loh, J., Lohar, D. P., Andersen, B., and Stacey, G. (2002). A two-component regulator mediates population-density-dependent expression of the *Bradyrhizobium* japonicum nodulation genes. *J. Bacteriol., 184,* 1759-1766.
Loh, J., Pierson, E. A., Pierson III, L. S., Stacey, G., and Chaterjee A. (2002b). Quorum sensing in plant-associated bacteria. *Curr. Op. Plant Biol., 5,* 285-290.
Lugtenberg B. J. J., Dekkers, L., and Bloemberg G. V. (2001). Molecular determinants of rhizosphere colonization by pseudomonads. *Ann. Rev. Phytopathol., 39,* 461-490.
MacConnell, J. T., and Bond, G. (1957). A comparison of the effect of combined nitrogen on nodulation in non-legumes and legumes. *Plant Soil, 8,* 378-388.
Mathesius, U. (2001). Flavonoids induced in cells undergoing nodule organogenesis in white clover are regulators of auxin breakdown by peroxidase. *J. Exp. Bot., 52,* 419-426.
Mathesius, U., Weinman, J. J., Rolfe, B. G., and Djordjevic, M. A. (2000). Rhizobia can induce nodules in white clover by "hijacking" mature cortical cells activated during lateral root development. *Mol. Plant-Microbe Interact., 13,* 170-182.
Mathesius, U., Mulders, S., Gao, M., Teplitski, M., Caetano-Anollés, G., *et al.* (2003). Extensive and specific responses of a eucaryote to bacterial quorum-sensing signals. *Proc. Natl. Acad. Sci. U.S.A., 100,* 1444-1449.
Maunuksela, L., Zeep, K., Koivula, T., Zeyer, J., Haatela, K., and Hahn, D. (1999). Analysis of *Frankia* populations in three soils devoid of actinorhizal plants. *FEMS Microbiol. Ecol., 28,* 11-22
Miller, I. M., and Baker, D. D. (1985). The initiation, development and structure of root nodules in *Elaeagnus angustifolia* L. (Elaeagnaceae). *Protoplasma, 128,* 107-119.
Miller, I. M., and Baker, D. D. (1986). Nodulation of actinorhizal plants by *Frankia* strains capable of both root hair infection and intercellular penetration. *Protoplasma, 131,* 82-91.
Miller, M. B., and Bassler, B. L. (2001). Quorum sensing in bacteria. *Annu. Rev. Microbiol., 55,* 165-99.
Nishimura, R., Hayashi, M., Wu, G., Kouchi, H., Imaizumi-Anraku, H., *et al.* (2002). HAR1 mediates systemic regulation of symbiotic organ development. *Nature, 420,* 426-430
Normand, P., Orso, S., Cournoyer, B., Jeannin, P., Chapelon, C., *et al.* (1996). Molecular phylogeny of the genus *Frankia* and related genera and emendation of the family *Frankia*ceae. *Int. J. Syst. Bacteriol., 46,* 1-9.

Okubara, P. A., Fujishige, N. A., Hirsch, A. M., and Berry, A. M. (2000). *Dg93*, a nodule-abundant mRNA of *Datisca glomerata* with homology to a soybean early nodulin gene. *Plant Physiol., 122,* 1073-1079.

Parsons, R., Stanforth, A., Raven, J. A., and Sprent, J. I. (1993). Nodule growth and activity may be regulated by a feedback mechanism involving phloem nitrogen. *Plant Cell Env., 16,* 125-136.

Pawlowski, K., and Bisseling, T. (1996). Rhizobial and actinorhizal symbioses: What are the shared features? *Plant Cell, 8,* 1899-1913.

Pawlowski, K., Swensen, S., Guan, C., Hadri, A.-E., Berry, A. M., and Bisseling, T. (2003). Distinct patterns of symbiosis-related gene expression in actinorhizal nodules from different plant families. *Mol. Plant-Microbe Interact., 16,* 796-807.

Penmetsa, R. V., and Cook, D. R. (1997). A legume ethylene-insensitive mutant hyperinfected by its rhizobial symbiont. *Science, 275,* 527-530.

Phillips, D. A. (2000). Biosynthesis and release of rhizobial nodulation gene inducers by legumes. In E. W. Triplett (Ed.), *Prokaryotic nitrogen fixation: A model system for the analysis of a biological process* (pp. 349-363). Wymondham, UK: Horizon Scientific Press.

Pizelle, G. (1965). L'azote minéral et la nodulation de l'aulne glutineux (*Alnus glutinosa*). Observations sur des plantes cultivées avec systémes racinaires compartimentés. *Bulletin de l'Ecole Nationale Supérieure Agronomique, 7,* 55-63.

Pizelle, G. (1966). L'azote minéral et la nodulation de l'aulne glutineux (*Alnus glutinosa*). II. Observations sur l'action inhibitrice de l'azote minéral à l'égard de la nodulation. *Annales de l'Institut Pasteur, 111,* 259-264.

Prin, Y., and Rougier, M. (1986). Cytological and histochemical characterization of the axenic root surface of *Alnus glutinosa. Can. J. Bot., 64,* 2216-2226.

Pueppke, S. G., and Broughton, W. J. (1999). *Rhizobium* sp. strain NGR234 and *R. fredii* USDA257 share exceptionally broad, nested host-ranges. *Mol. Plant-Microbe Interact., 12,* 293-318.

Ramirez-Saad, H., Janse, J. D., and Akkermans, A. D. L. (1998). Root nodules of *Ceanothus arboreus* contain the N_2-fixing *Frankia* endophyte and a phylogenetically related Nod^-/Fix^- actinomycete. *Can. J. Microbiol., 44,* 140-148.

Reddell, P., Yun, Y., and Shipton, W. A. (1997). Do *Casuarina cunninghamiana* seedlings dependent on symbiotic N_2 fixation have higher phosphorus requirements than those supplied with adequate fertilizer nitrogen? *Plant Soil, 189,* 213-219.

Redondo-Nieto, M., Rivilla, R., El-Hamdaoui, A., Bonilla, I., and Bolaños, L. (2001). Boron deficiency affects early infection events in the pea-*Rhizobium* symbiotic interaction. *Aust. J. Plant. Physiol., 28,* 819-823.

Ritchie, N. J., and Myrold, D. D. (1999b). Phylogenetic placement of uncultured *Ceanothus* microsymbionts using 16S rRNA gene sequences. *Can. J. Bot., 77,* 1208-1213.

Rodelas, B., Lithgow, J. K., Wisniewski-Dye, F., Hardman, A., Wilkinson, A., *et al.* (1999). Analysis of quorum-sensing-dependent control of rhizosphere-expressed (*rhi*) genes in *Rhizobium leguminosarum* bv. *viciae. J. Bacteriol., 181,* 3816-3823.

Scheres, B., van Engelen, F., van der Knaap, E., van de Wiel, C., van Kammen, A., and Bisseling, T. (1990). Sequential induction of nodulin gene expression in the developing pea nodule. *Plant Cell, 2,* 687-700.

Searle, I. R., Men, A. E., Laniya, T. S., Buzas, D. M., Iturbe-Ormaetxe, I., *et al.* (2003). Long-distance signaling in nodulation directed by a CLAVATA1-like receptor kinase. *Science, 299,* 109-112.

Sequerra, J., Capellano, A., Faure-Raynard, M., and Moiroud, A. (1993). Root hair infection process and myconodule formation on *Alnus incana* by *Penicillium nodositatum. Can. J. Bot., 72,* 955-962.

Smolander, A., and Sundman, V. (1987). *Frankia* in acid soils of forests devoid of actinorhizal plants. *Physiol. Plant., 70,* 297-303.

Soltis, D. E., Soltis, P. S., Morgan, D. R., Swensen, S. M., Mullin, B. C., *et al.* (1995). Chloroplast gene sequence data suggest a single origin of the predisposition for symbiotic nitrogen fixation in angiosperms. *Proc. Natl. Acad. Sci. U.S.A., 92,* 2647-2651.

Spaink, H. (2000). Root nodulation and infection factors produced by rhizobial bacteria. *Annu. Rev. Microbiol., 54,* 257-88.

Swensen, S. (1996). The evolution of actinorhizal symbioses: Evidence for multiple origins of the symbiotic association. *Am. J. Bot., 83,* 1503-1512.

Swensen, S. M., and Mullin, B. C. (1997). Phylogenetic relationships among actinorhizal plants. The impact of molecular systematics and implications for the evolution of actinorhizal symbiosis. *Physiol. Plant.*, *99*, 565-573.

Thomas, K. A., and Berry, A. M. (1989). Effects of continuous nitrogen application and nitrogen preconditioning on nodulation and growth of *Ceanothus griseus* var. *horizontalis*. *Plant Soil*, *118*, 181-187.

Tjepkema, J. D. (1978). The role of oxygen diffusion from the shoots and nodule roots in nitrogen fixation by root nodules of *Myrica gale* L. *Can. J. Bot.*, *61*, 2898-2909.

Torrey, J. G. (1990). Cross-inoculation groups within *Frankia* and host-endosymbiont associations. In C. R. Schwintzer and J. D. Tjepkma (Eds.), *The biology of* Frankia *and actinorhizal plants* (pp. 83-106). San Diego, CA: Academic Press.

Tsai, S. M., and Phillips, D. A. (1991). Flavonoids released naturally from alfalfa promote development of symbiotic *Glomus* spores in vitro. *Appl. Environ. Microbiol.*, *57*, 1485-1488.

Valverde, C. (2000). La simbiosis *Discaria trinervis–Frankia.* Regulación de la nodulación radicular. Ph.D. thesis. Facultad de Ciencias exactas, Universidad Nacional de La Plata, La Plata, Argentina.

Valverde, C., and Wall, L. G. (1999a). Time course of nodule development in *Discaria trinervis* (Rhamnaceae)–*Frankia* symbiosis. *New Phytol.*, *141*, 345-354.

Valverde, C., and Wall, L. G. (1999b). Regulation of nodulation in *Discaria trinervis* (Rhamnaceae)–*Frankia* symbiosis. *Can. J. Bot.*, *77*, 1302-1310.

Valverde, C., and Wall, L. G. (2002). Nodule distribution on the roots of actinorhizal *Discaria trinervis* (Rhamnaceae) in pots. *Environ. Exp. Bot.*, *47*, 95-100

Valverde, C., and Wall, L. G. (2003). The regulation of nodulation, nitrogen fixation and ammonium assimilation under a carbohydrate shortage stress in the *Discaria trinervis Frankia* symbiosis. *Plant Soil*, *254,* 139-153.

Valverde, C., Ferrari, A., and Wall, L. G. (2002). Phosphorus and the regulation of nodulation in the actinorhizal symbiosis between *Discaria trinervis* (Rhamnaceae) and *Frankia* BCU110501. *New Phytol.*, *153*, 43-52

Valverde, C., Wall, L. G., and Huss-Danell, K. (2000). Regulation of nodulation and nodule mass relative to nitrogenase activity and nitrogen demand in seedlings of *Discaria trinervis* (Rhamnaceae). *Symbiosis*, *28*, 49-62.

van Ghelue, M., Løvaas, E., Ringø, E., and Solheim, B. (1997). Early interactions between *Alnus glutinosa* and *Frankia* strain ArI3. Production and specificity of root hair deformation factor(s). *Physiol. Plant.*, *99*, 579-587.

Wall, L. G., and Huss-Danell, K. (1997). Regulation of nodulation in *Alnus Frankia* symbiosis. *Physiol. Plant.*, *99*, 594-600.

Wall, L. G. (2000). The actinorhizal symbiosis. *J. Plant Growth Reg.*, *19*, 167-182

Wall, L. G., Chaia, E., Valverde, C., and Lucki, G. (2000a). Specificity in *Discaria Frankia* symbioses. In F. O. Pedrosa, M. Hungria, M. G. Yates, and W. E. Newton (Eds.), *Nitrogen fixation: From molecules to crop productivity* (pp. 461-462). Dordrecht, The Netherlands: Kluwer Academic.

Wall, L. G., Hellsten, A., and Huss-Danell, K. (2000b). Nitrogen, phosphorous, and the ratio between them affect nodulation in *Alnus incana* and *Trifolium pratense*. *Symbiosis*, *29*, 91-105

Wall, L. G., Valverde, C., and Huss-Danell, K. (2003). Regulation of nodulation in the absence of N_2 is different in actinorhizal plants with different infection pathways. *J. Exp. Bot.*, *385*, 1253-1258.

Werner, D. (1992). *Symbiosis of plants and microbes*. London, UK: Chapman and Hall.

Wolters, D. J. (1998). *Ineffective* Frankia *in wet alder soil.* Ph.D. thesis. Wageningen Agricultural University, Wageningen, The Netherlands.

Wopereis, J., Pajuelo, E., Dazzo, F. B., Jiang, Q., Gresshoff, P. M., *et al.* (2000). Short root mutant of *Lotus japonicus* with a dramatically altered symbiotic phenotype. *Plant J.*, *23*, 97-114.

Yang, Y., Shipton, W. A., and Reddel, P. (1997). Effects of phosphorus supply on *in vitro* growth and phosphatase activity of *Frankia* isolates from *Casuarina*. *Plant Soil*, *189*, 75-79.

Yang, Y. (1995). The effect of phosphorus on nodule formation and function in the *Casuarina-Frankia* symbiosis. *Plant Soil*, *176*, 161-169.

Chapter 7

CARBON AND NITROGEN METABOLISM IN ACTINORHIZAL NODULES

C. VALVERDE[1] AND K. HUSS-DANELL[2]

[1]*Programa de Investigación sobre Interacciones Biológicas, Departamento de Ciencia y Tecnología, Universidad Nacional de Quilmes, Roque Saenz Peña 180, Bernal, B1876BXD, Argentina*

[2]*Department of Agricultural Research for Northern Sweden, Crop Science Section, Swedish University of Agricultural Sciences, SE-904 03 Umeå, Sweden*

1. INTRODUCTION

As a free-living organism, *Frankia* takes up nutrients from the soil and fixes N_2 for its own growth. Thus, the metabolism of saprophytic *Frankia* can be studied in pure cultures. However, when *Frankia* establishes symbiosis with actinorhizal plants and develops root nodules, the metabolism of both the host plants and the bacteria is modified. The shoot of the host plant always provides the root cells with C compounds, but in symbiosis, the shoot should also supply C compounds to nodules and the *Frankia* within them. In turn, the plant makes use of the N_2 fixed by *Frankia*. Root nodules are highly specialized organs in which *Frankia* and host cells together carry out the synthesis of organic N compounds that ultimately contribute to the support of plant growth and development. As a consequence, there is high metabolic activity in nodules and necessarily an important exchange of metabolites between *Frankia* and host cells.

Biological N_2 fixation must be followed by an efficient assimilation of the fixed-N within root nodules in order to make the whole nodulation process a good investment for the plant. It has been suggested that N assimilation in legumes could be enhanced by genetic engineering (Miflin and Habash, 2002; Vance and Lamb, 2001). However, such an approach faces the problem that N assimilation may require genetic modification of multiple steps to show an improvement in the N stored by the plant. It may not be as simple as enhancing the expression of one

K. Pawlowski and W. E. Newton (eds.), Nitrogen-fixing Actinorhizal Symbioses, 167–198.
© 2008 *Springer*.

enzyme. In fact, genetic engineering of metabolic routes needs the precise knowledge of how these routes work and how they are regulated in a given plant species. Actinorhizal plants are very diverse, and it may be inappropiate to extrapolate observations made in one plant genus to other plant genera or families.

Thus, it is necessary to understand how fixed-N arising from N_2 fixed by *Frankia* and C compounds supplied by the host plant are incorporated into organic molecules. This aspect of actinorhizal physiology has been only partially reviewed previoiusly (Benson and Silvester, 1993; Huss-Danell, 1990; 1997; Pawlowski *et al.*, 1996; Parsons and Sunley, 2001; Schubert, 1986). Here, we summarize and update information on C and N metabolism in actinorhizal nodules. Abbreviations used throughout this chapter are listed ater Table 1.

2. THE ACTINORHIZAL NODULE

2.1. Anatomy and General Pathways for C and N

The actinorhizal nodule is an organ developed from root tissue in response to infection by *Frankia* hyphae (see Chapter 6 in this volume). Because of its origin and anatomy, the nodule resembles a lateral root. The cortical tissue surrounds the central vascular bundles and an apical meristem directs nodule growth (Figure 1). With the exception of *Datisca*, actinorhizal plants and their nodules are perennial. In general, nodules from all actinorhizal plants are composed of meristematic, cortical, epidermal, endodermal, and vascular cells (Figure 1), although their relative distribution may vary among different plant families (see Chapter 5 in this volume).

In nodules like those of *Discaria*, there is a developmental gradient of cortical cells from the meristem of the nodule apex to the base of the nodule (Figure 1). Cortical cells grow and some of them become infected by *Frankia*, ultimately senescing. *Frankia* infects and extensively colonizes developing cortical cells, where it differentiates into the N_2-fixing cell type, the so-called vesicle (Figure 1). Both *Casuarina* and *Allocasuarina* nodules are exceptions because they produce no vesicles. Infected cortical cells are in contact with non-infected cells that are thought to play an important role in the metabolism and transport of the primary products of N_2 fixation and assimilation (Figure 1b). These processes take place in a rather defined region of the nodule tissue, the N_2-fixing zone, which can be easily identified under the microscope in the region of the nodule where *Frankia* produces vesicles (vc in Figure 1B).

The nodule can be regarded as an extension of the root system (Figure 1). It represents a strong sink for C during all stages of its development. Broadly speaking, nodules receive C sources through the phloem tissue and they export the products of N_2 fixation and assimilation through the xylem. To some extent, nodules can also fix CO_2 to sustain the pool of C compounds that are used to incorporate fixed-N. It is reasonable to think that nodules use part of the fixed-N for their own

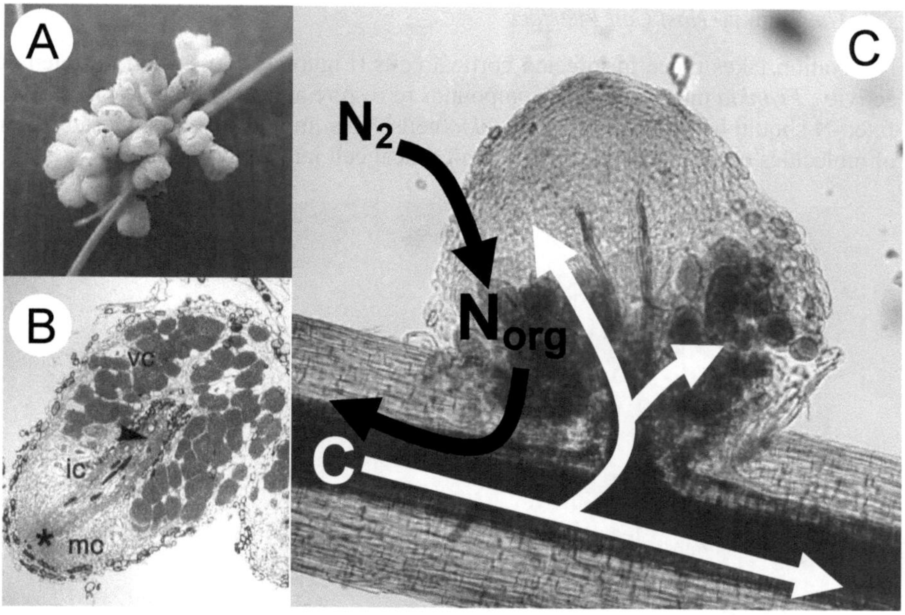

Figure 1. Actinorhizal nodules.
(A) Multilobed root nodules of Discaria trinervis. *(B) Longitudinal section of a* Discaria trinervis *nodule lobe: *, meristem; mc, mature uninfected cortical cells; ic, infected cortical cells; vc, infected cortical cells with* Frankia *vesicles; arrowhead, central vascular tissue.*
(C) Major flows of C and N through nodules.

needs but, in addition, receive fixed-N compounds from the phloem tissue. These latter fixed-N supplies are critical in at least two different stages of nodule development. First, an input of fixed-N is necessary to build up the nodule until it can begin to fix N_2. Second, once the nodule becomes active, fixed-N cycling through the nodule could be important both in providing information on the fixed-N status of the plant and in exerting control over N_2 fixation and assimilation (see section 7).

Herein, only current knowledge on C and N metabolism by mature active nodules will be discussed. Nodule N_2 fixation and assimilation, as well as any other metabolic activity, are drastically reduced during winter when nodules are dormant. Further, the description of C and N metabolism will be restricted to the N_2-fixing zone of mature nodules; towards the nodule base, the senescing zone may have a different C and N flow. In the senescent zone of 12-week-old *A. glutinosa* nodules, *Frankia* vesicles degenerate and nitrogenase transcripts are not detected (Pawlowski *et al.*, 1995), which may induce the remobilization of protein N into amino acids.

2.2. The Frankia*–Host Cell Interface*

N_2 fixation takes place in infected cortical cells (Figure 1). To sustain nitrogenase activity, *Frankia* must receive C compounds to respire and so produce ATP. In turn, fixed-N should be removed from *Frankia* cells. This implies that an important flow of molecules occurs through the *Frankia*-cortical cell interface (Figure 2).

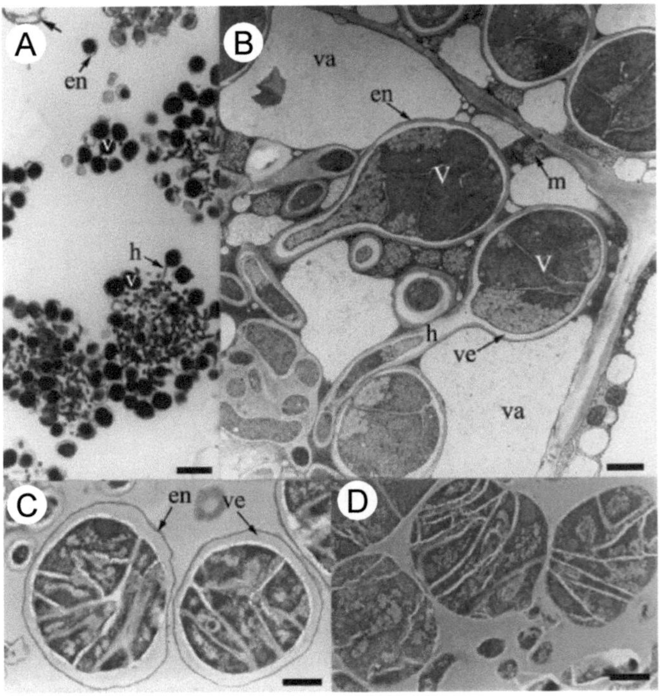

Figure 2. The Frankia*-host cell interface.*
(A) Purified nodule vesicle cluster preparation, stained with toluidine blue. The preparation is highly enriched in vesicle clusters, following two-step filtration. The clusters consist of Frankia *hyphae (h) and vesicles (v). The thin layer of host encapsulation is visible around several of the vesicles (en) and residual cell debris is present (arrow). Bar, 8 µm.*
(B) Transmission electron microscopy of Frankia *vesicles (v) and hyphae (h) within the* Alnus rubra *root nodule cortical cells. Immediately surrounding the vesicle cytoplasm is the* Frankia *cell wall and vesicle envelope (electron-translucent layer; ve). Outside the vesicle envelope, the host encapsulating wall layer can be seen as a thin, electron-dense layer (en). The vesicles are embedded within the host cytoplasm, which contains multiple mitochondria (m). The* Frankia*-infected cortical cells contain numerous small vacuoles (va). Bar, 1 µm.*
(C) Transmission electron microscopy of a nodule vesicle clusters preparation from Alnus incana *nodules; ve,* Frankia *vesicle envelope; en, host encapsulation. Bar, 1 µm.*
(D) Nodule vesicle clusters preparation as in (c), after pectinase treatment. Note the absence of the host encapsulation. Bar, 1 µm. This figure is reproduced from Berry et al. *(2002) with permission of New Phytologist.*

Such molecules face a complex array of layers. The host cell has its cell membrane and the modified plant-cell wall, the so-called "capsule". *Frankia* has its lipid envelope, its cell wall, and its cell membrane (Huss-Danell, 1990; Newcomb and Wood, 1987; Figure 2). Due to the presence of all these barriers, *Frankia* cells are physically separated from the cytoplasm of the plant cells in which they reside.

The *Frankia* lipid envelope consists of several layers, which are rich in hopanoids (Berry *et al.*, 1993). Most likely, the envelope thickness and relative hopanoid content of *Frankia* vesicles within nodules is regulated by the nodule O_2 concentration, which must be kept at a minimum close to nitrogenase to avoid nitrogenase inactivation (Kleeman *et al.*, 1994; Parsons *et al.*, 1987). In several actinorhizal genera, the vesicles are surrounded by a thick envelope (Huss-Danell, 1997).

Depending on the infection pathway, the formation of the capsule begins either during root-hair penetration by *Frankia* hyphae (as in *Alnus* and *Casuarina*) or upon penetration of cortical cells by intercellular hyphae that have already invaded the root cortex (as in *Elaeagnus* and *Dascaria*; see Chapter 6 in this volume). The characteristics of the capsule are not yet clear. Although the plant-cell wall of infected cells is quite similar to the wall of other plant cells, the capsule, which is continuous with host cell wall, is found only in infected cells. Electron micrographs show that it is an electron-dense matrix (Figure 2). It contains cellulose, hemicellulose, pectins, and "sticky" glycoproteins of the arabinogalactoprotein type (Berg, 1990; Berry *et al.*, 2002).

In summary, all the molecules that need to be translocated between the host cell and *Frankia* (and *vice versa*) have to overcome all these barriers. Because the *Frankia* lipid envelope is thinner around *Frankia* hyphae than around vesicles, transport processes might be favoured through the hyphal–host cell interface. *Frankia* forms a continuum of hyphae and vesicles (when present) among infected plant cells.

3. STUDYING C AND N METABOLISM OF NODULES

Experimental procedures have been applied to different levels of nodule complexity. The study of N compounds present in xylem exudates represents a picture of the nodule as a whole. Then, nodule compartments can be fractionated in order to observe the metabolic contribution of *Frankia* and host cells. To accomplish this, the metabolic performance of symbiotic *Frankia* has been studied with vesicle cluster preparations (see section 5) and metabolic aspects of nodule host cells have been analyzed in the soluble fraction of nodule homogenates. However, the cellular and sub-cellular compartmentalization of proteins and metabolites is lost in the preparation of nodule extracts. To overcome this problem, nodule sections have used to study the tissue localization both of proteins by immunohistochemistry and of mRNA by *in situ* hybridization (Table 1).

Table 1. Studies on C and N metabolism in nodules of actinorhizal symbioses.

Symbiosis	Compound	Methods	Reference
Alnus glutinosa	OCT	Purification and characterization	Scott *et al.*, 1981; Martin *et al.*, 1983
Alnus glutinosa	GS	Purification and characterization	Hirel *et al.*, 1982
Alnus glutinosa x *Frankia* AvcI1, CpI1 or crushed nodules	PEPC, CMPS	Purification and characterization, immunolocalization, pulse-chase of ^{14}C	Perrot-Rechenmann *et al.*, 1981; Mc Clure *et al.*, 1983
Alnus glutinosa	GS, ACOAT	Cloning, mRNA expression	Guan *et al.*, 1996
Alnus glutinosa	Polyamines	HPLC	Tonin *et al.*, 1991
Alnus glutinosa	SS, enolase	Cloning, mRNA expression	Van Ghelue *et al.*, 1996
Alnus glutinosa x crushed nodules	Amino acids	GC-MS, ^{15}N pulse	Baker *et al.*, 1997
Alnus incana x "local source of *Frankia*"	GS	Immunodetection	Lundquist and Huss-Danell, 1992
	GS	^{15}N-NMR	Lundberg *et al.*, 1996
	PEPC, PFK, GOGAT, CMPS	Cloning, mRNA expression	P.-O. Lundquist and C.P. Vance (unpublished data)
	SS, enolase	Cloning, mRNA expression	B. Solheim (unpublished data)
Casuarina glauca	GS	mRNA expression	Pawlowski *et al.*, 1996
C. cunninghamiana, C. glauca, C. equisetifolia x crushed nodules from *Frankia* KB or SI	Amino acids	HPLC	Sellstedt and Atkins, 1991
Datisca glomerata	GS	mRNA expression	Pawlowski *et al.*, 1996
Datisca glomerata x crushed nodules from *Ceanothus griseus*	RUBISCO activase	mRNA expression, immunodetection	Okubara *et al.*, 1999
Datisca glomerata	GS, amino acids	Immunolocalization, mRNA expression, HPLC	Berry *et al.*, 2003
Discaria trinervis x *Frankia* BCU110501	Amino acids GS, GOGAT, AAT, AS, MDH, PEPC, GDH	GC-MS, enzyme assays and immunodetection	Valverde, 2000; Valverde and Wall, 2002a; 2002b
Elaeagnus umbellata x *Frankia* EuIK1	AS	mRNA expression	Kim *et al.*, 1999
Myrica gale x crushed nodules	Amino acids	GC-MS, 15N pulse	Baker and Parsons, 1997; Baker *et al.*, 1997

Abbreviations used in Table 1 and throughout the chapter.

AAT: aspartate amino transferase (EC 2.6.1.1)
ACOAT: N^2-acetylornithine: glutamate *N*-acetyltransferase (EC 2.6.1.11)
ADP: adenosine diphosphate
Ala: alanine
Alc: allantoic acid
Aln: allantoin
AMP: adenosine monophosphate
Arg: arginine
AS: asparagine synthetase (EC 6.3.1.1)
Arg: arginine
Asn: asparagine
Asp: aspartate
ATP: adenosine triphosphate
cDNA: complementary DNA
Cit: citrulline
CMP: carbamyl phosphate
CMPS: carbamyl phosphate synthase (EC 6.3.5.5)
DNA: desoxyribonucleic acid
Frc: fructose
GABA: γ-amino butyric acid
GC-MS: gas chromatography coupled with mass spectrometry
GDH: glutamate dehydrogenase (EC 1.4.1.2)
Glc: glucose
Gln: glutamine
Glu: glutamate
Gly: glycine
GOGAT: glutamate synthase (EC 1.4.1.14)
GS: glutamine synthetase (EC 6.3.1.2)
HPLC: high performace liquid chromatography
IMP: inosine monophosphate
2-KG: 2-ketoglutarate
Mal: malate
MDH: malate dehydrogenase (EC 1.1.1.37)
mRNA: messenger RNA
NAD(P): nicotinamide adenine dinucleotide (phosphate)
NAD(P)H: reduced nicotinamide adenine dinucleotide (phosphate)
OA: oxalacetate
OCT: ornithine carbamyl transferase (EC 2.1.3.3)
Orn: ornithine
PEP: phosphoenolpyruvate
PEPC: phosphoenolpyruvate carboxylase (EC 4.1.1.31)
PFK: phosphofructokinase (EC 2.7.1.11)
pO_2: oxygen pressure
PP_i: inorganic pyrophosphate
PVPP: polyvinylpolypyrrolidone
Pyr: pyruvate
RNA: ribonucleic acid
Suc: sucrose
Succ: succinate
SS: sucrose synthase (EC 2.4.1.13)
TCA: tricarboxylic acids cycle
UDP: uridyl diphosphate

The analysis of nitrogenous compounds present in nodule extracts can provide a view of the main pools of organic N, but it does not necessarily reflect which compounds are the first produced as a direct result of fixed-N assimilation. Data on xylem composition must also be interpreted with care because the N pattern is a combination of recently assimilated NH_4^+ and recycled organic N. Use of $^{15}N_2$ is essential to establish whether or not N compounds present in root nodules and xylem sap are involved in the transport of recently fixed N (Baker *et al.*, 1997).

To investigate the primary assimilation of NH_4^+ in actinorhizal nodules, three major analytical approaches have been applied: (i) HPLC or GC-MS to study the composition of amino acids and other N-compounds in nodule extracts and xylem sap; (ii) quantification of enzyme activities related to primary NH_4^+ assimilation, and immunolocalization of the responsible enzymes; and (iii) analysis of gene expression with RNA-hybridization tools. All these approaches (Table 1) have advantages and disadvantages and are, therefore, complementary. Unfortunately, they have not been used systematically to study the same symbiotic system and so have resulted in a somewhat incomplete picture of the nitrogen-assimilation process in actinorhizal nodules; any comprehensive view thus reamins speculative. Given the great structural diversity between actinorhizal nodules from different plant families, physiological differences between these nodule types is also likely.

The detection of enzyme activities and protein in either nodule extracts or purified fractions (Table 1) provides biochemical evidence of a proposed metabolic pathway. However, enzyme activators and inhibitors can reversibly or irreversibly alter enzyme activities, so leading to misinterpretations of enzyme activities. Therefore, the use of phenolic-removing agents, like PVP (Lundquist and Huss-Danell, 1991a), and the desalting of nodule extracts, through molecular filters or exclusion chromatography columns, is recommended.

Several genes coding for proteins related to C and N metabolism have been cloned from cDNA libraries made from actinorhizal nodules; these include GS, ACOAT, SS, PEPC and GOGAT from *Alnus glutinosa* and *A. incana*, and AS from *Elaeagnus umbellata* (Table 1). Isolation of these cDNAs allowed the characterization of gene expression in symbiotic and non-symbiotic tissues by RNA hybridization. However, mRNA accumulation does not always correlate with enzyme or protein level and activity due to transcriptional, post-transcriptional or post-translational regulation of the expressed genes and gene products (Lam *et al.*, 1996). Thus, gene-expression studies must always be complemented with biochemical data on enzyme activities to produce a more complete view of the activity of a gene product.

4. CARBON SUPPLY TO NODULES

Although input of C from photosynthesis must feed several processes and compartments in nodules, at least in *Alnus incana*, there seems to be a correlation between shoot net CO_2 assimilation and nitrogenase activity under growth-chamber conditions (Huss-Danell, 1990) and a correlation between leaf area and N_2 fixation in the field (Huss-Danell *et al.*, 1992). Indirect evidence for such a relationship also comes from treatments (such as defoliation or darkening) that aim to reduce the C flow to nodules and which result in decreased nitrogenase activity (for a review see Huss-Danell, 1997). This result could indicate that decreased C flow results in a decresed respiration rate, which in turn causes a temporary elevation of pO_2. As a consequence, nitrogenase becomes at first inactivated and then degraded (Huss-Danell, 1997; Lundquist, 1993; Lundquist *et al.*, 2003).

Photoassimilates arrive in nodules of *Alnus glutinosa* within 10 minutes (Wheeler, 1971). There appears to be no correlation between nitrogenase activity and carbohydrate content of nodules as a whole, but high rates of nitrogenase activity require the delivery of new photoassimilates to nodules (Wheeler, 1971). Infected cells are likely to have a high respiratory capacity. This was demonstrated by supplying $^{14}CO_2$ to *Alnus glutinosa* shoots (Wheeler and Lawrie, 1976). At 5h after the pulse, microautoradiography showed that ^{14}C accumulated mainly in uninfected cells with no significant labeling in infected cells. A high respiratory capacity and an intense metabolite exchange with uninfected cells could explain this observation.

Starch granules are found in uninfected cells of nodule tissues (Berry and Sunell, 1990), although there is no direct correlation between such C reserves and nodule

N_2-fixation activity. The function of this starch stock is not clearly elucidated, but these C reserves are depleted in C-stress situations (Vikman *et al.*, 1990). In contrast, infected cells have few, if any, starch granules. Moreover, *Casuarina* nodules do not contain starch granules at all (E. Duhoux, unpublished data), which weakens any possible link between starch accumulation and N_2-fixation activity.

In most plants, the main phloem carbon-transport form is sucrose. Sucrose degradation can be initiated by invertase (either the apoplastic, vacuolar or cytosolic isoforms) or by symplastic sucrose synthase (SS). The latter appears to be primarily used in legume nodules (Gordon *et al.*, 1999). SS produces UDP-Glc and Frc. Where SS activity occurs, either starch biosynthesis or sucrose degradation should be taking place. If UDP-Glc is used for respiration, it has to be metabolized to PEP, which requires enolase in the last step. Thus, high enolase activities are expected in nodules that are metabolizing sucrose. At the sites of enolase activity, products of sucrose metabolism should be used either to fuel plant cells and/or bacterial cells or to assimilate fixed-N.

Alnus glutinosa SS and enolase mRNAs have been detected in infected nodule cells, in the pericycle of nodule vascular bundles, and in the nodule meristem (van Ghelue *et al.*, 1996). However, SS was not detected in uninfected cells where starch granules are present (van Ghelue *et al.*, 1996). Invertase may be responsible for sucrose degradation in the uninfected cells. The fact that SS was co-expressed with enolase makes it likely that the products of SS are channeled through glycolysis. In the same cells, the expression of *agthi1*, a thiamine synthetase-like protein, was also detected (Ribeiro *et al.*, 1996). This is reasonable because thiamine is a known cofactor of enzymes, like the pyruvate dehydrogenase complex and 2-ketoglutarate dehydrogenase, involved in C metabolism. The abundance of SS, enolase and *Agthi1* transcripts in pericycle cells might reflect the high energy requirement of these cells for transfer processes.

In addition to the C compounds that nodules receive from green tissues *via* the phloem, actinorhizal nodules can also fix CO_2 (Huss-Danell, 1990), but by a process that is different from the light-dependent CO_2 fixation in photosynthesis. CO_2 fixation was one of the first processes in C metabolism to be investigated in actinorhizal nodules. In *A. glutinosa*, the kinetics of ^{14}C labelling, after a pulse-chase with $^{14}CO_2$, demonstrated that C goes first into carbonated compounds and then moves to amino acids, particularly Glu, Asp and Cit (McClure *et al.*, 1983). However, the label in Cit was found primarily (80–89%) in the carbamyl group with only a minor part in C1. This result indicated that Cit labeling occurs through the activity of CMPS (see section 6.3; Figure 3). Mal was the organic acid that was mainly labeled with ^{14}C, but only a minor part of this label was used for amino-acid synthesis. Thus PEPC, which is the main carboxylating enzyme detected in *Alnus glutinosa* nodules, seems to be committed to create an important pool of Mal that is not directly used for Cit synthesis (McClure *et al.*, 1983). PEPC and CMPS are compensating enzymes for the C metabolism of the nodule cells; they are necessary to avoid a depletion of C substrates that are used to assimilate fixed-N.

5. CARBON UPTAKE AND METABOLISM BY SYMBIOTIC *FRANKIA*

It is not known which of the C compounds that are formed from the Suc fed to nodules *via* phloem are taken up and metabolized by *Frankia*. Sugars, TCA intermediates, and short fatty acids (like propionate) can be used by *Frankia* strains grown in culture (Benson and Schultz, 1990; Lechevalier and Lechevalier, 1990). However, such information can only serve as a suggestion for the C compounds used by symbiotic *Frankia*. There are differences in C metabolism between the free-living and symbiotic states for *Frankia* (Huss-Danell *et al.*, 1982; Lopez *et al.*, 1986) as well as for other bacteria (Huss-Danell, 1997). Therefore, symbiotic C uptake must be studied using *Frankia* is in its symbiotic form. This may be particularly important for those symbionts of actinorhizal species that belong to Rosaceae, Datiscaceae, Coriariaceae, and Ceanothus and cannot yet be cultured and, therefore, cannot be examined in their free-living state (see chapter 4 in this volume). In an attempt to obtain this type of insight, two types of studies have been performed. First, any stimulation of respiration that occurs when symbiotic *Frankia* has been fed with various compounds is taken to show which compounds can be taken up and which can be metabolized. Second, measurements of enzyme activities in crude extracts of symbiotic *Frankia* indicate the potential to metabolize the appropriate substrate compound.

Since the pioneering work by Akkermans *et al.* (1977) and van Straten *et al.* (1977), preparations of so-called vesicle clusters have been an essential tool for studies of symbiotic *Frankia*. Vesicle clusters have been used not only to study C metabolism but also nitrogenase, uptake hydrogenase, GS, and lipid composition (reviewed in Huss-Danell, 1997) and, more recently, arabinogalactans at the symbiotic interface (Berry *et al.*, 2002). To prepare vesicle clusters, root nodules are first disrupted by homogenization in buffer and the homogenate is sorted by particle size in step-wise filtrations. Vesicle clusters from *Alnus* are typically collected by centrifugation from the fraction between 20-100μm in size. A useful protocol is given by Lundquist and Huss-Danell (1991a).

The size of individual vesicle clusters varies and can equal the *Frankia* content of one infected cell or less if clusters are broken. A cluster can probably also comprise *Frankia* from more than one infected cell due to aggregation during preparation. Vesicle-cluster preparations have been made from *Hippophaë* and *Shepherdia* (Akkermans *et al.*, 1983; Huss-Danell *et al.*, 1982; van Straten *et al.*, 1977), where infected cells show similarities with *Alnus* in terms of vesicle size, shape, and distribution within host cells (Huss-Danell, 1997). In contrast, vesicle-cluster preparations from *Datisca,* which have a very different shape and distribution of vesicles in infected nodule cells (see chapter 5 in this volume), were difficult to obtain without contaminating mitochondria (Akkermans *et al.*, 1983). By definition, vesicle clusters cannot be obtained from *Casuarina* where *Frankia* does not produce vesicles.

Vesicle clusters from *Alnus* can be obtained in very high purity with more than 98% of the particle volume being vesicle clusters that are virtually free of mitochondria (Vikman and Huss-Danell, 1987b). Purity is essential when studying

C and N metabolism to ensure that measured activities are really those of *Frankia* and not of contaminating host material. The enzyme 6-phosphogluconate dehydrogenase can be used as a marker because it is NAD^+-dependent in *Frankia*, but $NADP^+$-dependent in plants (Vikman and Huss-Danell, 1987a). It is also critical to get physiologically active vesicle clusters for metabolite-exchange studies of symbiotic *Frankia* because uptake systems may be damaged during vesicle preparation and could lead to a false interpretation. For example, a negative result in respiration studies could be due to impaired uptake rather than an inability to metabolize the compound.

Vesicles clusters from *Alnus* show aerobic metabolism (Vikman, 1992). Their respiration, estimated by O_2 consumption, can be inhibited by cyanide (Vikman and Huss-Danell, 1987a). They also contain cytochromes of the a, b, c, and o types (Ching *et al.*, 1983). When supplied to vesicle clusters, several sugars (Table 2), such as Suc, maltose, trehalose, Glc, Frc, Glc-6-phosphate and 6-phosphogluconate in combination with NAD^+, enhanced respiration. However, this result does not necessarily mean that some or all of these compounds are used as C sources by symbiotic *Frankia*.

Dicarboxylic acids are also candidates as C sources for *Frankia* within infected cells (Table 2). They are the major source of C that rhizobial bacteroids use as fuel in legume nodules (Rosendahl *et al.*, 1990) and are taken up by *Rhizobium* through

Table 2. Carbon compounds that can be metabolized in symbiotic Frankia *from* Alnus.

Compound	Reference
Carbohydrates	
Glc	Lopez and Torres, 1985; Lopez *et al.*, 1986; Vikman and Huss-Danell,1987a
Frc, Suc, maltose	Lopez *et al.*, 1986
Glc-6-P, 6-P-gluconate	Lopez and Torres, 1985; Vikman and Huss-Danell,1987a
Trehalose	Lopez and Torres, 1985; Lopez *et al.*, 1986
Glycogen	Fontaine *et al.*, 1984
Dicarboxylic acids, aminoacids	
Succinate, isocitrate	Akkermans *et al.*, 1981
Mal + Glu	Akkermans *et al.*, 1981; Huss-Danell *et al.*, 1982; Vikman and Huss-Danell, 1991

specific dicarboxylate transporters (Engelke *et al.*, 1989). Succinate dehydrogenase activity can be measured in vesicle-cluster preparations from *Alnus* (Akkermans *et al.*, 1981). In addition to the sugars mentioned above, a combination of Mal and Glu, in the presence of NAD^+, also had a positive effect on respiration (Vikman and Huss-Danell, 1991), which implies the operation of a Mal-Asp shuttle at the *Frankia*-host cell interface. Although such carriers have not yet been reported (Vikman and Huss-Danell, 1991), their presence is strongly suggested by the high concentrations of Mal and Glu in *Alnus* nodules and because enzymes involved in Mal-Asp shuttles, *i.e.*, Glu-OA transaminase and MDH, have been detected both in symbiotic *Frankia* and host-cell cytoplasm (Akkermans *et al.*, 1981).

Amino acids are another alternative source of C that has not yet been sufficiently explored. However, Gln, when supplied as the sole source of fixed-N, promotes growth of the isolate CpI1 in the absence of detectable N_2-fixation activity, suggesting that it is used as C and N source (Zhang and Benson, 1992). The same strain can grow on Asp but, in this case, nitrogenase activity is observed. Other amino acids have less impact on cell growth (Zhang and Benson, 1992). In symbiosis, *Frankia* is apparently impaired in its ability to directly use the N that is generated by N_2 fixation (Lundquist and Huss-Danell, 1992) and, consequently, it needs organic N.

It is noteworthy that bacteroids of pea nodules have an absolute requirement for Glu, in addition to Mal, to fix N_2 efficiently. It appears that bacteroids cycle Glu to Asp, which is transported back to the plant cell and used for Asn synthesis. This process may contribute both to down regulate NH_4^+ assimilation and to provide a fixed-N source to symbiotic rhizobia (Lodwig *et al.*, 2003). The possibility that symbiotic *Frankia* uses amino acids as a C source, as well as an fixed-N source, deserves further attention.

Some *Frankia* isolates can grow on minimal medium supplemented with certain lipids as C source, whereas others require lipids in addition to the real C source, perhaps as precursors for vitamin or cofactor synthesis (Benson and Schultz, 1990; Lechevalier and Lechevalier, 1990). However, this situation may not apply to the symbiotic state. When *Frankia* was grown on lipids as a C source, the glyoxylate-cycle enzymes, isocitrate lyase and malate synthase, were detected, but this was not the case in vesicle clusters formed when the same isolate had infected *Alnus* (Huss-Danell *et al.*, 1982). It is possible that these enzymes are subject to metabolite repression within host cells when other C sources are available.

More effort is obviously needed to determine the C source that *Frankia* uses within infected cells. The fact that certain compounds do not stimulate *Frankia* respiration may not imply that they are not a good source of C, but instead could mean that the compound is not easily incorporated into symbiotic cells. Symbiotic *Frankia* may also profit not only from a single source of C, but also from a combination of metabolites, such as sugars and dicarboxylates, or dicarboxylates and amino acids, or even sugars and amino acids. We also need to learn where in vesicle clusters the uptake of C compounds occurs. Is it directly into vesicles or into the hyphae with further internal transport to vesicles? It should be very rewarding to examine the host-*Frankia* interface more closely.

Seen in retrospect, one problem with vesicle-cluster preparations is the presence in the preparations of barriers surrounding *Frankia*. Electron micrographs taken in earlier studies were interpreted as showing that the cytoplasm in *Frankia* vesicles was surrounded only by the cell membrane and cell wall of *Frankia* itself (Vikman and Huss-Danell, 1987b). However, it now seems clear that, in vesicle-cluster preparations from *Alnus, Frankia* is also surrounded by host membrane and the capsule of host origin (Figure 2A and C). By treating vesicle clusters with pectinase to remove the capsule (Figure 2D; Berry *et al.*, 2002), both fluorescence imaging and heavy-metal staining were improved (Kleemann *et al.*, 1994). It may be that this encapsulation of *Frankia* has obscured C-metabolism studies performed to date on vesicle clusters. One may argue that, in an infected cell, *Frankia* is indeed surrounded by this encapsulation as well but, when vesicle clusters are prepared, the integrity is broken and any gradients between host cytoplasm and *Frankia* are destroyed. If the encapsulation can be removed, *e.g.*, by pectinase treatment in such a way that vesicles and hyphae maintain their physiological activities, a new and better tool for metabolic studies becomes available.

The number of actinorhizal systems studied so far is rather limited with most of the work on the metabolism of symbiotic *Frankia* preformed on vesicle-cluster preparations from *Alnus* species. Other genera may turn out to be less difficult to work with. Studies on several genera will, of course, broaden the knowledge on C metabolism in symbiotic *Frankia*. Some ineffective strains of *Frankia* are known (Hahn *et al.*, 1988) and such strains may give useful information on links between N_2 fixation and C metabolism. The availability of established genetic protocols to obtain mutants in *Frankia* that are defective in key metabolic enzymes would be an important contribution to these kinds of studies. However, despite many efforts, genetic transformation of *Frankia* is still not possible. Once the sequencing and annotation of the *Frankia* genome are accomplished, it will be possible to screen for the presence in vesicle clusters of several enzymes and transporters using mRNA-expression techniques. It would also help if the precise tissue location of enzymes (by immunolocalization) and mRNAs (by *in situ*-hybridization) could be determined. Thus, a genomic approach, together with a widened list of actinorhizal genera, will indeed provide answers to many of the present questions on C metabolism in symbiotic *Frankia*.

6. NITROGEN METABOLISM

6.1. N_2 Fixation

In almost all actinorhizal plants, N_2 fixation likely takes place in the symbiotic vesicles, which occupy a large fraction of the volume of infected nodule cells. Both immunochemical detection and *in situ* hybridization demonstrate that nitrogenase is localized within vesicles of *Alnus* species (Huss-Danell and Bergman, 1990; Pawlowski *et al.*, 1995; Prin *et al.*, 1993). In the few genera where no vesicles are detected in nodule sections, *i.e., Casuarina* and *Allocasuarina*, nitrogenase expression and activity occur in the symbiotic hyphae (Laplaze *et al.*, 2000).

N_2 fixation produces NH_3, which becomes readily protonated at pH values near neutrality to give NH_4^+. Because neither GS activity (Blom *et al.*, 1981) nor GS protein (Lundquist and Huss-Danell, 1992) has been detected in symbiotic *Frankia* in *Alnus*, *Frankia* likely cannot readily profit from its own N_2 fixation. However, it might be that a very low GS activity (undetectable in the enzyme assay) still operates in symbiotic *Frankia*. It would be helpful to reinvestigate the presence of GS protein with immunolocalization using either different antibodies or by studies of mRNA expression. An alternative explanation would be that *Frankia* uses another NH_3-assimilating system, like GDH, which has not yet been detected.

If *Frankia* cannot directly profit from the N_2 fixed, it must obtain organic N from the host. A very attractive possibility for such N input would be the operation of an amino-acid cycle as recently described for pea nodules (Ludwig *et al.*, 2003). In that symbiosis, Glu is provided by the plant cell, it is transaminated to Asp in the bacteroid, and then cycles back to the plant for Asn synthesis. Similar cycles may operate in actinorhizal nodules and may explain the absence of NH_4^+ assimilation and the N dependency on the plant partner. In this respect, it is intriguing that Mal and Glu together with NAD^+ stimulated respiration of vesicle clusters from *A. incana* nodules (Vikman and Huss-Danell, 1991).

Nevertheless, it is generally accepted that NH_4^+ derived from symbiotic *Frankia* is exported to the plant-cell cytosol, where it is assimilated into organic N compounds. Either NH_4^+-specific channels or monovalent-cation transporters are likely to be responsible for NH_4^+ export. A high-affinity NH_4^+-uptake system is known to be activated after N starvation in cultured *Frankia* strain CpI1 (Mazzucco and Benson, 1984) and a similar channel might function to export NH_4^+. Alternatively, the small fraction of NH_3 present in equilibrium with NH_4^+ (due to the intracellular pH) could diffuse away from *Frankia* in response to a chemical gradient, be protonated in the plant cell, and then assimilated by host-cell enzymes.

The nature of the principal product of N_2 fixation, which is exported from legume-nodule bacteroids, has also been a matter of recent controversy. Although for many years NH_4^+ was accepted as the N form that bacteroids supplied to host cells, Ala has also been implicated as a possible N carrier from rhizobial bacteroids to nodule cells (Waters *et al.*, 1998). This proposal has been experimentally tested, and it has been concluded that ammonium should still be considered the primary product of N_2 fixation assimilated by soybean host cells (Li *et al.*, 2002). For this reason, and until experimental evidence is presented against it, we will assume that ammonium is also the N compound that actinorhizal nodule cells use as starting material for N assimilation.

6.2. Nitrogen Transport in Actinorhizal Plants

Estimates of energetic costs (in equivalents of ATP) for the synthesis of N export compounds show, at least in theory and without considering transport costs, that the more energy efficient compounds are Asn, Cit, Arg, Gly and Alc (less than 15 molecules of ATP per molecule of N compound excluding N_2 fixation) (Schubert,

1986). Actinorhizal plants can be separated into three groups based on the N-export compounds found in the xylem (Table 3). One group (of *Alnus* species) primarily export Cit; a second group (of members of the Elaeagnaceae, Myricaceae, and Rhamnaceae) are amide (Asn, Gln) exporters; and the third group (of *Casuarina*, *Coriaria* and *Datisca*) shows no clear preference for either type of N compound.

Table 3. Nitrogen compounds and their transport in actinorhizal plants.

Family	Plant species	Major xylem N compounds	Reference
Betulaceae	*Alnus crispa*	Cit	Schubert, 1986
	A. glutinosa	Cit	Gardner and Leaf, 1960
	A. incana	Cit	Miettinen and Virtanen, 1952
	A. inokumai	Gln, Asn, Glu, Cit	Wheeler and Bond, 1970
	A. nitida	Cit	Hafeez *et al.*, 1984
	A. rubra	Cit	Schubert, 1986
Casuarinaceae	*Casuarina cunninghamiana*	Cit, Asn, Arg	Sellstedt and Atkins, 1991
	C. equisetifolia	Asn	Bollard, 1957
		Cit	Walsh *et al.*, 1984
		Cit, Asn, Oth	Sellstedt and Atkins, 1991
	C. glauca	Asn, Cit, Oth	Sellstedt and Atkins, 1991
Coriariaceae	*Coriaria ruscifolia*	Gln, Asn, Cit	Bollard, 1957
		Glu, Gln, Arg, Asn	Wheeler and Bond, 1970; Hafeez *et al.*, 1984;
Datiscaceae	*Datisca glomerata*	Gln, Asn	Schubert, 1986
		Glu, Gln, Arg, Asn	Wheeler and Bond, 1970; Hafeez *et al.*, 1984
		Gln, Glu	Berry *et al.*, 2004
Elaeagnaceae	*Elaeagnus angustifolia* *E. umbellata*	Asn	Schubert, 1986
	Hippophaë rhamnoides	Asn	Schubert, 1986
		Asn	Wheeler and Bond, 1970
Myricaceae	*Comptonia spp*	Asn	Schubert, 1986
	Myrica cerifera	Asn, Gln	Schubert, 1986
	M. gale	Asn	Leaf *et al.*, 1959
Rhamnaceae	*Ceanothus americana*	Asn	Schubert, 1986
	Discaria trinervis	Asn	Valverde, 2000
Rosaceae		No information found	

There is no clear explanation for such differences in composition of N-transport compounds among plant species. However, the composition of amino compounds in the xylem stream can change significantly when either environmental or nutritional conditions are varied. For example, the relative Asn content dropped significantly, whereas both NH_3 and Gln concentrations increased, in *Casuarina glauca* plants fed with NH_4NO_3 (Sellstedt and Atkins, 1991). Similarly, the Cit content in the xylem of *A. glutinosa* nodulated plants increased more than 2-fold after NO_3^- fertilization (Baker *et al.*, 1997a). In *M. gale*, either NH_4^+ or NO_3^- fertilization caused a change in the main N compound (from Asn to Gln) of the xylem (Baker *et al.*, 1997b). Nevertheless, the pattern of N compounds that are mobilized in the xylem of actinorhizal plants seems to be a specific genetic trait.

In addition to the major N constituents of actinorhizal xylem sap (amides or citrulline), there are also minor components of the N pool that should be considered. At least in *A. glutinosa*, the polyamine (putrescine, spermine and spermidine) content of sap has been estimated to be 5 μM, which is *ca.* 300-fold lower than the normal Cit content (Tonin *et al.*, 1991). Within this polyamine pool, the putrescine level showed interesting variations with nutritional signals. It was much more abundant in non-nodulated roots fed with NH_4^+ than in roots of plants fed with NO_3^-, and its concentration increased 4-fold in nodules of plants fed with $(NH_4)_2SO_4$. After transfer of such plants from N-free to NH_4NO_3 nutrient solution, the sap putrescine content of nodulated plants increased 3- to 4-fold. So, it seems that the elevated putrescine response was largely due to the effect of NH_4^+ rather than NO_3^- (Tonin *et al.*, 1991). These observations imply a possible signaling role for these minor N compounds of sap because they could reflect the NH_4^+ concentration in the nodule cells.

Polyamine synthesis would be an efficient method for the plant to deal with large amounts of NH_4^+, which is obligately assimilated in roots and nodules, whereas NO_3^- reduction can occur in either roots or shoots, depending on the plant species, developmental age, and NO_3^- supply. Thus, the putrescine response could be a specific sensor of NH_4^+ levels. However, given the polybasic nature of polyamines, a role in the regulation of intracellular pH must not be neglected. Such a role might explain why the polyamine content responded more to NH_4^+ rather than to NO_3^- supply. Because pure cultures of *Frankia* strains can also produce several polyamines (Wheeler *et al.*, 1994), it is not clear yet if the polyamine pattern found in actinorhizal nodules is either entirely of plant origin or has a contribution from the symbiotic *Frankia*.

6.3. Biosynthesis of N Assimilatory Products in Nodules

Figure 3 presents the proposed relevant reactions that result in the synthesis of the main N compounds, which are normally found in the xylem sap of N_2-fixing plants, *i.e.*, amides (Asn, Gln), Cit, and ureides (Alc and Aln), the ureides being found only in tropical legumes that form determinate nodules.

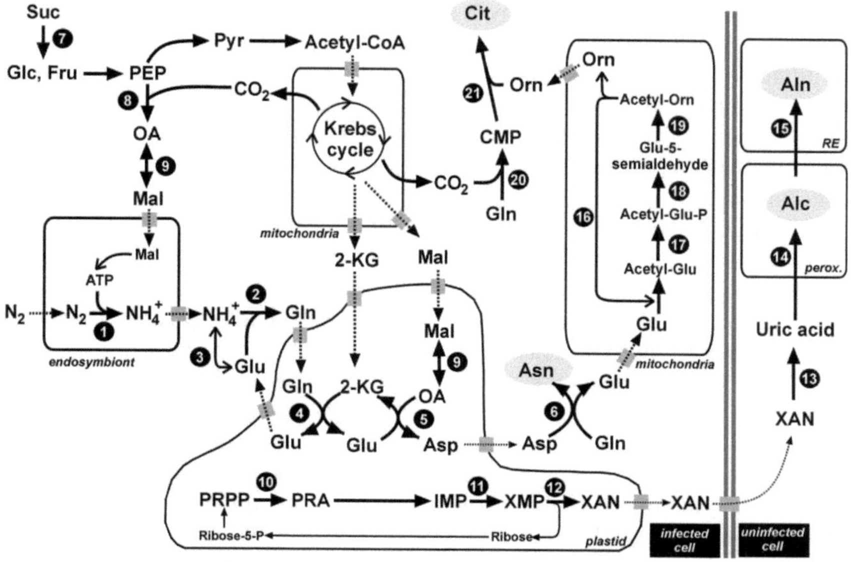

Figure 3. Nitrogen assimilation in nodules of N_2-fixing plants.
This scheme shows metabolic routes that result in the synthesis of the main N compounds found in the sap of N_2-fixing plants; amides (Asn, Gln), Cit, and ureides (Aln, Alc). Solid arrows represent single or multiple reactions catalyzed by enzymes. Dotted arrows indicate possible transport processes. Enzymes shown are (1) nitrogenase; (2) GS; (3) GDH; (4) GOGAT; (5) AAT; (6) AS; (7) SS or invertase; (8) PEPC; (9) MDH; (10) phosphoribosylpyrophosphate (PRPP) amidotransferase; (11) IMP dehydrogenase; (12) 5'-nucleotidase, 5'-nucleosidase; (13) xanthine (XAN) dehydrogenase; (14) uricase; (15) alantoinase; (16) ACOAT; (17) N-acetylglutamate 5-phosphotransferase; (18) N-acetylglutamyl-5-phosphate:NAD(P)$^+$ oxidoreductase; (19) N^2-acetylornithine: 2-oxoglutarate aminotransferase; (20) CMPS; (21) OCT.
Perox., peroxisomes; ER, endoplasmic reticulum; PRA, phosphoribosylamine; XMP, xanthine monophosphate.

Asn synthesis can proceed either by HCN incorporation *via* the β-cyanoalanine pathway or by direct amination of Asp in the Gln-dependent AS reaction. The latter pathway seems to be the primary reaction in plants and it has been completely described for alfalfa nodules (Vance, 2000). Ammonium derived from N_2 fixation is assimilated in the host-plant cells into Gln first and then into Glu by the combined action of GS and NADH-GOGAT. There are several GS isoforms, but cytosolic GS accounts for most of the nodule GS activity, meaning NH_4^+ assimilation occurs mainly in the plant cytosol. Gln and Glu are further metabolized to Asp and Asn by the combined action of AAT and AS. In alfalfa nodules, the C skeletons required for NH_4^+ assimilation are provided mainly in the form of Mal.

The reaction catalyzed by NADH-GDH offers another route for NH_4^+ assimilation into Glu. However, its high K_m for ammonia (>1-5mM), as determined in several plant species, has led to uncertainty over its role in N assimilation.

Rather, the main reaction catalyzed by NADH-GDH may be the deamination of Glu (Srivastava and Singh, 1987; Miflin and Habash, 2002).

Citrulline synthesis involves several enzymes of the urea cycle (Figure 3). However, the direct substrates are Orn and CMP, which OCT transforms into Cit. Ornithine is derived from Glu and CMP from Gln through the action of CMPS. The pathway needs an input of C compounds to sustain Glu consumption. Labeling experiments indicate that this is normally supported by the activity of PEPC, which channels C through the TCA cycle (Schubert, 1986; Figure 3).

Biosynthesis of the ureides Alc and Aln is more complicated because it is derived from the purine-degradation pathway, involves more than one cell type, and involves several subcellular compartments in legume nodules (Figure 3). So far, these ureides have not been detected in actinorhizal plants (Table 3).

It is quite clear that any route that is used to incorporate inorganic-N into any form of organic N needs a supply of C skeletons to avoid a depletion of C intermediates, like OA and 2-KG (Figure 3). Thus, both nitrogen fixation and N-assimilation are inseparable from C metabolism. In various legumes, estimates of the total cost of symbiotic N_2 fixation and assimilation are in the range of 6-12g C/g N fixed (Warenbourg and Roumet, 1989; Giller and Wilson, 1991). This C requirement can be approximately distributed into 36–39% for nodule growth and maintenance, 42-45% for nitrogenase activity, and 16–22% for NH_4^+ assimilation and transport (Giller and Wilson, 1991).

In legume nodules, *e.g.*, those of alfalfa, the complex interweaving of C and N metabolism involves integrated functioning of both cytosolic and organelle-targeted enzymes (for a review, see Vance, 2000). Efforts to improve N_2 fixation will demand an understanding of how plant enzymes involved in N and C pathways are regulated and whether they are amenable to genetic modification (Miflin and Habash, 2002; Vance and Lamb, 2001). Until we have more information from actinorhizal systems, we have to learn from legumes.

6.4. Studies on Nitrogen Metabolism in Different Actinorhizal Systems

6.4.1. Amide Exporter - Elaeagnus umbellata

Members of the Elaeagnaceae family are amide exporters, with Asn as the main N-containing compound found in the xylem (Table 3). Asn is commonly synthesized by AS, an enzyme that is Gln-dependent in all plants. AS is a frustrating enzyme to study because it is unstable, contaminated with asparaginase activity, and is affected by specific non-protein inhibitors (Romagni and Dayan, 2000). Thus, molecular tools have been used to shed light on the expression of this labile enzyme. Kim *et al.* (1999) isolated and characterized a cDNA clone that encodes a Gln-dependent AS from a nodule expression library. This was the first report of AS cloning from actinorhizal plants. Sequence analysis and comparisons with AS from other plants revealed a probable cytoplasmic localization and a Gln-binding domain. In nodule tissues, the mRNA was co-expressed with *nifH* mRNA, which suggests a role in N assimilation. In addition, the AS mRNA was only expressed in infected cells where

Frankia was differentiated into vesicles. These results provide strong evidence for metabolic control of AS mRNA expression, as observed for AS in alfalfa nodules (Shi *et al.*, 1997).

6.4.2. Amide Exporter - Myrica gale

In nodulated *Myrica gale*, Asn is the major amino acid both in nodule extracts and in xylem sap (Table 3). However, the predominant amino compounds in nodules or xylem sap are not necessarily directly involved in the assimilation and transport of recently fixed-N. The use of $^{15}N_2$ and GC-MS analysis of plant samples has provided relatively good information on the fate of assimilated NH_4^+ in *M. gale* (Baker and Parsons, 1997). The product of $^{15}N_2$ fixation was rapidly incorporated into Gln in detached nodules. In only 30s, $^{15}N_2$ had diffused into vesicles and NH_3 was transferred out of symbiotic *Frankia* to be assimilated in the plant cytoplasm. The pattern of amino-acid labeling in *M. gale* nodules suggested the operation of at least the GS-GOGAT cycle, AAT, and AS to synthesize Asn. ^{15}N was also exported in Gln, Glu, Asp, Ala, and GABA, in addition to Asn, so these compounds might be synthesized from the first product Gln (*via* glutamate decarboxylase and alanine aminotransferase). The high concentration of Asn and the low amount of Asp in the nodule suggest that the AS reaction operates exclusively in the Asn direction. The low levels of Asp may perhaps be useful in avoiding inhibition of PEPC, which might supply *Frankia* with organic-acid precursors. After 9h, less than 20% of the label that was present at 1 h remained in soluble amino acids in the nodule (Baker and Parsons, 1997). This ^{15}N transfer demonstrates the close linkage of N_2 fixation with the physiology of the whole plant.

6.4.3. Amide Exporter - Discaria trinervis

In xylem sap collected from root segments of symbiotic plants, Asn was the main N-compound exported from nodules. Asp, Glu, Ala, and Ser were also detected but at lower concentrations (Valverde, 2000). Such a pattern suggests that *D. trinervis* nodules produce Asn as the final product of N_2 fixation and assimilation. In agreement with this hypothesis, most of the enzymes involved in the synthesis of Asn from NH_4^+ and oxoacids are present in nodules, but not in roots, of *D. trinervis* seedlings that rely on N_2 as the sole source of N (Valverde and Wall, 2003a). Enzymatic assays on nodule extracts detected GS, AAT, and MDH activities. GOGAT activity could not be measured, in line with its tendency to become inactivated during preparation of nodule extracts (Blom, 1982). However, it was possible to detect the protein immunologically with alfalfa NADH-GOGAT antibodies. Immunoblots also detected the GS, MDH, AS, and PEPC proteins in nodule extracts. Even though heterologous antibodies were used on either nodule or root extracts, in all cases, the polypeptides detected had comparable molecular masses to those used to raise the antisera (Valverde and Wall, 2003a). All together, these observations suggest that the pathway for Asn synthesis depicted in Figure 3 also operates in *D. trinervis* nodules.

6.4.4. Citrulline exporter - Alnus glutinosa

Alnus glutinosa is probably the actinorhizal plant on which most studies have been carried out in terms of nodule N and C metabolism. For more than 50 years, *A. glutinosa* was known to be a Cit exporter (Miettinen and Virtanen, 1952; Bollard, 1957; Scott *et al.*, 1981). Pioneering studies of N assimilation showed that the first proteins of the N-assimilatory pathway in *A. glutinosa* were OCT (Martin *et al.*, 1983) and GS (Hirel *et al.*, 1982). OCT is the enzyme that produces Cit from Orn through the urea cycle and it was present in great amounts in *A. glutinosa* nodules (Martin *et al.*, 1983), most probably in mitochondria (Scott *et al.*, 1981). Initial assimilation of N, however, proceeds through GS, which was purified from root nodules as an octameric enzyme with K_m values of 0.9mM for Glu and 0.5mM for ATP (Hirel *et al.*, 1982). Nodule GS was activated by Mg^{2+} ions and strongly inhibited by AMP, ADP, and PP_i, which suggest a negative regulation of the protein activity by a low cell energy charge. In contrast, several amino compounds (Cit, Orn, Asn) did not influence GS activity directly *in vitro*. However, the same amino acids could have a regulatory effect *in vivo* through regulatory proteins that may have been lost in the purification process (see section 8.1.). Purified root GS behaved essentially as nodule GS (Hirel *et al.*, 1982). By using antibodies against GS from spinach chloroplasts, nodule GS was immunolocalized in the cytosol of the large inner cortical cells that surround *Frankia* vesicle clusters (Hirel *et al.*, 1982).

Differential screening of a nodule cDNA library with nodule and root cDNA resulted in the isolation of a nodule-specific clone that corresponded to the GS gene (AgGS1) with high homology to cytosolic isoforms of both alfalfa and soybean GS (Guan *et al.*, 1996). AgGS1 was expressed in all plant tissues, with the highest levels in nodules and the lowest in roots grown in the absence of combined-N. In nodules, GS mRNA was found both in the infected cortical cells and in cells identified as part of the multilayered pericycle of the central vascular bundle. The highest level of GS transcripts coincided with the highest expression of *nifH*. With the exception of the pericycle, no GS mRNA was detected in cells that did not express *nifH* (Guan *et al.*, 1996), which suggests that the expression of GS in *A. glutinosa* is under metabolic control, *i.e.*, by the presence of NH_4^+ to be assimilated, and not under developmental control as in legumes (Vance, 2000). Free NH_4^+ might be present in the pericycle of nodule vascular bundles because GS transcripts were detected there. In this case, it is possible that GS acts in the re-assimilation of N compounds prior to export (Guan *et al.*, 1996).

As part of the same cDNA library screening, a message encoding an acetylornithine transaminase (ACOAT; clone Ag118) was identified and characterized (Guan *et al.*, 1996). ACOAT catalyzes the fourth step in the Arg pathway that is used for the synthesis of Cit (Figure 3). The expression pattern of ACOAT was different from GS; no transcripts were found in the pericycle, but it was expressed only in infected cells. It is probable that assimilation of symbiotically derived NH_4^+ and the biosynthesis of the transport form, Cit, occur mainly in the infected cells of *A. glutinosa* nodules (Guan *et al.*, 1996). The deduced N-terminal sequence of the ACOAT cDNA clone suggests that it has a mitochondrial location, which would be consistent with the subcellular location of Cit biosynthesis in eukaryotic cells (Guan *et al.*, 1996).

6.4.5. Citrulline exporter - Alnus incana

Alnus incana has been the main species used for studies of the physiology and biochemistry of N_2 fixation of symbiotic *Frankia* (reviewed by Huss-Danell, 1997). The isolation of *Frankia* vesicle clusters has been very important for describing the activity of nitrogenase and its response to C stress, the respiratory activity of vesicles with different C substrates, and the occurrence of *Frankia* GS isoforms in the symbiotic state (Lundquist and Huss-Danell, 1991a; 1991b; 1992; Vikman and Huss-Danell, 1987a; 1987b; 1991; Vikman, 1992; Vikman *et al.*, 1990; see section 5). Fewer efforts have been made to characterize metabolic activities downstream of nitrogenase in nodules of *A. incana*.

It seems likely that GS is active in *A. incana* nodules because ^{15}N assimilation is specifically inhibited by methionine sulphoximine (Lundberg *et al.*, 1996). Similarly, Gln content of *A. incana* nodules is reduced by a dark stress (Lundquist *et al.*, 2003), which is known to affect drastically nitrogenase activity (Lundquist and Huss-Danell, 1991b). As a Cit exporter (Table 3), the inhibition of N_2 fixation and assimilation after a dark stress is also reflected in a lower level of Cit in *A. incana* nodules (Lundquist *et al.*, 2003). Several cDNAs have been identified by screening a nodule cDNA library with heterologous probes and by RT-PCR with degenerate oligonucleotides (Lundquist *et al.*, 1996; P. O. Lundquist and C. P. Vance, unpublished data as cited in Mullin and Dobritsa, 1996). By these means, messages for CMPS, PEPC, PFK, and GOGAT have been cloned and their expression detected only in the nodules. Using the same approach (B. Solheim, unpublished data as cited in Mullin and Dobritsa, 1996), cDNAs for *A. incana* SS and enolase have been isolated.

6.4.6. Mixed Exporter - Casuarina *spp.*

In general, in either ureide- or amide-transporting legumes, the principal N solute accounts for not less than 50% of translocated N. *Casuarina* species do not follow this trend because several N compounds, in similar proportions of the total N transported, can be found in the xylem sap. Asn, Cit, Arg, and Orn were the most abundant N compounds in three of the species studied (*Casuarina glauca, C. cunninghamiana*, and *C. equisetifolia*) (Sellstedt and Atkins, 1991). Citrulline was detected in xylem exudates of fully symbiotic *Casuarina cunninghamiana, C. equisetifolia*, and *C. glauca* but, with the exception of *C. equisetifolia* stems, it was not the main N compound. When plants were supplied with mineral-N, the proportion of Cit in relation to other N solutes did not change dramatically (Sellstedt and Atkins, 1991). Moreover, there was no clear change in the proportion of xylem amino acids in relation to the level of mineral-N. These results indicate that none of these products was formed predominantly from N_2 fixation.

Asn decreased with added mineral-N, but this happened with symbiotic as well as with non-symbiotic plants. In contrast, the relative concentration of Gln increased with addition of mineral-N. Xylem sap from *C. cunninghamiana* and *C. equisetifolia* contained up to 30% of Arg. The sum of Arg and Cit provides a more realistic measure of the proportion of exported N allocated to Cit synthesis (Sellstedt and Atkins, 1991). Little is known about NH_4^+ assimilation in *Casuarina*, but GS transcripts were detected in N_2-fixing infected cells of *C. glauca* nodules as well as in nodule pericycle cells (Pawlowski *et al.*, 1996).

6.4.7. Mixed Exporter - Datisca glomerata.

The herbaceous nature of *Datisca glomerata* and the low tissue levels of phenolics render it amenable to molecular biological studies. *Datisca glomerata* and *Coriaria spp.* are distinguished from other actinorhizal hosts in that the symbiotic N_2-fixing cells form a distinct dense sector within the nodule cortex, which is located to one side of the central vascular cylinder (see chapter 5 in this voulme). Whereas the xylem sap of *Datisca glomerata* plants is rich in Gln and Glu (Table 3), Arg is the most abundant amino acid after Gln in nodule extracts (Berry *et al.*, 2004). Such a difference points to Arg as the main product of N assimilation. If this is the case, it is expected that specific enzymes, like CMPS and OCT to provide the Cit needed to synthesize Arg and arginino-succinate synthetase and arginino-succinate lyase to generate Arg from Cit, are expressed at high levels in infected cells. These enzymes should then be expressed at comparatively lower levels in uninfected cells. Interestingly, GS transcripts and protein are detected only in uninfected cortical cells (Pawlowski *et al.*, 1996; Berry *et al.*, 2004). It has been proposed that Arg is synthesized in infected cells and then transported to uninfected cells, where it is catabolized by arginase, Arg iminohydrolase, and/or urease to release NH_4^+. This process could explain the expression of GS in uninfected cells. However, it remains unclear how NH_4^+ derived from N_2 fixation could be assimilated in infected cells, if not by the GS-GOGAT system. This hypothesis clearly deserves further.

6.5. Amide and Cit Synthesis in Actinorhizal Nodules

From the studies described in section 6.4, it is possible to depict a partial description of the pathways that lead to the export of Cit and amides in actinorhizal species. Whatever the final product of N assimilation would be, it seems likely that NH_4^+ is first assimilated by the action of GS and GOGAT (Figure 4). Both GS protein and mRNA have been detected in infected cells of both *Alnus glutinosa* (Guan *et al.*, 1996; Hirel *et al.*, 1983) and *Casuarina glauca* (Pawlowski *et al.*, 1996). Thereafter, pathways diverge in different species. Asn, the main amide exported by *Myrica*, *Elaeagnus*, *Ceanothus*, and *Discaria*, appears to be formed by the action of AAT and AS (Figure 4) as indicated by $^{15}N_2$-labeling studies in *Myrica gale* (Baker *et al.*, 1997) and by enzyme-activity assays and immunochemical detection in *Discaria trinervis* (Valverde and Wall, 2003a). Cit is normally synthesized from Orn and CMP, and Orn comes from Glu *via* the Orn cycle (Figure 4). In this case, the synthesis and export of Cit needs an anaplerotic reaction to replenish 2-KG. Both PEPC and CMPS can fix CO_2 in alder nodules; PEPC has been immunolocalized in

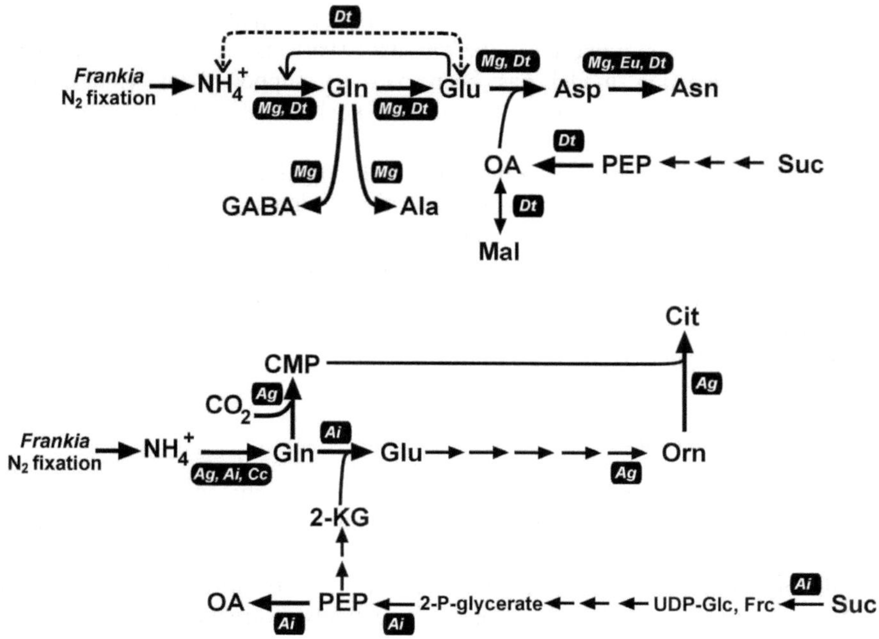

Figure 4. Amide and Cit synthesis in actinorhizal nodules.
The diagram shows the proposed pathways for the synthesis of amides (Asn, Gln) and Cit in nodules of actinorhizal plants. Those steps where an enzyme activity or its enzyme protein have been detected are highlighted with a box indicating the plant species in which the study has been carried out. Amide exporters: Dt, Discaria trinervis*; Eu,* Elaeagnus umbellata*; Mg,* Myrica gale. *Cit exporters: Ag,* Alnus glutinosa*; Ai,* Alnus incana*;* *Cc,* Casuarina cunninghamiana.

the host cytoplasm of *Alnus* nodule cells (McClure *et al.*, 1983), whereas OCT has been located in mitochondria (Martin *et al.*, 1983), which suggests that Cit synthesis occurs in this organelle (Figure 4).

7. THE REGULATION OF N ASSIMILATION

Nodules are organs with intense metabolic activity and an important flow-through of molecules and exchange of metabolites. Part of the recently fixed N_2 is used within the nodule, although nearly 50% of the N needs of the nodule are supplied by combined-N entering the nodule from the phloem (Schubert, 1986). Excess combined-N in the nodule is returned to the shoot *via* the xylem. Thus, xylem-sap composition reflects the result of the assimilation of combined-N as well as the the turnover and cycling of combined-N from distant organs, such as the leaves.

An integrative view of the physiology and biochemistry of N metabolism in nodules, within the context of the plant requirements, has led to the development of the concept of feedback regulation of N_2 fixation, nodule growth, and N assimilation, and the hypothesis of the "N stat" (Parsons *et al.*, 1993; Hartwig, 1998). This interesting model has been partially supported by experimental evidence in actinorhizal systems, although the biochemical signals involved still remain to be identified.

There is yet another level of complexity in the regulation of N metabolism in nodules. Environmental and particularly nutritional signals are either directly or indirectly transmitted to nodule cortical cells. Therein, changes in the availability of certain key nutrients affect both nitrogenase activity and its constituent proteins and also the NH_4^+-assimilatory pathway. For a detailed description of the mechanisms that modulate nitrogenase activity of *Frankia* in symbiosis, we refer the reader to Huss-Danell (1997) and to chapter 5 of this volume.

7.1. The Role of Internal N

All forms of combined-N uptake are regulated by the combined-N demand of the plant (Imsande and Touraine, 1994). For example, several amino acids applied directly to cotyledons of soybean seedlings had a negative impact on root NO_3^- assimilation (Muller and Touraine, 1992). Further, in *Arabidopsis*, the expression of the NH_4^+ transporter AMT1 in roots is negatively correlated with root Gln concentration (Rawat *et al.*, 1999). These two examples illustrate the inverse relationship that exists between NO_3^- and NH_4^+ uptake in roots and the amino-acid content of plant tissues.

Actinorhizal plants adjust the amount of symbiotic tissue to their N demand. For instance, nodulated plants of *Alnus glutinosa* and *Discaria trinervis* decrease the proportion of nodule mass in relation to plant weight when fertilized with mineral-N (Baker *et al.*, 1997; Valverde and Wall, 2002a). Also, the rate of nodule development in *Discaria trinervis* is decreased when leaf N content reaches its maximum level (Valverde *et al.*, 2000). These observations are in agreement with the theory that the N demand of the plant is the critical factor in regulating the symbiotic relationship and its performance (Hartwig, 1998). A negative feedback regulation by internal N is envisaged.

When combined-N was supplied to one side of a split-root system of seedlings of either *Hippophaë rhamnoides* or *Alnus incana* (Gentili and Huss-Danell, 2002; Wall *et al.*, 1998), it had inhibitory effects on nodule growth and activity of the untreated and physically separated side of the root system. Thus, incorporation of combined-N at sites that are distant from nodules can inhibit nodulation and N_2 fixation, which indirectly suggests long-distance signaling related to combined-N status.

In addition, feedback control may also work at the level of N assimilation within nodules and might be reflected in the amino-acid pattern of nodules and xylem sap. In fact, either NO_3^- or NH_4^+ supplied to *Myrica gale* plants induced a change in the amino-acid profile of xylem sap, namely, the Asn concentration decreased but both Gln and Glu concentrations increased (Baker *et al.*, 1997a). Similar treatments forced *Alnus glutinosa* nodulated roots to produce and export more Cit (Baker *et al.*, 1997b; Tonin *et al.*, 1990). The response of N assimilation at the level of the enzymes involved in C and N metabolism has also been studied in *Discaria trinervis* nodules (Valverde and Wall, 2003a). Application of combined-N to nodulated plants over a 4-week period resulted in a 50% decrease in GS activity and GS protein in nodule extracts (Valverde and Wall, 2003a), plus a concomitant decrease in both MDH activity and AS protein. Such behavior can be interpreted as feedback inhibition by N-satiety signaling.

The combined-N status of the plant must ultimately control N_2 fixation and assimilation, perhaps *via* signals transported in the phloem and an appropriate nodule sensor system. In actinorhizal plants, the final effect on N_2 fixation may be implemented through either decreasing carbohydrate availability or by altering the O_2-protection mechanisms (Parsons and Sunley, 2001; see also chapter 5 in this volume). Although either mechanism would reduce nitrogenase activity, the N-feedback signaling might also operate downstream of nitrogenase at the level of fixed-N metabolism.

More recently, the concept of an "N charge", as an indicator of the C and N status, has been proposed in relation to the combined-N feedback signaling in plants (equation (1), where [Gln], [Glu], and [2-KG] represent the concentrations of glutamine, glutamate and 2-ketoglutarate, respectively; Parsons and Sunley, 2001).

$$\text{N charge} = \frac{2\,[\text{Gln}] + [\text{Glu}]}{[\text{Gln}] + [\text{Glu}] + [\text{2-KG}]} \qquad (1)$$

Depending on the plant species, other N compounds (like Cit) would be more relevant in this ratio. It would be interesting to test this model in actinorhizal systems, where, in principle, total concentrations of Gln, Glu and 2-KG could be determined by biochemical methods in nodule extracts from plants subjected to different growth regimes that alter their C and N budgets. The biochemical significance of the "N charge" is still speculative, but it might operate through modulation of the activity of key enzymes, such as GS, that are involved in C and N metabolism. For example, Asp and Glu inhibit PEPC (Schuller *et al.*, 1990), whereas Glu and Gln act on pyruvate kinase (Podesta and Plaxton, 1994). If phosphorylation/dephosphorylation, a common regulatory mechanism of many enzymes, were involved, then there would also be a dependence on the AMP/ATP ratio (Hartwig, 1998).

7.2. Regulation of N Assimilation by Nutrients and Environmental Factors

If information on the regulation of N assimilation by internal combined-N is scarce in actinorhizal plant biology, it is even scarcer for the effects that external factors, such as either nutrients supply or light regime, may have on assimilation of fixed-N. This paucity of information may be due to nitrogenase being more sensitive to environmental stresses than the downstream enzymes. Thus, any effect observed on the enzymes of C and N metabolism would be related to the nitrogenase response. Several treatments that result in a reduced supply of carbohydrates to nodules of *Alnus* plants (like defoliation or darkening) are known to have a direct effect on nitrogenase activity, most probably due to a temporarily reduced respiration rate in infected cells and a concomitant increase in pO_2, which would lead to nitrogenase inactivation and degradation (Huss-Danell and Sellstedt, 1985; Lundquist and Huss-Danell, 1991a; 1991b; Lundquist *et al.*, 2003; Vikman *et al.*, 1990). The consequent decrease in NH_4^+ concentration could then have an impact on the activities of downstream enzymes.

In *Discaria trinervis*, combined-N acts as a negative regulator of nodulation and nodule growth, whereas P is a strong stimulator of nodule growth (Valverde *et al.*, 2000; Valverde *et al.*, 2002). When symbiotic seedlings were exposed to 2mM NH_4NO_3 over a four-week period, nodule GS, MDH and AS were inactivated and degraded, whereas GDH and AAT were activated; the amount of nitrogenase proteins was not affected by this treatment (Valverde and Wall, 2003a). In contrast, a 10-fold increase in P supply did not greatly affect the activity and amount of these enzymes, suggesting that N metabolism is not limited by the P level in nodules. On the other hand, suppression of the P supply induced a significant reduction in nodule GS, GOGAT, MDH and AS protein levels, although nitrogenase was again not affected. Only GDH activity was stimulated by a limited P supply (Valverde and Wall, 2003a). These responses might be explained by impaired C metabolism due to the lack of phosphate.

Plants that were shaded during the same period showed a complete degradation of nitrogenase and partial degradation of GS, AS and a nodule-specific MDH isoform, but GDH and AAT were activated (Valverde and Wall, 2003a). A similar pattern was observed when the darkening treatment was for a shorter period (6 days), except that nitrogenase activity was drastically decreased before its proteins were degraded (Valverde and Wall, 2003b). At least in *Discaria trinervis* nodules, dark stress and P limitation have strong and negative effects on enzymes downstream of nitrogenise, however, only dark stress induced a rapid degradation of nitrogenase proteins. These results suggest that, in the low P situation, symbiotic *Frankia* cells can respire at a rate high enough to keep the pO_2 below nitrogenise-inactivating levels. A prolonged darkness treatment also caused an important decrease in the Cit concentration of *Alnus incana* nodules (Lundquist *et al.*, 2003). Most probably, such decline in Cit content is due to the impaired N_2 fixation and decreased production of Gln necessary to sustain CMP synthesis to enter the Cit pathway (Lundquist *et al.*, 2003).

8. PROSPECTS

A survey of this exciting field of actinorhizal biology clearly reveals that we already have information from many actinorhizal systems, although no one system has been studied completely. This situation is probably a reflection of both the geographical location of research groups studying this aspect of actinorhizal biology and the different experimental approaches utilized (Table 1). Even so, many interesting questions remain. These include: (i) what C compounds do symbiotic *Frankia* cells receive from the nodule host cells?; (ii) what major pathways result in the synthesis of amides and Cit in nodules?; (iii) how are these pathways compartmentalized within the infected cells?; (iv) what is the metabolic role of uninfected cells?; (v) what determines the pattern of N molecules in the xylem?; and (vi) how are the pathways regulated by the N demand of the plant?

There is a clear need to concentrate research efforts to understand C and N metabolism in certain actinorhizal models, where different experimental approaches could be applied. However, we must keep in mind the biodiversity of actinorhizal plants and apply any differences to learn how nature has used several anatomical and biochemical solutions to solve the problem of N_2 fixation and assimilation.

With regards to genetic engineering techniques, transgenic actinorhizal plants have been established in only a few cases, namely *Casuarina* and *Allocasuarina* (Franche *et al.*, 1998), but these plants have such a long reproductive cycle that the idea of getting seeds is almost impractical. Only *Datisca* species would be useful in this endeavor due to its comparatively rapid growth. Alternatively, micro-propagation protocols could be envisaged to multiply genetically transformed plant-cell tissues. In any case, genetic tools are not yet available for other genera, although they are likely to be developed soon. On the bacterial side, it would be of value if the genome of *Frankia* was sequenced. This would provide a genetic basis to search for genes that are differentially expressed in symbiosis.

In summary, much work remains if we are to understand how these two organisms, *Frankia* and plant cells in root nodules, work together to share C sources and to redirect their metabolisms to provide the plant with combined-N for growth and development.

ACKNOWLEDGEMENTS

C. Valverde gratefully acknowledges the Universidad Nacional de Quilmes, CONICET and ANPCyT (all in Argentina) and IFS for their support through fellowships and grants. K. Huss-Danell acknowledges NFR and SJFR for their long-term financial support and STINT for financial support of collaboration with Universidad Nacional de Quilmes. Both authors are greatly indebted to C.T. Wheeler and L.G. Wall for their valuable comments on this work.

REFERENCES

Abeysekera, R. M., Newcomb, W., Silvester, W. B., and Torrey, J. G. (1990). A freeze-fracture electron microscopic study of *Frankia* in root nodules of *Alnus incana* grown at three oxygen tensions. *Can. J. Microbiol.*, *36*, 97-108.

Akkermans, A. D. L., Huss-Danell, K., and Roelofsen, W. (1981). Enzymes in the tricarboxylic acid cycle and the malate-aspartate shuttle in the N_2-fixing endophyte of *Alnus glutinosa*. *Physiol. Plant.*, *53*, 289-294.

Akkermans, A. D. L., Roelofsen, W., Blom, J., Huss-Danell, K., and Harkink, R. (1983). Utilization of carbon and nitrogen compounds by *Frankia* in synthetic media and in root nodules of *Alnus glutinosa, Hippophaë rhamnoides,* and *Datisca cannabina. Can. J. Bot., 61,* 2793-2800.

Akkermans, A. D. L., van Straten, J., and Roelofsen, W. (1977). Nitrogenase activity of nodule homogenates of *Alnus glutinosa*: a comparison with the *Rhizobium*-pea system. In W. E. Newton, J. R. Postgate and C. Rodriguez-Barrueco (Eds.), *Recent developments in nitrogen fixation* (pp. 591-603). London, UK: Academic Press.

Baker, A., Hill, G. F., and Parsons, R. (1997a). Evidence for N feedback regulation of N_2 fixation in *Alnus glutinosa* L. *J. Exp. Bot., 48,* 67-73.

Baker, A., Hill, G. F., and Parsons, R. (1997b). Alteration of N nutrition in *Myrica gale* induces changes in nodule growth, nodule activity and amino acid composition. *Physiol. Plant., 99,* 632-639.

Baker, A., and Parsons, R. (1997). Rapid assimilation of recently fixed N_2 in root nodules of *Myrica gale. Physiol. Plant., 99,* 640-647.

Benson, D. R., and Schultz, N. A. (1990). Physiology and biochemistry of *Frankia* in culture. In C. R. Schwintzer and J. D. Tjepkema (Eds.), *The biology of* Frankia *and actinorhizal plants* (pp. 107-127). San Diego, CA: Academic Press.

Benson, D. R., and Silvester, W. B. (1993). Biology of *Frankia* strains, actinomycete symbionts of actinorhizal plants. *Microbiol. Rev., 57,* 293-319.

Berg, R. H. (1990). Cellulose and xylans in the interface capsule in symbiotic cells of actinorhizae. *Protoplasma, 159,* 35-43.

Berry, A. M., Harriot, O. T., Moreau, R. A., Osman, S. F., Benson, D. R., and Jones, A. D. (1993). Hopanoid lipids compose the *Frankia* vesicle envelope, presumptive barrier of oxygen diffusion to nitrogenase. *Proc. Natl. Acad. Sci. U.S.A., 90,* 6091-6094.

Berry, A. M., Murphy, T. M., Okubara, P. A., Jacobsen, K. R., Swensen, S. M., and Pawlowski, K. (2004). Novel expression pattern of cytosolic glutamine synthetase in nitrogen-fixing root nodules of the actinorhizal host, *Datisca glomerata. Plant Physiol.,135,* 1849-1862.

Berry, A. M., Rasmussen, U., Bateman, K., Huss-Danell, K., Lindwall, S., and Bergman, B. (2002). Arabinogalactan proteins at the symbiotic interface in root nodules of *Alnus* spp. *New Phytol., 155,* 469-479.

Berry, A. M., and Sunell, L. A. (1990). The infection process and nodule development. In C. R. Schwintzer and J. D. Tjepkema (Eds.), *The biology of* Frankia *and actinorhizal plants* (pp. 61-81). San Diego, CA: Academic Press.

Blom, J. (1982). *Carbon and nitrogen metabolism of free-living* Frankia *and of* Frankia-Alnus *symbioses.* Ph.D. Thesis, Agricultural University, Wageningen, The Netherlands.

Blom J., Roelofsen, W., and Akkermans, A. D. L. (1981). Assimilation of nitrogen in root nodules of alder (*Alnus glutinosa*). *New Phytol., 89,* 321-326.

Bollard, E. G. (1957). Translocation of organic nitrogen in the xylem. *Aust. J. Biol. Sci., 10,* 292-301.

Ching, T. M., Monaco, P. A., and Ching, K. K. (1983). Energy status and cytochromes in isolated endophytic vesicle clusters of red alder root nodules. *Can. J. For. Res., 13,* 921-928.

Engelke, T., Jording, D., Kapp, D., and Pühler, A. (1989). Identification and sequence analysis of the *Rhizobium meliloti dctA* gene encoding the C4-dicarboxylate carrier. *J. Bacteriol., 171,* 5551-5560.

Fontaine, A. I., Lancelle, S. A., and Torrey, J. G. (1984). Initiation and ontogeny of vesicles in cultured *Frankia* strain HFPArI3. *J. Bacteriol., 160,* 921-927.

Franche, C., Laplaze, L., Duhoux, E., and Bogusz D. (1998). Actinorhizal symbioses: Recent advances in plant molecular and genetic transformation studies. *Crit. Rev. Plant Sci., 17,* 1-28.

Gentili, F., and Huss-Danell, K. (2002). Phosphorus modifies the effects of nitrogen on nodulation in split-root systems of *Hippophae rhamnoides. New Phytol, 153,* 53-61.

Giller, K. E., and Wilson, K. J. (1991). *Nitrogen fixation in tropical cropping systems.* 1st Ed. Wallingford, UK: CAB International.

Gordon, A. J., Minchin, F. R. James, C. L., and Komina, O. (1999). Sucrose synthase in legume nodules is essential for nitrogen fixation. *Plant Physiol., 120,* 867-877.

Guan, C., Ribeiro, A., Akkermans, A. D. L., Jing, Y., van Kammen, A., *et al.* (1996). Nitrogen metabolism in actinorhizal nodules of *Alnus glutinosa*: Expression of glutamine synthetase and acetylornithine transaminase. *Plant Mol. Biol., 32,* 1177-1184.

Hafeez, F., Akkermans, A. D. L., and Chaudhary, A.H. (1984). Observations on the ultrastructure of *Frankia* sp. in root nodules of *Datisca cannabina* L. *Plant Soil, 79,* 383-402.

Hahn, D., Starrenburg, M. J. C., and Akkermans, A. D. L. (1988). Variable compatibility of cloned *Alnus glutinosa* ecotypes against ineffective *Frankia* strains. *Plant Soil, 107*, 233-243.

Hartwig, U. A. (1998). The regulation of symbiotic N_2 fixation: A conceptual model of N feedback from the ecosystem to the gene expression level. *Persp. Plant Ecol. Evol. Syst., 1*, 92-120

Hirel, B., Perrot-Rechenmann, C., and Gadal, P. (1982). Glutamine synthetase in alder (*Alnus glutinosa*) root nodules. Purification, properties and cytoinmunochemical localization. *Physiol. Plant., 55*, 197-203.

Hsieh, M. H., Lam, H. M, van de Loo, F. J., and Coruzzi, G. (1998). A PII-like protein in *Arabidopsis*: Putative role in nitrogen sensing. *Proc. Natl. Acad. Sci. U.S.A., 95*, 13965-13970.

Huss-Danell, K. (1990). The physiology of actinorhizal nodules. In C. T. Schwintzer and J. D. Tjepkema (Eds.), *The biology of* Frankia *and actinorhizal plants* (pp. 129-156). San Diego, CA: Academic Press.

Huss-Danell, K. (1997). Actinorhizal symbioses and their N_2 fixation. *New Phytol., 136*, 375-405.

Huss-Danell, K., and Bergman, B. (1990). Nitrogenase in *Frankia* from root nodules of *Alnus incana* (L.) Moench: Immunolocalization of the Fe- and MoFe-proteins during vesicle differentiation. *New Phytol., 116*, 443-455.

Huss-Danell, K., Lundquist, P.-O., and Ohlsson, H. (1992). Distribution of biomass and nitrogen among plant parts and soil nitrogen in a young *Alnus incana* stand. *Can. J. Bot., 70*, 1545-1549.

Huss-Danell, K., Roelofsen, W., Akkermans, A. D. L., and Meijer, P. (1982). Carbon metabolism of *Frankia* spp. in root nodules of *Alnus glutinosa* and *Hippophae rhamnoides*. *Physiol. Plant., 54*, 461-466.

Huss-Danell, K., and Sellstedt, A. (1985). Nitrogenase activity in response to darkening and defoliation of *Alnus incana*. *Can. J. Exp. Bot., 36*, 1352-1358.

Imsande, J., and Touraine, B. (1994). N demand and the regulation of nitrate uptake. *Plant Physiol., 105*, 3-7.

Kim, H. B., Lee, S. H., and An, C. S. (1999). Isolation and characterization of a cDNA clone encoding asparagine synthetase from root nodules of *Elaeagnus umbellata*. *Plant Sci., 149*, 85-94.

Kleemann, G., Alskog, G., Berry, A. M., and Huss-Danell, K. (1994). Lipid composition and nitrogenase activity of symbiotic *Frankia* (*Alnus incana*) in response to different oxygen concentrations. *Protoplasma, 183*, 107-115.

Lam, H. M., Coschigano, K. T., Olivera, I. C., Melo-Oliveira, R., and Coruzzi, G. M. (1996). The molecular-genetics of nitrogen assimilation into amino acids in higher plants. *Annu. Rev. Plant Physiol. Plant Mol. Biol., 47*, 569-593.

Laplaze, L., Ribeiro, A., Franche, C., Duhoux, E., Auguy, F., Bogusz, D., *et al.* (2000). Characterization of a *Casuarina glauca* nodule-specific subtilisin-like protease gene, a homolog of *Alnus glutinosa ag12*. *Mol. Plant-Microbe Interact., 13*, 113-117.

Lechevalier, M. P., and Lechevalier, H. A. (1990). Systematics, isolation and culture of *Frankia*. In C. R. Schwintzer and J. D. Tjepkema (Eds.), *The biology of Frankia and actinorhizal plants* (pp. 35-60). San Diego, CA: Academic Press.

Li, Y., Parsons, R., Day, D. A., and Bergersen, F. J. (2002). Reassessment of major products of N_2 fixation by bacteroids from soybean root nodules. *Microbiol., 148*, 1959-1966.

Lodwig, E. M., Hosie, A. H. F., Bourdés, A., Findlay, K., Allaway, D., *et al.* (2003). Amino-acid cycling drives nitrogen fixation in the *Rhizobium*-legume symbiosis. *Nature, 422*, 722-726.

Lopez, M. F., and Torrey, J. G. (1985). Purification and properties of trehalase in *Frankia* ArI3. *Arch. Microbiol., 143*, 209-215.

Lopez, M. F., Young, P., and Torrey, J. G. (1986). A comparison of carbon source utilization for growth and nitrogenase activity in two *Frankia* isolates. *Can. J. Microbiol., 32*, 353-358.

Lundberg, P., Lundquist, P.-O., and Huss-Danell, K. (1996). ^{1}H, ^{13}C, ^{15}N and ^{31}P NMR methods for studying metabolites in an N_2-fixing root nodule symbiosis, *Alnus incana* x *Frankia*. *Bull. Magn. Reson., 18*, 131-132.

Lundquist, P.-O. (1993). *Dynamics of nitrogenase in the* Frankia-Alnus incana *symbiosis*. Ph.D. Thesis, Department of Plant Physiology, Umeå University, Umeå, Sweden.

Lundquist, P.-O., and Huss-Danell, K. (1991a). Nitrogenase activity and amounts of nitrogenase proteins in a *Frankia-Alnus incana* symbiosis subjected to darkness. *Plant Physiol., 95*, 808-813.

Lundquist, P.-O., and Huss-Danell, K. (1991b). Response of nitrogenase to altered carbon supply in a *Frankia-Alnus incana* symbiosis. *Physiol. Plant., 83*, 331-338.

Lundquist, P.-O., and Huss-Danell, K. (1992). Immunological studies of glutamine synthetase in *Frankia-Alnus incana* symbioses. *FEMS Microbiol Lett*, *91*, 141-146.

Lunquist, P.-O., Näsholm, T., and Huss-Danell, K. (2003). Nitrogenase activity and root nodule metabolism in response to O_2 and short-term N_2 deprivation in dark-treated *Frankia-Alnus incana* plants. *Physiol. Plant.*, *119*, 244-252.

Lundquist, P.-O., Pawlowski, K., and Vance, C. P. (1996). Carbamoyl phosphate synthase: Sequence analysis and nodule enhanced expression in nitrogen-fixing *Alnus incana* root nodules. *Plant Physiol.*, *111 (Suppl.)*, 90.

Martin F., Hirel, B., and Gadal, P. (1983). Purification and properties of ornithine carbamyl transferase I from *Alnus glutinosa* root nodules. *Z. Pflanz.*, *111*, 413-422.

Mazzucco C., and Benson, D. R. (1984). ^{14}C-Methylammonium transport by *Frankia* sp. strain CpI1. *J. Bacteriol.*, *160*, 636-641.

McClure, P. R., Coker, III, G. T., and Schubert, K. R. (1983). Carbon dioxide fixation in roots and nodules of *Alnus glutinosa*. *Plant Physiol.*, *71*, 652-657.

Miettinen, J. K., and Virtanen, A. I. (1952). The free amino acids in the leaves, roots and root nodules of the alder (*Alnus*). *Physiol Plant*, *5*, 540-557.

Miflin, B. J., and Habash, D. Z. (2002). The role of glutamine synthetase and glutamate dehydrogenase in nitrogen assimilation and possibilities for improvement in the nitrogen utilization of crops. *J. Exp. Bot.*, *53*, 979-987.

Muller, B., and Touraine, B. (1992). Inhibition of NO_3^- uptake by various phloem-translocated amino acids in soybean seedlings. *J. Exp. Bot.*, *43*, 617-623.

Mullin, B. C., and Dobritsa, S. V. (1996). Molecular analysis of actinorhizal symbiotic systems: Progress to date. *Plant Soil*, *186*, 9-20.

Newcomb, W., and Wood, S. M. (1987). Morphogenesis and fine structure of *Frankia* (Actinomycetales): The microsymbiont of nitrogen-fixing actinorhizal root nodules. *Int. Rev. Cytol.*, *109*, 1-89.

Okubara, P. A., Pawlowski, K., Murphy, T. M., and Berry, A. M. (1999). Symbiotic root nodules of the actinorhizal plant *Datisca glomerata* express Rubisco activase mRNA. *Plant Physiol.*, *120*, 411-420.

Parsons, R., Silvester, W. B., Harris, S., Gruitjers, W. T. M., and Bullivant, S. (1987). *Frankia* vesicles provide inducible and absolute oxygen protection for nitrogenase. *Plant Physiol.*, *83*, 728-731.

Parsons, R., Stanforth, A., Raven, J. A., and Sprent, J. I. (1993). Nodule growth and activity may be regulated by a feedback mechanism involving phloem nitrogen. *Plant Cell Env.*, *16*, 125-136.

Parsons, R., and Sunley, R. J. (2001). Nitrogen nutrition and the role of root-shoot nitrogen signalling particularly in symbiotic systems. *J. Exp. Bot.*, *52*, 435-443.

Pawlowski, K., Akkermans, A. D. L., van Kammen, A., and Bisseling, T. (1995). Expression of *Frankia nif* genes in nodules of *Alnus glutinosa*. *Plant Soil*, *170*, 371-376.

Pawlowski, K., Ribeiro, A., Guan, C., van Kammen, A., Berry, A. M., and Bisseling, T. (1996). Actinorhizal nodules from different plant families. In G. Stacey, B. Mullin, and P.M. Gresshoff (Eds.), *Biology of plant-microbe interactions*, *Vol. 1* (pp. 417-422). St Paul, MN: IS-MPMI.

Perrot-Rechenmann, C., Vidal, J., Maudinas, B., and Gadal, P. (1981). Immunocytochemical study of phosphoenolpyruvate carboxylase in nodulated *Alnus glutinosa*. *Planta*, *153*, 14-17.

Podesta, F. E., and Plaxton, W. C. (1994). Regulation of cytosolic carbon metabolism in germinating *Ricinus communis* cotyledons: II. Properties of phosphoenolpyruvate carboxylase and cytosolic pyruvate kinase associated with the regulation of glycolysis and nitrogen assimilation. *Planta*, *194*, 381-387.

Prin, Y., Mallein-Garin, F., and Simonet, P. (1993). Identification and localization of *Frankia* strains in *Alnus* nodules by *in situ* hybridization of *nifH* mRNA with strain-specific oligonucleotide probes. *J. Exp. Bot.*, *44*, 815-820.

Rawat, S. R., Silim, S. N., Kronzucker, H. J., Siddiqi, M. Y., and Glass, A. D. (1999). *AtAMT1* gene expression and NH_4^+ uptake in roots of *Arabidopsis thaliana*: Evidence for regulation by root glutamine levels. *Plant J.*, *19*, 143-152.

Ribeiro, A., Praekelt, U., Akkermans, A. D. L., Meacock, P. A., van Kammen, A., *et al.* (1996). Identification of *agthi1*, whose product is involved in biosynthesis of the thiamine precursor thiazole, in actinorhizal nodules of *Alnus glutinosa*. *Plant J.*, *10*, 361-368.

Romagni, J. G., and Dayan, F. E. (2000). Measuring asparagine synthetase activity in crude plant extracts. *J. Agri. Food Chem.*, *48*, 1692-1696.

Rosendahl, L., Vance, C. P., and Pedersen, W. B. (1990). Products of dark CO_2 fixation in pea root nodules support bacteroid metabolism. *Plant Physiol.*, *93*, 12-19.

Schubert, K. R. (1986). Products of biological nitrogen fixation in higher plants: Synthesis, transport, and metabolism. *Ann. Rev. Plant. Physiol.*, *37*, 539-574.

Schubert, K. R., and Boland, M. J. (1990). The ureides. In B. J. Miflin and P. J. Lea (Eds.), *The biochemistry of plants, Vol. 16* (pp.197-282). San Diego, CA: Academic Press.

Schuller, K. A., Turpin, D. H., and Plaxton, W. D. (1990). Metabolite regulation of partially purified soybean nodule phosphoenolpyruvate carboxylase. *Plant Physiol.*, *94*, 1429-1435.

Scott A., Gardner, I. C., and McNally, S. F. (1981). Localization of citrulline synthesis in the alder root nodule and its implication in nitrogen fixation. *Plant Cell Rep.*, *1*, 21-22.

Seguin, A., and Lalonde, M. (1989). Detection of pectinolytic activity and pel homologous sequences in *Frankia*. *Plant Soil*, *118*, 221-229.

Sellstedt, A., and Atkins, C. A. (1991). Composition of amino compounds transported in xylem of *Casuarina* sp. *J. Exp. Bot.*, *42*, 1493-1497.

Shi, L., Twary, S. N., Yoshioka, H., Gregerson, R. G., Miller, S. S., Samac, D. A., *et al.* (1997). Nitrogen assimilation in alfalfa: Isolation and characterization of an asparagine synthetase gene showing enhanced expression in root nodules and dark-adapted leaves. *Plant Cell*, *9*, 1339-1356.

Srivastava, H. S., and Singh, R. P. (1987). Role and regulation of L-glutamate dehydrogenase activity in higher plants. *Phytochem.*, 26, 597-610.

Tonin, G. S., Wheeler, C. T., and Crozier, A. (1990). Effect of nitrogen nutrition on amino acid composition of xylem sap and stem wood in *Alnus glutinosa*. *Physiol. Plant.*, *79*, *79*, 506-511.

Tonin, G. S., Wheeler, C. T., and Crozier, A. (1991). Effect of changes in nitrogen nutrition on the polyamine content of *Alnus glutinosa*. *Plant Cell Env.*, *14*, 415-421.

Valverde, C. (2000). *Regulación de la nodulación radicular en la simbiosis* Discaria trinervis–Frankia. Ph. D. Thesis, Facultad de Ciencias Exactas, Universidad Nacional de La Plata, La Plata, Argentina.

Valverde, C., Ferrari, A., and Wall, L. G. (2002). Phosphorus and the regulation of nodulation in the actinorhizal symbiosis between *Discaria trinervis* (Rhamnaceae) and *Frankia* BCU110501. *New Phytol.*, *153*, 43-52.

Valverde, C., and Wall, L.G. (2003a). Ammonium assimilation in *Discaria trinervis* root nodules. Regulation of enzyme activities and protein levels by the availability of macronutrients (N, P and C). *Plant Soil*, *254*, 139-153.

Valverde, C., and Wall, L.G. (2003b). The regulation of nodulation, nitrogen fixation and assimilation under a carbohydrate shortage stress in the *Discaria trinervis–Frankia* symbiosis. *Plant Soil*, *254*, 155-165.

Valverde, C., Wall, L. G., and Huss-Danell, K. (2000). Regulation of nodulation and nodule mass relative to nitrogenase activity and nitrogen demand in seedlings of *Discaria trinervis* (Rhamnaceae). *Symbiosis*, *28*, 49-62.

van Ghelue, M., Ribeiro, A., Solheim, B., Akkermans, A. D. L., Bisseling, T., and Pawlowski, K. (1996). Sucrose synthase and enolase expression in actinorhizal nodules of *Alnus glutinosa*: Comparison with legume nodules. *Mol. Gen. Genet.*, *250*, 437-446.

van Straten, J., Akkermans, A. D. L., and Roelofsen, W. (1977). Nitrogenase activity of endophyte suspensions derived from root nodules of *Alnus*, *Hippophaë*, *Shepherdia* and *Myrica* spp. *Nature*, *266*, 257-258.

Vance, C. P. (2000). Amide biosynthesis in root nodules of temperate legumes. In E. W. Triplett (Ed.), *Prokaryiotic nitrogen fixation: A model system for analysis of a biological process* (pp. 589-607). Wymondham, UK: Horizon Scientific Press.

Vance C. P., and Lamb, J. F. S. (2001). Application of biochemical studies to improving nitrogen fixation. *Aust. J. Exp. Agri.*, *41*, 403-416.

Vikman, P.-A. (1992). The symbiotic vesicle is a major site for respiration in *Frankia* from *Alnus incana* root nodules. *Can. J. Microbiol.*, *38*, 779-784.

Vikman, P.-A., and Huss-Danell, K. (1987a). Capacity for hexose respiration in symbiotic *Frankia* from *Alnus incana*. *Physiol. Plant.*, *70*, 349-354.

Vikman, P.-A., and Huss-Danell, K. (1987b). Purity of *Frankia* preparations from root nodules of *Alnus incana*. *Physiol. Plant.*, *71*, 489-494.

Vikman, P.-A., and Huss-Danell, K. (1991). Respiration of malate and glutamate in symbiotic *Frankia* prepared from *Alnus incana*. *J. Exp. Bot.*, *42*, 221-228.

Vikman, P.-A., Lundquist, P.-O., and Huss-Danell, K. (1990). Respiratory capacity, nitrogenase activity and structural changes of *Frankia*, in symbiosis with *Alnus incana*, in response to prolonged darkness. *Planta*, *182*, 617-625.

Wall, L. G., Hellsten, A., and Huss-Danell, K. (1998). P alters N effects on nodulation in *Alnus incana* and *Trifolium pratense*. In C. Elmerich, A. Kondorosi and W. E. Newton (Eds.), *Biological nitrogen fixation for the 21st century* (pp. 363-364). Dordrecht, The Netherlands: Kluwer Academic Publishers.

Wall, L. G., Hellsten, A., and Huss-Danell, K. (2000). Nitrogen, phosphorus, and the ratio between them affect nodulation in *Alnus incana* and *Trifolium pratense*. *Symbiosis*, *29*, 91-105.

Walsh, K. B., Ng, B. H., and Chandler, G. E. (1984). Effects of nitrogen nutrition on xylem sap composition of Casuarinaceae. *Plant Soil*, *81*, 291-293.

Warenbourg, F. R., and Roumet, C. (1989). Why and how to estimate the cost of symbiotic N_2 fixation? A progressive approach based on the use of ^{14}C and ^{15}N isotopes. *Plant Soil*, *115*, 167-177.

Waters, J. K., Hughes, B. L., Purcell II, L. C., Gerhardt, K.O., Mawhinney, T. P., and Emmerich, D. W. (1998). Alanine, not ammonia, is excreted from N_2-fixing soybean nodule bacteroids. *Proc. Natl. Acad. Sci. U.S.A.*, *95*, 12038-12042.

Wheeler, C. T. (1971). The causation of the diurnal changes in nitrogen fixation in the nodules of *Alnus glutinosa*. *New Phytol.*, *70*, 487-495.

Wheeler, C. T., and Bond, G. (1970). The amino acids of non-legume root nodules. *Phytochem.*, *9*, 705-708.

Wheeler, C. T., and Lawrie, A. C. (1976). Nitrogen fixation in root nodules of alder and pea in relation to the supply of photosynthetic assimilates. In P.S. Nutman (Ed.), *Symbiotic nitrogen fixation in plants* (pp. 497-510). Cambridge, UK: Cambridge University Press.

Wheeler, C. T., Tonin, G. S., and Sutcliffe, A. (1994). Polyamines of *Frankia* in relation to nitrogen nutrition. *Soil Biol. Biochem.*, *26*, 577-581.

Zhang, X., and Benson, D. R. (1992). Utilization of amino acids by *Frankia* sp. strain CpI1. *Arch. Microbiol.*, *158*, 256-261.

Chapter 8

ECOLOGY OF ACTINORHIZAL PLANTS

J. O. DAWSON

Departments of Natural Resources and Environmental Sciences and Plant Biology, University of Illinois at Urbana-Champaign, IL, U.S.A.

1. IMPORTANCE OF ACTINORHIZAL PLANTS

The actinorhizal plants that are nodulated by the symbiotic, N_2-fixing actinomycete *Frankia* include 25 genera in 8 families of angiospermous plants (Table 1). Actinorhizal plant families, together with all legumes and the rhizobially-nodulated genus *Parasponia*, have been placed in the rosid clade, which contains plants with a predisposition to nodular symbiosis with diazotrophs (Soltis *et al.,* 1995). The symbiotic organ is a coralloid-root nodule formed on primary roots infected by *Frankia*. Many important ecological interactions, patterns, and functions of actinorhizal plants are specifically associated with this N_2-fixing symbiosis but, in this chapter, only the symbiosis-related ecological features will be discussed.

Fixed nitrogen (fixed-N) availability is commonly limiting to primary productivity and other processes of ecosystems throughout the world. Nitrogen fixation by actinorhizal plants is a major source of fixed-N in diverse and widespread terrestrial ecosystems, including forests, bogs, swamps, coastal dunes, landslides, glacial deposits, riparian zones, shrub lands, prairies, and deserts (Dawson, 1986). Actinorhizal plants play important roles in wild-land ecosystem function and are used in land reclamation, range management, forestry, agro-forestry, and horticulture. It is probable that actinorhizal plants contribute a high proportion of the total amount of N fixed globally, primarily in wild or extensively managed lands. Dixon and Wheeler (1986) estimated that the contribution of actinorhizal plants to terrestrial global nitrogen fixation could be as great as 25% of the total.

Estimated rates of actinorhizal nitrogen fixation are comparable to those of legumes. Nitrogen-fixation rates vary widely within and among actinorhizal

K. Pawlowski and W. E. Newton (eds.), Nitrogen-fixing Actinorhizal Symbioses, 199–234.
© 2008 *Springer.*

Table 1. Actinorhizal plant taxa, species number, number of species known to be nodulated, and native distribution.

Family	Genus	Number of species	Known nodulated	Native distribution
Betulaceae	*Alnus*	47	47	Europe, Asia, N. America, Andes Mtns.
Casuarinaceae	*Allocasuarina*	59	54	Australia
	Casuarina	18	18	Australia, Tropical Asia, S-W Pacific
	Ceuthostoma	2	2	Oceania
	Gymnostoma	18	18	Australia, New Caledonia, Sumatra
Coriariaceae	*Coriaria*	16	16	Mediterranean, Asia, New Zealand, N America
Datiscaceae	*Datisca*	2	2	Asia, N America, Europe
Elaeagnaceae	*Elaeagnus*	45	35	Europe, Asia, N America
	Hippophae	3	2	Europe, Asia
	Shepherdia	3	2	N America
Myricaceae	*Comptonia*	1	1	N America
	Myrica	60	28	All continents except Australia
Rhamnaceae	*Adolphia*	1	1	N America
	Ceanothus	55	31	N America
	Colletia	17	4	S America
	Discaria	10	5	S America, Australia, New Zealand
	Kentrothamnus	2	2	S America
	Talguenea	1	1	S America
	Trevoa	6	2	S America
Rosaceae	*Cercocarpus*	20	4	Mexico, S-W United States
	Chamaebatia	2	1	Sierra Nevada Mtns.
	Cowania	25	1	Mexico, S-W United States
	Dryas	3	1	Arctic
	Purshia	4	2	W-N America

Adapted from Baker and Schwintzer (1990), Bond (1983), Dawson (1986), Huguet (2003), and Swensen (1996).

species depending on methodology used as well as ecological and genetic factors. Thus, it follows that a wide range of values has been reported for the major taxa. Estimates of the annual N_2-fixation rate for: (i) *Alnus rubra* (red alder trees) range from 22 to over 300kg/ha (summarized in Hibbs and Cromack, 1990); (ii) *Ceanothus velutinus* (snowbrush ceanothus) range from 24-101 kg/ha (McNabb and Cromack, 1985; Rose and Youngberg, 1981); and (iii) *Myrica faya* in Hawaii is 18kg/ha (Vitousek and Walker, 1989). Actual N_2-fixation values for *Casuarina equisetifolia* range from 15-94kg $ha^{-1}y^{-1}$ (summarized by Dommergues, 1997).

Much of what is known about actinorhizal associations has been derived from controlled studies in laboratories, growth chambers, and greenhouses. The physiology and biochemistry of both actinorhizal and other N_2-fixing systems is relatively well characterized, whereas our knowledge of both the amounts of N_2 fixed and the ecological controls of nitrogen fixation in nature remains less developed (Vitousek *et al.*, 2002). There are considerable difficulties inherent in ecological studies of actinorhizal plants, particularly the subterranean ecology. These difficulties include: (i) precisely estimating annual rates of fixation; (ii) dealing with the great complexity and heterogeneity of soil and associated ecosystems; and (iii) the necessity of excavating actinorhizal nodules, often in rocky soil and at considerable depth.

The host plants in actinorhizal symbioses are all perennial and all, but those of the genus *Datisca*, are trees or shrubs. Perennial plants have complex seasonal, annual, and multi-year reproductive, growth, and nutritional cycles, all of which renders them less amenable to study than annual plants. Reproductive capacity and related aspects of physiological maturity may require years to develop in trees, and the results of controlled studies that employ seedlings cannot be directly applied to mature plants.

Despite the difficulties associated with actinorhizal studies in general and ecological approaches in particular, greater ecological understanding is a prerequisite for managing and domesticating actinorhizal plants and their communities. Ecological inference in controlled experiments is strengthened by introducing some of the ecological complexities of nature, such as the multiple variables encountered in the field. However, there is no substitute for studying the actinorhizal symbioses *in situ*.

In nature, interactions occur among actinorhizal organisms and a dynamic range of biotic, physical and chemical agents. Such complexity is the norm and not the exception. A complex of environmental factors determines the actual survival, growth, nodulation, and nitrogen fixation of the symbiotic partners at any given site on the Earth's surface. So, ecological studies of actinorhizal plants, ranging from the global to the micro-ecosystem scale, are important and necessary complements to molecular and organism studies.

Major anthropogenic and natural changes in the environment have the potential to impact both the occurrence of actinorhizal plant and critical functions locally and globally. Evidence for climatic and human impact on regional actinorhizal plant occurrence and abundance has been obtained from the archaeologically rich Patacancha Valley in the mountains of southern Peru (Chepstow-Lusty *et al.,* 1998). An 8-meter core taken from a small, dry lake revealed an enormous increase in pollen of *Alnus acuminata* (Andean alder) occurring *ca.* A.D. 1000. At about this time, a warming trend reversed a 900-year period of cooler temperatures, affording a climate more suitable for Andean alder. Evidence from the core suggests that the cool period was accompanied by decreased pre-Incan farming, but continued degradation and erosion of the soil.

At the time of warming, the Inca took over the valley and the signature of soil erosion in the core dropped dramatically. The Inca employed soil-sparing techniques, such as terraces and irrigation canals, and both seeds and pollen of maize and other crops appeared in sediments and dated archaeological finds of the period. As the Incan population quadrupled and stabilized, the occurrence of alder pollen persisted, even as the growing population used alder trees for firewood, roof beams, and door lintels. The increase in population was accompanied by increased agricultural activity, but apparently with reduced soil erosion, which suggests a system for conserving alder trees that would have stabilized soil on steep slopes.

Speculation that Andean alders were used in agro-forestry practices, possibly even in association with terraces where they could both stabilize and improve the soil, is not unreasonable. Contemporary indigenous peoples of Central America still use alders, intermingled with corn, on steep slopes to stabilize soil and to increase fertility by lopping off branches and scattering them as a green manure. Additional evidence for alder management by the Inca appeared in historical records shortly after the Spanish conquest in the early 1500s. Chroniclers among the conquistadors wrote that the Inca had a strong tradition of tree planting. Alder cultivation was overseen by the emperor, who punished both illegal woodcutters and those who burned the trees by putting them to death (Krajick, 1998).

After the Spanish conquest, terraces, canals and tree plantings were abandoned, and alder stands gradually disappeared. Today, alder is restricted to a few remote ravines in the valley. These observations illustrate how changes in climate and civilization might have shifted the distribution to higher elevations and greatly increased the regional occurrence and use of an actinorhizal tree. In the city of Manizales, in Colombia, Andean alders are currently planted and protected in the municipal watershed to guarantee a clean and seasonally reliable source of water.

Today, humankind faces the probability of new, widespread, anthropogenic changes in the global environment as well as cyclical and episodic changes that occur on a global scale. According to geological patterns, massive volcanic eruptions with drastic global cooling, ice ages, asteroid collisions, and warming cycles are inevitable in the long term. More immediately, increasing atmospheric carbon-dioxide levels, fixed-N pollution of air and water, global warming, ozone pollution, acid rain, and other problems associated with industrial and agricultural activity are increasing. Plants, animals, and microorganisms cross geographic barriers routinely as global travel, commerce, and transport continue to increase. Introductions and escapes of biotic organisms into new territories, which lack the co-evolved checks and balances, can result in disease epidemics, harmful insect infestations, population explosions of animal pests that disrupt balanced ecosystems, and the invasion of biotic communities by exotic plants, including several actinorhizal species that could displace natives.

Ultimately, such continent hopping by living organisms can result in the extinction of native species and the loss of biological diversity. In the context of dynamic change, the occurrence, distribution, abundance, biotic associations, genetic makeup, and ecological function of actinorhizal plants will change accordingly. Understanding the ecological basis for change can facilitate effective and sustainable management of these important actinorhizal plant resources.

2. OCCURRENCE AND DISTRIBUTION OF ACTINORHIZAL PLANT TAXA AND THEIR MICROSYMBIONTS

2.1. Global Distribution of Actinorhizal Plant Taxa

Most actinorhizal plant species occur in boreal and temperate ecosystems of both the Northern and Southern Hemispheres. A widespread representative of an actinorhizal plant genus with many species of cold and temperate climates in the Northern Hemisphere is *Alnus* (alder). The genus *Alnus* dates back in the fossil record to the Upper Cretaceous, 70-80 million years ago. A basal position on the alder phylogenetic tree, a slow evolutionary rate, and morphologically primitive characteristics indicate that the subgenus Alnaster, represented by the widespread subspecies of *Alnus viridis* (green alder) that occurs in cold climates and at the margins of glaciers, is closest to the ancestral form of the alders (Navarro *et al.,* 2003). Northeast Asia is the probable center of origin for this genus because this region has the greatest number of alder species and the highest level of alder endemism (Murai, 1964; Navarro *et al.,* 2003).

A. viridis ssp. *sinuata*, *Shepherdia Canadensis*, and *Dryas drummondii* are post-glacial primary successional plants in the Northern Hemisphere (Kohls *et al.,* 2003), whereas *Coriaria* plays a similar role on post-glacial deposits in the mountains of New Zealand in the Southern Hemisphere. A sudden cooling started about 12,700 years ago, in the midst of the last global warming cycle, and lasted 1,300 years. This cold period, known as the Younger Dryas, is named for the pollen of this actinorhizal tundra plant that turned up in a lakebed in Denmark when it should not have. Two earlier sudden cold periods, known as the oldest Dryas (13,800 years ago) and the older Dryas (13,450 years ago), have also been revealed by sudden increases in the pollen of this actinorhizal plant at more southerly locations (Roberts, 1998). *Shepherdia* in Canada and *Hippophae* in Scandinavia were apparently post-glacial precursors of boreal forests at the end of the last Quaternary glaciation together with *Alnus* on both continents (Baker and Miller, 1980; Silvester, 1977). *Myrica gale* is widespread in bogs of boreal regions in the Northern Hemisphere, whereas *Alnus*, *Shepherdia*, and *Hippophaë* are actinorhizal plant genera present in boreal and cool temperate regions of the Northern Hemisphere today.

Many *Casuarina* and *Allocasuarina* species occur in warm temperate regions of Australia in a variety of habitats. Unlike most taxa of actinorhizal plants with distributions restricted to higher latitudes and altitudes, some *Casuarina* and all *Gymnostoma* spp. are endemic to tropical forests in the south-west Pacific region. The tropical genus *Gymnostoma* may represent an ancestral stock of the family Casuarinaceae. This genus was widespread in Gondwanaland at least by the beginning of the Tertiary. It predates *Casuarina* and *Allocasuarina* in the fossil record and is least specialized with the simplest morphology. It has small chromosomes with the lowest chromosome number ($n = 8$), and represents the only actinorhizal family of Gondwanian origin with generic differentiation occurring in early Tertiary times (Barlow, 1983). The genus *Gymnostoma* may have diverged into *Casuarina* and *Allocasuarina* species having both tropical and temperate adaptations. All other actinorhizal plant families are of northern Laurasian origin.

In the tropics, some genera of actinorhizal plants occur at high elevations with cool moist climates that are similar to those of temperate regions. Examples include: the *Myrica* belt around Mount Kilimanjaro in Africa; *Myrica* in the mountains of the Canary Islands, the Caribbean and Central America; and the alders (*Alnus* spp.) of the highlands of tropical regions in South America and Asia.

In cooler regions of the globe, actinorhizal plants, with the exceptions noted previously, are the predominant tree and shrub forms of N_2-fixers. In the tropics, tree legumes replace actinorhizal plants as the predominant N_2-fixing trees and shrubs. The reasons for this dichotomy are unknown. Many late-successional tropical forests contain canopy legumes capable of N_2 fixation, whereas N_2-fixers are absent from most late-successional temperate and boreal forests. Vitousek *et al.* (2002) suggest that the relatively high fixed-N availability in lowland tropical rainforests could permit legumes to maintain a fixed-N-demanding lifestyle (McKey, 1994) without the necessity of expending considerable energy to fix atmospheric N_2. Conversely, in temperate and boreal forests, early-successional actinorhizal species are typically replaced by tree species that are less demanding of available fixed-N, perhaps at some threshold level below that necessary to support actinorhizal plants without costly N_2 fixation.

In warm-temperate transitional zones between the tropics and temperate regions, such as the monsoonal arid regions of Australia and North America, both leguminous and actinorhizal species occur together. In the Sonoran Desert and other semiarid regions of western North America, shrubby species of *Cercocarpus, Cowania,* and *Ceanothus* occur together with a variety of woody legumes, including *Prosopis* and *Acacia.* In the canyons of desert mountains, the cool moist sky islands that are separated by deserts and grasslands, *Alnus oblongifolia* (Arizona alder), a relatively tall tree of the riparian zone, occurs along perennial streams extending out into the upper Sonoran Desert.

Similarly in Australia, species of *Acacia* and other tree legumes occur in scrublands and forests of arid and semiarid zones together with the casuarinas. *Allocasuarina* spp. are prominent together with shrubby legumes in drier regions, including the central and northern desert monsoonal regions, whereas larger *Casuarina* species occur primarily along streams, in coastal areas, and in the moist tropical rainforests of northeastern Australia (Midgley *et al.,* 1983). *Allocasuarina decaisneana* is striking for its large size among trees of the desert of central Australia. It is found in sandy soils and has a system of deeply penetrating roots, most likely allowing it to access permanent water in deep soil strata.

Among the actinorhizal genera, *Myrica,* in the family Myricaceae, has the widest geographic distribution. They are small trees or shrubs common to nearly all major landmasses and with species occurring from tropical to cold regions. According to morphological features, fossil and pollen records, and a recent molecular study based on the *rbcL* gene (summarized in Huguet, 2003), the family Myricaceae is considered to be the most ancient actinorhizal family with *Myrica* appearing at least as early as 85-83 million years before the present.

Apart from the Casuarinaceae and Myricaceae, actinorhizal plants seem to be restricted to cold and temperate climates, perhaps reflecting a global trend of continental cooling during their evolutionary development. They seem to be distributed accordingly within either the Northern or Southern Hemispheres. There is hemispheric separation of cold- and temperate-climate actinorhizal plants because of the tropical equatorial barrier to temperate-plant migration.

How did actinorhizal plants of Laurasian origin reach the Southern Hemisphere? An example of a Laurasian actinorhizal genus that occurs on both sides of the equator is *Coriaria*. *Coriaria* species are distributed in four separate areas in the world; the most conspicuous disjunct distribution in flowering plants. The phylogenetic relationships of 12 *Coriaria* species collected from the representative disjunct areas were inferred by comparing the combined data set of *rbcL* and *matK* (maturase K) genes (Hasebe *et al.,* 1998). The divergence time between the Eurasian species and other species distributed in the Southern Hemisphere was estimated as between 63 and 59 million years ago, using *rbcL* and *matK* molecular clocks, respectively. These results do not support the hypotheses that continental drift created the disjunct distribution pattern by separating earlier contiguous populations of ancestral *Coriaria*. Rather they suggest that the distribution pattern was formed by geographical migrations and separations in the Cenozoic. Other actinorhizal plant genera and their geographic distributions are included in Table 1.

2.2. Global Distribution of Frankia *Taxa*

Infective *Frankia* are known to occur on all continents, except Antarctica, and on many islands, but vary spatially (Paschke *et al.,* 1994; Simonet *et al.,* 1999) and temporally (Wollum *et al.,* 1968) in occurrence. Nodulation of an actinorhizal plant may not occur either because *Frankia* is absent or, more likely, because specific strains, which are able to nodulate a given host, are not present in a soil or are unable to nodulate a host under existing soil conditions.

Within their native ranges, most actinorhizal plant species nodulate with *Frankia* strains capable of symbiotic nitrogen fixation, although nodulation with ineffective strains also occurs in nature (Wolters *et al.,* 1997). Soil near actinorhizal hosts generally has greater nodulation capacity than surrounding soils (Jeong and Myrold, 2001; Smolander, 1990; Zimpfer *et al.,* 1999). However, infective *Frankia* can be found in a variety of soils both within and outside the immediate influence of actinorhizal hosts as well as outside of the native range of actinorhizal plant species (Burleigh and Dawson, 1994; Lawrence *et al.,* 1967; Maunuksela *et al.,* 1999, 2000; Paschke and Dawson, 1992a; Zimpfer *et al.,* 1997). *Frankia* able to nodulate *Alnus*, *Myrica*, *Dryas* and *Elaeagnus* are found widespread outside the native range of their host plants, suggesting that they have the capacity to persist as a saprophyte (Kohls *et al.,* 1994; Maunuksela *et al.,* 1999; 2000; Nickel *et al.,* 1999; 2000).

The presence of plant species of genera, which are not actinorhizal but that are closely related to actinorhizal genera, can increase *Frankia*-nodulation capacity of soil for actual host species (Gauthier *et al.,* 2000; Paschke and Dawson, 1992b; Smolander *et al.,* 1990). Increased rhizosphere soil-nodulation capacity of actinorhizal hosts and some other closely related plant species suggests the release of compounds that stimulate *Frankia* growth or infectious capacity or both (Zimpfer *et al.,* 2002; 2003). A compound from roots of an actinorhizal plant has been shown to stimulate *Frankia* spore germination (Krumholz *et al.,* 2003). In order to determine the distribution of *Frankia* in nature, they must first be retrieved from soil or nodules and described to allow differentiation among genotypes. Since the first confirmed isolation of the actinomycete genus *Frankia* (Callaham *et al.,* 1978), a variety of methods have been used to obtain a coherent classification of this bacterial symbiont.

Frankia can be difficult, or impossible in the case of some actinorhizal host genera, to isolate in pure culture from nodules. There are few reports of successful isolations of *Frankia* from soil (Baker and O'Keefe, 1984). Furthermore, not all cultures have been derived from single spores and some isolates could represent multiple strains. Many "atypical" strains, isolated from nodules but incapable of reinfecting a host, have been catalogued, perhaps as a result of the mistaken isolation of rhizosphere *Frankia* from inadequately sterilized nodule surfaces (Huguet *et al.,* 2001; see also chapter 2 in this volume). If the actinomycetous partner cannot be identified or isolated, the descriptive and predictive value of ecological inference is weak.

Methods for describing and comparing *Frankia* strains include soluble protein patterns, isoenzyme patterns, fatty acids, serology, DNA-DNA relatedness, and genome and plasmid restriction analyses. Recently, ecological studies of the symbiotic interactions between *Frankia* and its hosts have benefited from advances in molecular biology, which allow identification of nodular microsymbionts without first obtaining isolates (Hahn *et al.,* 1999; see also chapter 2 in this volume). Thus, ecological and distributional features can be studied with respect to specific rather than generic *Frankia*. Nodules are usually collected in the field, but *Frankia* in a soil are extracted, using actinorhizal bait plants to capture individuals from infective populations.

The earliest studies indicated that *Frankia* strains are heterogeneous and clustered within two main phylogenetic groups: the *Alnus-Myrica-Casuarina* group and the Elaeagnaceae group (reviewed by Hahn *et al.,* 1999). Strains that infect *Myrica* spp. often grouped with *Alnus*-infective strains. Additional work (Navarro *et al.,* 1997) placed *Gymnostoma-Frankia* in the *Elaeagnus* group. Clawson *et al.* (1998) and Hahn *et al.* (1999) defined a third "*Dryas* group", which encompassed *Coriaria*, *Datisca*, *Ceanothus*, and the actinorhizal plants in the family Rosaceae (Table 2; see also chapter 4 in this volume).

Maggia and Bousquet (1994) have proposed that host-plant evolution has proceeded toward more specificity and, therefore, the most primitive actinorhizal plants, such as *Myrica* and *Gymnostoma*, should be able to host the broadest ranges of *Frankia* genotypes. Navarro *et al.* (1997) disagree and suggest that *Gymnostoma* and *Casuarina-Allocasuarina* constitute two divergent plant lineages associated with two divergent Frankia lineages. Simonet *et al.* (1999) have suggested that *Casuarina* species from Australia are less specific and have a broader host range than more recently evolved *Allocasuarina* spp. They further suggest that the more

Table 2. Dominant phylogenetic groups of Frankia *naturally associated with host genera.*

		Phylogenetic groups of *Frankia*		
Family	Genus	"Alnus"	"Dryas"	"Elaeagnus"
Betulaceae	*Alnus*	X		
Casuarinaceae	*Allocasuarina*	X		
	Casuarina	X		
	Ceuthostoma			?
	Gymnostoma			X
Coriariaceae	*Coriaria*		X	
Datiscaceae	*Datisca*		X	
Elaeagnaceae	*Elaeagnus*			X
	Hippophae			X
	Shepherdia			X
Myricaceae	*Comptonia*	X		
	Myrica	X		
Rhamnaceae	*Adolphia*		?	
	Ceanothus		X	
	Colletia			X
	Discaria			X
	Kentrothamnus			X
	Retanilla			X
	Talguenea			X
	Trevoa			X
Rosaceae	*Cercocarpus*		X	
	Chamaebatia		X	
	Cowania		X	
	Dryas		X	
	Purshia		X	

Adapted from Clawson et al. (1998), Hahn et al. (1999), Navarro et al. (1997), and Torrey (1990).

recently divergent and less specialized species of *Casuarina*, such as *Casuarina equisetifolia*, likewise symbiose with less-specialized *Frankia* strains possessing greater saprophytic abilities.

Myrica spp. have been described as promiscuous because some species, such as *M. pensylvanica* and *M. californica,* can host a diversity of nodular *Frankia* in the field that is sufficient to allow them to serve as a reservoir host for *Frankia* strains that infect plants from other actinorhizal families (Clawson *et al.,* 1999; Huguet, 2003). Other species of Myricaceae, such as *M. gale*, are not promiscuous in the field and have greater field specificity for microsymbionts (Huguet *et al.,* 2001). Actinorhizal plants in the family Elaeagnaceae have also been described as promiscuous and species, such as *Shepherdia canadensis*, can support diverse nodular *Frankia* (Huguet *et al.,* 2001).

It is noteworthy that green alder has been found nodulated in New Zealand (Benecke, 1969) and *Alnus nepalensis* (Himalayan alder) nodulates spontaneously in Canberra, Australia; in both cases, separated by thousands of miles and isolated by oceans from their native ranges. In contrast, *Casuarina* and *Allocasuarina* spp. from the same region do not nodulate in New Zealand, in other places outside their native ranges, or indeed even in some soils within their native ranges, unless the appropriate *Frankia* strains have been inoculated or otherwise introduced (Simonet *et al.,* 1999).

A diversity of *Frankia* genotypes able to nodulate *Casuarina* and *Allocasuarina* plant spp. seem to occur naturally only in soils within the host-genus native range (Simonet *et al.,* 1999). Elsewhere, exotic casuarinas must be inoculated in order to nodulate when planted in new locations (Diem and Dommergues, 1990). It has been postulated that the source of nodular *Frankia* in older introduced casuarinas that occur outside their native range may have been nodules and soils from seedlings planted by seafarers beginning at least 4 centuries before the present (Simonet *et al.,* 1999). Virtually all of the nodular *Frankia* in casuarinas found outside of Australia are of one phylogenetic group, which is also the only Casuarinaceae group of *Frankia* to successfully yield isolates, all of which suggests greater saprophytic capacity. In one Jamaican location where casuarinas have been introduced, this same group of *Casuarina*-infective *Frankia* is highly localized in soil around host trees. This may be a result of substances released from the host that increase infective *Frankia* populations in soil (Zimpfer *et al.,* 2002; 2003).

As noted previously, *Myrica, Alnus, Dryas,* and *Elaeagnus* spp. nodulate readily either outside their native ranges or with soils and nodule inocula obtained from locations distant from their native ranges (Kohls *et al.,* 1994; Paschke *et al.,* 1994). The explanation generally offered for this capacity is that the host species are promiscuous and can consort symbiotically with a heterogeneous mix of widespread microsymbiont genotypes. However, their ability to nodulate as newcomers in soils far from their place of origin could also be a result of: (i) the widespread occurrence of certain genotypes that include genera within their specific host-compatibility ranges; (ii) a greater dispersal capacity of their compatible microsymbionts; (iii) microsymbionts that possess greater competitive or saprophytic capabilities; or (iv) *Frankia* that are supported in the rhizospheres of many non-actinorhizal plants.

Some *Frankia* strains might be more readily dispersed over long distances because of differences in the amount of sporulation, spore longevity, spore resistance to harsh environmental conditions, or greater potential for wind dispersal. Additionally, the ability of spores to survive in soil particles, which are ingested by or adhering to animals, and to survive in the digestive tracts of soil arthropods and the birds and other animals that consume them, might magnify their dispersal. Likewise, there could be both biotic and abiotic agents in different soil types and conditions that influence nodulation and host/microsymbiont specificity.

Our knowledge of global distribution patterns of *Frankia* is far from complete. Even though we know that *Frankia,* which are able to infect a given host, may or may not occur in soils outside of native-host influence and range, we cannot yet comprehensively characterize non-infective *Frankia* soil populations. It also remains to be determined whether the lack of infectivity of soil-born *Frankia* in bioassays might be due to either genetic incompatibility or a number of environmental triggers that alter the physiological capacity of the microsymbiont and its plant partner to nodulate. Direct determination of the nature, dynamics and distribution of *Frankia* populations in soil and the range and occurrence of non-infective *Frankia* strains would allow more detailed determination of the range of *Frankia* groups globally and regionally.

2.3. Niches of Actinorhizal Plant Species

A common ecological niche for actinorhizal plants is where soil nitrogen availability is low. Low levels of available combined-N can critically limit the establishment and growth of plants unable to fix N_2 symbiotically. In such a niche, actinorhizal and other plants possessing symbiotic associations with N_2-fixing bacteria are often favored. Available combined-N is severely limiting to plant growth in soils low in organic matter, the major pool of soil N.

Examples of low-N soils include recent volcanic deposits, landslide areas on steep slopes, sand and gravel deposits of rivers and streams, sand dunes, beaches, recent glacial deposits, and eroded soils. Combined-N can also be limited in availability in grasslands, where an extensive system of fine roots competes strongly with the roots of other plants for water and nutrients. Many arid ecosystems are N-limited due both to extreme moisture and temperature conditions that inhibit organic matter mineralization and to patchy accumulation of wind-distributed soil particles containing nitrogenous substances. Wetlands may be N-limited because of slow mineralization of organic-N. In wetlands, low rates of N mineralization can result from low O_2 concentrations, sub-optimal temperatures for mineralization of N, and loss of both organic and mineral forms of N from soil through leaching, dilution, and transport by water. Furthermore, NO_3^- present in soil solution can be lost to plants as a result of denitrification in anaerobic strata and in flooded soils.

The early stage of plant succession following a disturbance, which creates a N-limited situation, is a common temporal niche for actinorhizal plants. In common with other pioneer plant species, actinorhizal plants are generally light demanding and do not persist under the closed canopies of later successional forest species (Côté *et al.,* 1988). Combined-N accumulates in soil with the shedding and decomposition of actinorhizal plant roots and litter. At a later point in successional development, the threshold of available soil-N, which allows other plants to become established, is crossed and other plants eventually displace the actinorhizal colonizers.

Alnus rubra in western North America and *A. glutinosa* (European alder) in Europe occur in riparian zones, such as gravel and sand bars, as well as in N-poor wetlands by virtue of their abilities both to fix N_2 and to tolerate flooding. *A. rubra* also occurs on landslides, volcanic mudflows, and after fire in the cool wet temperate regions of the Pacific Northwest of North America. These plants are light demanding and colonize disturbed sites. Species of *Alnus*, *Shepherdia*, *Hippophaë* and *Dryas* probably played a major role in soil formation immediately following the retreat of continental ice sheets over vast areas of Europe, Asia, and North America

(Baker and Miller 1980; Bond, 1983; Silvester, 1977), just as some of them do today as primary successional plants of recent glacial deposits (Kohls *et al.,* 2003).

Some actinorhizal plants of savannas, grasslands, chaparral, and shrublands differ from early successional actinorhizal plant species by persisting as stable components of plant communities. The ability to fix N_2 in these ecosystems probably allows the plant, *e.g.*, *Ceanothus* species in open grasslands and chaparral, to compete favorably with plants more efficient at either acquiring or using scarce available soil N. *Ceanothus americanus* (New Jersey Tea) occurs as a stable plant component of tallgrass prairies and oak savannas of central North America. This species has a massive main root and reproduces readily after fire from seeds and sprouts. These prairie and savanna ecosystems accumulate abundant organic-N in soil, but competition for nutrients and water by grasses, together with droughts that reduce N mineralization, may limit the availability of N to other plants. These examples illustrate typical niches and similar patterns of niche differentiation occur for other actinorhizal plant species globally.

2.4. Dispersal of Actinorhizal Plants and Frankia

The dispersal mechanisms of actinorhizal plants vary among taxa, and generally involve the movement of seeds. Alders and casuarinas produce many small winged seeds that are dispersed by wind and water. The rosaceous *Cercocarpus*, *Cowania*, and *Dryas* species have achenes with a persistent feathery style that facilitates wind dispersal. Actinorhizal plants in the Coriariaceae, Elaeagnaceae, and Myricaceae families produce seeds in fruits, which for some species of actinorhizal plants are ingested and the seeds likely dispersed by birds and mammals (Martin *et al.,* 1961). In the Rhamnaceae, *Ceanothus* species produce seeds in capsules. On drying, these capsules rupture in such a way as to forcefully eject the seeds a short distance. None of these mechanisms is either unusual or unique to actinorhizal plants.

Unlike host-plant dispersal methods, the mechanisms of dispersal of *Frankia* propagules have not been empirically established. Infective *Frankia* are present in newly deposited glacial till and young sand dunes prior to colonization by host plants (Kohls *et al.,* 2003; Young *et al.,* 1992). The likely mechanisms for *Frankia* dispersal are wind (anemochoric dispersal), water (hydrochoric dispersal), and biological vectors (zoochoric dispersal).

Frankia occurs in soils that lack host plants (Huss-Danell and Frej, 1986; Paschke and Dawson, 1992a; Zimpfer *et al.,* 1997) and on fresh substrates newly colonized by actinorhizal plants (Lawrence *et al.,* 1967; Schramm, 1966), which indicates that propagules of *Frankia* are mobile in nature. *Frankia* produces spores that are able to withstand desiccation (Burleigh and Torrey, 1990) for at least 18 months (Righetti and Munns, 1981; Tortosa and Cusato, 1991).

One possible mode of *Frankia* colonization of new habitats is anemochoric dispersal. However, unlike fungi, *Frankia* does not produce aerial sporangia for dispersal of spores and has a slow growth rate relative to soil saprobes. Aerial dispersal of *Frankia* would occur most readily where wind erosion and wind transport of soil particles containing *Frankia* propagules are prevalent. These conditions are widespread in arid regions, such as those of Australia, where there are many endemic *Casuarina* and *Allocasuarina* spp., as well as in glacial regions during the winter, when silt deposits are exposed with the recession of summer meltwaters. Of potential importance here is the induction of melanin by tyrosine, which is a prevalent trait of *Frankia* isolates from *Casuarina* spp. is (Lai, 1996). This induced trait, which is accompanied by increased sporulation in some strains, may afford a means of protection of airborne spores or hyphal particles of Casuarinaceae-infective *Frankia* from ultraviolet radiation.

Hydrochoric dispersal over long distances might be possible in riparian or lake ecosystems. Su and Lin (1989) speculated that *Frankia* moves in stream water in Asia; Huss-Danell *et al.* (1997) demonstrated that the action of streams, waves and changing water levels transport infective *Frankia* attached to superficial sediment particles in Alaska and Sweden; and Arveby and Huss-Danell (1988) suggested that *Frankia* moves laterally in the soil solution of peat.

Zoochoric dispersal is also very likely. McIlveen and Cole (1976) found that organisms that transport soil can act as dispersal vectors for spores of Endogonaceae fungi and it is probable that organisms, including humans, which move soil, may also passively transport *Frankia*. The transcontinental shipping of domestic livestock with soil particles adhering to fur and within folds of tissue is an example. Furthermore, the introduction of European sheep to New Zealand may have introduced *Frankia* from Europe to this isolated island nation.

Casuarina-infective *Frankia* can pass through the digestive tracts of earthworms, which undoubtedly disperse *Frankia* vertically together with large volumes of soil (Reddell and Spain, 1991). Birds, many of which consume earthworms and other soil invertebrates, also ingest large soil particles that function as grit for grinding food in their gizzards. Some bird species also transport mud that contains infective *Frankia* for nest construction as well as soil invertebrates as a food source for nestlings (Paschke and Dawson, 1993). *Frankia* spores are known to survive and maintain the ability to infect host plants after passage through the digestive tracts of birds (Burleigh and Dawson, 1995).

Infective *Frankia* is present in soils of tropical lowland forests of Costa Rica, which lack known actinorhizal hosts (Paschke and Dawson, 1992a). Central America is a funnel for the migratory routes of many bird species moving annually between North and South America. If birds are a major dispersal agent for *Frankia* and, if *Frankia* strains have some degree of saprophytic capacity, then the presence of *Frankia* in soils of regions where migratory birds congregate, but that lack host plants, would be easily explainable. Such findings provide support for the proposal that migratory bird species transport *Frankia* over long distances.

The possibility of codispersal of actinorhizal hosts and their *Frankia* microsymbionts exists for actinorhizal genera, such as *Myrica* and *Elaeagnus*, which have avian-dispersed seeds (Martin *et al.,* 1961). Birds tend to defecate when leaving perches in trees and shrubs, thus possibly more frequently inoculating soils near actinorhizal plants that are attractive as food sources and provide perches. Mammals that consume actinorhizal fruit and seeds could likewise codisperse the microsymbiont through soil adhering to their fur and other external tissues (Martin *et al.,* 1961). Seafaring traders, who transported *Casuarina* seedlings in soil for planting in distant lands, may also have introduced both soil and nodular *Frankia* throughout the tropical and warm temperate regions of the world (Simonet *et al.,* 1999).

What we know for certain about the distribution of actinorhizal plants and *Frankia* indicates the likelihood of many undiscovered interrelationships among actinorhizal partners and a variety of biotic and abiotic dispersal agents.

3. ECOLOGICAL FACTORS INFLUENCING INFECTIVE *FRANKIA* POPULATIONS

Many factors affect the abundance of soil-borne *Frankia* that are able to infect their host plants. Samples from uniform soils exhibit considerable variation in infective units of *Frankia* (Oremus, 1980; Zimpfer *et al.,* 1997; 1999), perhaps because of patchy favorable soil microhabitats, the localized release of *Frankia* from degrading nodules, localized stimulation of infective capacity by release of compounds from host plants and related genera, or the localized deposition by dispersal agents. Intense agricultural cultivation diminishes *Frankia*-nodulation capacity of soils (Paschke *et al.,* 1994; Zimpfer *et al.,* 1997), whereas the presence of actinorhizal host plants is a major factor in promoting *Frankia* presence and abundance in the rhizosphere (Zimpfer *et al.,* 1999), possibly as the result of compounds released from the host plants.

Research on *Frankia* ecology and physiology has some practical barriers not encountered in the parallel symbiosis between legumes and rhizobial bacteria. *Frankia* isolates grow slowly relative to rhizobial bacteria, and their filamentous nature precludes standard plate counts for enumeration. Furthermore, colonies of *Frankia* grown in liquid culture form coherent masses that cannot be directly assayed using photospectrometry, however, new methods to amend *Frankia* cultures to produce dispersed cells in liquid culture have recently been published (Harriott and Bourret, 2003). Despite this improvement in methodology, ecophysiological studies of the response of *Frankia* to physical and chemical variables *in vitro* remain more laborious than those of many other bacteria. Additionally, methods for genetic transformation of *Frankia* still remain elusive, so that strains with markers are not available for ecological studies in the field. Nonetheless, much has been learned about the ecology of this important microsymbiont despite these inherent difficulties.

3.1. Biotic Factors Influencing Frankia *in Soil*

As described in the preceeding section, zoochoric dispersal is probably a major ecological determinant of the occurrence of infective *Frankia* populations. Several studies have indicated that the presence of *Frankia* in young soils is not dependent on host-plant presence. In soils of the barrier islands off the coast of Virginia, colonization by *Frankia*, which is able to infect actinorhizal *Myrica cerifera,* occurs prior to colonization by host plants (Young *et al.,* 1992). On the islands of Java

(Becking, 1970) and Hawaii (Vitousek and Walker, 1989; Burleigh and Dawson, 1994), *Myrica* is nodulated on volcanic soils. On the island of Jamaica, one-year-old reclaimed bauxite mines harbored low levels of *M. cerifera*-infective *Frankia* (Zimpfer *et al.,* 1997). In Alaska's Glacier Bay, *Dryas drummondii* and *Alnus viridis ssp. sinuata* (Sitka alder) nodulate readily on newly-formed glacial soils (Lawrence *et al.,* 1967), indicating the early presence in glacial deposits of infectious *Frankia.*

Host plants apparently release compounds that increase the numbers of infective *Frankia* in soil (Krumholz *et al.,* 2003; Zimpfer *et al.,* 1999; 2002; 2003). Most actinorhizal plants are nodulated in their native habitats and *Frankia* is usually found in abundance in soils beneath nodulated host plants (Arveby and Huss-Danell, 1988; Oremus, 1980; Smolander, 1990; Van Dijk, 1979; Zimpfer *et al.,* 1999). However, some actinorhizal genera are often sparsely nodulated or not nodulated at all even within their native habitats (Lawrie *et al.,* 1982) and, therefore, probably contribute little to either host-plant success or the overall accumulation of soil nitrogen. In addition, there are numerous reports of *Frankia* in soils lacking actinorhizal host plants (Bermudez de Castro *et al.,* 1976; Dawson and Klemp, 1987; Huss-Danell and Frej, 1986; Paschke and Dawson, 1992b; Rodríguez-Barrueco, 1968; Smolander, 1990; Smolander and Sundman, 1987; Young *et al.*, 1992; Zimpfer *et al.,* 1997), providing indirect evidence that *Frankia* can grow and survive in soils in the absence of symbiotic hosts.

Compounds present in *Casuarina cunninghamiana* cladode tissue extracts increase the nodulation capacity of soil containing naturally occurring *Frankia,* whereas the addition of organic-binding agents to the host rhizosphere decreases nodulation (Zimpfer *et al.,* 2002). The increase in nodulation capacity caused by compounds released into plant rhizospheres may result from one or more of many possibilities, including: (i) the stimulation of *Frankia* growth; (ii) the stimulation of "helper" bacteria; (iii) the inhibition of competitors of *Frankia*; (iv) the enhancement of signaling by molecules important in the nodulation process; and (v) the stimulation of root-hair curling or fine-root production (Benoit and Berry, 1997; Gauthier *et al.,* 2000; Kapulnik *et al.,* 1987; Knowlton *et al.,* 1980). In addition, *Alnus* and *Elaeagnus* root extracts increase *Frankia* spore germination (Krumholz *et al.,* 2003). Further, *Alnus*-infective *Frankia* can grow as saprophytes in incubated soils and the addition of *Alnus* leaves to incubated soils can increase both growth and infectivity of *Alnus*-infective strains of *Frankia* (Nickel *et al.,* 1999; 2000).

Important compounds that may influence the interaction between plants and soil microorganisms include the phenolics in plant tissue (Harborne, 1973) and soils (Li *et al.,* 1970; Shindo *et al.,* 1978; Whitehead, 1964; Whitehead *et al.,* 1983). Various phenolics significantly stimulate or inhibit the *in vitro* growth of *Frankia* strains and also alter their morphological development (Perradin *et al.*, 1983; Vogel and Dawson, 1986). An isolate of *Frankia* from *A. viridis* ssp *crispa*, which has a high level of the phenolics, pinosylvin and pinosylvin methyl ether, to discourage browsing by arctic hares (Clausen *et al.,* 1987), was least impaired by a variety of plant phenolics tested *in vitro*. This result hints at the possibility of both intracellular and rhizosphere influences of plant phenolics on the growth, physiology, and morphology of *Frankia.*

The growth in pure culture of *Casuarina*-associated *Frankia* can be promoted with various fatty acids (Selim *et al.*, 1996). Flavonoids also influence the actinorhizal symbiosis (Benoit and Berry, 1997; Laplaze *et al.*, 1999). Because *Casuarina* trees are infected intracellularly following root-hair curling, which is induced by an unknown *Frankia* signal (Laplaze *et al.*, 1999), this and other similarities with the infection process in legumes has led to the hypothesis that flavonoids act as actinorhizal plant signals, that activate the production of *Frankia* root hair-deforming factor (Prin and Rougier, 1987; Van Ghelue *et al.*, 1997). Thus, it is plausible that actinorhizal plant hosts and their relatives produce compounds that either mediate host/symbiont signaling, stimulate *Frankia* growth, or otherwise promote nodulation (Benoit and Berry, 1997; Gauthier *et al.*, 2000; Kapulnik *et al.*, 1987; Knowlton *et al.*, 1980).

"Helper" bacteria are known that increase axenic nodulation of *Alnus* plants (Knowlton *et al.*, 1980), maybe by inducing root-hair curling and possible amelioration of rhizosphere pH (Knowlton and Dawson, 1983). The presence of soil biota can synergistically increase *Frankia* nodulation of a *Casuarina* host (Zimpfer *et al.*, 2003). Other soil organisms, therefore, seem to play a critical ecological role in enhancing the actinorhizal nodulation process in nature, however, the organisms and processes underlying this enhancement remain to be fully elucidated.

Frankia's ability to proliferate in *A. glutinosa* rhizospheres (Vergnaud *et al.*, 1985) and its localization near *C. cunninghamiana* (Zimpfer *et al.*, 1999) suggest that it may be favored in host rhizospheres. *Frankia* nodulation capacity is increased in the rhizosphere of *Betula pendula* Roth., a member of the same family as *Alnus* spp. (Smolander *et al.*, 1990). Similarly, *Frankia* nodulation is stimulated in soil beneath *Betula nigra* (Paschke and Dawson, 1992a). Using *Gymnostoma poissonianum* as a bait plant, soil from the rhizosphere of a non-nodulated endemic rhamnaceous species, *Alphitonia neocaledonica,* was shown to harbor more infective *Frankia* than the rhizospheres of either *Pinus caribea* or bare soil (Gauthier *et al.*, 2000). Evidently, stimulation of *Frankia*'s infectious capacity is independent of the regulation of host specificity. In addition, different populations of *Frankia* were found in alder capture plants depending on whether the soil originated from birch, pine, or spruce stands (Maunuksela *et al.*, 1999). It may be that non-actinorhizal genera able to stimulate *Frankia* had symbiotic ancestors, but have since lost the genetic capacity to symbiose (Simonet *et al.*, 1999).

Non-actinorhizal plants that stimulate the infectious capacity of rhizosphere *Frankia* may derive some benefit from a rhizospheric association with *Frankia*, or may represent plant clade members with a predisposition to the evolution of actinorhizal symbioses. Whatever the reason, the ability of non-actinorhizal plants to stimulate *Frankia* infectivity has important ecological consequences for the maintenance of actinorhizal microsymbionts in the absence of specific hosts. What is more, this situation raises many questions about the evolutionary and functional significance of rhizosphere *Frankia*.

In sympatric associations of *M. gale, Alnus incana* ssp. *Rugosa*, and *Shepherdia canadensis* in a Lake Michigan sand dune ecosystem in Wisconsin, USA, the likelihood of host-plant nodulation by soil-borne *Frankia* was increased by the presence of actinorhizal plants in general, but not the presence of specific host plants (Huguet, 2003). Infectious *Frankia* was detected in 82% of 120 plots located in sand dune communities at one study location, even though only 14% of the plots supported actinorhizal host plant species. Infective *Frankia* were present in soils of young dunes prior to the establishment of any actinorhizal hosts. Submerged soils had no infectious capacity whatsoever, whereas soils with greater *in-situ* moisture contents or subject to intermittent saturation had low infectious capacities overall.

In the same study, PCR-RFLP of the 16S-23S IGS indicated that *Frankia* nodular strains resulting from soil inoculations were in distinct host-specific groups (Huguet, 2003), which showed greater diversity among *Shepherdia*-infective strains than among the closely related *Alnus-Myrica* strains. Further, ecological patterns of nodulation corresponded with genetic differences of groups of host-specific nodular strains. All *Shepherdia*-infective *Frankia* were more abundant in soil samples from drier earlier-successional sites, whereas *Alnus*- and *Myrica*-infective *Frankia* strains were more abundant in moister soils of later-successional communities.

There were two strongly divergent phylogenetic groups of *Shepherdia*-infective *Frankia*. One group was dominant in soils of the early stages of sand dune formation and succession, where nodulation capacity was greatest, whereas the other group dominated elsewhere in later successional stages particularly near *Shepherdia* host plants (Huguet, 2003). These results suggest that soil-borne infective *Frankia* genotypes occurring together in the same sand dune soils are not only host-specific, but also are specifically associated with different ecological conditions in their infective state. They also indicate that host specificity is a dominant determinant of symbiotic partnerships in nature, even where host plants occur in close proximity in the same ecosystem. However, ecological factors can influence both the amount of nodulation in different host-specific *Frankia* groups and the infectivity of different strains within a host-specificity group.

Allelopathy can also influence nodulation and growth of actinorhizal partners. Actinorhizal plants have been successfully interplanted as N_2-fixing nurse trees with valuable black walnut trees (Friedrich and Dawson, 1984). Native Arizona alder (*A. oblongifolia*) and Arizona walnut (*Juglans major*) occur together in isolated stretches of riparian forest in Arizona as do various other combinations of walnut species and actinorhizal plants in forests and plantations throughout the world. Walnut's allelochemical, juglone (Dawson and Seymour, 1983), is a respiration inhibitor that was probably evolved as a chemical defense against insect herbivory, but it is also toxic in varying degrees to plants and microorganisms. Both *Frankia* strains and actinorhizal plants are inhibited by juglone, but both host species and microsymbiont strains vary in their response to this allelochemical (Dawson and Seymour, 1983; Neave and Dawson, 1989). This exemplifies just one of many such possible biotic agents with the capacity to alter the composition and growth of associated *Frankia* and actinorhizal plants.

3.2. Abiotic Factors

Abiotic factors that influence infective populations of *Frankia* in soil include time, moisture, pH, aeration, temperature, inorganic chemicals, clay content, organic matter, salinity, and movement by water and wind. For example, the nodulation of *C. velutinus* Dougl. is inversely proportional to the length of time that this shrub has been absent from a site (Wollum *et al.,* 1968). This light-demanding actinorhizal shrub colonizes coniferous forests in the Pacific Northwest of North America after either forest fires or logging. It is typically replaced over a 50-year period by either Douglas fir or mixed conifers in forest succession. In forest stands up to 100 years old, *Ceanothus* plants, which were established from soil seed banks after harvesting, nodulated well. As the age of the stands, which were harvested, increased to 350 years, nodulation declined. Thus *Ceanothus*-infective *Frankia* can apparently persist in soil for many years in the absence of a host. Dry storage of a soil that contained *Frankia* did not change its infective capacity after drying after 3.5 years (Zimpfer *et al.,* 1997). Whatever the means, *Frankia* can persist in an infective form in soils over a long period of time.

Myrold and Huss-Danell (1994) found that a soil's nodulation units changed seasonally, even though the number of genomic units obtained by PCR amplification of *Frankia*-specific DNA sequences remained constant in the same soil. This result is evidence that the physiological response of soil-borne *Frankia* to environmental stimuli is a primary determinant of infectious capacity.

Frankia infectious capacity increases with moisture availability in soils (Dawson *et al.,* 1989; Righetti *et al.,* 1986), although infective *Frankia* is not found in permanently submerged soils (Huss-Danell *et al.,* 1997). Waterlogged soils are generally deficient in O_2 and inhibit activity of free-living microaerophiles, such as *Frankia.* Some *Frankia* strains of wetlands cannot fix N_2 and *Alnus glutinosa* bearing these ineffective bacteria are stunted and have yellow leaves, characteristic of combined-N deficiency (van Dijk and Sluimer-Stolk, 1990). Ineffective *Frankia* strains are widely distributed in waterlogged soils that support *Alnus* spp. and can reach high densities at some sites (Wolters *et al.,* 1997). The actinomycete can survive desiccation both as spores and as hyphae that are induced to form cellular trehalose as an osmotic protectant (Burleigh and Dawson, 1994).

The nodulation of actinorhizal plants is inhibited in acid soils (Bond, 1951; Dixon and Wheeler, 1983; Griffiths and McCormick, 1984; Hensley and Carpenter, 1984) and growth studies of various *Frankia* strains have established a variety of optimal pH values, which are generally neutral to slightly acidic (Lechevalier and Lechevalier, 1990). Even so, infective *Frankia* occur in strongly acidic soils due either to saprophytic growth of acid-tolerant *Frankia* strains or to the persistence of spores of acid-sensitive strains (Faure-Renaud *et al.,* 1986). However, under natural conditions, in which soluble Al^{3+} is inherent to acid soils worldwide (Martin, 1988), Al^{3+} may partially overcome the inhibitory effects of low pH, possibly owing to cell-surface potential phenomena (Igual and Dawson, 1999). Thus, the classification of *Frankia* strains as either acid-tolerant or acid-sensitive, based solely on *in-vitro* growth capacity at low pH, is probably flawed and the inhibition of nodulation of actinorhizal plants in acid soils may not be explainable solely on the basis of low pH.

Frankia colonies in soil apparently behave as a free-living microaerophiles, requiring O_2 for growth and, possibly, N_2 fixation. So, not surprisingly, the greatest *Frankia* infectivity occurs in soil types and strata affording the least resistance to O_2

diffusion (Dawson *et al.,* 1989). Infectious capacity is usually greatest near the soil surface provided that moisture and other microhabitat requirements for *Frankia* are adequate (Paschke and Dawson, 1992b). Infectious capacity may be greater in deeper strata in sandy and other coarse-textured soils, which have greater macropore space to facilitate O_2 diffusion (Dawson *et al.,* 1989). It is possible that O_2 demands are greatest at high temperatures, which induce higher rates of respiration, and when other stresses induce rapid physiological responses, which are typically accompanied by higher rates of respiration. Insofar as free-living *Frankia* in soil might be able to fix N_2, its adaptive O_2-control system, which consists of multiple monolyaers of lipids located in the vesicle, is able to protect the nitrogenase enzyme from O_2 (Silvester and Harris, 1990; see chapter 5 in this volume).

The optimal temperature range for nodulation of *C. velutinus* from western North America is 22-26°C (Wollum and Youngberg, 1969). The optimal temperature range for nodulation of most other actinorhizal host-microsymbiont combinations is unknown. For a given pair of actinorhizal symbionts, the intersection of optimal temperature ranges in soil for growth of the microsymbiont and primary root development of the host plant would likely be the optimal range of temperatures for nodulation. Most isolated strains of *Frankia* grow fastest in culture at temperatures between 25°C and 35°C (Lechevalier and Lechevalier, 1990). The optimal temperatures for growth and development in soil of roots of actinorhizal plant species undoubtedly vary widely according to climate but are virtually unknown in detail. Among *Frankia* isolates, those derived from nodules of Casuarinaceae spp. generally have higher optimal culture temperatures than isolates from cooler temperate regions, perhaps reflecting adaptations to warmer climates in their places of origin. It would be reasonable to assume that Casuarinaceae host species would likewise be adapted to optimize root growth at higher soil temperatures. Freezing of soil containing *Frankia* for several years at –20°C did not alter the structure of nodulating *Frankia* populations, but did decrease the nodulation capacity of frozen soil (Maunuksela *et al.,* 2000).

Frankia strains interact with inorganic chemicals in soil in many ways likely to influence their free-living populations. For example, some strains produce siderophores in iron-limited situations to aid in iron uptake (Boyer *et al.,* 1999). The trace element molybdenum is an essential micronutrient for nitrogenase function in *Frankia,* and must be present in soils in order for symbiotic nitrogen fixation to occur (Hewitt and Bond, 1961). Martin *et al.* (2003) found, in red alder stands of varying ages, that a decline in nodulation capacity was most closely related to lower pH and higher nitrate concentrations. As alder litter and sloughed roots increase organic nitrogen in the soil, mineralization of the N-rich tissue also increases and results in higher nitrate concentrations and lower pH. The authors speculated that low pH resulting from nitrification and the stimulation by nitrate of competing soil bacteria may decrease *Frankia* populations in soil.

Other inorganic chemicals are potentially toxic to *Frankia* in soil ecosystems, particularly pollutants resulting from mining and industrial activity. *A. glutinosa* and *A. rubra* nodulation are reduced by cadmium in soils (Hensley and Carpenter, 1987). *Casuarina*-infective *Frankia* was able to grow and nodulate host plants at high nominal Al concentrations of 800μM (279μM monomeric Al) (Igual *et al.,*

1997). At the same time, soluble Al^{3+} overcomes the inhibitory effects of low pH on *Frankia* growth (see earlier in this section).

Nickel is required for hydrogenase synthesis and function in *Frankia* and amounts of this element greater than those typically found in soils results in greater nodulation and N_2 fixation in *A. glutinosa*. However, even higher Ni^{2+} levels apparently become toxic to the host plant before they become toxic to *Frankia* (Wheeler *et al.,* 2001).

Twelve *Frankia* strains were sensitive to low concentrations (<0.5mM) of Ag^{+}, AsO_2^{-}, Cd^{2+}, SbO_2^{-}, and Ni^{2+} (Richards *et al.* 2002). Most of the strains were less sensitive to Pb^{2+} (6-8mM), CrO_4^{2-} (1.0-1.75mM), AsO_4^{3-} (>50mM), and SeO_2^{2-} (1.5-3.5mM). Although most strains were sensitive to 0.1mM Cu^{2+}, four strains were resistant to elevated levels of Cu^{2+} (2-5mM and concentrations as high as 20mM). The mechanism of SeO_2^{2-} resistance may involve reduction of soluble selenite to insoluble elemental selenium, whereas Pb^{2+} resistance and Cu^{2+} resistance may involve sequestration or binding mechanisms. Some *Frankia* strains, therefore, have the capacity to resist heavy metal toxicity in soil.

Frankia infectious capacity increases with the amount of clay in substrates (Smolander *et al.,* 1988; Zimpfer *et al.,* 2002). Clay and organic matter particles in soil can increase its water-holding capacity and the amount of clay or organic matter in soil correlates positively with *Frankia* infectious capacity (Burleigh and Dawson, 1994; Dawson *et al.,* 1989; Righetti *et al.,* 1986; Smolander *et al.,* 1988). In Jamaica for example, a moist, fertile soil with a substantial clay component and neutral pH appeared particularly supportive of *Myrica*-infective *Frankia* (Zimpfer *et al.,* 1997).

Frankia strains isolated from *Casuarina* spp. are more tolerant of high concentrations of NaCl (approaching that of seawater) than isolates from North American hosts not normally exposed to sodic soils (Dawson and Gibson, 1987). *C. equisetifolia, M. pensylvanica,* and *Hippophaë rhamnoides* all nodulate in coastal soils. Infective populations of *Frankia* are present in soils of coastal sand dunes occupied by *M. pensylvanica* in Virginia, USA (Young *et al.,* 1992). This natural variability in salt tolerance of *Frankia* strains could be exploited to produce inocula for use with salt-tolerant host plants to increase the productivity of saline soils.

Soil organic material most likely supports the saprophytic grpwth of *Frankia* and its physiological status in soils. Evidence for this suggestion comes from enzymes secreted and the metabolic activities of *Frankia* strains. *Frankia* strains secrete proteinases, aminopeptidases, and proteasome-like corpuscles (Benoit *et al.,* 1992; Girgis and Schwenke, 1993; Muller *et al.,* 1991), all of which can degrade proteins in soil organic matter to give amino acids that could be used as N sources by *Frankia* (Zhang and Benson, 1992). Extracellular pectinases and cellulases could enable *Frankia* to break down complex organic substances into simple organic substrates to support growth in soil (Safo-Sampath and Torrey, 1988; Seguin and Lalonde, 1989). Also, some *Frankia* spp. are capable of metabolizing complex phospholipids (Girgis and Schwenke, 1993).

In addition to supplying nutrients, organic matter also improves the water-holding capacity of soils and supports beneficial as well as antagonistic microorganisms. *Frankia*'s ability to produce antibiotics could enhance the ability of this slow-growing actinomycete to compete as a saprophyte with faster-growing soil microorganisms (Haansu *et al.*, 2001).

At a given time during their growth and infective phases, any one factor might be the overriding determinant of growth and nodulation capacity of *Frankia*. Undoubtedly, however, many of the biotic and abiotic factors above would typically interact in a complex manner to influence nodulation.

4. ECOLOGY OF ACTINORHIZAL PLANTS

4.1. Ecophysiology

4.1.1. Light Responses

Light intensity and duration are important determinants of the occurrence and function of actinorhizal plants. Actinorhizal plants tend to be light demanding and many function physiologically as early successional shade-intolerant species (Côté *et al.*, 1988), which typically can take photosynthetic advantage of high light intensity. Other actinorhizal plants occur as stable components of ecosystems that feature scattered trees and shrubs in stands kept open to light penetration by either aridity or fires. Their intolerance of shade generally precludes their persistence or regeneration in the understory of dense forest stands. Evolutionary pressure for this type of photosynthetic adaptation could have arisen from the typically open nature of its disturbed or N-limited niches plus the considerable energy requirements for nitrogen fixation.

Because actinorhizal plants are capable of symbiotic nitrogen fixation, they can maintain high leaf-N levels that are associated with high rates of photosynthesis (Krueger and Ruth, 1968). Several species of alder, including *A. glutinosa*, retain green foliage longer than other temperate deciduous trees and do not resorb foliar N as efficiently as other temperate deciduous species in the autumn (Côté and Dawson, 1986). This prolonged leaf and nitrogen retention is associated with prolonged photosynthesis of *A. glutinosa* and probably other alders (Neave and Dawson, 1989). Prolonged autumnal photosynthesis may give alders a competitive growth advantage over other temperate deciduous species. It may also provide photosynthate for seasonally high rates of nodule growth, for maintenance of high rates of nitrogen fixation observed for alders in early autumn (Kaelke and Dawson, 2003), and for the energy demands for enzymatic processes inducing dormancy as well as the needs for carbohydrate reserves to maintain dormant tissue over the winter. Apparently, temperature, not photoperiod, eventually decreases photosynthesis and induces senescence in alder leaves (Côté and Dawson, 1986), illustrating an important difference in light response of alders in comparison with temperate deciduous trees that initiate foliar autumnal senescence processes in precise synchrony with day-length changes.

Photosynthesis is the primary driver of symbiotic nitrogen fixation (Gordon and Wheeler, 1978) and it is likely that the greatest gains in actinorhizal nitrogen fixation can be obtained by genetic manipulation of the host to increase photosynthetic capacity rather than through selection or alteration of the microsymbiont alone (Dawson and Gordon, 1979). Because whole-plant photosynthesis rates are linearly correlated with whole-plant nitrogenase activities in alder (Gordon and Wheeler, 1978), the search for strains of *Frankia* and host genotypes more efficient at nitrogen fixation might practically be evaluated on the basis of the ratio of nitrogenase activity to photosynthesis on a whole-plant basis.

The impact of elevated carbon-dioxide levels on plants has been the focus of many recent studies that were initiated in response to the alarm over the increasing levels of atmospheric carbon dioxide from the widespread use of fossil fuels. Vogel and Curtis (1995) found with *A. glutinosa* that a doubling of ambient CO_2 in open-top chambers for 160 days in the field resulted in a 16% greater rate of CO_2 fixation, a 50% increase in specific leaf N, and a 46% increase in specific nitrogenase activity of detached nodules relative to controls at ambient levels of atmospheric CO_2. Unlike other plants, actinorhizal trees and shrubs on N-limited soils can maintain high net CO_2 assimilation with minimal negative adjustment of photosynthetic capacity following prolonged exposure to elevated CO_2.

4.1.2. Moisture

Actinorhizal plants have a broad range of ecological amplitudes with respect to soil moisture. Species such as *Cowania mexicana* and *Purshia tridentata* occur in semi-arid temperate shrub ecosystems in North America and are able to tolerate drought. Dalton and Zobel (1977) found that the xerophytic shrub *P. tridentata* can tolerate severe water stress, but nitrogenase activity of nodulated plants declines sharply under these conditions and requires 12 days of watering to recover. In contrast, the wetland tree *A. glutinosa* is much more sensitive to water stress (Seiler and Johnson, 1984). *A. incana* ssp. *incana* (grey alder), an alder of Europe that occurs on drier soils than those occupied by *A. glutinosa*, exhibited declines in nitrogenase activity with decreasing water potentials from -0.6 to -1.4 Mpa (Sundström and Huss-Danell, 1987). Stomatal closure occurred with water stress and there was a strong relation between stomatal closure, which reduces photosynthesis, and the inhibition of nitrogenase activity. *A. viridis* ssp. *crispa, M. pennsylavanica, M. cerifera, Ceanothus* spp., and *Comptonia peregrina* occur on dry soils in North America and apparently have adaptations to drought.

The *Casuarina* spp. are riparian or coastal species in nature, but can develop a long tap root to exploit deeper sources of soil moisture (Yadav, 1983) and are commonly planted in uplands worldwide in tropical, subtropical, and warm temperate regions. The *Allocasuarina* spp. are associated with more arid environments in Australia and are more drought tolerant than *Casuarina* spp. (Subbarao and Rodríguez-Barrueco, 1995). Actinorhizal plants of riparian zones and wetlands have physiological and morphological adaptations that enable them to tolerate flooding and soil waterlogging. Wetland *M. gale* and riparian *Casuarina* spp. produce negatively geotropic "nodule roots" under waterlogged soil conditions. These serve as alternate O_2-delivery pathways to enable sufficient respiratory activity in nodules to meet the energy requirements of N_2 fixation under low soil O_2 conditions (Silvester *et al.,* 1988a; Wheeler *et al.,* 1979).

A. glutinosa and *A. rubra* are flood tolerant alders that develop nodule lenticels when flooded (Silvester *et al.,* 1988b; Tjepkema, 1978). In response to flooding, *Alnus rubra* forms hypertrophied nodule lenticels with subsequent full restoration of nitrogenase activity after 50 days, when there was a 10-fold increase in intercellular space around nodule cells containing *Frankia* vesicles (Batzli and Dawson, 1999). *A. incana* ssp. *rugosa* is a wetland alder that is not actually tolerant of soil waterlogging, but that inhabits the margins of wetlands with a stable water level where it maintains a superficial root system, thus avoiding the hypoxia of waterlogged soils (Kaelke and Dawson, 2003).

Soil moisture may also influence the occurrence of actinorhizal plants because the germinants may be particularly sensitive to desiccation, such as for *A. glutinosa* (McVean, 1956), which limits their occurrence to seasonally moist sites.

4.1.3. Fire and Heat

Species of the actinorhizal plant genus *Ceanothus* in chaparral vegetation of California have leaves that are coated with flammable resins. The leaves and branches of these plants are small, adding to their flammability. Thus, wildfires pass quickly with limited heating of soil, which prevents damage to seeds and roots important in supporting growth of sprouts after fire. *Ceanothus* roots are specially adapted to enable the plant to grow in areas that were recently burned. The plants' ability to produce heat-resistant seeds that remain dormant yet viable in ground litter for long periods of time contributes to the ability of *Ceanothus* to recover quickly following fire (Keeley, 1989). Fire also insures germination under open light conditions and releases nutrient to support post-fire growth. *Comptonia peregrina* seeds respond to a heat signal generated by direct sunlight, so stimulating germination under open conditions created by a disturbance that simultaneously assures the high light levels necessary to establish the seedlings (Dow and Schwintzer, 1999).

4.1.4. Seasonality

Actinorhizal nodules are perennial structures that become dormant along with the whole plant during the winter in temperate and boreal host species. During the winter, there are few symbiotic nodule vesicles, no detectable nitrogenase activity, reduced numbers of hyphae, and, in some instances, an increase in spores (Schwintzer *et al.,* 1982; Wheeler *et al.,* 1983). To maintain viability in the dormant state, nodules and other root tissues increase reserve carbohydrates prior to dormancy, which subsequently decrease over winter, perhaps by supporting the metabolism of the overwintering nodules (Wheeler and Bowes, 1974; Wheeler *et al.,* 1983). Trehalose and glycogen produced by *Frankia* may be reserve carbohydrates for overwintering in the nodular microsymbiont (Benson and Eveleigh, 1979; Lopez *et al.,* 1984).

An unusual aspect of actinorhizal alder physiology is a delay in foliar senescence in the autumn, associated with prolonged photosynthesis (Neave *et al.,* 1989) and a lack of resorption of foliar N prior to leaf drop. In contrast, the net resorption and conservation of foliar phosphorus, carbohydrates, and other nutrients during alder leaf senescence is apparently equal in efficiency to that of other temperate deciduous trees and shrubs (Côté and Dawson, 1986). One might speculate that there would be little evolutionary selection pressure for efficient resorption and reuse of foliar N in a N_2-fixing tree, however, the lack of resorption of foliar N leads to a consequent drop of leaves with high concentrations of N. This accelerated shedding of internal N by alder accelerates the rate at which recently fixed N is added to soil relative to the expected rate if alder resorbed and recycled N similarly to other temperate deciduous trees.

4.1.5. Plant Nutrients

Of the essential mineral nutrients, molybdenum, cobalt, phosphorus, and nitrogen are of special interest with respect to nitrogen fixation in actinorhizal plants. Molybdenum is required for nitrogenase and root nodules of *A. glutinosa* have a nodular Mo concentration 6-times greater than roots (Becking, 1970). Cobalt is the only mineral element exclusively essential for N_2-fixing plants because the microsymbionts require it for vitamin B_{12} production (Dixon and Wheeler, 1983).

Nitrogen-fixing plants have a relatively high demand for phosphorus because of the high energy demands for nitrogen fixation and, hence, for the phosphorus compounds important in biochemical energetics (Huss-Danell, 1990). Phosphorus deficiency limits nodulation and nitrogen fixation in actinorhizal plants (Diem and Dommergues, 1990). *Casuarina* windbreaks on infertile coastal sand dunes in southeast China become phosphorus limited after litter harvesting for fuel, resulting in loss of vigor and increased disease problems in subsequent plantings. The presence of mycorrhizae on nodulated alders in a tripartite symbiosis improves phosphate uptake by *A. viridis* ssp. *viridis* (Mejstrik and Benecke, 1969). In a tripartite association between the ectomycorrhizal fungus *Alpova diplophloeus*, *Frankia*, and *Alnus tenuifolia*, apparently the ectomycorrhizal partner increased the solubilization of ground basalt, enhancing mineral acquisition and consequent growth and nitrogen fixation of the host plant (Yamanaka *et al.*, 2003).

Combined-N at concentrations necessary for optimal plant growth not only inhibits nodulation but also inhibits nitrogen fixation in nodules already formed. These two inhibitory influences of combined-N occur concurrently under both field and laboratory conditions (Bond and Mackintosh, 1975; Ingsted, 1980; Rodríguez-Barrueco *et al.,* 1970). Theoretically, it is energetically more efficient for actinorhizal plants to take up combined-N in the form of NH_4^+ or NO_3^- from the soil solution than to fix N_2 symbiotically, hence the observed self-imposed inhibition of nodulation and nitrogen fixation in an early successional actinorhizal alder forest.

It is possible that the uptake and metabolism of NO_3^- complement nitrogen fixation seasonally by supplying N to plants during the autumn in temperate regions. Nitrate reductase activity has a lower optimal temperature than that of nitrogenase activity. At lower temperatures typically occurring in temperate-region soils during autumn, *A. glutinosa* nodules switch the expenditure of energy on nitrogenase activity to nitrate reductase activity, yielding temporally distinct and complementary mechanisms for N acquisition in an actinorhizal system (Vogel and Dawson, 1986).

Alder forest soils typically increase in acidity, total N, and organic matter content while decreasing in cation-exchange capacity with increasing stand age (Bormann *et al.,* 1994). With bacterial mineralization of its litter and sloughed roots, alder releases NH_4^+ into the soil, which is then converted to NO_3^- under aerobic conditions, sometimes at very high rates (Paschke *et al.,* 1989). The latter process of nitrification is the cause of acidification in alder soils (Van Miegrot and Cole, 1985). Increases in NO_3^- concentration and lower pH values were most closely associated with a decline in nodulation capacities of soils from red alder stands of different ages in the Pacific Northwest of North America (Martin *et al.,* 2003). This observation suggests that alder, *via* feedback mechanisms triggered by high NO_3^- levels, switches from nitrogen fixation to the more efficient uptake of mineral-N in response to its own enrichment of soil-N fertility. Acidification displaces nutrient cations from soil exchange sites by protons, followed by leaching that decreases fertility. With red alder, successional coniferous species replace the red alder and act as base recyclers (Van Miegrot and Cole, 1985), counteracting the decrease in fertility resulting from acidification.

One anthropogenic change evident today is an increase in atmospheric N pollution, which creates the potential for widespread increases in background levels of N in soil when delivered in precipitation. Such a change could decrease the biological diversity of plants and associated animals (Tilman *et al.,* 1996). Does this mean that fixed-N provided to terrestrial ecosytems by actinorhizal plants add to the threat to biodiversity? Biological diversity with respect to nitrogen fertility would predictably depend on patches with a wide range of available combined-N levels to support organisms with differing requirements for combined-N. An increase in soil-N resulting from atmospheric pollution would tend to first homogenize and eliminate patches at the low end of N-fertility levels together with their adapted organisms. However the N-rich patches associated with actinorhizal plants would remain important to organisms dependent upon high levels of N-fertility and hence would support biological diversity, barring an extreme increase in soil-N levels due to pollution.

It is a mistake to assume that actinorhizal plants are ecologically important only for the amount of nitrogen fixed in a given terrestrial ecosystem. Where the total or relative amount of nitrogen fixed by actinorhizal plants is unknown or relatively small, it can nonetheless be sufficient to allow hosts to establish competitively and assume other critical roles in the functioning of their particular terrestrial ecosystems. This could be the case for actinorhizal plants of sand dune ecosystems, deserts, grasslands and savannas.

Salt concentration can also differentially influence the survival and symbiotic performance of actinorhizal host plants (El-Lakany and Luard, 1982). In Australia, three *Casuarina* species occur along rivers according to their respective salt tolerances (Subbarao and Rodríguez-Barrueco, 1995). *C. cunninghamiana* occurs along freshwater portions of rivers, *C. glauca* occurs near brackish water and in lower reaches of rivers influenced by tidal surges, whereas *C. equisetifolia* occurs in coastal areas directly influenced by sea spray and incursions of seawater (Midgley *et al.,* 1983). At the boundaries between these zones of salinity, hybrid forms of these tree species are found.

In addition to the factors influencing the ecology and physiology of host plants summarized above, symbiotic performance can be compromised by the absence of appropriate mycorrhizal and N_2-fixing symbionts (Diem and Gauthier, 1982; Reddell *et al.,* 1988), low soil temperatures (Reddell *et al.,* 1985), deficiencies of other nutrients, and low soil pH (Bond, 1957; Coyne, 1973). Climate change, soil pollutants, human-altered fire regimes, increased ozone levels, sulfur pollution and acid rain, pests, diseases, soil erosion, hydrological changes, volcanic activity, desertification, logging, plantation establishment, and many other factors too numerous to list impact the occurrence, health, ecological and physiological function of actinorhizal species. The identification and study of ecological and physiological responses of actinorhizal plants to environmental change will be important in optimizing human stewardship of these valuable natural resources.

4.2. Interactions among Actinorhizal and Associated Plant Species

The preponderant open N-limited niches for actinorhizal plants as proscribed by their light-demanding nature and capacity to fix N_2 symbiotically have been described. The dynamics in time and space that create and eliminate these niches involves competition for plant resources. In many cases, actinorhizal plants ultimately facilitate the establishment and growth of other plant species that replace them in predictable patterns of plant succession. In other cases, the introduction of exotic actinorhizal plants can displace other plant species and alter ecological processes, so creating novel ecosystems and successional processes.

Crocker and Major were among the first to note a correlation between nutrient accumulation associated with early primary successional stands of *A. viridis* ssp. *sinuata* at Glacier Bay, Alaska, and the development of subsequent forest communities. However, the relative timing and importance of fixed-N supplied to associated plants by actinorhizal N_2 fixers and the balance of their facilitative *versus* their competitive influences over glacial chronosequences has seldom been determined precisely in ecological studies. Direct evidence for facilitation of associated plants by provision of fixed-N during primary succession is lacking at Glacier Bay and other areas of recent glacial deposits (Walker, 1993).

Using δN^{15} values in a chronosequence of deglaciation at Glacier Bay, Kohls *et al.* (2003) found that the amount of N supplied to willow trees by actinorhizal nitrogen fixation increased over a 40-year period to an equilibrium value, suggesting that fixed N_2 was the dominant source of N at that time. This time coincided with the occurrence of dense thickets of alder, which undoubtedly provided competitive pressure on the establishment of other plants. But, at the earliest stages of primary succession, actinorhizal nitrogen fixation was not the major source of N in associated plants; it would obviously not be advantageous to a N_2-fixing plant, which was establishing itself on a nutrient-limited site, to leak any substantial amount of the N fixed at the expenditure of considerable energy. Of the actinorhizal plants, *Shepherdia* depended most, *Dryas* least, and *Alnus* at an intermediate level on atmospheric nitrogen fixation. Further, the small colony-forming *Dryas* shrub seemed to decrease its reliance on N_2 fixation over time and with increased levels of available soil-N. It appears then that actinorhizal species may not be as important as fixed-N sources during the earliest stages of primary succession as once thought and that other sources of combined-N can be important during early primary succession (Vitousek and Walker, 1989). Even so, actinorhizal plants are ultimately a dominant source of N available to plants during a prominent phase of primary succession on glacial deposits at Glacier Bay.

To capture the benefits of nitrogen fixation in either balanced natural stands or inter-plantings of actinorhizal trees or shrubs with companion trees, it is important that the growth rate and density of the N_2-fixing tree or shrub not exceed that of associated trees (Dawson, 1990). In the Pacific Northwest of North America, *A. rubra* and *C. velutinus* are vigorous colonizers of logged and burned areas of Douglas-fir forest. Although they provide benefits from nitrogen fixation, a high density of colonizing alder or snowbrush slows the regeneration of the desired Douglas fir. Similarly, where the growth rate or density of a companion tree in plantations exceeds that of the actinorhizal nurse plant, or where allelopathy or other competitive influences inhibit the actinorhizal plant, the actinorhizal plant will be displaced, so causing the loss of both a supply of fixed-N and all other inter-planting benefits that were afforded the companion crop. If the companion crop is shade loving, such as shade-coffee, the growth rate and stature of the N_2-fixing shade planting are less important than its density and ultimate crown structure in providing optimal light conditions to the shade-requiring companion crop.

The impact of the introduction of exotic actinorhizal plants on terrestrial ecosystems can be dramatic, especially because of the competitive advantage afforded them by their ability to fix N_2 symbiotically (Richardson *et al.,* 2000). The best documented example of a major impact of an actinorhizal exotic on ecosystem composition, structure and function is that of the invasion of *M. faya* from the Canary Islands on volcanic deposits in Hawaii Volcanoes National Park (Vitousek and Walker, 1989). Dense stands of this exotic plant have established themselves in a novel community that greatly increases combined-N inputs from symbiotic fixation. Similarly, in south Florida, plantings of *Casuarina* species have given rise to widely naturalized offspring that displace native vegetation, particularly on sandy soils and near wetlands in this low-elevation peninsula. Frequent hurricanes disperse its small seeds over great distances and the exposure of bare soil by the uprooting of other trees creates an ideal seedbed and open site for its establishment in native plant communities. Elsewhere, exotic *Elaeagnus umbellata* and *Elaeagnus angustifolia* bear fruit, which is consumed by birds that disperse the plants widely. *E. umbellata* is a common weed of the eastern U.S.A., whereas *E. angustifolia* is a common invader of valuable riparian forests in semi-arid regions of western North America.

4.3. Actinorhizal Plant Interactions with Heterotrophic Organisms and Humans

There are some interactive relatiohships with heterotrophic organisms that are particular to actinorhizal plants and that relate to their N_2-fixing capacity. A few examples are given to illustrate the general nature of these unique relationships.

Spittlebugs (Homoptera: Cercopoidea) are common insects that feed on xylem sap. Many show a preference for N_2-fixing plants that afford an abundant and reliable supply of xylem-borne organic-N compounds (Thompson, 1994). At least 20 species have a primary association with actinorhizal *Alnus*, *Casuarina*, *Ceanothus, Comptonia, Elaeagnus*, and *Myrica* hosts (Thompson, 1999).

In plantations of *A. acuminata* in the cloud forest regions of Colombia, the avian fauna is more diverse and representative of native forest birds than those of plantations of either *Eucalyptus* or *Pinus* spp. Most common are native birds that feed on soil invertebrates. Most likely, the high rates of soil-N accretion associated with Andean alder plantations (Carlson and Dawson, 1985) increase the biomass of soil invertebrates, thus making a situation attractive to these birds.

Nitrogen-fixing plants are beneficial and attractive to herbivores because of their abundant tissue nitrogen. So, N_2-fixing trees typically invest resources in chemical and other defenses against herbivory (Haukioja, 1991), with phenolics being a one such deterrant. As mentioned above, *A. viridis* ssp. *crispa* produces phenolic pinosylvin methyl ether to discourage browsing by voracious arctic hares (Bryant *et al.,* 1983). Similarly, the wide-ranging actinorhizal shrub, *P. tridentata* (bitterbrush), which is the most important browse species for wildlife and livestock in the intermountain west of North America, has populations with varying phenolic contents, particularly in response to browsing (Paschke, 1997). Alder root tissue is also high in phenolics, and alders have been observed to decrease the incidence of fungal root pathogens in associated Douglas fir (Li *et al.,* 1970; 1972).

The cold streams that support trout and salmon, particularly the headwaters of rivers where anadromous salmon spawn, are often infertile. It is possible that alders commonly and abundantly associated with these freshwater streams are important sources of combined-N for the planktonic organisms at the beginning of the food chain for salmonids and their fry (Goldman, 1961).

The management and use of actinorhizal plants by humans has been documented extensively in several reviews (Dawson, 1990; 1992; Diem and Dommergues, 1990; Dixon and Wheeler, 1986; Gordon and Dawson, 1979; Hibbs and Kromack, 1990; Hibbs *et al.,* 1994; Huss-Danell, 1997; Paschke, 1997; Schwencke and Carú, 2001; Silvester, 1977; Wheeler and Miller, 1990). The potential for domestication and use of actinorhizal plants for food, medicine, soil improvement, bioremediation, soil stabilization, aesthetic plantings, agro forestry, wood, fiber and chemicals remains largely untapped. Although actinorhizal plants in general are not highly susceptible to pests or diseases (Dommergues, 1997), their use in large-scale monocultures on poor sites and under stressful conditions increases the likelihood of serious damage by insects and diseases. Some actinorhizal plants, such as Elaeagnaceae genera including *Hippophaë,* are receiving renewed attention because they produce high concentrations of antioxidant compounds, such as flavonols, which are important in human nutrition (Rösch *et al.,* 2003).

Human encroachment on the wild lands, where actinorhizal plants are key functional ecosystem components, continues unabated and accelerates with the growth of human populations and increasing standards of living. It is primarily through an increased understanding of both the importance of actinorhizal plants to ecosystem function and to our natural heritage, and their capacity to provide valuable products and by-products that they will be effectively managed, preserved, and used sustainably.

REFERENCES

Arveby, A. S., and Huss-Danell, K. (1988). Presence and dispersal of infective *Frankia* in peat and meadow soils in Sweden. *Biol. Fertil. Soils*, *6*, 39-44.

Baker, D., and Miller, N. G. (1980). Ultrastructural evidence for the existence of actinorhizal symbioses in the late Pleistocene. *Can. J. Bot.*, *58*, 1612-1620.

Baker, D., and O'Keefe, D. (1984). A modified sucrose fractionation procedure for the isolation of frankiae from actinorhizal root nodules and soil samples. *Plant Soil*, *78*, 23-28.

Baker, D. D., and Schwintzer, C. W. (1990). Introduction. In C. R. Schwintzer and J. D. Tjepkema (Eds.), *The biology of* Frankia and *actinorhizal plants* (pp. 3-13). New York, NY: Academic Press.

Barlow, B. A. (1983). Casuarinas, a taxonomic and biogeographic review. In S. J. Midgley, J. W. Turnbull, and R. D. Johnston (Eds.), *Casuarina ecology, management* and *utilization* (pp. 10-18). Melbourne, Australia: CSIRO.

Batzli, J. M., and Dawson, J. O. (1999). Development of flood-induced lenticels in red alder nodules prior to the restoration of nitrogenase activity. *Can. J. Bot.*, *77*, 1373-1377.

Becking, J. H. (1970). Frankiaceae, fam. Nov. (Actinomycetales) with one new combination and six new species of the genus *Frankia* Brunchorst 1886. *Int. J. Syst. Bacteriol.*, *20*, 201-220.

Benecke, U. (1969). Symbionts of alder nodules in New Zealand. *Plant Soil*, *30*, 149-149.

Benoit, L. F., and Berry, A. M. (1997). Flavonoid-like compounds from seeds of red alder (*Alnus rubra*) influence host nodulation by *Frankia* (Actinomycetales). *Physiol. Plant.*, *99*, 588-593.

Benoit, P., Müller, A., Diem, H. G., and Schwenke, J. (1992). High-molecular-mass multi-catalytic proteinase complexes produced by the nitrogen-fixing actinomycete *Frankia* strain BR. *J. Bacteriol.*, *174*, 1495-1504.

Benson, D. R., and Eveleigh, D. E. (1979). Nitrogen fixing homogenates of *Myrica pensylvanica* (bayberry) non-legume root nodules. *Soil Biol. Biochem.*, *11*, 331-334.

Bermudez de Castro, F., Miguel, C., and Rodriguez-Barrueco, C. (1976). A study of the capacity of soil to induce nodules of *Alnus glutinosa* (L.) Gaertn. and *Myrica gale* L. with special reference to the specificity of the endophytes. *Ann. Microbiol.*, *127A*, 307-315.

Bond, G. (1951). The fixation of nitrogen associated with the root nodules of *Myrica gale* L., with special reference to its pH relation and ecological significance. *Ann. Bot.*, *15*, 447-459.

Bond, G. (1957). The development and significance of the root nodules of *Casuarina*. *Ann. Bot.*, *21*, 373-380.

Bond, G. (1983). Taxonomy and distribution of non-legume nitrogen-fixation systems. In J. C. Gordon and C. T. Wheeler (Eds.), *Biolgical nitrogen fixation in forest ecosystems: Foundations* and *applications* (pp. 55-87). The Hague, The Netherlands: Martinus Nijhoff/Dr. W. Junk.

Bond, G., and Mackintosh, A. H. (1975). Effect of nitrate nitrogen on the nodule symbioses of *Coriaria* and *Hippophae*. *P. Roy. Soc. Lond. B Bio.*, *190*, 199–209.

Bormann, B. T., Cromack, K., and Russel, W. O. (1994). Influences of red alder on soils andlong-term ecosystem productivity. In D. E. Hibbs, D. S. DeBell and R. F. Tarrant (Eds.), *The biology* and *management of red alders* (pp. 47-56). Corvallis, OR: Oregon State University Press.

Boyer, G. L., Kane, S. A., Alexander, J. A., and D. B. Aronson. (1999). Siderophore formation in iron-limited cultures of *Frankia* sp. 52065 and *Frankia* sp. CeSI5. *Can. J. Bot.*, **77**, 1316-1320.

Bryant, J. P., Wieland, G. D., Reichardt, P. B., Lewis, V. E., and McCarthy, M. C. (1983). Pinosylvin methyl ether deters snowshoe hare feeding on green alder. *Science*, *222*, 1023-1025.

Burleigh, S. H., and Dawson, J. O. (1994). Occurrence of *Myrica*-nodulating *Frankia* in Hawaiian volcanic soils. *Plant Soil*, *164*, 283-289.

Burleigh, S. H., and Dawson, J. O. (1995). Spores of *Frankia* strain HFPCcI3 nodulate *Casuarina equisetifolia* after passage through the digestive tracts of captive parakeets (*Melopsittacus undulatus*). *Can. J. Bot.*, *75*, 1527-1530.

Burleigh, S., and Torrey, J. G. (1990). Effectiveness of different *Frankia* cell types as inocula for the actinorhizal plant *Casuarina*. *App. Environ. Microbiol., 56,* 2565-2567.

Callaham, D., Del Tredici, P., and Torrey, J. G. (1978). Isolation and cultivation *in vitro* of the actinomycete causing root nodulation in *Comptonia*. *Science, 199,* 899-902.

Carlson, P. J., and Dawson, J. O. (1985). Soil nitrogen changes, early growth, and response to soil internal drainage of a plantation of *Alnus jorullensis* in the Colombian highlands. *Turrialba, 35,* 141-150.

Chepstow-Lusty, A. J., Bennett, K. D., Fjeldsa, J., Kendall, A., Galliano, W., and Tupayachi Herrera, A. (1998). Tracing 4,000 years of environmental history in the Cuzco area, Peru, from the pollen record. *Mt. Res. Dev., 18,* 159-172.

Clausen, T. P., Reichardt, P. B., and Bryant, J. P. (1987). Pinosylvin and pinosylvin methyl ether as feeding deterrents in green alder. *J. Chem. Ecol., 12,* 2117-132.

Clawson, M. L., Carú, M., and Benson, D. R. (1998). Diversity of *Frankia* strains of root nodules of plants from the families Elaeagnaceae and Rhamnaceae. *Appl. Environ. Microbiol., 64,* 3539-3543.

Clawson, M. L., Gawronski, J., and Benson, D. R. (1999). Dominance of *Frankia* strains instands of *Alnus incana* subsp. *rugosa* and *Myrica pensylvanica*. *Can. J. Bot., 77,* 1203-1207.

Côté, B., Carlson, R. W., and Dawson, J. O. (1988). Leaf photosynthetic characteristics of seedlings of actinorhizal *Alnus* ssp. and *Elaeagnus* spp. *Photosynth. Res., 16,* 211-218.

Côté, B., and Dawson, J. O. (1986). Autumnal changes in total nitrogen, salt-extractedproteins, and amino acids in leaves and adjacent bark of black alder, eastern cottonwood, and white basswood. *Physiol. Plant., 67,* 102-108.

Coyne, P. D. (1973). (Ph.D. thesis, Australian National University, Canberra, Australia.). Summarized by Torrey, J. G. *Casuarina*: Actinorhizal nitrogen-fixing tree of the tropics. In *Biological nitrogen fixation technology for tropical agriculture* (pp. 427-439). Cali, Colombia: Centro Internacional de Agricultura Tropical.

Dalton, D. A., and Zobel, D. B. (1977). Ecological aspects of nitrogen fixation by *Purshia tridentata*. *Plant Soil, 48,* 57-80.

Dawson, J. O. (1986). Actinorhizal plants: Their use in forestry and agriculture. *Outlook Agr., 15,* 202-208.

Dawson, J. O. (1990). Interactions among actinorhizal and associated species. In C. R. Schwintzer and J. D. Tjepkema (Eds.), *The biology of* Frankia and *actinorhizal plants* (pp. 288-316). New York, NY: Academic Press.

Dawson, J. O. (1992). Nitrogen fixation in forestry and agroforestry. In F. B. Metting, Jr. (Ed.), *Soil microbial ecology* (pp. 227-253). Basel, Switzerland: Marcel Decker.

Dawson, J. O., and Gibson, A. H. (1987). Sensitivity to sodium chloride of selected *Frankia* isolates from *Casuarina, Allocasuarina* and North American host plants. *Physiol. Plant., 70,* 272-278.

Dawson, J. O., and Gordon, J. C. (1979). Nitrogen fixation in relation to photosynthesis in *Alnus glutinosa*. *Bot. Gaz., 140,* S70-S75.

Dawson, J. O., and Klemp, M. T. (1987). Variation in the capacity of black alder to nodulate in central Illinois soils. In R. L. Hay, F. W. Woods, and H. DeSelm (Eds.), *Sixth central hardwood forest conference* (pp. 255-260). Knoxville, TN: Department of Forestry, Wildlife, and Fisheries, University of Tennessee.

Dawson, J. O., Kowalski, D. G., and Dart, P. J. (1989). Variation with soil depth, topographicposition and host species in the capacity of soils from an Australian locale to nodulate *Casuarina* and *Allocasuarina* seedlings. *Plant Soil, 118,* 1-13.

Dawson, J. O., and Seymour, P. E. (1983). Effects of juglone concentration on growth *in vitro* of *Frankia* ArI3 and *Rhizobium japonicum* strain 71. *J. Chem. Ecol., 9,* 1175-1183.

Diem, H. G., and Dommergues, Y. R. (1990). Current and potential uses and management of Casuarinaceae in the tropics and subtropics. In C. R. Schwintzer and J. D. Tjepkema (Eds.), T*he biology of* Frankia and *actinorhizal plants* (pp. 317-342). New York, NY: Academic Press.

Diem, H. G., and Gauthier, D. (1982). Effet de l'infection endomycorrhizienne (*Glomus mosseae*) sur la nodulation et la croissance de *Casuarina equisetifolia*. *C.R. Acad. Sci. III -Vie 294,* 215-218.

Dixon, R. O. D., and Wheeler, C. T. (1983). Biochemical, physiological, and environmental aspects of symbiotic nitrogen fixation. In J. C. Gordon and C. T. Wheeler (Eds.), *Biological nitrogen fixation in forest ecosystems: foundations* and *applications* (pp. 107-171). The Hague, The Netherlands: Nijhoff/Dr. W. Junk.

Dixon, R. O. D., and Wheeler, C. T. (1986). *Nitrogen fixation in plants*. New York, NY: Chapman and Hall.

Dommergues, Y. R. (1997). Contribution of actinorhizal plants to tropical soil productivity and rehabilitation. *Soil Biol. Biochem.*, *29*, 931-941.

Dow, M. A., and Schwintzer, C. R. (1999). Seed germination, seedling emergence, and seed bank ecology of sweet-fern (*Comptonia peregrina* (L.) Coult.). *Can. J. Bot.*, *77*, 1378-1386.

El-Lakany, M. H., and Luard, J. E. (1982). Comparative salt tolerance of selected *Casuarina* species. *Austr. Forest Res.*, *13*, 11-20.

Faure-Raynaud, M., Bonnefoy-Poirier, M. A., and Moiroud, A. (1986). Influence de pH acides sur la viabilité d'isolats de *Frankia*. *Plant Soil*, *96*, 347-358.

Friedrich, J. M., and Dawson, J. O. (1984). Soil nitrogen concentration and *Juglans nigra* growth in mixed plots with nitrogen-fixing *Alnus*, *Elaeagnus*, *Lespedeza*, and *Robinia* species. *Can. J. Forest Res.*, *14*, 864-868.

Gauthier, D., Jaffre, T., and Prin, Y. (2000). Abundance of *Frankia* from *Gymnostoma* spp.in the rhizosphere of *Alphitonia neocaledonica*, a non-nodulated Rhamnaceae endemic to New Caledonia. *Eur. J. Soil Biol.*, *36*, 169-175.

Girgis, M. G. Z., and Schwencke, J. (1993). Differentiation of *Frankia* by their electrophoretic patterns of intracellular esterases and aminopeptidases. *J. Gen. Microbiol.*, *139*, 2225-2232.

Goldman, C. R. (1961). The contribution of alder trees (*Alnus tenuifolia*) to the primary productivity of Castle Lake, California. *Ecol.*, *42*, 282-288.

Gordon, J. C., and Dawson, J. O. (1979). Potential uses of nitrogen-fixing trees and shrubs in commercial forestry. *Bot. Gaz.*, *140*, 88-90.

Gordon, J. C., and Wheeler, C. T. (1978). Whole plant studies on photosynthesis and acetylene reduction in *Alnus glutinosa*. *New Phytol.*, *80*, 179-186.

Griffiths, A. P., and McCormick, L. H. (1984). Effects of soil acidity on nodulation of *Alnus glutinosa* and viability of *Frankia*. *Plant Soil*, *79*, 429-434.

Haansu, J. P., Klika, K. D., Söderholm, P. P., Ovcharenko, V. V., Pihlaja, K., *et al.* (2001). Isolation and biological activity of frankiamide. *J. Ind. Microbiol. Biotechnol.*, *27*, 62-66.

Hahn, D., Nickel, A., and Dawson, J. O. (1999). Assessing *Frankia* populations in plants and soil using molecular methods. *FEMS Microbiol. Ecol.*, *29*, 215-227.

Harborne, J. (1973). *Phytochemical methods*. London, UK: Chapman and Hall.

Harriott, O. T., and Bourret, A. (2003). Improving dispersed growth of *Frankia* using Carbopol. *Plant Soil*, *254*, 69-74.

Hasebe, M., Kofuji, R., Shindo, S., Hiwatashi, Y., Kobayashi-Arakawa, S., *et al.* (1998). Division of speciation mechanisms II: Annual report 1998. Japanese National Institute for Basic Biology. Retrieved October 10, 2003 from http://www.nibb.ac.jp/annual_report/1998/24.html.

Haukioja, E. (1991). Cyclic fluctuations in density: Interactions between a defoliator and its host tree. *Acta Oecol.*, *12*, 77-88.

Hensley, D. L., and Carpenter, P. L. (1984). Effect of lime additions to acid strip-mine spoil on survival, growth and nitrogen fixation (acetylene reduction) of several woody legume and actinomycete-nodulated species. *Plant Soil*, *79*, 353-367.

Hensley, D. L., and Carpenter, P. L. (1987). The effect of cadmium on growth and acetylene reduction (N_2 fixation) by *Alnus glutinosa*. *Hort. Science*, *22*, 69-70.

Hewitt, E. J., and Bond, G. (1961). Molybdenum and the fixation of nitrogen in *Casuarina* and *Alnus* root nodules. *Plant Soil*, *14*, 159-175.

Hibbs, D. E., and Cromack, Jr., K. (1990). Actinorhizal plants in Pacific Northwest forests. In C. R. Schwintzer and J. D. Tjepkema (Eds.), *The biology of* Frankia and *actinorhizal plants* (pp. 343-363). New York, NY: Academic Press.

Hibbs, D. E., DeBell, D. S., and Tarrant, R. F. (Eds.) (1994). *The biology and management of red alder*. Corvallis, OR: Oregon State University Press.

Huguet, V. (2003). Spécificité d'association du couple symbiotique *Frankia/Myrica* (Ph.D. thesis, Université Claude Bernard Lyon I, France).

Huguet, V., Batzli, J. M., Zimpfer, J. F., Normand, P., Dawson, J. O., and Fernandez, M. P. (2001). Diversity and specificity of *Frankia* strains in nodules of sympatric *Myrica gale*, *Alnus incana*, and *Shepherdia canadensis* determined by *rrs* gene polymorphism. *Appl. Environ. Microbiol.*, *67*, 2116-2122.

Huss-Danell, K. (1990). The physiology of actinorhizal nodules. In C. R. Schwintzer and J. D. Tjepkema (Eds.), *The biology of* Frankia and *actinorhizal plants* (pp. 129-156). New York, NY: Academic Press.

Huss-Danell, K. (1997). Actinorhizal plants and their N_2 fixation. *New Phytol.*, *136*, 375-405.

Huss-Danell, K., and Frej, A. (1986). Distribution of *Frankia* in soils from forests and afforestation sites in northern Sweden. *Plant Soil*, *90*, 407-418.

Huss-Danell, K., Uliassi, D., and Renberg, I. (1997). River and lake sediments as sources of infective *Frankia* (*Alnus*). *Plant Soil*, 197, 35-39.

Igual, J. M., and Dawson, J. O. (1999). Stimulatory effects of aluminum on *in vitro* growth of *Frankia*. *Can. J. Bot.*, *77*, 1321-1326.

Igual, J. M., Rodríguez-Barrueco, C., and Cervantes, E. (1997). The effects of aluminum on nodulation and symbiotic nitrogen fixation in *Casuarina cunninghamiana* Miq. *Plant Soil*, *190*, 41-46.

Ingsted, T. (1980). Growth, nutrition and nitrogen fixation in grey alder at varied rate of nitrogen addition. *Physiol. Plant.*, *50*, 353-364.

Jeong, S. C., and Myrold, D. D. (2001). Population size and diversity of *Frankia* in soils of *Ceanothus velutinus* and Douglas-fir stands. *Soil Biol. Biochem.*, *33*, 931- 941.

Kaelke, C. M., and Dawson, J. O. (2003). Seasonal flooding regimes influence survival, nitrogen fixation, and the partitioning of nitrogen and biomass in *Alnus incana* ssp. *rugosa*. *Plant Soil*, *254*, 167-177.

Kapulnik, Y., Joseph, C. M., and Phillips, D. A. (1987). Flavone limitations to root nodulation and symbiotic nitrogen fixation in alfalfa. *Plant Physiol.*, *84*, 1193-1196.

Keeley, S. C. (1989). The California chaparral: Paradigms reexamined. *Natural History Museum of Los Angeles County, CA, Science Series* Number 34.

Knowlton, S., Berry, A., and Torrey, J. G. (1980). Evidence that associated soil bacteria may influence root hair infection of actinorhizal plants by *Frankia*. *Can. J. Microbiol.*, *26*, 971-977.

Knowlton, S., and Dawson, J. O. (1983). Effects of *Pseudomonas cepacia* and cultural factors on the nodulation of *Alnus rubra* roots by *Frankia*. *Can. J. Bot.*, *61*, 2877-2882.

Kohls, S. J., Baker, D. D., van Kessel, C., and Dawson, J. O. (2003). An assessment of soil enrichment by actinorhizal N_2 fixation using $\delta^{15}N$ values in a chronosequence of deglaciation at Glacier Bay, Alaska. *Plant Soil*, *254*, 11-17.

Kohls, S. J., Thimmapuram, J., Bushena, C. A., Paschke, M. W., and Dawson, J. O. (1994). Nodulation patterns of actinorhizal plants in the family Rosaceae. *Plant Soil*, *162*, 229-239.

Krajick, K. (1998). Green farming by the Incas? *Science*, *281*, 323.

Krueger, K. W., and Ruth, R. H. (1968). Photosynthesis of red alder, Douglas-fir, Sitka spruce, and western hemlock seedlings. In *Biology of alder: Proceedings of a symposium; 1967 April 14-15*. Portland, OR: U.S. Department of Agriculture, Forest Service, Pacific Northwest Forest and Range Experiment Station.

Krumholz, G. D., Chval, M. S., McBride, M. J., and Tisa, L. S. (2003). Germination and physiological properties of *Frankia* spores. In P. Normand, K. Pawlowski, and J. O. Dawson (Eds.), Frankia *symbiosis* (pp. 57-68). Dordrecht, The Netherlands: Kluwer Academic Publishers.

Lai, Chia-Chin. (1996). The effect of tyrosine on pigmentation, growth, and infectivity of *Frankia* (M.Sc. thesis, University of Illinois at Urbana-Champaign).

Laplaze, L., Gherbi, H., Frutz, T., Pawlowski, K., Franche, C., *et al.* (1999). Flavan-containing cells delimit *Frankia*-infected compartments in *Casuarina glauca* nodules. *Plant Physiol.*, *121*, 113-122.

Lawrence, D. B., Schoenike, R. E., Quispel, A., and Bond, G. (1967). The role of *Dryas drummondii* in vegetation development following ice recession at Glacier Bay, Alaska, with special reference to its nitrogen fixation by root nodules. *J. Ecol.*, *55*, 793-813.

Lawrie, A. C. (1982). Field nodulation in nine species of *Casuarina* in Victoria. *Aust. J. Bot.*, *30*, 447-460.

Lechevalier, M. P., and Lechevalier, H. A. (1990). Systematics, isolation, and culture of *Frankia*. In C. R. Schwintzer and J. D. Tjepkema (Eds.), *The biology of* Frankia and *actinorhizal plants* (pp. 35-60). New York, NY: Academic Press.

Li, C. Y., Lu, K., Trappe, J., and Bollen, W. (1970). Separation of phenolic compounds in alkali hydrolysates of forest soil by thin-layer chromatography. *Can. J. Soil Sci.*, *50*, 458-460.

Li, C. Y., Trappe, J. M., and Bollen, W. B. (1972). *Poria weirii*-inhibiting and other phenolic compounds in roots of red alder and Douglas fir. *Microbios*, *5*, 65-68.

Lopez, M. F., Fontaine, M. S., and Torrey, J. G. (1984). Levels of trehalose and glycogen in *Frankia* sp. HFPArI3 (Actinomycetates). *Can. J. Microbiol.*, *30*, 746-752.

Maggia, L., and Bousquet, J. (1994). Molecular phylogeny of the actinorhizal Hamamelidae and relationships with host promiscuity towards *Frankia*. *Mol. Ecol.*, *3*, 459-467.

Martin, A. C., Zim, H. S., and Nelson, A. L. (1961). *American wildlife & plants: A guide to wildlife food habitats*. New York, NY: Dover Publications.

Martin, K. J., Posavatz, N. J., and Myrold, D. D. (2003). Nodulation potential of soils from red alder stands covering a wide age range. *Plant Soil, 254*, 187-192.

Martin, R. B. (1988). Bioinorganic chemistry of aluminum. In H. Sigel and A. Sigel (Eds.), *Metal ions in biological systems: Aluminum* and *its role in biology* (pp. 1-57). New York, NY: Marcel Dekker.

Maunuksela, L., Hahn, D., and Haahtela, K. (2000). Effect of freezing of soils on nodulation capacities of total and specific *Frankia* populations. *Symbiosis, 29*, 107-120.

Maunuksela, L. Zepp, K., Koivula, T., Zeyer, J., Haahtela, K., and Hahn, D. (1999). Analysis of *Frankia* populations in three soils devoid of actinorhizal plants. *FEMS Microbiol. Ecol., 28*, 11-21.

McIlveen, W. D., and Cole, H. (1976). Spore dispersal of Endogonaceae by worms, ants, wasps, and birds. *Can. J. Bot., 54*, 1486-1489.

McKey, D. (1994). Legumes and nitrogen: The evolutionary ecology of a nitrogen-demanding lifestyle. In J. I. Sprent and D. McKey (Eds.), *Advances in legume systematics: Part 5 – the nitrogen factor* (pp. 211-228). Kew, UK: Royal Botanic Gardens.

McNabb, D. H., and Cromack, K. (1985). Dinitrogen fixation by a mature *Ceanothus velutinus* (Dougl.) stand in the Western Oregon Cascades. *Can. J. Bot., 29*, 1014-1021.

McVean, D. N. (1956). Ecology of *Alnus glutinosa* (L.) Gaertn. III. Seedling establishment. *J. Ecol., 44*, 195-218.

Mejstrik, V., and Benecke, U. (1969). The ectotrophic mycorrhizas of *Alnus viridis* (Chaix) DC and their significance in respect to phosphorous uptake. *New Phytol.*, 68, 141-149.

Midgley, S. J., Turnbull, J. W., and Johnson, R. D. (1983). Casuarina *ecology, management and utilization*. Melbourne, Austrralia: CSIRO.

Müller, A., Benoit, P., Diem, H. G., and Schwencke, J. (1991). Age-dependent changes in extracellular protein, aminopeptidase activities in *Frankia. J. Gen. Microbiol., 137*, 2787-2796.

Murai, S. (1964). Phytotaxonomical and geobotanical studies on genus *Alnus* in Japan (III).Taxonomy of whole world species and distribution of each section. *Bull. Gov. Forest Exp. Sta.* (Tokyo, Japan), *171*, 1-107.

Myrold, D. D., and Huss-Danell, K. (1994). Population dynamics of *Alnus*-infective *Frankia* in a forest soil with and without host trees. *Soil Biol. Biochem., 26*, 533-540.

Navarro, E., Bousquet, J., Moiroud, A., Munive, A., Piou, D., and Normand, P. (2003). Molecular phylogeny of *Alnus* (Betulaceae), inferred from nuclear ribosomal DNA ITS sequences. *Plant Soil, 254*, 207-217.

Navarro, E., Nalin, R., Gauthier, D., and Normand, P. (1997). The nodular microsymbionts of *Gymnostoma* spp. are *Elaeagnus*-infective *Frankia* strains. *Appl. Environ. Microbiol., 63*, 1610-1616.

Neave, I. A., and Dawson, J. O. (1989). Juglone reduces growth, nitrogenase activity, and root respiration of actinorhizal black alder seedlings. *J. Chem. Ecol., 15*, 1823-1836.

Neave, I. A., Dawson, J. O., and DeLucia, E. H. (1989). Autumnal photosynthesis is extended in nitrogen-fixing European black alder compared with white basswood: Possible adaptive significance. *Can. J. Forest Res., 19*, 12-17.

Nickel, A. (2000). Population dynamics of *Frankia* in soil (Ph.D. thesis, Swiss Technical University (ETH), Zürich, Switzerland).

Nickel, A., Hahn, D., Zepp, K., and Zeyer, J. (1999). *In situ* analysis of introduced *Frankia* populations in root nodules obtained on *Alnus glutinosa* grown under different water availability. *Can. J. Bot., 77*, 1231-1238.

Oremus, P. A. I. (1980). Occurrence and infective potential of the endophyte of *Hippophaë rhamnoides* L. ssp. *rhamnoides* in coastal sand dune areas. *Plant Soil, 56*, 123-139.

Paschke, M. W. (1997). Actinorhizal plants in rangelands of the western United States. *J. Range Manage., 50*, 62-72.

Paschke, M. W., and Dawson, J. O. (1992a). The occurrence of *Frankia* in tropical forest soils of Costa Rica. *Plant Soil, 142*, 63-67.

Paschke, M. W., and Dawson, J. O. (1992*b*). *Frankia* abundance in soils beneath *Betual nigra* and other non-actinorhizal woody plants. *Acta Oecol., 13*, 407-415.

Paschke, M. W., and Dawson, J. O. (1993). Avian dispersal of *Frankia. Can. J. Bot., 71*, 1128-1131.

Paschke, M. W., Dawson, J. O., and Condon, B.M. (1994). *Frankia* in prairie, forest, and cultivated soils of central Illinois, U.S.A. *Pedobiologia, 38*, 546-551.

Paschke, M. W., Dawson, J. O., and David, M. B. (1989). Soil nitrogen mineralization in mixed plantations of black walnut with actinorhizal autumn olive or black alder. *Plant Soil, 118*, 33-42.

Perradin, Y., Mottet, M., and LaLonde, M. (1983). Influence of phenolics on *in vitro* growth of *Frankia* strains. *Can. J. Bot., 61*, 2807-2814.

Prin, Y., and Rougier, M. (1987). Preinfection events in the establishment of the *Alnus-Frankia* symbiosis: Study of the root hair deformation step. *Plant Physiol., 6*, 99-106.

Reddell, P., Bowen, G. D., and Robson, A. D. (1985). The effects of soil temperature on plant growth, nodulation and nitrogen fixation in *Casuarina cunninghamiana*. Miq. *New Phytol., 101*, 441-450.

Reddell, P., Rosbrook, P. A., Bowen, G. D., and Gwaze, D. (1988). Growth responses in *Casuarina cunninghamiana* plantings to inoculation with *Frankia*. *Plant Soil, 108*, 79-86.

Reddell, P., and Spain, A. V. (1991). Earthworms as vectors of viable propagules of mycorrhizal fungi. *Soil Biol. Biochem., 23*, 767-774.

Richards, J. W., Krumholz, G. D., Chval, M. S., and Tisa, L. S. (2002). Heavy metal resistance patterns of *Frankia* strains. *Appl. Environ. Microbiol., 68*, 923-927.

Richardson, D. M., Allsopp, N., D'Antonio, C. M., Milton, S. J., and Rejmanek, M. (2000). Plant invasions - the role of mutualisms. *Biol. Rev., 75*, 65-93.

Righetti, T. L., Chard, C. H., and Backhause, R. A. (1986). Soil and environmental factors related to nodulation in *Cowania* and *Purshia*. *Plant Soil, 91*, 147-160.

Righetti, T. L., and Munns, D. N. (1981). Soil factors limiting nodulation and nitrogen fixation in *Purshia*. In J. M. Lyons, R. C. Valentine, D. A. Phillips, D. W. Rains, and R. C. Huffaker (Eds.), *Genetic engineering of symbiotic nitrogen fixation* and *conservation of fixed nitrogen* (pp. 395-407). New York, NY: Plenum Press.

Roberts, N. (1998). *The Holocene: An environmental history* (Second Edition). Oxford, UK: Blackwell.

Rodriguez-Barrueco, C. (1968). The occurrence of the root nodule endophytes of *Alnus glutinosa* and *Myrica gale* in soils. *J. Gen. Microbiol., 52*, 189-194.

Rodriguez-Barrueco, C., Mackintosh, A. H., and Bond, G. (1970). Some effects of combined nitrogen on the nodule symbioses of *Casuarina* and *Ceanothus*. *Plant Soil, 33*, 129-139.

Rösch, D., Bergmann, M., Knorr, D., and Kroh, L. W. (2003). Structure-antioxidant efficiency relationships of phenolic compounds and their contribution to the antioxidant activity of sea buckthorn juice. *J. Agr. Food Chem., 51*, 4233-4239.

Rose, S. L., and Youngberg, C. T. (1981). Tripartite association of snowbrush (*Ceanothus velutinus*): Effect of vesicular-arbuscular mycorrhizae on growth nodulation and nitrogen fixation. *Can. J. Bot., 59*, 34-39.

Safo-Sampath, S., and Torrey, J. G. (1988). Polysaccharide-hydrolysing enzymes of *Frankia* (Actinomycetales). *Plant Soil, 112*, 89-97.

Seguin, A., and Lalonde, M. (1989). Detection of pectolytic activity and *pel* homologous sequences in *Frankia*. *Plant Soil, 118*, 221-229.

Schramm, J. R. (1966). Plant colonization studies on black wastes from anthracite mining in Pennsylvania. *T. Am. Philos. Soc., 56*, 1-194.

Schwencke, J., and Carú, M. (2001). Advances in actinorhizal symbiosis: Host plant-*Frankia* interactions, biology, and applications in arid land reclamation. A review. *Arid Land Res. Manag., 15*, 285-327.

Schwintzer, C. R., Berry, A. M., and Disney, L. D. (1982). Seasonal pattern of root nodule growth, endophyte morphology, nitrogen activity and shoot development in *Myrica gale*. *Can. J. Bot., 60*, 746-757.

Seiler, J. R., and Johnson, J. D. (1984). Growth and acetylene reduction of black alder seedlings in response to water stress. *Can. J. Forest Res., 14*, 477-480.

Selim, S., Delacour, S., and Schwencke, J. (1996). Specific long-chain fatty acids promote optimal growth of *Frankia*: Accumulation and intracellular distribution of palmitic and propionic acid. *Arch. Microbiol., 165*, 252-257.

Shindo, H., Ohta, S., and Kuwatsuka, S. (1978). Behavior of phenolic substances in the decaying process of plants. IX. Distribution of phenolic acids in soils of rice paddy fields and forests. *Soil Sci. Plant Nutr., 24*, 233-243.

Silvester, W. B. (1977). Dinitrogen fixation by plant associations excluding legumes. In R. Hardy and W. Silvester (Eds.), *A treatise of dinitrogen fixation* (Vol. 4, pp. 141-190). New York, NY: Academic Press.

Silvester, W. B., and Harris, S. L. (1990). Oxygen regulation and hemoglobin. In C. R. Schwintzer and J. D. Tjepkema (Eds.), *The biology of* Frankia and *actinorhizal plants* (pp. 157-176). New York, NY: Academic Press.

Silvester, W. B., Whitbeck, J., Silvester, J. K., and Torrey, J. G. (1988a). Growth, nodule morphology, and nitrogenase activity of *Myrica gale* with roots grown at various oxygen levels. *Can. J. Bot., 66*, 1762-1771.

Silvester, W. B., Silvester, J. K., and Torrey, J. G. (1988b). Adaptation of nitrogenase to varying oxygen tension in root nodules of *Alnus incana* ssp. *rugosa*. *Can. J. Bot., 66*, 1772-1779.

Simonet, P., Navarro, E., Rouvier, C., Reddell, P., Zimpfer, J., *et al.* (1999). Co-evolution between *Frankia* populations and host plants in the family Casuarinaceae and consequent patterns of global dispersal. *Env. Microbiol., 1*, 525-533.

Smolander, A. (1990). *Frankia* populations under different tree species – with a special emphasis on soils under *Betula pendula*. *Plant Soil, 121*, 1-10.

Smolander, A., Rönkkö, R., Nurmiaho-Lassila, E.-L., and Haahtela, K. (1990). Growth of *Frankia* in the rhizosphere of *Betula pendula*, a nonhost tree species. *Can. J. Microbiol., 36*, 649-656.

Smolander, A., and Sundman, V. (1987). *Frankia* in acid soils of forests devoid of actinorhizal plants. *Physiol. Plant., 70*, 297-303.

Smolander, A., van Dijk, C., and Sundman, V. (1988). Survival of *Frankia* strains introduced into soil. *Plant Soil, 106*, 65-72.

Soltis, D. E., Soltis, P. S., Morgan, D. R., Swensen, S. M., Mullin, B. C., *et al.* (1995). Chloroplast gene sequence data suggest a single origin of the predisposition for symbiotic nitrogen fixation in angiosperms. *Proc. Natl. Acad. Sci. U.S.A., 92*, 2647-2651.

Su, H., and Lin, L.-P. (1989). The dissemination of endophytes and existence of endomycorrhiza on the rhizoplane of *Alnus formosana*. *J. Chin. Agric. Chem. Soc., 27*, 513-523.

Subbarao, N. S., and Rodríguez-Barrueco, C. (1995). *Casuarinas*. Lebanon, NH: Science Publishers.

Sundström, K.-R., and Huss-Danell, K. (1987). Effects of water stress on nitrogenase activity in *Alnus incana*. *Physiol. Plant., 70*, 342-348.

Swensen, S. M. (1996). The evolution of actinorhizal symbioses: Evidence for multiple origins of the symbiotic association. *Am. J. Bot., 83*, 1503-1512.

Thompson, V. (1994). Spittlebug indicators of nitrogen-fixing plants. *Ecol. Entomol., 19*, 391-398.

Thompson, V. (1999). Spittlebugs associated with actinorhizal host plants. *Can. J. Bot., 77*, 1387-1390.

Tilman, D., Wedin, D., and Knops, J. (1996). Productivity and sustainability influenced by biodiversity in grassland ecosystems. *Nature, 379*, 718.

Tjepkema, J. (1978). The role of oxygen diffusion from the shoots and nodule roots in nitrogen fixation by root nodules of *Myrica gale*. *Can. J. Bot., 56*, 1365-1371.

Torrey, J. G. (1990). Cross-inoculation groups within *Frankia* and host-endosymbiont associations. In C. R. Schwintzer and J. D. Tjepkema (Eds.), *The biology of* Frankia *and actinorhizal plants* (pp. 83-106). New York, NY: Academic Press.

Tortosa, R. D., and Cusato, M. (1991). Effective nodulation of rhamnaceous actinorhizal plants induced by air dry soils. *Plant Soil, 131*, 229-233.

van Dijk, C. (1979). Endophyte distribution in the soil. In J. C. Gordon, C. T. Wheeler, and D. A. Perry (Eds.), *Symbiotic nitrogen fixation in the management of temperate forests* (pp. 84-94). Corvalis, OR: Oregon State University Press.

van Dijk, C., and Sluimer-Stolk, A. (1990). An ineffective strain type of *Frankia* in the soil of natural stands of *Alnus glutinosa* (L.) Gaertn. *Plant Soil, 127*, 107-121.

Van Ghelue, M., Lovaas, E., Ringo, E., and Solheim, B. (1997). Early interactions between *Alnus glutinosa* and *Frankia* strain ArI3: Production and specificity of root hair deformation factor(s). *Physiol. Plant., 99*, 579-587.

Van Miegroet, H., and Cole, D. W. (1985). Acidification sources in red alder and Douglas-fir soils: Importance of nitrification. *Soil Sci. Soc. Am. J., 49*, 1274- 1279.

Vergnaud, L., Chaboud, A., Prin, Y., and Rougier, M. (1985). Preinfection events in the establishment of *Alnus-Frankia* symbiosis: Development of a spot inoculation technique. *Plant Soil, 87*, 67-68.

Vitousek, P. M., Cassman, K., Cleveland, C., Crews, T., Field, C. B., *et al.* (2002). Towards an ecological understanding of biological nitrogen fixation. *Biogeochemistry, 57/58*, 1-45.

Vitousek, P. M., and Walker, L. R. (1989). Biological invasion by *Myrica faya* in Hawaii: Plant demography, nitrogen fixation, and ecosystem effects. *Ecol. Monogr., 59*, 247-265.

Vogel, C. S., and Curtis, P. S. (1995). Leaf gas and nitrogen dynamics of N_2-fixing field-grown *Alnus glutinosa* under elevated atmospheric CO_2. *Glob. Change Biol., 1*, 55-61.

Vogel, C., and Dawson, J. O. (1986). *In vitro* growth of five *Frankia* isolates in the presence of four phenolic acids. *Soil Biol. Biochem., 18*, 227-231.

Walker, L. R. (1993). Nitrogen fixers and species replacements in primary succession. In J. Miles and D. W. H. Walton (Eds.), *Primary succession on land* (pp. 249-272). London, UK: Blackwell Scientific.

Wheeler, C. T., and Bowes, B. G. (1974). Effects of light and darkness on nitrogen fixation by root nodules of *Alnus glutinosa* in relation to their cytology. *Z. Pflanzenphysiol., 71*, 71-75.

Wheeler, C. T., Henson, I. E., and McLaughlin, M. E. (1979). Hormones in plants bearing actinomycete nodules. *Bot. Gaz.* (Chicago), Suppl., *140*, S52-S57.

Wheeler, C. T., Hughes, L. T., Oldroyd, J., and Pulford, I. D. (2001). Effects of nickel on *Frankia* and its symbiosis with *Alnus glutinosa* (L) Gaertn. *Plant Soil, 231*, 81-90.

Wheeler, C. T., and Miller, I. M. (1990). Current and potential uses of actinorhizal plants in Europe. In C. R. Schwintzer and J. D. Tjepkema (Eds.), *The biology of* Frankia and *actinorhizal plants* (pp. 365-389). New York, NY: Academic Press.

Wheeler, C. T., Watts, S. H., and Hillman, J. R. (1983). Changes in carbohydrates and nitrogenase compounds in the root nodules of *Alnus glutinosa* in relation to dormancy. *New Phytol., 95*, 209-218.

Whitehead, D. (1964). Identification of *p*-hydroxybenzoic, vanillic, *p*-coumaric, and ferulic acid in soils. *Nature, 202*, 417-419.

Whitehead, D., Dibb, H., and Hartley, R. (1983). Bound phenolic compounds in water extracts of soils, plant roots and leaf litter. *Soil Biol. Biochem., 15*, 133-136.

Wollum, C. T., and Youngberg, A. G. (1969). Effect of soil temperatures on nodulation of *Ceanothus velutinus* Dougl. *Proc. Soil Sci. Soc. of America, 33*, 801-803.

Wollum, C. T., Youngberg, A. G., and Chichester, F. W. (1968). Relation of previous timber stand age to nodulation of *Ceanothus velutinus. Forest Sci., 14*, 114-118.

Wolters, D. J., Akkermans, A. D. L., and Van Dijk, C. (1997). Ineffective *Frankia* strains in wet stands of *Alnus glutinosa* L. Gaertn. in the Netherlands. *Soil Biol. Biochem., 29*, 1702-1712.

Yadav, J. S. P. (1983). Soil limitations for successful establishment and growth of *Casuarina* plantations. In S. J. Midgley, J. W. Turnbull, and R. D. Johnston (Eds.), *Casuarina ecology, management* and *utilization* (pp. 138-157). Melbourne, Austrralia: CSIRO.

Yamanaka, T., Li, C.-Y., Bormann, B. T., and Okabe, H. (2003). Tripartite associations in alder: Effects of *Frankia* and *Alpova diplophloeus* on the growth, nitrogen fixation and mineral acquisition of *Alnus tenuifolia. Plant Soil, 254*, 179-186.

Young, D. R., Sande, E., and Perters, G. A. (1992). Spatial relationships of *Frankia* and *Myrica cerifera* on a Virginia, U.S.A., barrier island. *Symbiosis, 112*, 209-220.

Zhang, X., and Benson, D. R. (1992). Utilization of amino acids by *Frankia* strain CpI1. *Arch. Microbiol., 158*, 256-261.

Zimpfer, J. F., Kaelke, C. M., Smyth, C. A., Hahn, D., and Dawson, J. O. (2003). *Frankia* inoculation, soil biota, and host tissue amendment influence *Casuarina* nodulation capacity of a tropical soil. *Plant Soil, 254*, 1-10.

Zimpfer, J. F., Kennedy, G. J., Smyth, C. A., Hamelin, J., Navarro, E., and Dawson, J. O. (1999). Localization of *Casuarina*-infective *Frankia* near *Casuarina cunninghamiana* trees in Jamaica. *Can. J. Bot., 77*, 1248-1256.

Zimpfer, J. F., McCarty, B., Kaelke, C. M., Mulongwe, L., Igual, J. M., Smyth, C. A., and Dawson, J. O. (2002). *Casuarina cunninghamiana* cladode extracts increase the *Frankia* infectious capacity of a tropical soil. *Symbiosis, 33*, 73-90.

Zimpfer, J. F., Smyth, C. A., and Dawson, J. O. (1997). The capacity of Jamaican mine spoils, agricultural and forest soils to nodulate *Myrica cerifera, Leucaena leucocephala* and *Casuarina cunninghamiana. Physiol. Plant., 99*, 664-672.

Chapter 9

MOLECULAR BIOLOGY OF ACTINORHIZAL SYMBIOSES

L. LAPLAZE, S. SVISTOONOFF, C. SANTI, F. AUGUY, C. FRANCHE AND D. BOGUSZ

Equipe Rhizogenèse, UMR 1098,
Institut de Recherche pour le Développement (IRD), BP 64501,
911 Avenue Agropolis, 34394 Montpellier Cedex 5, France

1. INTRODUCTION

Two N_2-fixing root-nodule symbioses between soil bacteria and plants have been described, one involves rhizobia and legumes (and also *Parasponia*) and the other involves *Frankia* and actinorhizal plants. The rhizobia/legume symbioses include more than 1700 plant species of the Fabaceae (Leguminosae) family, whereas the *Frankia*/actinorhizal plant symbioses cover about 260 species belonging to 8 angiosperm families. Legume and actinorhizal nodules differ in their ontogeny and structure. However, recent phylogenetic studies, based on *rbcL* gene-sequence analysis, have shown that all plants able to enter a root-nodule symbiosis belong to the same clade, which suggests that they share a predisposition for symbiosis (Doyle, 1998; Soltis *et al.*, 1995). The molecular bases of this predisposition are unknown. In this respect, a comparison of the genetic program of legume and actinorhizal symbioses is of great interest, especially if we are ever to transfer the ability to fix N_2 to crop plants, such as cereals.

In legumes, knowledge of the molecular biology of the symbiotic interaction has progressed considerably during the last decade (for review, see Schultze and Kondorosi, 1998). Actinorhizal plants are mostly woody plants, trees or shrubs and are, therefore, recalcitrant to molecular-biology techniques. However, progress in nucleic-acid isolation allowed the first actinorhizal nodulin gene in *Alnus glutinosa* to be characterised in 1994 (Goetting-Minesky and Mullin, 1994). Since then, several putative symbiotic genes have been isolated from different actinorhizal species (for reviews, see Franche *et al.*, 1998b; Pawlowski, 1997). The study of actinorhizal symbiotic genes has greatly benefited from the recent development of a

K. Pawlowski and W. E. Newton (eds.), Nitrogen-fixing Actinorhizal Symbioses, 235–259.
© 2008 *Springer*.

transformation procedure for actinorhizal trees of the Casuarinaceae family (Franche *et al.*, 1997; for reviews, see Franche *et al.*, 1998b; Smouni *et al.*, 2002). This technical breakthrough opened new options for the study of genes involved in actinorhizal symbioses. For instance, it paved the way for the study of the regulatory sequence within the promoters of symbiotic genes (Laplaze *et al.*, 2002). Transgenic plants are useful tools to study gene function by providing a means to modulate either the level or pattern of expression. Further, the expression conferred by promoters, such as the cauliflower virus 35S promoter, which might be useful for these kinds of experiments, has been characterised in transgenic Casuarinaceae (Franche *et al.*, 1998b; Smouni *et al.*, 2002). With these tools now available, our understanding of the molecular mechanisms of actinorhizal symbioses has increased and will continue to do so.

In this chapter, the contribution of plant molecular-biology approaches to our understanding of actinorhizal symbioses is described first. Then, the molecular mechanisms of infection, nodule development, and nodule function are analysed, followed by a discussion of the evolution of symbiotic genes in light of recent experiments on the expression of heterologous genes in transgenic plants. Finally, new and exciting approaches to study the molecular biology of actinorhizae are examinied.

2. INFECTION PROCESS

2.1. Interface between Frankia *and the Plant Cell*

During the infection of actinorhizal plants by *Frankia*, the bacteria come in close contact with the plant cell. The interface between the two symbiotic partners is an important zone of exchange of both signals and nutrients. Thus, this structure, which being derived from the plant cell wall has some very specific properties (see chapter 6 this volume), and the symbiotic genes that might be involved in its formation and/or functioning have been studied and described.

The symbiotic genes, *ag12/cg12*, are from *A. glutinosa* (Ribeiro *et al.*, 1995) and *Casuarina glauca* (Laplaze *et al.*, 2000b), respectively. They both encode proteases of the subtilisin family and show 85% similarity at the amino-acid level. These genes are specifically expressed during plant-cell infection, but expression is turned down when plant cells differentiate to fix N_2. When the *cg12* promoter-reporter gene fusions were introduced into *Allocasuarina verticillata*, its expression was observed during the first steps of the infection process, *i.e.*, when *Frankia* is invading deformed root hairs, but it is not induced by *Frankia* diffusible factors (Svistoonoff *et al.*, 2003a). Therefore, expression of these genes is correlated with plant-cell invasion by the endosymbiont from the very start of the symbiotic process.

The function of these proteins is still unknown. The presence of a putative signal peptide in the sequence of both proteins indicates that the corresponding enzymes are probably secreted into the extracellular compartment. Our immunolocalisation experiments showed that CG12 localizes in the cell wall and the matrix surrounding *Frankia* in infected cells of *Casuarina glauca* nodules (our unpublished data).

Sequence alignments with other subtilases revealed that these genes belong to the pyrolisin subfamily and are closely related to the LeSbt3/4 tomato family (Laplaze *et al.*, 2000b). Subtilases belong to one of two classes, *i.e.*, they are either processing or degradative subtilases. Degradative subtilases have poor substrate specificity and are involved in the degradation of a wide range of proteins. Most bacterial subtilases belong to this class as do some plant subtilases, such as the well-known melon fruit cucumisin (Yamagata *et al.*, 1994; for a review, see Siezen and Leunissen, 1997). If AG12/CG12 belongs to this class, it could be involved in general protein digestion, which is associated with the cell-wall remodelling that occurs in response to *Frankia* infection.

In contrast, processing subtilases have high levels of substrate specificity and are generally involved in maturation of inactive proteins or peptide hormones. Mammalian proprotein convertases, which cleave their substrates at paired dibasic residues to produce active hormones or neuropeptides, and the yeast KEX-2 subtilase, involved in the maturation of the mating pheromone, are examples of well-known processing subtilases. In plants, subtilases likely belong to this latter class and include P69B, a potential tomato LRR-protein maturation subtilase (Jorda *et al.*, 1999), and SBP50, which may be involved in prosystemin maturation (Schaller and Ryan, 1994). If AG12/CG12 belongs to this class of subtilase, it might be involved in the maturation of unknown proteins or propeptides at the interface between the plant and the bacteria. Biochemical characterisation of CG12 should help to determine the function of these subtilases in the infection process.

The two genes, *agNt84/ag164*, were isolated from *A. glutinosa* (Pawlowski *et al.*, 1997). These genes are strongly expressed in cells invaded by *Frankia* that do not yet fix N_2. They code respectively for a 10.57-kDa and 9.19-kDa glycine- and histidine-rich protein. Both have a signal peptide that probably targets them to the extracellular compartment, presumably to the interface between the two symbiotic partners. Both proteins have several phosphorylation sites and an N-terminal glycine that is potential target for myristylation. Fragments of AgNt84 and Ag164 produced in *Escherichia coli* can bind metal ions, like Ni^{2+}, Co^{2+}, and Zn^{2+}, suggesting that they play such a role (Gupta *et al.,* 2002; Pawlowski *et al.,* 1997). Of these, Co^{2+} was suggested as their likely target because it is known to be essential for N_2-fixing symbiosis (Pawlowski *et al.*, 1997). Zinc and nickel are less likely candidates but they are required for efficient nitrogen fixation. Alternatively, these proteins, the so-called metallohistins, could play a role in limiting the growth of symbiotic Frankia by sequestering metal ions (Gupta *et al.,* 2002).

2.2. Prenodule Formation

In actinorhizal plants that are infected intra-cellularly, *Frankia* infection triggers cell divisions in the cortical cell adjacent to the infection site. These cell divisions give rise to a small protuberance, called a prenodule (Figure 1D). The endosymbiont invades some of the prenodule cells that subsequently enlarge (Figure 1E; Callaham and Torrey, 1977; see chapter 6 in this volume). The prenodule is an obligatory step in intracellular infection but is not directly involved in nodule formation.

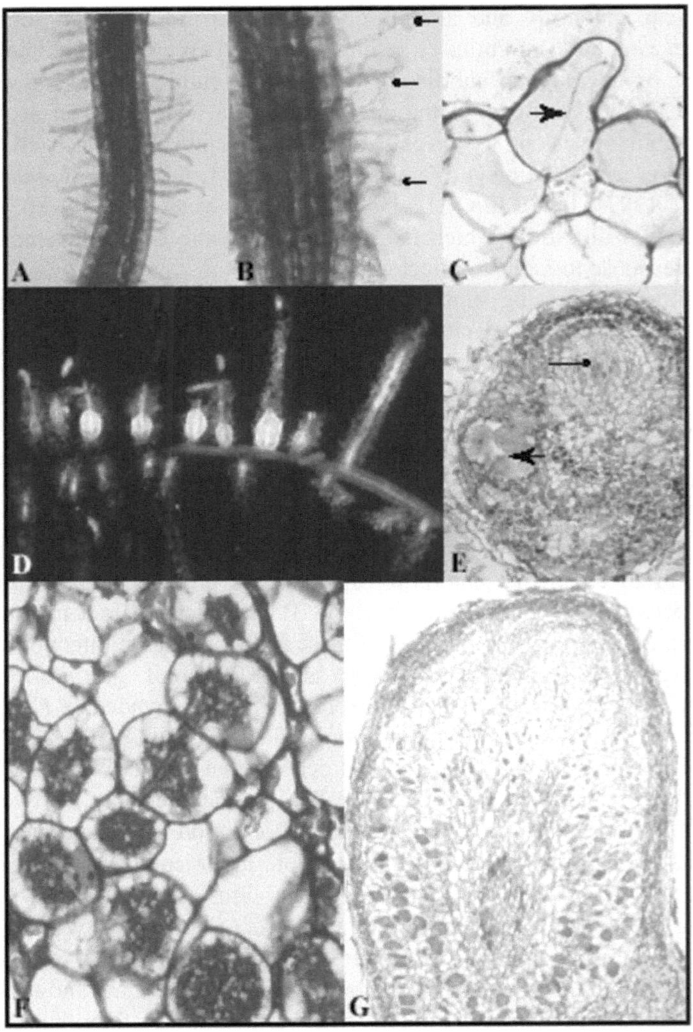

Figure 1. Infection of actinorhizal plants by Frankia.
A and B: Roots of Casuarina glauca *infected by* Frankia. *Root-hair deformation occurs (black arrows). C: Detail of a transverse section of a* C. glauca *root showing a* Frankia *filament (arrow) in a root hair. D: Infected* Allocasuarina verticillata *root with white protuberances corresponding to prenodules. E: Transverse section of an infected* C. glauca *root with a prenodule (thick arrow) and a young nodule primordium (thin arrow). F: Detail of a longitudinal section of a* C. glauca *nodule showing an infection thread progressing between two cortical cells. G: Longitudinal section of a* C. glauca *nodule lobe.*

Recently, prenodule physiology and function studies, using molecular techniques (Laplaze *et al.*, 2000a), showed that *Frankia* can fix N_2 in prenodule infected cells. The formation of vesicles that were associated with a strong reducing potential (Angulo Carmona, 1974) and the expression of the nitrogenase *nifH* structural gene (Laplaze *et al.*, 2000a) were clearly demoonstrated. As was the differentiation of these plant cells to allow nitrogen fixation by both the expression of *cghb*, a symbiotic hemoglobin gene, and cell-wall lignification (Laplaze *et al.*, 2000a). Moreover, expression of molecular markers and starch accumulation in uninfected prenodule cells suggest that they have the same characteristics as their nodule counterparts (Callaham and Torrey, 1977; Laplaze *et al.*, 2000a). Taken together, these results suggest that the prenodule is formed of two cell types, infected and uninfected cells, which undergo the same differentiation towards symbiotic nitrogen fixation as the corresponding nodule cells. The prenodule is, therefore, a very simple symbiotic organ and it might be a significant remnant of the evolution of endophytic N_2-fixing symbioses in plants (Gualtieri and Bisseling, 2000; Laplaze *et al.*, 2000a; see chapter 10 in this volume).

3. NODULE DEVELOPMENT

3.1. Nodule Formation and Structure

After prenodule development and infection, cell divisions are induced in the pericycle opposite to a protoxylem pole that will give rise to a nodule lobe primordium. An apical meristem is responsible for primordium growth toward the root surface in regions not infected by *Frankia* (Figure 1E). The primordium does not incorporate the prenodule but gets infected by hyphae coming from the prenodule (Duhoux *et al.*, 1996; see chapter 6 in this volume).

Mature actinorhizal nodules are indeterminate and multilobed structures. Each nodule lobe contains a central vascular bundle surrounded by an endoderm, an expanded cortex, and a periderm (Figures 1G and 2). Only certain cortical cells are infected by *Frankia* (Figure 1F). Two types of actinorhizal nodule can be defined; one is the *Myrica* type that exhibits a so-called nodule root at the apex of each lobe, the other is the *Alnus* type that does not have this feature (Duhoux *et al.*, 1996). Nodule roots lack root hairs, have a smaller root cap, and are not infected. They show negative geotropism, have an extensive aerenchyma, and facilitate the diffusion of gases (O_2 in particular) in and out of the nodule lobe (Callaham and Torrey, 1977; Schwintzer and Lancelle, 1983; Tjepkema, 1978).

3.2. Comparison with Lateral Root Development

Because of their origin and structure, actinorhizal nodules are often regarded as modified lateral roots. The formation of a nodule root at the apex of some actinorhizal nodules reinforces this view, however, some important differences exist. First, in *Comptonia*, some cortical cells close to the nodule-lobe primordium divide and are incorporated in the growing young lobe, whereas lateral-root

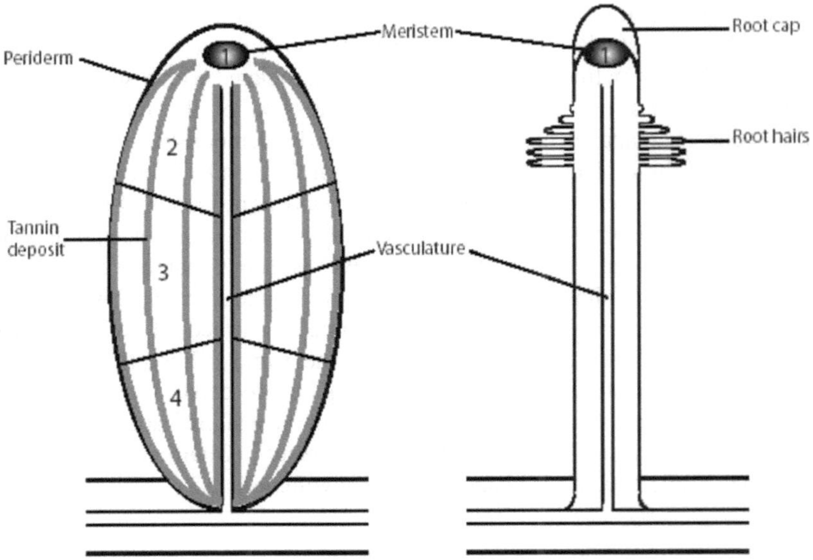

Figure 2. Comparison of the structures of an actinorhizal nodule lobe and a lateral root.
(1), meristem ; (2), infection zone; (3), fixation zone; (4), senescence zone.

primordia originate only in the pericycle (Callaham and Torrey, 1977). Moreover, the distribution of lateral roots is not changed in nodulated plants, thus suggesting that the formation of these two types of organs is regulated independently (Angulo Carmona, 1974; Valverde, 2000). Finally, differences exist in gene expression (Franche *et al.*, 1998b; Pawlowski, 1997).

One exciting question is the extent to which the development of a lateral root and an actinorhizal nodule share common steps. In an attempt to answer this question, the *HRGPnt3*-gene promoter fused to the β-glucuronidase (*gus*) gene was introduced into *A. verticillata* plants. The *Nicotiana tabacum HRGPnt3* gene encodes a plant cell-wall protein expressed during early stages of lateral-root development and is a very good molecular marker of lateral-root initiation (Keller and Lamb, 1989; Vera *et al.*, 1994). Unfortunately, *HRGPnt3-gus* was expressed during neither lateral root nor nodule development in *A. verticillata*, suggesting that the specificity of expression is not maintained in a heterologous environment (our unpublished data). The isolation from actinorhizal plants of homologues of genes known to be involved in lateral-root development should help answer this question.

Another exciting challenge for the future is to identify the genes responsible for the specific developmental features of nodules. One candidate gene is dg93, a nodule-specific gene from *Datisca glomerata* (Okubara et al., 2000), which encodes a 105 amino-acid protein with 74% similarity to the soybean early nodulin ENOD93. This gene, plus EuSAMS1 (Lee *et al.*, 2001), are the only actinorhizal symbiotic genes known to be expressed in the nodule-lobe meristem so far. It is also expressed in infected cells and in the vascular cylinder. Its function is unknown but, because it is nodule-specific and present in the nodule-lobe meristem, it might be involved in setting up the specific characteristic of nodule lobes as opposed to lateral roots. Further studies on DG93 function will be needed to clarify this point.

3.3. Role of Plant Hormones

Even though the molecular mechanisms responsible for actinorhizal nodule development are poorly understood, it is likely that nodulation is linked to a local change in hormonal balance. Auxins, which are known to play a central role in lateral root initiation (for a review, see Malamy and Benfey, 1997), are obvious candidates for regulatory molecules. Actinorhizal nodules contain large quantities of auxin and cytokinins (Dullaart, 1970; Henson and Wheeler, 1977; Wheeler *et al.*, 1979) and some *Frankia* strains can secrete auxins and cytokinins in culture (Berry *et al.*, 1989; Stevens and Berry, 1988). Moreover, phenylacetic acid (PAA) is known to be released by *Frankia* strains *in vitro* (Hamad *et al.* (2003). This molecule is auximimetic and was able to induce the formation of nodule-like structures on *A. glutinosa* roots. These results suggest that *Frankia* might induce nodule formation by secreting an auximimetic molecule.

Recently, auxin-responsive genes, such as *GH3* from soybean (Li *et al.*, 1999), have been used as markers to visualise auxin accumulation *in situ*. This molecular marker has been successfully used to study the changes in auxin level or sensitivity during legume nodulation (Mathesius *et al.*, 1998). The soybean *GH3* promoter–*gus* fusion was introduced into *Allocasuarina verticillata* but no expression was detected in the transgenic plants, even after incubation with auxin. Consequently a homologue of *GH3* was isolated from a *Casuarina glauca* ESTs library (our unpublished data). Its promoter contains putative auxin-response elements. Transfer of a *cgGH3* promoter-*gus* fusion in *A. verticillata* is underway and will hopefully be a useful molecular marker to study auxin accumulation *in situ* during actinorhizal nodule development.

Like lateral roots and legume nodules, actinorhizal nodules are formed opposite to protoxylem poles. In legumes, a regulator that is emitted by the stele is responsible for nodule positioning (Libbenga *et al.*, 1973). Other studies suggest that ethylene, which is produced in the region of the pericycle opposite to the phloem poles, inhibits cell divisions in the neighbouring cortical cells, thus controlling the position of nodule initiation (Heidstra *et al.*, 1997; Penmetsa and Cook, 1997). The mechanisms responsible for both lateral-root and actinorhizal-nodule positioning are not known but it is tempting to speculate that gradients of hormones, such as ethylene, play a role.

3.4. Is there a Role for ENOD40 *in Actinorhizal Nodulation?*

The early nodulin gene, *ENOD40*, was first isolated from soybean (Yang *et al.*, 1993). In legumes, *ENOD40* is a key gene for nodule organogenesis and a limiting factor in nodule development (Charon *et al.*, 1999). It also plays a role in mycorrhizal symbiosis (Staehelin *et al.*, 2001). *ENOD40* is induced by nodulation factors, its expression precedes the first cortical-cell divisions (Fang *et al.*, 1998), and it is expressed in the vascular system of roots, stems, mature nodules, and in the developing nodule primordia (Crespi *et al.*, 1994). Over expression of *ENOD40* results in a significant increase in cortical-cell division, suggesting that ENOD*40* action may play a role in initiating nodule morphogenesis (Charon *et al.*, 1999).

Identification of homologues in non-leguminous plants suggests that this gene may have a more general biological function. Recent work (Röhrig *et al.*, 2002) has revealed that *ENOD40* encodes two peptides that bind to sucrose synthase. Thus, a function in phloem unloading and/or sink strength determination would be consistent with the effects of its over- and under-expression in *Medicago* (Charon *et al.*, 1999; Staehelin *et al.*, 2001).

Homologues of *ENOD40* were isolated from two actinorhizal plants, *A. glutinosa* and *C. glauca*, and were named *agENOD40* and *cgENOD40*, respectively. A Southern blot showed that *cgENOD40* is encoded by a single gene and contains no introns (Santi *et al.*, 2003). In the legume genes, two highly conserved regions were distinguished; box I in the 5'-end, spanning a conserved ORF, and box II in the central part of the gene that corresponds to non-coding RNA. Neither actinorhizal gene encodes the conserved peptides found in legumes; there are insertions, deletions, and frame shifts in box I, which contrasts with the high conservation in box II (U. von Groll, A. Ribeiro and K. Pawlowski, personal communication).

Expression of a *cgENOD40-gus* fusion in transgenic *A. verticillata* and *C. glauca* was found in vascular tissue of the roots, shoots and nodules. No expression was found at earlier stages of infection by *Frankia*; neither in prenodules and nodule primordia nor in response to Nod factors (Santi *et al.*, 2003).

Contrary to legumes, all actinorhizal plants have a lignified root system, and the mechanisms of carbon transport must consequently be different. A comparison of phloem unloading in *C. glauca* and *Medicago truncatula*, using fluorescent tracers, indicates that, in *Casuarina,* unloading is mostly symplastic whereas, in *M. truncatula*, it is mostly apoplastic (Santi *et al.*, 2003). These results suggest that ENOD*40* is involved in increasing apoplastic phloem unloading to induce nodulation in legumes, whereas in actinorhizal plants, like *C. glauca* with mostly symplastic phloem unloading mechanisms, *cg*ENOD*40* plays no role in nodule induction.

4. NODULE FUNCTIONING

4.1. Actinorhizal Nodule Compartmentation

Actinorhizal nodule lobes display two levels of compartmentation. Because of the presence of a meristem at the apex in a mature nodule lobe, the first level of compartmentation, which involves the different steps of the symbiotic interaction, occurs longitudinally. Four zones (Figure 2) have been defined based on morphological (Angulo Carmona *et al.*, 1974; Duhoux *et al.*, 1996) and gene-expression studies (Gherbi *et al.*, 1997; Ribeiro *et al.*, 1995).

Zone (1) is the *apical meristem*, which is free of *Frankia*. In the *Myrica*-type nodule, after some time, the meristem undergoes a change leading to the formation of the nodule root, whereas in the *Alnus*-type nodule, meristem activity stops. Nothing is known about either the signal or molecular mechanisms associated with these different outcomes.

Zone (2) is adjacent to the meristem and is an *infection zone*, where some of the young cortical cells that result from the meristem activity are infected by *Frankia*. The bacteria start to proliferate while remaining encapsulated in a plant-derived matrix and the plant cell enlarges.

Zone (3) is the *fixation zone* and contains both infected and uninfected cortical cells. Infected cells are hypertrophied and are filled with *Frankia* filaments that differentiate vesicles, where nitrogen fixation takes place. The appearance and shape of these vesicles are controlled by the plant. In some genera, *e.g.*, *Casuarina*, infected cells have a lignified cell wall. Uninfected cells are smaller and, in some species, contain amyloplast and phenolic compounds; these cells might be involved in nitrogen and carbon metabolism (see below).

The final zone, zone (4), is a basal *senescence zone*, where both plant cells and bacteria degenerate and nitrogen fixation is switched off.

An additional level of compartmentation has been described in *C. glauca* nodules (Laplaze *et al.*, 1999). An accumulation of flavans (a class of flavonoids) occurs in uninfected cells in the endodermis, below the periderm and in the cortex. These cells form layers that delimit *Frankia*-infected compartments in the nodule lobe (Figure 2). Gene-expression studies (Laplaze *et al.*, 1999; Smouni *et al.*, 2002) confirm that these cells represent a third specialised cell type in the cortex of *C. glauca* nodules. Further, an accumulation of tannins occurs in early steps of both intra-cellular infection (Angullo Carmona, 1974; Callaham and Torrey, 1977; Laplaze *et al.*, 1999; Duhoux *et al.*, 2001) and inter-cellular infection of actinorhizal plants (Torrey, 1976; Miller and Baker, 1985) and *Parasponia rigida* (Lancelle and Torrey, 1984), the only non-legume nodulated by rhizobia. Interestingly, these deposits occur neither in pseudonodules induced by auxin-transport inhibitors (Laplaze *et al.*, 1999) nor in nodules induced by ineffective *Frankia* strains (Guan *et al.*, 1996b). Because phenolic compounds influence *Frankia* growth *in vitro* (Perradin *et al.*, 1982) and because *Frankia* hyphae never cross these layers of flavan-containing cells, they might be involved in restricting bacterial infection to certain parts of the nodule. Alternatively, they might contribute to limiting O_2 penetration in the nodule cortex.

4.2. Late Actinorhizal Nodulin Genes

Actinorhizal genes that are expressed in nodules are listed in Table 1 at the end of this section. Two cDNAs that encode genes involved in nitrogen metabolism, glutamine synthetase (GS) and acetylornithine transaminase (AOTA) have been characterized in *A. glutinosa* (Guan *et al.*, 1996a). GS is responsible for assimilating NH_4^+ derived from bacterial nitrogen fixation and AOTA is involved in the synthesis of citrulline, which is the form in which fixed-N is transported in *Alnus*

Table 1. Actinorhizal genes expressed in nodules.

Name	Description	Expression profile in nodules
Dg93 [a]	Similar to a soybean early nodulin gene (*GmENOD93*)	Meristem, early infection zone, periderm, cells of vascular cylinder
Ag12 [b]/ ***Cg12*** [c,d]	Similar to plant subtilisin-like proteases	Root hairs post-inoculation, strong in young infected cells of the infection zone; weak in infected cells of the fixation zone
AgNt84 [e]/ ***Ag164*** [e]	Glycine- and histidine-rich protein, possibly a metal-binding protein	Young infected cells of the infection zone
Ag11 [f]	Glutamine synthetase	High level in infected cells of the fixation zone and in pericycle
Ag118 [f]	Acetylornithine transaminase	High level in infected cells of the fixation zone
EuNOD-AS1 [g]	Asparagine synthetase	Similar expression as *Ag118*
AgPgh1 [h]	Enolase	Infected cells of the N_2 fixation zone and the pericycle; weak in infected cells of the infection zone
AgSus1 [h]	Sucrose synthase	Similar expression as *AgPgh1*
Agthi1 [i]	Involved in thiazole biosynthesis	Similar expression as *AgPgh1* and *AgSus1*
EuSAMS1 [j]/ ***EuSAMS2*** [j]	S-adenosyl-L-methionine synthetase	Meristem, infected cells of the N_2 fixation zone, central vascular system; also *EuSAMS2* is expressed in the infection zone
EuNOD-CHT1 [k]/ ***EuNOD-CHT2*** [k]	Chitinase	*EuNOD-CHT1*: elevated in meristem, weak in outer cortex layer and uninfected cells of the fixation zone *Eu-CHT2*: elevated in infected cells of the N_2 fixation zone and central vascular system, weak in the senescence zone
Dgrca [l]	Rubisco activase	High level in nuclei of infected cell, weak in uninfected cortical cells adjacent to the periderm and vascular cylinder
CgCHS1 [m]	Chalcone synthase	Apex of young nodule lobes; also in phenolic containing cells of cortex
hb-Sym [n] ***hb-Cg1F*** [o]	Hemoglobin	High levels in infected cells of the fixation zone, weak in infected cells of infection zone
CgMT1 [p]	Metallothionein	High levels in infected cells of the fixation zone and in the pericycle
Ag40 [q]/ ***Cg40*** [r]	Similar to legume early nodulin gene *ENOD40*	High level in the vascular bundles of mature nodule lobes
AgNOD-CP1 [s]	Cystein protease	Nodule-specific
Ag13 [t]	Proline- and glutamic acid-rich protein	Infected cells of the senescence zone and pericycle

a, Okubara *et al.*, 2000 (*Datisca glomerata*); b, Ribeiro *et al.*, 1995 (*Alnus glutinosa*); c, Laplaze *et al.*, 2000b (*Casuarina glauca*); d, Svistoonoff *et al.*, 2003a (*C. glauca*); e, Pawlowski *et al.*, 1997 (*A. glutinosa*); f, Guan *et al.*, 1996a (*A. glutinosa*); g, Kim *et al.*, 1999 (*Eleagnus umbellata*); h, van Ghelue *et al.*, 1996 (*A. glutinosa*); i, Ribeiro *et al.*, 1996 (*A. glutinosa*); j, Lee *et al.*, 2001 (*E. umbellata*); k, Kim and An, 2002, (*E. umbellata*); l, Okubura *et al.*, 1999 (*D.glomerata*); m, Laplaze *et al.*, 1999 (*C glauca*); n, Jacobson-Lyon *et al.*, 1995 (*C. glauca*); o, Gherbi *et al.*, 1996 (*C. glauca*); p, Laplaze *et al.*, 2002 (*C. glauca*); q, U. von Groll, A. Ribeiro and K. Pawlowski, personal communication (*A. glutinosa*); r, Santi *et al.*, 2003 (*C. glauca*); s, Goetting-Minesky and Mullin, 1994 (*A. glutinosa*
t, Guan *et al.*, 1997 (*A. glutinosa*).

(Miettinen and Virtanen, 1952). These two genes were shown by *in situ* hybridisation to be expressed both in the infected cells of the fixation zone and in the pericycle of the vascular system (GS only). Thus, it has been suggested that ammonium assimilation and synthesis occur in these cells (Guan *et al.*, 1996a).

A cDNA encoding asparagine synthetase (AS), which is an enzyme linked with NH_4^+ assimilation, was isolated from *Eleagnus umbellata* (Kim *et al.*, 1999). Asparagine is the major N-compound of *Eleagnus* nodules (Wheeler and Bond, 1970), which is consistent with the high level of AS expression in root nodules (Kim *et al.*, 1999). Furthermore, AS mRNA was confined to fully infected cells of the fixation zone, which suggests that, in *E. umbellate* and unlike in alfalfa nodules, AS expression is under metabolic control.

A good carbon flux to the nodule is essential to provide energy, reductant, and acceptor molecules for fixed-N. A sucrose synthase (SuSy) cDNA and an enolase cDNA were isolated from *A. glutinosa* and the corresponding genes were expressed both in infected cells and in the pericycle of the nodules (van Ghelue *et al.*, 1996). Because SuSy mRNA was not detected in starch-containing non-infected cells, it is likely that apoplastic invertase was responsible for starch biosynthesis, not SuSy.

A similar expression pattern occurs for *agthi1*, which shares homology with yeast *thi4*, encoding an enzyme involved in the biosynthesis of the thiamine precursor thiazol (Ribeiro *et al.*, 1996). Thiamine is a co-factor of both glycolysis and the Calvin cycle. Because *agthi1* is expressed in the same cells as both sucrose synthase and enolase, it is probably correlated with the high energy-demanding processes taking place in infected cells and in the pericycle (Ribeiro *et al.*, 1996).

A RuBisCo activase (RCA) cDNA was identified from a *Datisca glomerata* cDNA library (Okubara *et al.*, 1999). Although RCA transcripts were detected by *in situ* hybridisation in nuclei of infected cells, in some uninfected cells, and in the vascular cylinder of nodules, the corresponding protein did not accumulate to a detectable level. Okubara *et al.* (1999) suggested that inefficient splicing of mRNA and translation of the message were responsible for the absence of protein. RuBisCo activase is involved in photosynthetic carbon reduction *via* the action of RubisCo, so the significance of an mRNA that encodes RCA in nonphotosynthetic symbiotic root nodules of *D. glomerata* still remains to be clarified (Okurbara *et al.*, 1999).

In a free-living state, at atmospheric pO_2, all *Frankia* strains can fix N_2 because they can form vesicles that limit O_2 diffusion. In cultured *Frankia*, the increased number of lipid laminae provides an adaptative barrier to O_2 penetration (Parsons *et al.*, 1987). In all actinorhizal plants, except *Casuarina* and *Allocasuarina, Frankia* in symbiosis forms vesicles. Various actinorhizal nodule structures are also involved in protecting nitrogenase from O_2 (Huss-Danell, 1997; Silvester *et al.*, 1990). However, in contrast to legume nodules, no barrier to gas diffusion through the inner cortical zone has been identified in actinorhizal nodules. In *C. glauca* nodules, an O_2-diffusion barrier is created by lignification of the cell wall of the infected and adjacent uninfected cortical cells (Berg and McDowell, 1988).

In *Casuarina*, a large quantity of the O_2-transport protein hemoglobin (hb) has been found in nodules (Fleming *et al.*, 1987). The purified protein was shown to be similar to the legume leghemoglobin, which implies a similar function (Gibson *et al.*, 1989). This large amount of hb is consistent with the absence of *Frankia* vesicles in *Casuarina* nodules. Symbiotic *hb* genes (Jacobson-Lyon *et al.*, 1995) and a corresponding cDNA (Gherbi *et al.*, 1997) were isolated from *Casuarina.* Localisation of *hb* mRNA in nodules by *in situ* hybridisation showed that the corresponding *hb* symbiotic genes are induced in young infected cells prior to the detection of *Frankia nifH* mRNA, suggesting that hb contributes to reducing O_2 tension before *nif*-gene expression (Gherbi *et al.*, 1997). With *C. glauca* nodules, immunogold localisation showed that hb is confined to the cytoplasm and nuclei of infected host cells and is not associated with the *Frankia* membrane (Goodchild and Miller, 1997). Thus, in *Casuarina* as in the nodules of legumes, it seems that O_2 regulation is mediated by a host-derived O_2-diffusion barrier and O_2-transport proteins. Somewhat surprisingly, hb was found in nodules of both *Myrica gale* (Pathirana and Tjepkema, 1995) and *A. glutinosa* (Suharjo and Tjepkema, 1995), where *Frankia* vesicles are also present. This observation suggests that, even in the presence of vesicles, symbiotic hb assures the flow of O_2 within infected cells. It would be instructive to see if the location of hb is similar in nodules with *Frankia* vesicles and in nodules without *Frankia* vesicles.

Metallothioneins (MTs) are a group of low-molecular-weight cysteine-rich proteins, which in animals are believed to play roles in several different processes, such as detoxification of heavy metals, homeostasis of intracellular metal (Kägi, 1991), defence against intracellular oxidants, and regulation of metal-containing enzymes (Andrews, 2000). Although the exact function of plant MTs is not understood, the diversity of MT-gene responses suggests that plant MTs might be involved in the defence reaction to pathogens, apoptosis, growth and development, and heavy-metal metabolism. A clone for a type 1 metallothionein (*cgMT1*) was isolated from a *C. glauca* nodule cDNA library. *In situ* hybridisation indicated that, in nodules, *cgMT1* transcripts were present both in mature *Frankia*-infected cells and in the pericycle. The promoter region of *cgMT1* was isolated and fused to the β-glucuronidase (*gus*) gene. Transgenic Casuarinaceae showed that the *cgMT1* promoter was most active in large *Frankia*-infected cells of the N_2-fixing zone of nodules, in roots and in the oldest region of the shoot (Laplaze *et al.*, 2002). It has been suggested that cgMT1 might be involved in metal-ion transport required for nitrogenase function and might also be part of the antioxidant defences against reactive oxygen species (ROS) induced during nodulation (Laplaze *et al.*, 2002). Further studies are needed to identify cgMT1 function. Analysis of transgenic plants over expressing *cgMT1* is underway and should help to understand the physiological function of the product of this gene.

As mentioned previously, in old nodule lobes, a senescence zone is observed, where both plant cytoplasm and bacteria undergo degradation. The *ag13* and *Ag-NOD-CP1* cDNAs were found to encode a glutamic acid/proline-rich protein and a cysteine proteinase, respectively (Goetting-Minesky and Mullin, 1994; Pawlowski, 1997). Both genes were expressed in *Frankia* infected cells. The deduced proteins have a putative signal peptide, suggesting an extracellular function probably in the compartment surrounding the endophyte. A role as a defence-related protein and in senescence has been proposed for *ag13* and *Ag-Nod-CP1*, respectively.

Two cDNAs encoding putative proteins structurally related to defence proteins have been isolated from an *E. umbellata* root-nodule cDNA library (Kim and An, 2002). The two clones, *EuNOD-CHT1* and *Eu-CHT2*, encode chitinases. Based on their spatial-temporal expression patterns in non-symbiotic tissues and during nodule differentiation, it has been suggested that *Eu*NOD-CHT1 is involved in defence response, whereas *Eu*-CHT2 might be involved in normal plant development and in defence response to external pathogens (Kim and An, 2002).

Acyl carrier protein (ACP) is a component of plant fatty-acid synthase, which is located in chloroplasts. A cDNA corresponding to an ACP was isolated from a *C. glauca* nodule cDNA library (Laplaze *et al.*, 1998). The corresponding protein showed all the characteristic features of plant ACP, including a putative chloroplast transit-peptide cleavage-site motif and a putative phosphopantotheine-attachment site (Laplaze *et al.*, 1998). ACP may participate in the fatty-acid biosynthesis that occurs during plant-cell infection (Laplaze *et al.*, 1998).

Two cDNAs (*EuSAMS1* and *EuSAMS2*) that encode S-adenosyl-L-methionine synthetase were isolated from a nodule cDNA library of *E. umbellata* (Lee *et al.*, 2001). *SAMS* are housekeeping genes that encode the enzyme responsible for the major methyl-group donor in most trans-methylation processes. *In situ* hybridisation showed that these *SAMS* genes were differentially expressed; both *EuSAMS1* and *EuSAMS2* in the meristematic zone; only *EuSAMS2* in the infection zone; both *EuSAMS1* and *EuSAMS2* in the infected cells of the fixation zone; and *EuSAMS1* and *EuSAMS2* in the central vascular system of *E. umbellata* nodules (Lee *et al.*, 2001). A role in nitrogen metabolism and in methylation of cell-wall constituents has been postulated for *EuSAMS1* and *EuSAMS2*, respectively (Lee *et al.*, 2001).

5. EVOLUTIONARY ORIGIN OF SYMBIOTIC GENES

The genetic transformation procedures developed for Casuarinaceae trees (Diouf *et al.*, 1995; Franche *et al.*, 1997; Smouni *et al.*, 2002) provide valuable tools to investigate the conservation of the mechanisms for nodule-specific expression between legumes and actinorhizal plants. Using this approach, the *gus*-reporter gene under the control of promoters from either early or late nodulin genes from legumes was introduced in transgenic Casuarinaceae and the regulation of *gus* expression was investigated during the ontogenesis of actinorhizal nodules.

The *ENOD12* gene, which encodes a (hydroxy)proline-rich protein, is a well-characterized early nodulin gene. Two *ENOD12* genes (*ENOD12A* and *B*) exist in pea (Govers *et al.*, 1991). These two genes are expressed in roots in response to inoculation with either *Rhizobium* or purified Nod factors (Horvath *et al.*, 1993). Expression is found in root hairs of infected plants, in root cells containing the infection thread, and in cortical cells immediately in front of the infection thread. In the mature pea nodule, expression is confined to the distal part of the infection zone, suggesting that ENOD12 is a cell-wall protein involved in the infection process (Bauer *et al.*, 1994). In actinorhizal plants, no homologue of this symbiotic gene has been identified so far.

The *gus* gene, under the control of the promoter region from the early pea *PsENOD12B* nodulin gene (kindly provided by T. Bisseling; Vijn *et al.*, 1995), was introduced into *A. verticillata* and *C. glauca*. The expression pattern of the *PsENOD12B-gus* construct was established in transgenic plants regenerated from 13 and 6 transformed calli of *A. verticillata* and *C. glauca*, respectively, which were obtained after *Agrobacterium tumefaciens*-mediated gene transfer. In nodulated Casuarinaceae plants, no blue staining was observed in roots; but in nodules, *Frankia*-infected cells of the nitrogen-fixation zone expressed reporter-gene activity in both *Casuarina* and *Allocasuarina*. A kinetic analysis of ß-glucuronidase activity in *Frankia*-infected roots established that the *PsENOD12B-gus* construct was not expressed during the early stages of the symbiotic process (our unpublished data). Even though no homologue of *ENOD12* has been found in Casuarinaceae, these results indicate that *PsENOD12* drives nodule-specific expression in actinorhizal plants. The specificity of expression conferred by this sequence appears to be different in actinorhizal plants and legumes. *PsENOD12* directs gene expression in the infection zone of legume nodules, whereas it is expressed only in the nitrogen-fixation zone in actinorhizae, indicating that the signals responsible for the early expression are not recognized in this heterologous host plant.

The promoters of plant hemoglobin genes were also introduced in *A. verticillata* and *C. glauca*. Hemoglobins are widely distributed throughout higher plants and belong to two different families, symbiotic and non-symbiotic hemoglobins. Symbiotic hemoglobin is expressed at high concentration in the N_2-fixing nodules of both legumes and non-legumes, where it facilitates O_2 diffusion to N_2-fixing endosymbiotic bacteria (Appleby, 1992). Non-symbiotic hemoglobins are widespread and have been identified in both symbiotic and non-symbiotic plants (Bogusz *et al.*, 1988; Hunt *et al.*, 2001; Taylor *et al.*, 1994; Trevaskis *et al.*, 1997). These non-symbiotic proteins are expressed at a low level, and their pattern of expression and biochemical properties suggest that they have other functions besides O_2 transport that remain still to be determined.

Three different hemoglobin sequences were studied in transgenic Casuarinaceae; they were the promoter regions of the hemoglobin genes from soybean *(lbc3)* (Lauridsen *et al.*, 1993), *Parasponia andersonii*, and *Trema tomentosa* (Bogusz *et al.*, 1990). The symbiotic gene, *Lbc3*, is expressed at a high level in soybean nodules (Lauridsen *et al.*, 1993). *P. andersonii*, a non-legume belonging to the Ulmaceae family, enters a symbiotic association with *Rhizobium* (Trinick, 1979). The *Parasponia* hemoglobin sequence is expressed both in the N_2-fixing nodules and at low level in the root tissue (Bogusz *et al.*, 1988). *T. tomentosa* is a non-nodulated relative of *P. andersonii* (Akkermans *et al.*, 1978) and the corresponding hemoglobin gene belongs to the non-symbiotic family.

In transgenic *C. glauca* and *A. verticillata,* both the soybean and *P. andersonii* hemoglobin promoters directed *gus*-gene expression in *Frankia* infected cells; some blue staining was also observed in the root tip of the *Parasponia* construct, indicating recognition of the sequence conferring the non-symbiotic expression.

The *T. tomentosa hb* promoter was mainly expressed in the root system (Franche *et al.*, 1998a). Because these different patterns of expression were similar to those of the endogenous soybean, *P. andersonii* and *T. tomentosa hb* genes, it was concluded that these promoters retain their cell-specific expression in transgenic Casuarinaceae. Conversely, the symbiotic *C. glauca hb* promoter retains its nodule-specific expression in legumes (Jacobsen-Lyon *et al.*, 1995). These findings suggest that, although root nodulation has evolved independently in legumes, *Parasponia*, and actinorhizal plants, *hb* genes have maintained their regulatory mechanisms. Moreover, in agreement with other results (Andersson *et al.*, 1997; Jacobsen-Lyon *et al.*, 1995), our studies show that the *Parasponia* symbiosis seems more closely related to actinorhizal symbioses than to legume symbioses, although both legumes and *P. andersonii* are nodulated by the same endosymbiont (rhizobia). Altogether, the fact that legume and actinorhizal symbiotic *hb*-gene promoters retain their specific expressions in endophyte-infected cells of heterologous nodules suggests that similar transcription factors and DNA regulatory elements are used to regulate theses genes. This hypothesis is in agreement with the proposal that legume and *Casuarina hb* genes belong to the same class 2 group of *hb* genes (Hunt *et al.*, 2001).

6. FUTURE TRENDS

6.1. Looking for an Actinorhizal Model System

In legume research, two species, *Medicago truncatula* and *Lotus japonicus*, have been proposed as model systems to develop the same tools that have fueled breakthroughs in the understanding of *Arabidopsis* plant growth and development. The choice of these species was based on a number of criteria, including their diploid autogamous nature, short generation times, genome size (only three-to-four times that of *Arabidopsis*), and the possibility of genetically transforming these species with *Agrobacterium tumefaciens* (Barker *et al.*, 1990; Cook *et al.*, 1997; Handberg and Stougaard, 1992). So far, suitable models have not been selected for actinorhizal plants (Pawlowski, 1999). Nevertheless, the characteristics of three species, *D. glomerata*, *A. glutinosa* and *C. glauca*, will be reviewed to assess whether one of them may be considered as a model.

Datisca glomerata is the only herbaceous species among actinorhizal plants. The major advantage of *D. glomerata* is its short life cycle (about six months). In addition, these plants are diploid, self pollinating, and produce abundant progeny (Wang and Berry, 1996). Compared to *Arabidopsis*, which grows vegetatively as a ground rosette about 2-4cm in diameter with a flowering stem that is 20-30cm long, more space is needed to cultivate *Datisca* plants, which can grow to a height of 60 cm. So far, gene transfer has not been reported for *Datisca*. Nevertheless, plant regeneration from leaf segments of *D. glomerata* has been reported (Wang and Berry, 1996) and successful transient expression of a *35S-gus* construct has been obtained after particle bombardment of *Datisca* leaves (C. Franche, unpublished data). The major drawback with *Datisca* is that, so far, its microsymbiont *Frankia* has not been isolated and grown in pure culture.

Genetic studies of *Frankia* have been difficult for a variety of reasons, including their low growth rates, multicellular nature, poor germination, and lack of genetic markers. Most of the efforts to develop shuttle vectors necessary for the genetic analysis of *Frankia* and for the production of mutants have been focused on *Alnus* microsymbionts (Benson and Silvester, 1993; Mullin and An, 1990). To enhance the development of specific cloning vectors, several plasmids isolated from *Frankia alni* have recently been sequenced (Lavire *et al.*, 200; John *et al.*, 2001; Xu *et al.*, 2002). The analysis of the ORFs might lead to new possibilities for the genetic manipulation of the actinomycete *Frankia alni*. The most valuable characteristics of the host plant, *A. glutinosa*, is that it is a diploid tree with a small genome (2C = 1.1pg) (Pawlowski, 1999). *In vitro* micropropagation of *Alnus* has been described (*e.g.*, Hendrickson *et al.*, 1995; Simon *et al.*, 1985) and the susceptibility of *A. glutinosa* and *A. acuminata* to four strains of *A. rhizogenes* has been established (Savka *et al.*, 1992) but, to our knowledge, transgenic *Alnus* trees have never been obtained. The major drawback of *Alnus* includes its generation time, which is about ten years.

In the Casuarinaceae family, transgenic plants have been obtained for two species, *A. verticillata* and *C. glauca,* after gene transfer by either *A. rhizogenes* or disarmed strains of *A. tumefaciens* (for a review, see Smouni *et al.*, 2001). It is easier to produce transgenic plants with *Allocasuarina* because of the time required (six months) to obtain rooted plants, the large number of transgenic plants produced per transformed callus, and finally the good rooting ability of the regenerated shoots (Franche *et al.*, 1997). However, *C. glauca* has the smallest genome among actinorhizal plants at 2C = 0.7pg compared with that of *A. verticillata*, which is 2C = 1.9pg (Schwencke *et al.* 1998). Even so, Casuarinaceae are small trees, and the production of seeds takes two-to-five years (National Research Council, 1984). Random sequencing of expressed sequence tags (ESTs) from roots and nodules of *C. glauca* is currently in progress in our laboratory (D. Bogusz, unpublished data). Pure cultures of infective *Frankia* strains are available for Casuarinaceae (Diem *et al.*, 1982), but no notable effort has been made so far to develop a shuttle vector the genetic analysis of these strains.

6.2. Use of Model Organisms

Symbiotic genes are assumed to have been recruited from non-symbiotic genes. Some of them belong to gene families that have similar properties in different organisms. When homologues of the symbiotic genes can be found in model organisms, such as either *Arabidopsis* or even *E. coli* and yeast, the general properties of these genes can be studied much more easily.

6.2.1. Use of Actinorhizal Nodulin-Gene Homologues in Arabidopsis

Ribeiro *et al.* (1995) identified in *Arabidopsis* a homologue, named *ARA12*, of the early actinorhizal nodulin genes, *ag12/cg12*. The corresponding ARA12 protein shares 61% similarity at the amino-acid level with AG12. In order to understand the role of *ARA12*, an *ARA12* promoter-*gus* fusion was introduced into *Arabidopsis* and promoter activity was detected in young developing tissues, suggesting a role for ARA12 in protein or polypeptide processing during development (Svistoonoff *et al.*, 2003b). Other subtilases in *Arabidopsis* have been studied, *e.g.*, SDD1, which is involved in the distribution of stomata (Berger and Altmann, 2000). Like these *Arabidopsis* subtilases, AG12/CG12 could also be involved in the regulation of some developmental processes induced by *Frankia* penetration.

6.2.2. Characterization of CgENOD40 *in* A. thaliana

The *ENOD40* genes are involved in nodule organogenesis in legumes but they have also been isolated from non-N_2-fixing plants, like tobacco and rice (Kouchi *et al.*, 1999). A general function has been proposed for ENOD40, probably in sucrose metabolism (Röhrig *et al.*, 2002). Further, the soybean *ENOD40-2* promoter can drive the reporter-gene expression in transgenic *A. thaliana* (Mirabella *et al.*, 1999).

As described previously, the *C. glauca cgenod40* promoter drives reporter-gene expression under non-symbiotic conditions in actinorhizal plants. As a prelude to understanding the role played by CgENOD40 in non-symbiotic development, the *cgENOD40* promoter has been fused with *gus* and introduced in the non-N_2-fixing plant, *Arabidopsis.* Future studies of these transgenic *A. thaliana* should help improve our understanding of the non-symbiotic role of CgENOD40 and, in particular, to provide a simple way of checking induction by either certain hormones, such as cytokinin and auxin, or by bacterial factors or inhibitors of auxin transport. In parallel, the sense construct for over-expressing *cgENOD40* has also been introduced in the model plant to check for a putative effect on development.

6.2.3. Use of Arabidopsis Root-Development Genes

As described previously, lateral roots and actinorhizal nodules share similar initiation and structure. A very exciting study would be to determine to what extent both of these developmental processes share common molecular mechanisms. This should help us understand the evolution of the actinorhizal nodule, which is a highly specialised root organ. In recent years, some key genes, such as *ALF4* (Celenza *et al.*, 1995), that are involved in lateral-root development have been identified using the model plant *A. thaliana.* This growing knowledge can be exploited to compare lateral-root and nodule development in actinorhizal plants by isolating homologues of these important genes from actinorhizal plants, using cDNA libraries, RT-PCR or ESTs sequences.

6.3. EST Sequencing and Genomics

For the past few years, several international research consortia have collaborated on projects to provide a full set of genomic tools for the model legumes *Medicago truncatula* and *Lotus japonicus* (Jiang and Gresshoff, 1997; Oldroyd and Geurts, 2001). A similar international effort is required in actinorhizal genomics to provide a comparison with model legumes. A relatively rapid way to study the complexity of genes expressed during symbiosis is partial sequencing of cDNAs. Our recent EST project, using mRNA isolated from roots and young nodules of *C. glauca*, has produced several hundred ESTs, which correspond to novel actinorhizal nodulin genes (our unpublised data). Comparisons between *C. glauca* and legume EST databases will be of great interest and should reveal which molecular mechanisms are common to the two endophytic root-nodule symbioses and which are not. Furthermore, studies using micro- or macro-arrays should help to improve our overall understanding of the changes in gene expression induced by the symbiotic interaction.

6.4. Exploration of the Expression of Actinorhizal Nodulin Genes in Rice

Rice has recently become a model for cereals because of the accumulation of molecular information for this species, the efficiency of its transformation, its small genome, and the economic importance of this crop, which feeds about half of the world's population (Shimamoto, 1998). Research on biological nitrogen fixation and on plant molecular genetics has progressed to the point where it is not unrealistic to design strategies aimed at developing N_2-fixing capacity in cereals. Among the strategies already tested, the introduction of rhizobia into plant roots failed to give significant results, suggesting that the induction of a nodule is necessary to confer the proper environment for nitrogen fixation (Gough *et al.*, 1997). However, nodule-like structures, called paranodules, can be induced in a number of cereals, including rice, by 2,4-D treatment (Ridge *et al.*, 1993).

The possibility of non-legumes recognizing LCOs (lipochitooligosaccharides) produced by rhizobia has been studied using transgenic plants that contain the *MsENOD12A* and *MsENOD12B* promoters from the early nodulin gene of *Medicago sativa* fused to the *gus*-reporter gene. Terada *et al.* (2001) demonstrated that the microballistic application of the Nod factor, NodRm-IV (C16:2,S) from *Rhizobium meliloti*, changed the ß-glucuronidase activity in transgenic roots exposed to 2,4-D. This result suggests that rice possesses receptors that recognize some components of the Nod factor tested.

No sequence from an actinorhizal symbiotic gene has ever been expressed in rice. In collaboration with E. Guiderdoni (CIRAD Biotrop, Montpellier, France), we examined the possibility of expressing in rice the ß-glucuronidase gene under the control of the promoter of the *cgMT1* metallothionein gene from *C. glauca* (Laplaze *et al.*, 2002). Analysis of transgenic rice plants revealed consistent GUS histochemical staining in root tissues. Staining was mainly observed in root tips, in the elongation zone of the primary and secondary roots, and in lateral roots, whereas no GUS activity was detected in the root differentiation zone. Histological investigation of longitudinal and transverse sections of primary, secondary and lateral roots detected the presence of GUS crystals in the endodermis and pericycle cell layers as well as in the vascular system (phloem and xylem cells). As previously observed in transgenic *PcgMT1-gus A. verticillata* plants (Laplaze *et al.*, 2002), the root meristems and the lateral roots exhibited the most intense staining.

Histochemical assay of shoot sections of rice plants demonstrated that the immature blade of the innermost rolled leaf did not exhibit detectable staining, whereas blade and sheath tissues of leaves of higher rank stained deep blue with a more intense GUS signal in the vascular system. The specificity of staining in the vascular system, as compared with other hypodermal parenchyma and sclerified leaf tissues, appears to increase as the leaf matures. In the aerial part of the transgenic *PcgMT1-gus A. verticillata plants*, reporter-gene activity was also mostly restricted to the oldest region of the shoots (Laplaze *et al.*, 2002).

These data establish that the promoter from the actinorhizal metallothionein gene (*cgMT1*) can drive the expression of a reporter gene in *Oryza sativa*. The specificity of expression observed in transgenic rice plants is similar to that observed in transgenic *cgMT1-gus A. verticillata* trees. The possibility of obtaining GUS expression in 2,4-D-induced paranodules of rice has not yet been investigated.

7. CONCLUSIONS

Nodule development is largely under the control of the plant genome and is only triggered by bacterial factors. Hence, the mechanisms controlling nodule development might be derived from processes common to all plants. Studies of how these common processes have been altered might provide new means to design strategies by which non-legume plants can be given the ability to establish a symbiosis with N_2-fixing bacteria. Because actinorhizal nodule lobes are modified lateral roots, it is important to use all the tools that have become available from studies of *Arabidopsis* root development to investigate nodule formation. The large range of developmental mutants and their corresponding genes in *Arabidopsis* now makes it possible to screen for homologues in actinorhizal plants. The availability of both promoter probes (to sense auxins levels) and transgenic Casuarinaceae should allow the question of auxin requirements in the nodule-developmental process to be addressed.

Finally, it is now a priority to develop genomic approaches to actinorhizal symbioses. Strong international cooperation among groups willing to generate actinorhizal programmes should help to raise funds dedicated to genomics. In this respect, it is important that a model actinorhizal plant is recognized and adopted soon by actinorhizal biologists.

REFERENCES

Akkermans, A. D. L., Abdulkadir, S., and Trinick, M. J. (1978). N_2-fixing root nodules in *Ulmaceae*: *Parasponia* or (and) *Trema* spp. *Plant Soil, 49,* 711-715.

Andersson, C. R., Llewellyn, D. J., Peacock, W. J., and Dennis, E. S. (1997). Cell-specific expression of the promoters of two nonlegume hemoglobin genes in transgenic legume, *Lotus corniculatus*. *Plant Physiol., 113,* 45-57.

Andrews, G. K. (2000). Regulation of metallothionein gene expression by oxydative stress and metal ions. *Biochem. Pharmacol., 59,* 95-104.

Angulo Carmona, A. F. (1974). La formation des nodules fixateurs d'azote chez *Alnus glutinosa* (L.). *Acta Bot. Neerl., 23,* 257-303.

Appleby, C. A. (1992). The origin and functions of haemoglobin in plants. *Science in Progress, 76,* 365-398.

Barker, D. G., Bianchi, S., London, F., Dattee, Y., Essad, S., *et al.* (1990). *Medicago truncatula*, a model plant for studying the molecular genetics of the *Rhizobium*-legume symbiosis. *Plant Mol. Biol., 8,* 40-49.

Bauer, P., Crespi, M. D., Szecsi, J., Allison, L. A., Schultze, M., *et al.* (1994). Alfalfa *enod12* genes are differentially regulated during nodule development by Nod factors and *Rhizobium* invasion. *Plant Physiol., 105,* 585-592.

Benson, D., and Silvester, W. B. (1993). Biology of *Frankia* strains, actinomycete symbionts of actinorhizal plants. *Microbiol. Rev., 57,* 293-319.

Berg, R. H., and McDowell, L. (1988). Cytochemistry of the wall of infected cells in *Casuarina* actinorhizae. *Can. J. Bot., 66,* 2038-2047.

Berger D., and Altmann T. (2000). A subtilisin-like serine protease involved in the regulation of stomatal density and distribution in *Arabidopsis thaliana. Genes Dev., 14,* 1119-1131.

Berry, A. M., Kahn, R. K. S., and Booth, M. C. (1989). Identification of indole compounds secreted by *Frankia* HFPArI3 in defined culture medium. *Plant Soil, 118,* 205-209.

Bogusz, D., Appleby, C. A., Landsmann, J., Dennis, E., Trinick, M. J., and Peacock, W. J. (1988). Functioning haemoglobin genes in non-nodulating plants. *Nature, 331,* 178-180.

Bogusz, D., Llewelyn, D. J., Craig, S., Dennis, E. S., Appleby, C. A., and Peacock, W. J. (1990). Non legume hemoglobin genes retain organ specific expression in heterologous transgenic plants. *Plant Cell, 2,* 633-641.

Callaham, D., and Torrey, J. G. (1977). Prenodule formation and primary nodule development in roots of *Comptonia* (*Myricaceae*). *Can. J. Bot., 51,* 2306-2318.

Celenza, J. L., Grisafi, P. L., and Fink, G. R. (1995). A pathway for lateral root formation in *Arabidopsis thaliana. Genes Dev., 9,* 2131-42.

Charon, C., Sousa, C., Crespi, M., and Kondorosi, A. (1999). Alteration of *enod40* expression modifies *Medicago truncatula* root nodule development induced by *Sinorhizobium meliloti. Plant Cell, 11,* 1953-1965.

Cook, R. C., VandenBosch, K., de Bruijn, F., and Huguet, T. (1997). Model Legumes get the Nod. *Plant Cell, 9,* 275-281.

Crespi, M., Jurkevitch, E., Poiret, M., d'Aubenton-Carafa, Y., Petrovics, G., *et al.* (1994). *enod40,* a gene expressed during nodule organogenesis, codes for a non-translatable RNA involved in plant growth. *EMBO J., 13,* 5099-5112.

Diem, H. G., Duhoux, E., Zaid, H., and Arahou, M. (2000). Cluster roots in Casuarinaceae: Role and relationship to soil nutrient factors. *Ann. Bot., 85,* 929-936.

Diem, H. G., Gauthier, D., and Dommergues, Y. R. (1982). Isolation of *Frankia* from nodules of *Casuarina equisetifolia. Can. J. Microbiol., 28,* 526-530.

Diouf, D., Gherbi, H., Prin, Y., Franche, C., Duhoux, E., and Bogusz, D. (1995). Hairy root nodulation of *Casuarina glauca*: A system for the study of symbiotic gene expression in an actinorhizal tree. *Mol. Plant-Microbe Interact., 8,* 532-537.

Doyle, J. J. (1998). Phylogenic perspectives on nodulation: Evolving views of plants and symbiotic bacteria. *Trends Plant Sci., 3,* 473-478.

Duhoux, E., Diouf, D., Gherbi, H., Franche, C., Ahée, J., and Bogusz, D. (1996). Le nodule actinorhizien. *Acta Bot. Gal., 143,* 593-608.

Duhoux, E., Rinaudo, G., Diem, H. G., Auguy, F., Fernandez, D., *et al.* (2001). Angiosperm *Gymnostoma* trees produce root nodules colonized by arbuscular mycorrhizal fungi related to *Glomus. New Phytol., 149,* 115-125.

Dullaart, J. (1970). The auxin content of root nodules and roots of *Alnus glutinosa* (L.) Vill. *J. Exp. Bot., 21,* 975-984.

Fang, Y. W., and Hirsch, A. M. (1998). Studying early nodulin gene *enod40* expression and induction by nodulation factor and cytokinin in transgenic alfalfa. *Plant Physiol., 116,* 53-68.

Fleming, A. I., Wittenberg, J. B., Dudman, W. F., and Appleby, C. A. (1987). The purification, characterization and ligand-binding kinetics of hemoglobins from root nodules of the non-leguminous *Casuarina glauca-Frankia* symbiosis. *Biochim. Biophys. Acta, 911,* 209-220.

Franche, C., Diouf, D., Le, Q. V., N'Diaye, A., Gherbi, H., *et al.* (1997). Genetic transformation of the actinorhizal tree *Allocasuarina verticillata* by *Agrobacterium tumefaciens. Plant J., 11,* 897-904.

Franche, C., Diouf, D., Laplaze, L., Auguy, F., Frutz, T., *et al.* (1998a). Soybean (*lbc3*), *Parasponia*, and *Trema* hemoglobin gene promoters retain symbiotic and nonsymbiotic specificity in transgenic

Casuarinaceae: Implications for hemoglobin gene evolution and root nodule symbioses. *Mol. Plant-Microbe Interact., 11,* 887-894.

Franche, C., Laplaze, L., Duhoux, E., and Bogusz, D. (1998b). Actinorhizal symbioses: Recent advances in plant molecular and genetic transformation studies. *Crit. Rev. Plant Sci., 17,* 1-28.

Gherbi, H., Duhoux, E., Franche, C., Pawlowski, K., Nassar A., *et al.* (1997). Cloning of a full-length symbiotic hemoglobin cDNA and *in situ* localization of the corresponding mRNA in *Casuarina glauca* root nodule. *Physiol. Plant., 99,* 608-616.

Gibson, Q. H., Wittenberg, J. B., Wittenberg, B. A., Bogusz, D., and Appleby, C. A. (1989). The kinetics of ligand binding to plant hemoglobins. *J. Biol. Chem., 264,* 100-107.

Goetting-Minesky, M. P., and Mullin, B. C. (1994). Differential gene expression in an actinorhizal symbiosis: Evidence for a nodule-specific cysteine proteinase. *Proc. Natl. Acad. Sci. U.S.A., 91,* 9891-9895.

Goodchild, D. S., and Miller, C. (1997). Immunogold localisation of hemoglobin in *Casuarina* root nodules. *Protoplasma, 198,* 130-134.

Gough, C., Galera, C., Vasse, J., Webster, G., Cocking, E. C., and Denarie J. (1997). Specific flavonoids promote intercellular root colonization of *Arabidopsis thaliana* by *Azorhizobium caulinodans* ORS571. *Mol. Plant-Microbe Interact., 10,* 560-570.

Govers, F., Harmsen, H., Heidstra, R., Micheielsen, P., Prins, M., *et al.* (1991). Characterization of the pea *enod12B* gene and expression analyses of the two *enod12* genes in nodule, stem and flower tissue. *Mol. Gen. Genet., 228,* 160-166.

Gualtieri, G., and Bisseling, T. (2000). The evolution of nodulation. *Plant Mol. Biol., 42,* 181-194.

Guan, C., Akkermans, A. D. L., van Kammen, A., Bisseling, T., and Pawlowski K. (1997). *ag13* is expressed in *Alnus glutinosa* nodules in infected cells during endosymbiont degradation and in the nodule pericycle. *Physiol. Plant., 99,* 601-607.

Guan, C., Ribeiro, A., Akkermans, A. D. L., Jing, Y., van Kammen, A., *et al.* (1996a). Nitrogen metabolism in actinorhizal nodules of *Alnus glutinosa:* Expression of glutamine synthetase and acetylornithine transaminase. *Plant Mol. Biol., 32,* 1177-1184.

Guan, C., Wolters, D. J., van Dijk, C., Akkermans, A. D. L., van Kammen, A., *et al.* (1996b). Gene expression in ineffective actinorhizal nodules of *Alnus glutinosa. Acta Bot. Gall., 143,* 613-620.

Gupta, R. K., Dobritsa, S. V., Stiles, C. A., Essington, M. E., Liu, Z. Y., *et al.* (2002). Metallohistins: A new class of plant metal-binding proteins. *J. Prot. Chem., 21,* 529-536.

Hammad, Y., Nalin, R., Marechal, J., Fiasson, K., Pepin, R., *et al.* (2003). A possible role for phenyl acetic acid (PAA) on *Alnus glutinosa* nodulation by *Frankia. Plant Soil, 254,* 193-205.

Handberg, K., and Stougaard, J. (1992). *Lotus japonicus,* an autogamous, diploïd legume species for classical and molecular genetics. *Plant J., 2,* 487-496.

Heidstra, R., Yang, W. C., Yalcin, Y., Peck, S., Emons, A., *et al.* (1997). Ethylene provides positional information on cortical cell division but is not involved in Nod factor-induced root hair tip growth in *Rhizobium*-legume interaction. *Development, 124,* 1781-1787.

Hendrickson, O. Q., Burgess, D., Perinet, P., Tremblay, F., and Chatatpaul, L. (1995). Effects of *Frankia* on field performance of *Alnus* clones and seedlings. *Plant Soil, 150,* 295-302.

Henson, I. E., and Wheeler, C. T. (1977). Hormones in plants bearing nitrogen-fixing root nodules: Partial characterization of cytokinins from root nodules from *Alnus glutinosa* (L.) Gaertn. *J. Exp. Bot., 28,* 1076-1086.

Horvath, B., Heidstra, R., Lados, M., Moerman, M., Spaink, H. P., *et al.* (1993). Lipo-oligosaccharides of *Rhizobium* induce infection-related early nodulin gene expression in pea root hairs. *Plant J., 4,* 727-733.

Hunt, P. W., Watts, R. A., Trevaskis, B., Llewelyn, D. J., Burnell, J., *et al.* (2001). Expression and evolution of functionaly distinct haemglobin genes in plants. *Plant Mol. Biol.,* 47, 677-692.

Huss-Danell, K. (1997). Actinorhizal symbioses and their N2 fixation. *New Phytol., 136,* 375-405.

Jacobsen-Lyon, K., Jensen, E. O., Jorgensen, J.-E., Marcker, K. A., Peacock W. J., and Dennis, E. S. (1995). Symbiotic and non-symbiotic hemoglobin genes of *Casuarina glauca. Plant Cell, 7,* 213-222.

Jiang, Q., and Gresshoff, P. M. (1997). Classical and molecular genetics of the model legume *Lotus japonicus. Mol. Plant-Microbe Interact., 10,* 59-68.

John, T. R., Rice, J. M., and Johnson, J. D. (2001). Analysis of pFQ12, a 22.4-kb *Frankia* plasmid. *Can. J. Microbiol., 47,* 608-617.

Jorda, L., Coego, A., Conejero, V., and Vera, P. (1999). A genomic cluster containing four differentially regulated subtilisin-like processing protease genes is in tomato plants. *J. Biol. Chem., 274,* 2360-2365.

Kägi, J. H. R. (1991). Metallothioneins. *Methods Enzymol., 205,* 613-626.

Keller, B., and Lamb, C. J. (1989). Specific expression of a novel cell-wall hydroxyproline-rich glycoprotein gene in lateral root initiation. *Genes Dev., 3,* 1639-1646.

Kim, H. B., and An, C. S. (2002). Differential expression patterns of an acidic chitinase and a basic chitinase in the root nodule of *Elaeagnus umbellata. Mol. Plant-Microbe Interact., 15,* 209-215.

Kim, H. B., Lee, S. H., and An, C. S. (1999). Isolation and characterization of a cDNA clone encoding asparagine synthetase from root nodules of *Elaeagnus umbellata. Plant Sci., 149,* 85-94.

Kouchi, H., and Hata, S. (1993). Isolation and characterization of novel nodulin cDNAs representig genes expressed at early stages of soybean nodule development. *Mol. Gen. Genet., 238,* 106-119.

Kouchi, H., Takane, K., So, R. B., Ladha, J. K., and Reddy, P. M. (1999). Rice *ENOD40*: Isolation and expression analysis in rice and transgenic soybean root nodules. *Plant J., 18,* 121-129.

Lancelle, S. A., and Torrey, J. G. (1984). Early development of *Rhizobium*-induced root nodules of *Parasponia rigida* I. Infection and early nodule initiation. *Protoplasma, 123,* 26-37.

Laplaze, L., Duhoux, E., Franche, C., Frutz, T., Svistoonoff, S., *et al.* (2000a). *Casuarina glauca* prenodule cells display the same differentiation as the corresponding nodule cells. *Mol. Plant-Microbe Interact., 13,* 107-112.

Laplaze, L., Gherbi, H., Duhoux, E., Pawlowski, K., Auguy, F., *et al.* (2002). Symbiotic and non-symbiotic expression of *cgMT1*, a metallothionein-like gene from the actinorhizal tree *Casuarina glauca. Plant Mol. Biol., 49,* 81-92.

Laplaze, L., Gherbi, H., Franche, C., Duhoux, E., and Bogusz, D. (1998). cDNA sequence for an Acyl Carrier Protein from actnorhizal nodules of *Casuarina glauca* (PGR98-066). *Plant Physiol., 116,* 1605.

Laplaze, L., Gherbi, H., Frutz, T., Pawlowski, K., Franche, C., *et al.* (1999). Flavan-containing cells delimit *Frankia*-infected compartments in *Casuarina glauca* nodules. *Plant Physiol., 121,* 113-122.

Laplaze, L., Ribeiro, A., Franche, C., Duhoux, E., Auguy, F., *et al.* (2000b). Characterization of a *Casuarina glauca* nodule-specific subtilisin-like protease gene, a homologue of *Alnus glutinosa ag12. Mol. Plant-Microbe Interact.,* 13, 113-117.

Lauridsen, P., Franssen, H., Stougaard, J., Bisseling, T., and Marker, K. A. (1993). Conserved regulation of the soybean early nodulin *ENOD2* gene promoter in determinate and indeterminate transgenic root nodules. *Plant J., 3,* 483-492.

Lavire, C., Louis, D., Perriere, G., Briolay, J., Normand, P., and Cournoyer, B. (2001). Analysis of pFQ31, a 8551-bp cryptic plasmid from three symbiotic nitrogen-fixing actinomycete *Frankia. FEMS Microbiol. Letters, 197,* 111-116.

Lee, S. H., Kim, H. B., and An, C. S. (2001). Structures and expression patterns of two cDNA clones encoding S-adenosyl-L-methionine synthetase from the root nodule of *Elaeagnus umbellata. Aus. J. Plant Physiol., 28,* 951-957.

Li, Y., Wu, Y. H., Hagen, G., and Guilfoyle, T. (1999). Expression of the auxin-inducible *GH3* promoter/GUS fusion gene as a useful molecular marker for auxin physiology. *Plant Cell Physiol., 40,* 675-682.

Libbenga, K. R., van Iren, F., Bogers, R. J., and Schraag-Lamers, M. F. (1973). The role of hormones and gradients in the initiation of cortex proliferation and nodule formation in *Pisum sativum* L. *Planta, 114,* 29-39.

Malamy, J. E., and Benfey, P. N. (1997). Organization and cell differentiation in lateral roots of *Arabidopsis thaliana. Development, 124,* 33-44.

Mathesius, U., Schlaman, H. M., Spaink, H. P., Sautter, C., Rolfe, B. G., *et al.* (1998). Auxin transport inhibition precedes root nodules formation in white clover roots and is regulated by flavonoids and derivatives of chitin oligosaccharides. *Plant J., 14,* 23-34.

Miettinen, J. K., and Virtanen, A. I. (1952). The free amino acids in the leaves, roots, and root nodules of the alder (*Alnus*). *Physiol. Plant., 5,* 540-557.

Miller, I. M., and Baker, D. D. (1985). Initiation, development and structure of root nodules in *Eleagnus angustifolia* L. (*Eleagnaceae*). *Protoplasma, 128,* 107-119.

Mirabella, R., Martirani, L., Lamberti, A., Iaccarino, M., and Chiurazzi M. (1999). The soybean *enod40(2)* promoter is active in *Arabidopsis thaliana* and is temporally and spatially regulated. *Plant Mol. Biol., 39,* 177-181.

Mullin, B. C., and An, C. S. (1990). The molecular genetics of *Frankia*. In C. R. Schwintzer and J. D. Tjepkema (Eds.), *The biology of* Frankia *and actinorhizal plants* (pp. 195-214). San Diego, CA: Academic Press.

National Research Council. (1984). Casuarinas: Nitrogen-fixing trees for adverse sites. Washington, DC: National Academy Press.

Okubara, P. A., Fujishige, N. A., Hirsch, A. M., and Berry, A. M. (2000). *Dg93*, a nodule-abundant mRNA of *Datisca glomerata* with homology to a soybean early nodulin gene. *Plant Physiol., 122,* 1073-1079.

Okubara, P. A., Pawlowski, K., Murphy, T. M., and Berry, A. M. (1999). Symbiotic root nodules of the actinorhizal plant *Datisca glomerata* express rubisco activase mRNA. *Plant Physiol., 120,* 411-420.

Oldroyd, G. E., and Geurts, R. (2001). *Medicago truncatula*, going where no plant has gone before. *Trends Plant Sci., 6,* 552-4.

Parsons, R., Silvester, W. B., Harris, S., Gruijters, W. T. M., and Bullivant, S. (1987). *Frankia* vesicles provide inducible and absolute oxygen protection for nitrogenase. *Plant Physiol., 83,* 728-731.

Pathirana, S. M., and Tjepkema, J. D. (1995). Purification of hemoglobin from the actinorhizal root nodules of *Myrica gale* L. *Plant Physiol., 107,* 827-831.

Pawlowski, K. (1997). Nodule-specific gene expression. *Physiol. Plant., 99,* 617-631.

Pawlowski, K. (1999). Actinorhizal plants: Time for genomics. In J. G. M de Wit., T. Bisseling. and W. J. Stiekema (Eds.), *Biology of plant-microbe interactions, Vol. 2* (pp. 368-372). St. Paul, MN: International Society for Molecular Plant-Microbe Interactions.

Pawlowski, K., Twigg, P., Dobritsa, S., Guan, C. H., and Mullin, B. (1997). A nodule-specific gene family from *Alnus glutinosa* encodes glycine-and histidine-rich proteins expressed in the early stages of actinorhizal nodule developoment. *Mol. Plant-Microbe Interact., 10,* 656 -664.

Penmetsa, R. V., and Cook, D. R. (1997). A legume ethylene-insensitive mutant hyperinfected by its rhizobial symbiont. *Science, 275,* 527-530.

Perradin, Y., Mottet, M. J., and Lalonde, M. (1982). Influence of phenolics on *in vitro* growth of *Frankia* strains. *Can. J. Bot., 61,* 2807-2814.

Ribeiro, A., Akkermans A. D. L, van Kammen, A., Bisseling, T., and Pawlowski, K. (1995). A nodule-specific gene encoding a subtilisin-like protease is expressed in early stages of actinorhizal nodule development. *Plant Cell, 7,* 785-794.

Ribeiro, A., Praekelt, U., Akkermans, A. D. L., Meacock , P. A., van Kammen, A., *et al.* (1996). Identification of *agthi1*, whose product is involved in biosynthesis of the thiamine precursor thiazole, in actinorhizal nodules of *Alnus glutinosa*. *Plant J., 10,* 361-368.

Ridge, R. W., Ridge, K. M., and Rolfe, B. G. (1993). Nodule-like structures induced on the roots of rice seedlings by addition of the synthetic auxin 2,4-dichloro-phenoxy-acetic acid. *Aus. J. Plant Physiol., 20,* 705-717.

Röhrig, H., Schmidt, J., Miklashevichs, E., Schell, J., and John, M. (2002). Soybean *enod40* encodes two peptides that bind to sucrose synthase. *Proc. Natl. Acad. Sci. U.S.A., 99,* 1915-1920.

Roussis, A., van de Sande, K., Papadopoulou, K., Drenth, J., Bisseling, T., *et al.* (1995). Characterization of the soybean gene *Gmenod40-2*. *J. Exp. Bot., 46,* 719-724.

Santi, C., von Groll, U., Ribeiro, A., Chiurazzi, M., Auguy, F., *et al.* (2003). Comparison of nodule induction in legume and actinorhizal symbioses: The induction of actinorhizal nodules does not involve *enod40*. *Mol. Plant-Microbe Interact., 16,* 808-816.

Savka, M. A., Liu, L., Farrand, S. K., Berg, R. H., and Dawson, J. O. (1992). Induction of hairy roots or *pseudoactinorhizae* on *Alnus glutinosa*, *A. acuminata* and *Eleagnus angustifolia* by *Agrobacterium rhizogenes*. *Acta Oecol., 13,* 423-431.

Schaller, A., and Ryan, C. A. (1994). Identification of a 50-kDa systemin binding protein in tomato plasma membranes having Kex2p-like properties. *Proc. Natl. Acad. Sci. U.S.A., 91,* 11802-11806.

Schultze, M., and Kondorosi, A. (1998). Regulation of symbiotic nodule development. *Ann. Rev. Genet., 32,* 33-57.

Schwencke, J., Bureau, J. M., Crosnier, M. T., and Brown, S. (1998). Cytometric determination of genome size and base composition of three species of three genera of Casuarinaceae. *Plant Cell Rep., 18,* 346-349.

Schwintzer, C. R., and Lancelle, S. A. (1983). Effect of water-table depth on shoot growth, root growth, and nodulation of *Myrica gale* seedlings. *J. Ecol., 71,* 489-501.

Shimamoto, K. (1998). Molecular biology of rice (304 pp.). Berlin, Germany: Springer.

Siezen, R. J., and Leunissen, J. A. M. (1997). Subtilases: The subtilisin-like serine proteases. *Prot. Sci., 6,* 501-523.

Silvester, W. B., Harris, S. L., and Tjepkema, J. D. (1990). Oxygen regulation and hemoglobin. In C. R. Schwintzer and J. D. Tjepkema (Eds.), *The biology of* Frankia *and actinorhizal plants* (pp. 157-176). San Diego, CA: Academic Press.

Simon, L., Stein, A., Côté, S., and Lalonde, M. (1985). Performance of *in vitro* propagated *Alnus glutinosa* (L.) Gaertn. clones inoculated with *Frankiae*. *Plant Soil, 87,* 125-133.

Smouni, A., Laplaze, L., Bogusz, D., Guermache, F., Auguy, F., *et al.* (2002). The 35S promoter is not constitutively expressed in the transgenic tropical actinorhizal tree, *Casuarina glauca. Funct. Plant Biol., 29,* 649-656.

Soltis, D. E., Soltis, P. S., Morgan, D. R., Swensen, S. M., Mullin, B. C., *et al.* (1995). Chloroplast gene sequence data suggest a single origin of the predisposition for symbiotic nitrogen fixation in angiosperms. *Proc. Natl. Acad. Sci. U.S.A., 92,* 2647-2651.

Staehelin, C., Charon, C., Boller, T., Crespi, M., and Kondorosi, A. (2001). *Medicago truncatula* plants overexpressing the early nodulin gene *enod40* exhibit accelerated mycorrhizal colonization and enhanced formation of arbuscules. *Proc. Natl. Acad. Sci. U.S.A., 98,* 15366-15371.

Stevens, G. A., and Berry, A. M. (1988). Cytokinin secretion by *Frankia* sp. HFPArI3 in defined medium. *Plant Physiol., 87,* 15-16.

Suharjo, U. K. J., and Tjepkema, J. D. (1995). Occurence of hemoglobin in the nitrogen-fixing root nodules of *Alnus glutinosa. Physiol. Plant., 95,* 247-252.

Svistoonoff, S., Laplaze, L., Auguy, F., Runions, J., Duponnois, R., *et al.* (2003a). *cg12* expression is specifically linked to infection of root hairs and cortical cells during *Casuarina glauca* and *Allocasuarina verticillata* actinorhizal nodule development. *Mol. Plant-Microbe Interact., 16,* 600-607.

Svistoonoff, S., Laplaze, L., Auguy, F., Santi, C., Fontanillas, E., *et al.* (2003b). Expression pattern of *ara12*, an Arabidopsis homologue of the nodule-specific actinorhizal subtilases *cg12/ag12*. *Plant Soil, 254,* 239-244.

Taylor, E. R., Nie, X. Z., MacGregor, A. W., and Hill, R. D. (1994). A cereal haemoglobin gene is expressed in seed and root tissues under anaerobic conditions. *Plant Mol. Biol., 24,* 853-862.

Terada, R., Ignacimuthu, S., Bauer, P., Kondorosi, E., Schultze, M., *et al.* (2001). Expression of early nodulin promoter gene in transgenic rice. *Cur. Sci., 81,* 270-276.

Tjepkema, J. (1978). The role of oxygen diffusion from the shoots and nodule roots in nitrogen fixation by root nodules of *Myrica gale. Can. J. Bot., 56,* 1365-1371.

Torrey, J. G. (1976). Initiation and development of root nodules of *Casuarina* (*Casuarinaceae*). *Amer. J. Bot.,* 63, 335-345.

Trevaskis, B., Watts, R. A., Andersson, C. R., Llewelyn, D. J., Hargrove, M. S., *et al.* (1997). Two haemoglobin genes in *Arabidopsis thaliana*: the evolutionary origins of leghemoglobins. *Proc. Natl. Acad. Sci. U.S.A., 94,* 12230-12234.

Trinick, M. L. (1979). Structure of nitrogen-fixing nodules formed by *Rhizobium* on roots of *Parasponia andersonii. Appl. Environ. Microbiol., 55,* 2046-2055.

Valverde, C. (2000). La symbiosis *Discaria trinervis-Frankia*. Regulacion de la nodulacion radicular. Ph.D. thesis, Facultad de Ciencias Exactas, Universidad National de La Plata, La Plata, Argentina.

van Ghelue, M., Ribeiro, A., Solheim, B., Akkermans, A. D. L., Bisseling, T., and Pawlowski, K. (1996). Sucrose synthase and enolase expression in actinorhizal nodules of *Alnus glutinosa*: Comparison with legume nodules. *Mol. Gen. Genet., 250,* 437-46.

Vera, P., Lamb, C., and Doerner, P. W. (1994). Cell-cycle regulation of hydroxyproline-rich glycoprotein HRGPnt3 gene expression during the initiation of lateral root meristem. *Plant J., 6,* 717-727.

Vijn, I., Yang, W. C., Pallisgard, N., Jensen, E. Ø., van Kammen, A., and Bisseling, T. (1995). *Vsenod5, Vsenod12* and *Vsenod40* expression during *Rhizobium*-induced nodule formation on *Vicia sativa* roots. *Plant Mol. Biol., 28,* 1111-1119.

Wang, H. Y., and Berry, A. M. (1996). Plant regeneration from leaf segments of *Datisca glomerata. Acta Bot. Gall., 143,* 609-612.

Wheeler, C. T., and Bond, G. (1970). The amino acids of non-legume root nodules. *Phytochemistry, 9,* 705-708.

Wheeler, C. T., Henson, I. E., and Mc Laughlin, M. E. (1979). Hormones in plants bearing actinomycetes nodules. *Bot. Gaz. (Chicago), 140,* 52-57.

Xu, X. D., Kong, R. Q., de Bruijn, F. J., He, S. Y., Murry, M. A., *et al.* (2002). DNA sequence and genetic characterization of plasmid pFQ11 from *Frankia alni* strain CpI1. *FEMS Microbiol. Lett., 207,* 103-107.

Yamagata, H., Masuzawa, T., Nagaoka, Y., Ohnishi, T., and Iwasaki, T. (1994). Cucumisin, a serine protease from melon fruits, shares structural homology with subtilisin and is generated from a large precursor. *J. Biol. Chem., 269,* 32725-32731.

Yang, W. C., Katinakis, P., Hendriks, P., Smolders, A., de Vries, F., *et al.* (1993). Characterization of *Gmenod40,* a gene showing novel patterns of cell-specific expression during soybean nodule development. *Plant J., 3,* 573-585.

Chapter 10

COMPARISON BETWEEN ACTINORHIZAL AND LEGUME SYMBIOSIS

K. PAWLOWSKI[1] AND J. I. SPRENT[2]
[1]*Department of Botany, Stockholm University, 10691 Stockholm, Sweden.*
[2]*32 Birkhill Avenue, Wormit, Fife, DD6 8PW UK*

1. INTRODUCTION

Two types of root-nodule symbioses exist between higher plants and N_2-fixing soil bacteria; these are the legume-rhizobia and actinorhizal symbioses. In both interactions, bacteria induce the formation of special plant organs, nodules, on the plant roots. Bacteria fix N_2 while being hosted within plant cells and are provided with carbon sources by the plant hosts. In the case of rhizobial symbioses, a diverse group of eleven genera of Gram-negative unicellular soil bacteria (*Rhizobium, Allorhizobium, Azorhizobium, Bradyrhizobium, Mesorhizobium, Sinorhizobium, Methylobacterium, Blastobacter, Devosia, Burkholderia and Ralstonia;* Chen *et al.*, 2001; Moulin *et al.*, 2001; Rivas *et al.*, 2002; van Berkum *et al.*, 2002*;* Vandamme *et al.*, 2002; Young *et al.*, 2001) induce nodules on the roots of legumes and one non-legume, *Parasponia* sp. (Ulmaceae). In the case of actinorhizal symbioses, Gram-positive actinomycetous soil bacteria of the genus *Frankia* induce nodules on the roots of dicotyledonous plants from eight different families, mostly trees or woody shrubs. Legume nodule primordia are formed in the root cortex and develop into stem-like organs with a peripheral vascular system and infected cells in the central tissue. In contrast, actinorhizal nodule primordia are formed in the root pericycle, like lateral root primordia. Mature actinorhizal nodules are coralloid organs composed of multiple lobes, each of which represents a modified lateral root without root cap, a superficial periderm, and infected cells in the expanded cortex.

K. Pawlowski and W. E. Newton (eds.), Nitrogen-fixing Actinorhizal Symbioses, 261–288.
© 2008 *Springer.*

Interestingly, nodules of the only non-legume (*Parasponia*) that is able to enter a rhizobial symbiosis structurally and developmentally resemble actinorhizal nodules (Lancelle and Torrey, 1984; 1985).

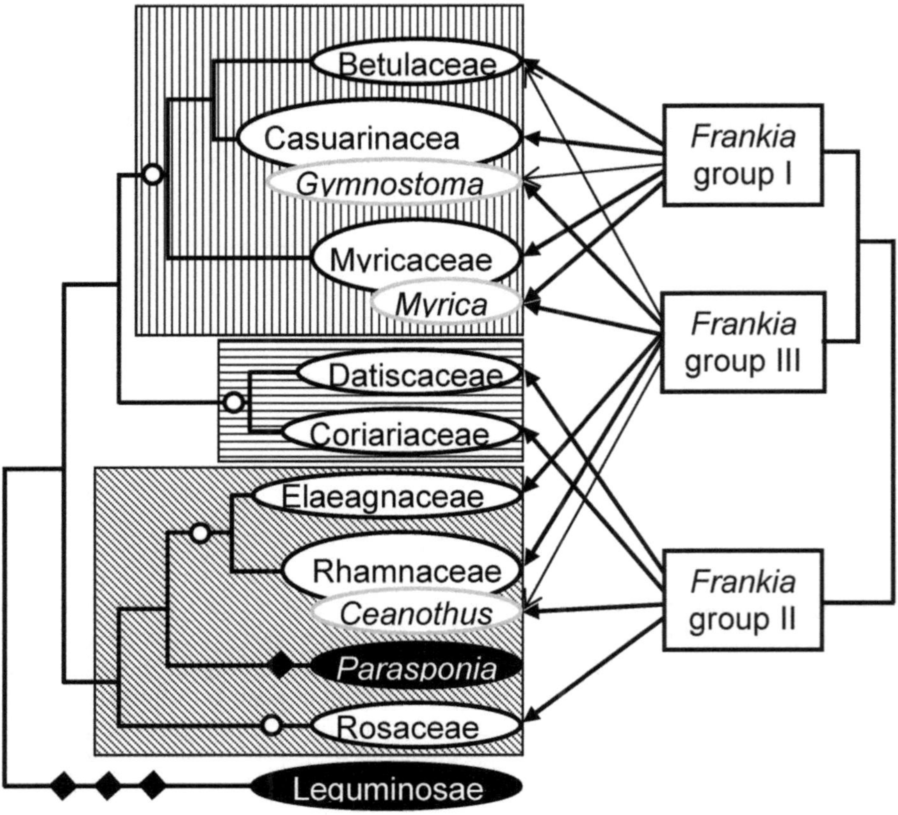

Figure 1: Simplified scheme of the phylogenetic relationship between actinorhizal plants as explained by Swensen and Benson (this volume) and Benson and Clawson (2000). Groups of plants infected by rhizobia are labeled by inverse print. Boxes indicate the three main groups of actinorhizal plants: higher Hamamelidae (vertical stripes); Cucurbitales (horizontal stripes); and Rosales (diagonal stripes). The phylogenetic relationship between the three clades of symbiotic Frankia *is included. Some actinorhizal genera (*Gymnostoma, Myrica, Ceanothus*) that differ in microsymbiont specificity from the rest of the family are indicated. Thick arrows connect* Frankia *clades with the members and plant groups that the clades are commonly associated with. Thin arrows indicate that members of that clade have been either isolated from or detected in an effective or ineffective nodule of a member of the plant group at least once. Host specificity exists within the* Frankia *clades,* i.e., *not all members of a* Frankia *clade can nodulate all plants associated with that clade. Circles indicate the four putative origins of the ability to enter into an actinorhizal symbiosis (Swensen, 1996); black rhombs indicate the four putative origins of the ability to enter a symbiosis with rhizobia (Doyle, 1998).*

Evolutionary analysis indicates that N_2-fixing root-nodule symbioses evolved 50-100 million years ago (Kistner and Parniske, 2002). As a comparison, plant symbioses with arbuscular mycorrhizal fungi are much older and fossil evidence dates them back for at least 400 million years (Remy *et al.,* 1994). Thus, mechanisms that were evolved for the fungal symbioses may have been exploited in the evolution of N_2-fixing root-nodule symbioses, which will be discussed later. Phylogenetic analysis has shown that all plants able to enter a root-nodule symbiosis belong to a single clade (Rosid I; see Figure 1), *i.e.*, they go back to a common ancestor (Soltis *et al.,* 1995). Within the Rosid I clade, rhizobial symbioses are proposed to have evolved four times independently, namely three times within the legume family and once for *Parasponia* (Doyle, 1998). Similarly, four independent origins have been hypothesized for actinorhizal symbioses (Swensen, 1996; Figure 1), although another proposal suggests that present-day actinorhizal symbioses go back to an ancestral symbiosis that emerged before the divergence of extant actinorhizal plants (Clawson *et al.,* 2004). At any rate, the phylogenetic data suggest that the common ancestor of the Rosid I clade had acquired a property upon which a root-nodule symbiosis could develop.

2. NODULE STRUCTURE

2.1. Legume Nodules

Two types of legume nodules are known, determinate and indeterminate (see Figure 2). The type of nodule depends on the host plant (Trinick and Galbraith, 1980). Indeterminate nodules have an apical meristem, the activity of which leads to the formation of a developmental gradient in the inner tissue. Close to the meristem, cells are infected by infection threads. More basal cells internalize rhizobia in an endocytosis-like process, surrounding them with a peribacteroid membrane (PBM) that is derived from the plant plasma membrane. Internalized rhizobia along with their PBMs are called symbiosomes. Within symbiosomes, rhizobia differentiate into bacteroids, express nitrogenase genes, and begin to fix N_2. In determinate nodules, the meristem stops its activity early in nodule development, and new infected cells mostly do not arise by infection, but by division of infected cells (Newcomb, 1981; Rolfe and Shine, 1984). Indeterminate nodules can be unbranched (caesalpinoid type) or lobed (mucunoid and crotalarioid type; Corby 1988). There are two types of determinate nodules, the desmodioid type, which occurs in the Phaseoleae and Loteae, and the aeschynomenoid type (Doyle, 1998; Sprent, 1995; 2001). The determinate nodules of *Sesbania rostrata*, ostensibly of the aeschynomenoid type, show some phenotypic plasticity in that they retain an inactive meristem that can be reactivated by ethylene leading to indeterminate growth (Fernandez-Lopez *et al.,* 1998). Phylogenetic analysis and the distribution of primitive features, like persistent infection threads, suggest that indeterminate caesalpinioid nodules represent the basal form and that determinate nodules are a phenomenon which evolved independently at different points during legume evolution (Sprent, 2001).

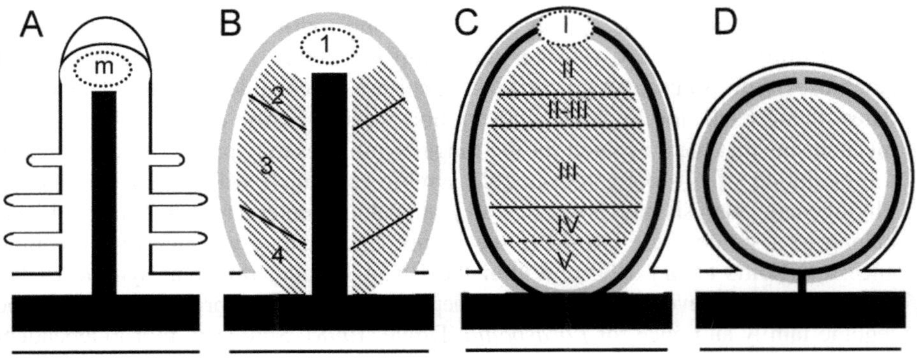

Figure 2: Comparison between lateral root (A), actinorhizal nodule lobe (B), indeterminate legume nodule (C), and determinate legume nodule (D). The vascular system is given in black. The actinorhizal nodule lobe is surrounded by a periderm (dark grey). The vascular system of legume nodules (C, D) is embedded in the nodule parenchyma that forms a turgor-controlled O_2-diffusion barrier. (A), (B), and (C) are indeterminate organs with an apical meristem (m, 1, I), the activity of which leads to the formation of a developmental gradient of infected cells in the tissue containing the infected cells (hatched). Zone of infection (2, II); interzone (II-III); zone of nitrogen fixation (3, III); and zone of senescence (4, IV) (Vasse et al., *1990; Ribeiro* et al., *1995). A zone of saprophytic growth (V) is found so far only in alfalfa nodules, where bacteria are released from remaining infection threads, re-invade senescent plant cells, and live saprophytically in a unique ecological niche (Timmers* et al., *2000).*

In some tropical legumes that form caesalpinioid nodules, *e.g.*, *Andira* and *Chamaecrista*, as well as in *Parasponia* sp., rhizobia are not internalized in infected cells in an endocytosis-like process, but they develop into their N_2-fixing form within branching infection threads (reviewed by Sprent, 2001). On the other hand, there are some aeschynomenoid nodules where no infection threads are formed at any point, but where rhizobia colonize the root cortex intercellularly and are taken up into nodule primordium cells from the apoplast (for *Arachis*, see Chandler 1978; and for *Stylosanthes,* see Chandler *et al.,* 1982).

2.1.1. Oxygen-diffusion Pathways

Nitrogenase is very sensitive to O_2, but the nitrogen-fixation process requires large amounts of ATP that have to be obtained by respiration. This situation leads to the so-called oxygen dilemma of nitrogen fixation - nitrogenase has to be protected from O_2, but O_2 has to be brought to the sites of respiration. In principle, because O_2 diffusion is far more rapid in air than in water, several different mechanisms are available to help solve this dilemma. These include: (a) restriction of O_2 diffusion by cell layers devoid of intercellular spaces, which could be achieved either by occlusion of the intercellular spaces or by turgor control; (b) metabolic protection, *i.e.*, the rapid removal of O_2 by highly active respiratory processes; and (c) the presence of a high-affinity O_2-binding protein that transports O_2 to the sites of respiration and not those of nitrogen fixation. In legumes as well as *Parasponia*, infected cells contain large amounts of nodule-specific O_2-transport proteins, the leghemoglobins (Appleby, 1984; Appleby *et al.,* 1983). Furthermore, an O_2-diffusion barrier, the so-called nodule endodermis, surrounds the inner tissue (reviewed by Minchin, 1997). In legumes, this is possible because the vascular system is located in the periphery of the nodule, so the nodule endodermis can protect the infected cells without restricting the access of O_2 to the vascular system where it is needed to generate energy (ATP) for transport processes. In *Parasponia* nodules, however, with their actinorhizal-like morphology, this problem seems to be solved by having infected cells form two separate regions in the cortex, each of which is surrounded by its own O_2-diffusion barrier. Between these two areas, O_2 can pass to the central vascular system (see chapter 5 in this volume; Tjepkema and Cartica, 1982).

2.2. Actinorhizal Nodules

As described above, actinorhizal nodules are composed of multiple lobes each of which represents a modified lateral root. Due to the activity of the apical meristem, the infected cells in the expanded cortex are arranged in a developmental gradient (see Figure 2). In the infection zone, they become gradually filled with branching *Frankia* hyphae. In the nitrogen-fixation zone, *Frankia* vesicles have developed and bacterial nitrogen fixation takes place. In the zone of senescence, bacterial material is degraded by the plant. In most actinorhizal nodules, infected and uninfected cells are interspersed in the cortex of the nodule lobe. In the actinorhizal Cucurbitales, however, the infected cells form a continuous patch, kidney-shaped in cross-section, at one side of the acentric stele (Hafeez *et al.,* 1984; Newcomb and Pankhurst, 1982; see chapter 5 in this volume).

2.2.1. Oxygen-diffusion Pathways and Oxygen Protection in Actinorhizal Nodules

In contrast to most legume nodules, actinorhizal nodule lobes are surrounded by a superficial periderm that can be more or less impermeable to gas. To provide the nodules with O_2 as required to produce energy for nitrogen fixation, the nodule periderm may be disrupted by lenticels (as in *Alnus, Datisca,* and *Coriaria*) or nodule roots can be formed (as in Casuarinaceae, Myricaceae, and *Datisca*). Lenticels are also found on the surface of desmodioid legume nodules (Sprent, 2001). Nodule roots are agraviotropically growing roots with reduced root caps, without root hairs, and with large air spaces in the cortex that are formed at the tip of nodule lobes, *i.e.*, by a change in the activity of the nodule lobe meristem (Torrey, 1976). Subsequently, new nodule lobe meristems can be induced next to the origin of the nodule root. Nodule roots provide access to O_2 for nodules formed on roots of plants grown in wetlands, and their length depends on the pO_2 level with the length increasing with decreased pO_2 (Silvester *et al.,* 1988; Sprent and Scott, 1979). *Datisca glomerata* can form either nodule roots (in liquid culture) or lenticels (in soil).

Diverse protection systems for the O_2-sensitive process of bacterial nitrogen fixation have developed in different actinorhizal plant genera (see chapter 5 in this volume). The morphology of actinorhizal nodule lobes does not allow for a legume-like O_2-diffusion control system, except for *Coriaria* nodules. These nodules have developed a comparable system, which uses a long lenticel at the non-infected side of the nodule lobe, with presumably turgor-based control in the cell layers between stele and periderm to regulate the access of O_2 to the infected cells (Silvester and Harris, 1989). It is likely to be due to this system that *Coriaria* is the actinorhizal plant, which shows the fastest adaptation to changes in external pO_2 as determined by nitrogen-fixation rates (Silvester and Harris, 1989).

Frankia can contribute to O_2 protection by forming vesicles. Furthermore, *Frankia* vesicles are sites of major respiratory activity, thereby achieving the metabolic removal of O_2 (Vikman, 1992). However, the differences in vesicle morphology among actinorhizal plants make it unlikely that *Frankia* vesicles contribute to O_2 protection in all cases. For instance, in *Casuarina* nodules, *Frankia* does not form vesicles (Berg and McDowell, 1987) and the plant seems to be solely responsible for O_2 protection of nitrogenase, which involves the production of hemoglobin in infected cells as is the case with legume nodules. Recently, bacteriohemoglobins have been discovered in *Frankia* and these may be involved in the protection of nitrogenase as well (Beckwith *et al.,* 2002).

In summary, O_2-protection mechanisms in actinorhizal nodules are diverse and may involve both the host and microsymbiont in exerting metabolic control to decrease O_2 levels, in physically restricting O_2 diffusion, and/or in regulating O_2 levels by the presence of hemoglobin.

3. NODULE-INDUCTION MECHANISMS

In both legume and actinorhizal symbioses, infection can take place either intra-cellularly *via* root hairs or inter-cellularly *via* penetration of the root epidermis and bacterial colonization of the root cortex. The pathway by which the bacteria enter the plant depends on the host-plant species. Mechanisms employed in both symbioses are summarized in Figure 3.

3.1. Legume Nodules

Components of the plant-root exudate induce the expression of the rhizobial *nod* genes, the products of which catalyse the production and export of rhizobial signal molecules, lipochitooligosaccharides, the so-called Nod factors (reviewed by Pueppke, 1996). Nod factors are essential for both inter- and intra-cellular infection. In most legumes examined, rhizobia enter the plant intra-cellularly *via* root hairs. However, different inter-cellular infection mechanisms have been characterized for some tropical legumes, most of them woody. In some cases, intermediates exist, *i.e.*, plants where both intra-cellular and inter-cellular infection is possible depending on the growth conditions (Subba-Rao *et al.,* 1995). During inter-cellular infection, rhizobia exploit gaps in the epidermis to enter the root; they cannot penetrate the epidermis directly.

During intracellular infection, Nod factors induce the deformation of root hairs in the root-hair extension zone by inducing the blockage and re-initiation of root-hair growth (Heidstra *et al.,* 1994). Nod factors also induce both the formation of pre-infection thread structures (PITs) in cortical cells, and cell divisions in the inner cortex (in plants forming indeterminate nodules) or in the outer cortex (in plants forming determinate nodules). When a rhizobium is trapped in a root-hair curl, local hydrolysis of the cell wall takes place (van Spronsen *et al.,* 1994), the plasma membrane invaginates, and new cell-wall material is deposited, leading to the formation of a tubular structure, called the infection thread (for reviews, see Brewin, 1991; Kijne, 1992; Newcomb, 1981). The ultrastructure of the infection thread wall is similar to that of the cell wall. Within the infection thread, rhizobia enter the plant root while still embedded in the infection-thread matrix that seems to consist of plant and bacterial compounds.

Concomitant with infection-thread formation, cortical cells are mitotically reactivated, leading to the formation of the nodule primordium. Infection threads grow toward primordium cells and release rhizobia into their cytoplasm. In plants that form indeterminate nodules, primordia are formed in the inner cortex and infection threads have to traverse outer cortical cells to reach them. Infection threads follow a tip-growth mechanism that requires the polarization of the cytoplasm. Therefore, cortical cells undergo morphological changes to support infection-thread growth and rearrange their cytoplasm to form a radially oriented conical structure, the cytoplasmic bridge that resembles a preprophase band, the so-called pre-infection thread (PIT; Kijne, 1992; van Brussel *et al.,* 1992). PIT formation can be induced by Nod factors alone, but infection thread formation, *i.e.*, invagination of the root-hair plasma membrane and synthesis of the infection-thread wall and matrix, occurs only in presence of rhizobia. PITs, but no infection threads, are formed in response to rhizobial lipopolysaccharide mutants (Spaink, 2000). PIT-forming cells re-enter the cell cycle and most likely become arrested in the G2 phase, *i.e.*, they undergo DNA re-duplication but get re-arrested before mitosis (Yang *et al.,* 1994).

During inter-cellular infection, bacteria enter the roots *via* gaps in the root epidermis, *e.g.*, through the cracks at the junctions of emerging lateral or adventitious roots (for *Arachis*, see Chandler, 1978; Boogerd and van Rossum, 1997; for *Stylosanthes,* see Chandler *et al.,* 1982; for *Aeschynomene*, see Alazard and Duhoux, 1990; for *Neptunia*, see James *et al.,* 1992; and for *Sesbania rostrata*, see Ndoye *et al.,* 1994). For *Lotus uliginosus*, infection occurs *via* enlarged epidermal cells (James and Sprent, 1999), whereas with *Mimosa,* rhizobia enter directly between epidermal cells (de Faria *et al.,* 1988). Concomitantly, cell divisions, which lead to the development of a nodule primordium, are induced in the root cortex. In many cases, once rhizobia begin colonizing the apoplast, infection-thread formation can be induced within the root cortex (James *et al.,* 1992; James and Sprent, 1999; Ndoye *et al.,* 1994; Schaede *et al.,* 1940; Subba-Rao *et al.,* 1995). The infection of nodule primordium cells can occur *via* infection threads; rhizobia are released into the cytoplasm from an unwalled portion of the infection thread and surrounded by PBMs. In those cases where no infection thread formation takes place at all, inter-cellular rhizobia move through the cortex by abortive infection and the resulting collapse of cortical cells, leading to the formation of infection strands (in *Stylosanthes*, see Chandler *et al.*, 1982; in *Mimosa*, see de Faria *et al.*, 1988; in *Aeschynomene*, see Alazard and Duhoux, 1990). They enter primordium cells through a structurally altered, probably partly degraded, part of the cell wall and become surrounded by PBMs (for *Arachis*, see Chandler, 1978; for *Aeschynomene*, see Alazard and Duhoux, 1990; for *Stylosanthes*, see Chandler *et al.*, 1982; for *Chamaecytisus*, see Vega-Hernández *et al.*, 2001).

Parasponia is also infected inter-cellularly. Here, the colonization of the root surface by rhizobia leads to the induction of cortical-cell divisions and the formation of a so-called prenodule, the presence of which leads to ruptures in the root epidermis (Bender *et al.*, 1987). Rhizobia enter the root cortex through these ruptures and colonize it inter-cellularly. This process is similar to the infection of the leguminous tree *Chamaecytisus* (Vega-Hernández *et al.*, 2001), where an abortive intra-cellular infection *via* root hairs leads to the formation of a nodule primordium. This, in turn, causes ruptures in the root epidermis through which rhizobia enter the root cortex and colonize it inter-cellularly. In *Chamaecytisus*, infection threads are not formed, but rhizobia move through the cortex by the collapse of cortical cells and are taken up directly into nodule primordium cells (Vega-Hernández *et al.*, 2001). In contrast, in *Parasponia*, infection threads are formed within the root at a later stage. They start from inter-cellular rhizobial colonies, when the formation of the nodule primordium has been induced in the root pericycle (Bender *et al.*, 1987; Lancelle and Torrey, 1984; 1985).

3.2. Actinorhizal Nodules

Like rhizobia, *Frankia* strains can enter the roots of their host plants either intra-cellularly *via* root hairs or inter-cellularly, and the mode of infection depends on the host plant species. Intra-cellular infection takes place in the higher Hamamelidae, *i.e.*, in plants of the families Betulaceae, Casuarinaceae and Myricaceae (Figure 1). The infection mechanisms are described in detail in chapter 6 of this volume, so they are only briefly summarized here (Figure 3).

Intra-cellular infection is similar to the corresponding process described for rhizobia. *Frankia* culture supernatants contain a factor of unknown chemical nature that induces the deformation of root hairs. When a hypha is trapped in a root hair curl, an infection thread-like structure ("encapsulation") develops by which the hypha enters the plant root. Within this tubular ingrowth, which grows by cell-to-cell passage like infection threads in legumes, the hypha is embedded in a cell wall-like matrix, the equivalent of the infection thread wall. There is no equivalent of the legume infection thread matrix in actinorhizal symbioses.

Concomitantly, cell divisions are induced in the root cortex. The infection thread-like structure grows to the dividing cortical cells and infects some of them by intense branching within the cells, filling them with *Frankia* hyphae from the inside outward. Once the cell is filled with branched hyphae in infection thread-like structures, vesicles develop and nitrogen fixation starts. This cortex-based structure is called the prenodule. While the prenodule develops, the formation of the nodule

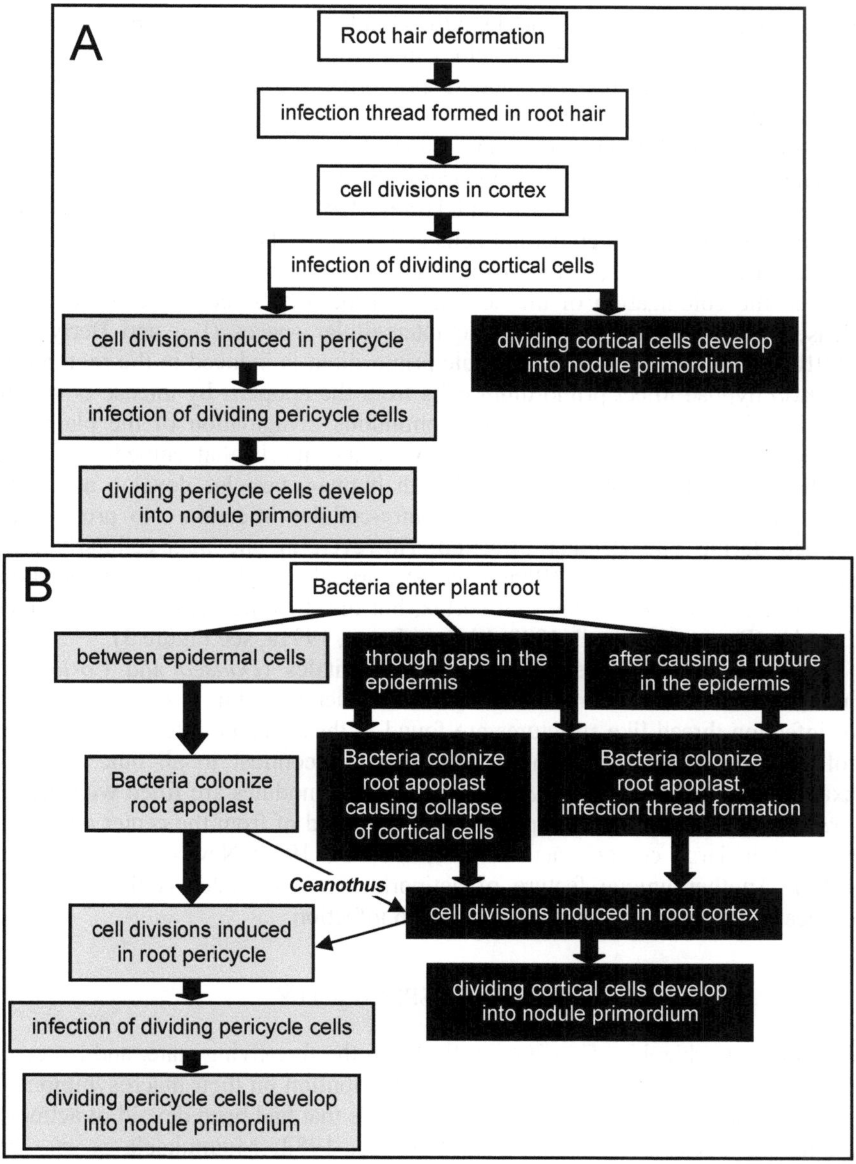

Figure 3. Comparison of nodule-induction mechanisms in actinorhizal plants and legumes. (A) Intracellular infection (higher Hamamelidae and legumes).(B) Intercellular infection (Rosales, Cucurbitales, legumes). Gray background indicates mechanisms specific for actinorhizal plants; black background indicates mechanisms specific for legumes; and white background indicates mechanisms present in both groups.

primordium is induced in the pericycle of the root vascular system, like in the case of lateral root primordia. Hyphae in infection thread-like structures ("encapsulated hyphae") grow from the prenodule to the nodule primordium and infect primordium cells. The nodule primordium develops into a nodule lobe. Depending on the host-plant species, more than one nodule primordium can be formed per prenodule (Torrey and Callaham, 1979).

During inter-cellular infection, which has been described for actinorhizal Rosales, *i.e.,* Rhamnaceae, Elaeagnaceae and Rosaceae (Berry and Sunell, 1990; Miller and Baker, 1986; Racette and Torrey, 1989), *Frankia* hyphae enter the root by penetration between epidermal cells and colonize the root cortex inter-cellularly. In contrast to rhizobia, *Frankia* does not depend on gaps in the root epidermis. During the colonization of the cortex, the root cortical cells secrete an electron-dense protein-rich material into the intercellular spaces (Liu and Berry, 1991a; 1991b), and the formation of a nodule primordium is induced in the root pericycle. *Frankia* hyphae infect primordium cells from the apoplast by intense branching of invading hyphae, concomitant with continuous invagination of the plant plasma membrane and the synthesis of a cell wall-like matrix that embeds the invading hyphae. Once the host cell is filled with hyphae, vesicles develop and nitrogen fixation starts, as is the case during intra-cellular infection. No prenodules are formed during inter-cellular infection. However, in one inter-cellularly infected species, *Ceanothus griseus* (Rhamnaceae), cortical cell divisions are induced during bacterial colonization of the root cortex, but the dividing cells are not infected by *Frankia* (Berry and Sunell, 1990; Liu and Berry, 1991a; see Figure 3).

In host plants of the actinorhizal Cucurbitales (*Datisca* and *Coriaria*), the infection mechanism has not been examined in detail yet but, because no prenodules or infection thread-like structures are found in these plants, infection is assumed to follow the inter-cellular pathway. However, in contrast to all other host plants examined, infected cells of *Datisca* and *Coriaria* nodules are filled with branching *Frankia* hyphae from the periphery inward, instead of from the center outward, and they retain large central vacuoles (Mirza *et al.,* 1994; Newcomb and Pankhurst, 1982). Another unique feature of actinorhizal Cucurbitales is that their nodule cortical cells become multinucleate prior to infection.

4. HOST SPECIFICITY

In contrast to rhizobia, *Frankia* can fix N_2 in the free-living state, and hence could be expected to be less dependent in their distribution on their macrosymbionts than are rhizobia. *Frankia* has been detected in soils that had been devoid of actinorhizal plants for decades (Normand and Lalonde, 1982; Maunuksela *et al.,* 1999), but rarely in soils outside the normal area of distribution of actinorhizal plants (Huss-Danell *et al.,* 1999, but see chapter 8 in this volume). Although several rhizobial strains depend on their host plants for persistence (Woomer *et al.,* 1988), others are better adapted to saprophytic growth (Hirsch, 1996). Evidence suggests that non-symbiotic rhizobia persist in soils in the absence of legumes and can acquire symbiotic genes from inoculant strains upon introduction of host legumes (Sullivan *et al.,* 1995; 1996; Sullivan and Ronson, 1998). However, for several actinorhizal genera (*Datisca, Coriaria, Ceanothus* and those within the Rosaceae), it has not yet been possible to isolate the microsymbiont (Swensen and Benson, this volume), which raises the suspicion that these *Frankia* strains might represent obligate symbionts.

4.1. Signal Exchange during Legume-nodule Induction

Flavonoids excreted by plant roots activate the rhizobial transcription factor NodD, which activates the transcription of the *nod/nol/noe* genes, the products of which produce specific lipochitooligosaccharides, the Nod factors (reviewed by Carlson *et al.,* 1994). Various substitutions at the terminal chitin monomers, as well as variations in the acyl group at the non-reducing end, lead to a high degree of structural variability among Nod factors (reviewed by Mergaert *et al.,* 1997). Host specificity is determined mostly by Nod factors (reviewed by Mergaert *et al.,* 1997), although it can be influenced by mechanisms increasing the concentration of rhizobia at the infection site (reviewed by Hirsch, 1999). Host range can vary between the extremes of one bacterial strain with one host plant, *i.e., Azorhizobium caulinodans* ORS571/*Sesbania rostrata* (de Bruijn 1989), and one bacterial strain that can infect more than 112 genera of legumes as well as the non-legume *Parasponia andersonii, i.e., Rhizobium* NGR234 (Pueppke and Broughton, 1999).

Nod factors induce the following effects on host plants: (i) root-hair deformation; (ii) PIT formation; and (iii) cortical cell divisions, *i.e.*, the formation of a nodule primordium. On some plants, Nod factors are sufficient to induce the formation of either nodule primordia or whole bacteria-free nodules (for *Medicago sativa,* see Truchet *et al.,* 1991; for *Glycine soja,* see Stokkermans and Peters, 1994).

It is interesting that among rhizobia, *nod* and *nif* (nitrogen fixation) gene phylogeny can differ from rRNA phylogeny (Haukka *et al.,* 1998), which indicates that lateral transfer of *nod* genes has taken place and has allowed the recipient strains to infect new hosts. An extreme case was shown by the finding of "symbiotic islands" in a *Mesorhizobium loti* strain (Sullivan *et al.,* 1995; Sullivan and Ronson, 1998). Analogous to the "pathogenicity islands" of bacterial animal pathogens (Lee, 1996), this strain contains a large DNA region with all genes required for an efficient root-nodule symbiosis, which was transferable into other soil bacteria.

4.2. Signal Exchange during Actinorhizal Nodule Induction

The *Frankia* Nod factor equivalent has not yet been characterized, partially due to the fact that, in contrast to legume symbioses, no convenient bioassay is available. Root-hair deformation on actinorhizal plants can also be induced by several non-symbiotic soil bacteria (Knowlton *et al.,* 1980). However, a partial purification of the root hair-deforming factor from the supernatant of *Frankia* cultures led to the conclusion that no lipochitooligosaccharides are involved in the infection of actinorhizal plants (Cérémonie *et al.,* 1999).

Host specificity of *Frankia* strains has been described in chapter 4 of this volume. Several factors hamper comparisons between the host ranges of rhizobia and *Frankia* strains. First, the number of symbiotic genera among legumes by far exceeds the total number of actinorhizal genera, *e.g.*, there are 25 actinorhizal genera altogether, whereas the wide host-range *Rhizobium* strain NGR234 itself can infect more than 112 different leguminous genera. On the other hand, actinorhizal genera come from different plant families, making it difficult to compare the host specifity of NGR234 with that of a *Frankia* strain, which can infect *Myrica gale* (Myricaceae) and *Casaurina glauca* (Casuarinaceae)! To complicate matters further, some actinorhizal genera are more promiscuous than others. However, this property is dependent on the ecological context because strong differences were found between greenhouse and field trials (see chapter 4 of this volume).

Second, the infection of actinorhizal plants seems to be less strictly controlled than does that of legumes. Actinorhizal nodules can accommodate not only different *Frankia* strains, but also, in their outer cortex, non-N_2-fixing related actinomycetes that cannot re-infect the plant on their own (Mirza *et al.,* 1992; Ramirez-Saad *et al.,* 1998). In some cases, *Frankia* strains isolated from actinorhizal nodules could not re-infect the host plants from which they were isolated, but were instead able to infect other actinorhizal plants (*e.g.*, Torrey, 1990; see chapter 4 in this volume). At any rate, no *Frankia* strain specific to a single host-plant species has ever been characterized.

For *Alnus glutinosa*, it is known that the plant's degree of resistance against infection by ineffecient *Frankia* strains (strains that cannot fix N_2 in symbiosis) is genetically determined (Wolters *et al.,* 1999). However, it is not known at which stage infection by these strains is blocked.

5. ROOT NODULES AND OTHER ROOT SYMBIOSES

More than 80% of all terrestrial plants can enter arbuscular mycorrhizal (AM) symbioses with fungi (Remy *et al.,* 1994). These interactions go back at least 400 million years and involve the stable uptake of hyphae into root cortical cells, but the fungal structures ("arbuscules") are degraded by the infected plant cell after a few days (Harrison, 1999). In most cases, AM does not involve the formation of special lateral root-like structures. However, a few cases are known where AM fungi induce the formation of so-called myconodules on the roots of some tropical trees. Myconodules resemble single-lobed actinorhizal nodules but have only a short life-time, which might be due to the short life-time of arbuscule-containing cells in general. Myconodules have been found on some legume trees and on the actinorhizal *Gymnostoma* sp., but also on trees that do not enter N_2-fixing root nodule symbioses (Béreau and Garbaye, 1994; Duhoux *et al.,* 2001). Hence, also fungi can trigger the formation of lateral root-like structures on higher plants.

Legume genetics have shown that rhizobial and AM symbioses involve common components in plant-signal transduction. Several legume mutants have been identified that are affected in both symbioses (Duc *et al.,* 1989; Stougaard, 2001). In all cases, these mutants are affected in early steps of the interaction, *e.g.* in *symRK* mutants of *Lotus japonicus* or *nork* mutants of alfalfa, the response to rhizobial Nod factors is blocked before root-hair deformation, and the penetration of the root epidermis by hyphae of AM fungi is blocked (Endre *et al.,* 2002; Stracke *et al.,* 2002). These results indicate that Nod-factor perception and signal transduction involves modules developed for the interaction of plants with AM fungi.

Another type of myconodule is known. It involves *Penicillium nodositatum*, which can use the actinorhizal intra-cellular infection pathway of *Alnus* to form parasitic myconodules (Pommer, 1956; Sequerra *et al.*, 1994; 1995). Again, these myconodules resemble single-lobed actinorhizal nodules, showing that *Frankia* Nod-factor equivalents can also be formed by fungi.

5.1. Which Phytohormones are Involved in Nodule Induction?

Rhizobial Nod factors seem to affect plant morphogenesis *via* auxin transport. Nod factor application results in a block in auxin transport in the root vascular system, which in turn causes an accumulation of auxin at the site of Nod factor application (Mathesius *et al.*, 1998a). Eventually, the auxin that accumulates in the stele at the Nod factor-application site leaks into the cortex and cell divisions occur. Once in the root tip, auxin from the stele is redirected and transported upward in the root cortex (Friml and Palme, 2002). Nod factors seem first to cause auxin depletion, *i.e.*, a decrease of the auxin/cytokinin ratio, and then an auxin excess, *i.e.*, an increase of the auxin/cytokinin ratio, in the root cortex. It is unclear which of these effects causes the induction of cell division. The effect of Nod factors on auxin concentrations might be mediated *via* plant flavonoids (Mathesius *et al.,* 1998b), which regulate enzymes involved in auxin degradation (Mathesius, 2001).

Support for the suggestion that nodule induction occurs *via* manipulation of plant-auxin transport arises from the fact that, in some legumes, exogenous application of auxin-transport inhibitors can lead to the induction of so-called pseudonodules, *i.e.*, rhizobia-free nodules (Hirsch *et al.,* 1989). The induction of nodule-like structures on alfalfa roots by *E. coli*, which produce a cytokinin (Cooper and Long, 1994), might be seen in the same context, *i.e.*, a localized change of the auxin-cytokinin balance seems to cause legume nodule induction. For actinorhizal plants, induction of pseudonodules by exogenously applied cytokinins has also been shown (Rodriguez-Barrueco and Bermudez de Castro, 1973). Pseudo-actinorhizal nodules also occur on plants in whose roots the auxin/cytokinin ratio has been changed by infection with *Agrobacterium rhizogenes* (Berg *et al.,* 1992). The differences in the localization of the pseudonodule primordium in legumes *versus* actinorhizal plants might be explained by differences in phytohormone balance in the two systems.

Another symbiotic interaction between plant roots and microorganisms that leads to morphological changes in the root system is the ectomycorrhizal symbiosis, which leads to the formation of short highly branched lateral roots (see, *e.g.*, Tagu *et al.,* 2002). Like nodule formation, ectomycorrhization requires the manipulation of plant development *via* auxin produced by the fungi (reviewed by Barker and Tagu, 2000; Martin *et al.,* 2001). One ectomycorrhizal fungus has been also shown to produce an auxin antagonist, hypaphorine, a betaine of tryptophan (Ditengou and Lapeyrie 2000). Interestingly, through interfering with the auxin response, hypaphorine blocks the extension of elongating root hairs and leads to the swelling of root hairs (Ditengou *et al.,* 2000), an effect similar to the initial response of elongating root hairs to rhizobial Nod factors (Heidstra *et al.,* 1994). It also affects membrane depolarization and ion currents similarly to Nod factors (Reboutier *et al.,* 2002), implying that the same signal-transduction pathway may be involved. In contrast to the effects of Nod factors, hypaphorine also has this effect on root hairs of non-host plants (Reboutier *et al.,* 2002). Obviously, symbiotic bacteria and fungi target the same plant signal-transduction pathway *via* different mechanisms.

Using plant mutants, ethylene has been implicated in the mechanism by which legumes control the number of nodules formed (called autoregulation; reviewed by Guinel and Geil, 2002). Due to the lack of actinorhizal plant mutants, the ethylene effect has not been examined in this system. However, in legumes, the repression of nodulation in response to soil-N is controlled by the same receptor kinases as autoregulation (Searle *et al.,* 2002), and nitrogen inhibition of nodulation in actinorhizal plants is well known and seems to involve similar combinations of root- and shoot-derived factors as in legumes (see chapter 6 in this volume). The receptor kinases involved in both combined-N inhibition and autoregulation of legume nodulation are closely related to CLAVATA1 from *Arabidopsis*. This receptor kinase is involved in cell fate determination in shoot apical meristems (De Young and Clark, 2001) and controls not only nodule, but also lateral-root, meristems. Therefore, it is likely that the same basic mechanisms that were recruited in legumes to control nodulation were also adapted by actinorhizal plants. Interestingly, the induction of pseudonodules on alfalfa roots by cytokinin is under autoregulatory control, whereas auxin transport inhibitor-dependent nodule formation is not (Cooper and Long, 1994; Hirsch *et al.,* 1989). This result might imply that autoregulation works *via* an ethylene effect of auxin transport or degradation.

6. EVOLUTION OF ROOT-NODULE SYMBIOSES

Generally speaking, root-nodule symbioses require: (i) the controlled penetration by a microsymbiont of the plant root, and the concomitant suppression of plant defense; (ii) the stable internalization of the microsymbiont into plant cells; and (iii) the induction of the formation of an organ with a vascular connection to the root stele. These processes have to be brought about *via* signal exchange between host and microsymbiont.

The first two requirements are already established for AM fungi. In exceptional cases, AM fungi even induce the formation of lateral root-like structures (Béreau and Garbaye, 1994; Duhoux *et al.,* 2001). The chitinaceous nature of the rhizobial Nod factors, in combination with the overlap in the signal-transduction pathways for legume-nodule formation and AM formation, has led to the hypothesis that perhaps N_2-fixing bacteria have copied fungal signal molecules. On the other hand, chitins occur in plant secondary cell walls (Benhamou and Asselin, 1989) and plant chitinases, which are involved in plant development, have been identified (Domon *et al.,* 2000; Zhong *et al.,* 2002). There is also evidence that chitooligosaccharide-like signal molecules exist in plants (de Jong *et al.,* 1995; Dyachok *et al.,* 2002;

Schmidt *et al.,* 1993) and that chitinases produce them by acting on apoplastic arabinogalactan proteins (Domon *et al.,* 2000; Passarinho *et al.,* 2001). Hence, rhizobial Nod factors may instead resemble endogenous plant-signal molecules.

6.1. Ways for Symbiotic Bacteria to enter Plant Cells

There are two types of internalization of bacteria within plant cells in root nodule symbioses. Infection-thread formation occurs in both actinorhizal and legume symbioses, whereas symbiosome formation occurs in legumes only. Bacterial nitrogen fixation in persistent infection threads occurs in actinorhizal symbioses, in *Parasponia*, and in some tropical legumes (*e.g.*, *Andira, Chamaecrista*; reviewed by Sprent, 2001), but endocytotic uptake into symbiosomes takes place in all other nodule-forming legumes. There are some legume symbioses in which infection threads are never formed and in which the bacteria enter the roots inter-cellularly and become internalized in symbiosomes (*e.g.*, for *Arachis*, see Chandler, 1978; for *Stylosanthes*, see Chandler *et al.,* 1982), but phylogenetic analysis implies that these are evolved cases (Sprent, 2001). It seems likely that the basal method of internalizating N_2-fixing bacteria was *via* infection threads.

As a mechanism for bacteria to enter plant roots, infection threads are formed only in root hairs. This is not surprising because infection threads are expanding by tip growth, which requires a polarization of the cytoplasm. Trichoblasts are already organized to support tip growth in that root hairs arising from them do so by a tip-growth mechanism. It seems possible that the corresponding reorganization of the cytoplasm cannot be achieved in atrichoblasts.

There is evidence implying that root-hair infection by rhizobia requires a specific structure of the epidermal cell wall (Matthysse and McMahan, 2001; Mort and Grover, 1988). This requirement, together with the fact that not all plants form numerous root hairs, might be taken to imply that root hair-independent infection-thread formation was the basal way of bacterial entry into the root. However, the uniformity in root-hair infection, when contrasted with the diversity of intercellular-infection mechanisms (reviewed, for legumes, by Sprent, 2001), points to the former mechanism as basal. Further, root-hair infection would be advantageous to the plant because it allows control of bacterial entry into the root.

For legumes, infection-thread growth through cortical cells requires the formation of pre-infection thread (PIT) structures formed by root cortical cells re-entering, but not completing, the cell cycle, which leads to endoreduplicaton and the formation of a cytoplasmic bridge through which the infection thread will grow (Yang *et al.,* 1994). PIT-like structures have also been observed in the infection zone of actinorhizal nodules, indicating that they form a general part of infection-thread growth (Berg, 1999). It is interesting that, in Cucurbitales, where *Frankia* is internalized from the apoplast in infection thread-like structures, nodule cells become multinucleate prior to and during the infection process (Hafeez *et al.,* 1984; Newcomb and Pankhurst, 1982). These data indicate that, also in plants with an intercellular infection mechanism, some cell-cycling is required before a cell can support the internalization of the microsymbiont. Because endoreduplication of nuclear DNA has also been observed in cells infected by AM fungi (Berta *et al.*, 2000), it is possible that such a mechanism for the internalization of microsymbionts in plant cell was already in place (if only for fungi), when N_2-fixing symbioses evolved (Parniske, 2000). However, infection thread-like structures do not exist in AM symbioses, because the intracellular growth of AM fungal hyphae does not involve PIT formation. Clearly, the internalization mechanism for AM fungi had to be changed significantly to allow the internalization of bacteria and some new functions had to be recruited. For instance, the promoter of the nodule-specific subtilase, *cg12*, is active in infection thread-containing cells of *Allocasuarina verticillata* roots as well as in infected cells of nodules and prenodules, but not in cells that contain fungal arbuscules (Svistoonoff *et al.*, 2003).

The host range of *Frankia* is ostensibly wider than that of rhizobia and includes plants from eight different families, but there are far more legume genera that are able to enter rhizobial symbioses. Was an actinorhizal symbiosis an impediment to further evolution of the host-plant species? Interestingly, the only legumes, where bacteroids still fix N_2 in infection threads and where no complete endocytotic process takes place, are woody species; this situation is particularly obvious in the genus, *Chamaecrista* (Naisbitt *et al.*, 1992; Sprent, 2001). The structure of *Frankia* necessitates a symbiosis with persistent infection threads. The internalization of microsymbionts by continuous invagination of the plasma membrane without complete endocytosis has to counteract the turgor of the plant cell, putting high demands on the cytoskeleton, and perhaps requiring some stabilization of the cell walls of infected cells (as by lignified tissues). In contrast, the endocytotic internalization of rhizobia in symbiosomes allows turgor control by aquaporins in the peribacteroid membranes (Dean *et al.*, 1999). In the only intracellular symbiosis between a higher plant and the filamentous cyanobacterium *Nostoc,* the N_2-fixing *Gunnera* symbiosis, the microsymbiont is internalized by a complete endocytotic process (Bergman *et al.*, 1992). Although in arbuscular mycorrhizal symbioses, branching fungal hyphae are internalized in root cortical cells by continuing invagination of the plasma membrane as in the case of *Frankia* hyphae, arbuscules only have a life time of a few days (Harrison, 1999). It is tempting to speculate that this restriction, *i.e.*, the necessity for persistent infection threads requiring a stabilization of the walls of infected cells, impaired the distribution of the actinorhizal symbiosis.

6.2. Signal-transduction Pathways for the Differentiation of Nodule Cells

Rhizobia and *Frankia* are only stably internalized (in contrast to infection-thread growth through cells) in cells formed after signal exchange with the microsymbiont, whereas AM fungi can form arbuscules in pre-existing root cortical cells. Thus, some novel differentiation appears necessary for stable internalization of a bacterial microsymbiont. This suggestion is supported by the fact that there are several fungal pathogens that enter plant cells (see, *e.g.* Ellis *et al.*, 2000), but no known bacterial pathogens (Kistner and Parniske, 2002). This new differentiation required for the internalization of symbiotic bacteria could share common features in legumes and actinorhizal plants and, indeed, evidence exists that infected cell-specific transcription factors are conserved among legumes and actinorhizal Hamamelidae.

This evidence is based on the observations that a transcriptional fusion between the promoter of the gene encoding the symbiotic hemoglobin of *Casuarina glauca* and the ß-glucuronidase (*GUS*) reporter gene was found to be expressed specifically in the infected cells of nodules of the legume *Lotus corniculatus* (Jacobsen-Lyon *et al.,* 1995). Furthermore, a soybean leghemoglobin promoter-*GUS* fusion was expressed specifically in the infected cells of actinorhizal *Allocasuarina verticillata* nodules (Franche *et al.,* 1998), even though the infected cells are not at morphologically equivalent positions in both types of nodules. However, although the expression pattern of a *GUS* fusion with the hemoglobin promoter from *Parasponia andersonii* was conserved in the actinorhizal *A. verticillata* (Franche *et al.,* 1998), it was not conserved in the legume *L. corniculatus* (Andersson *et al.,* 1997). These results suggest that legumes and higher Hamamelidae are more closely related to each other than to the other N_2-fixing plant groups. On the other hand, the hemoglobin gene expressed in infected cells of *Parasponia* nodules, in contrast to the symbiotic (leg-)hemoglobin genes of *Casuarina* and legumes, is not nodule-specific (Landsmann *et al.,* 1988; Trinick *et al.,* 1989), so the discrepancy may be due to the promoter sequences responsible for expression during non-symbiotic development.

Another type of infection-related transcription factor may also be conserved among legumes and actinorhizal plants. The promoter of the nodule-specific subtilase gene from *Casuarina glauca, cg12* (Laplaze *et al.,* 2000b; Svistoonoff *et al.,* 2003) is active in infection thread-containing cells (beginning with infected root hairs) in both legumes and actinorhizal plants, as shown by *cg12* promoter-marker gene fusions in *Allocasuarina verticillata* and *Medicago truncatula* (S. Svistoonoff, L. Laplaze, C. Franche and D. Bogusz, unpublished data). In actinorhizal plants, *cg12* expression occurs throughout the nitrogen-fixation zone, whereas, in legume nodules, it is restricted to the distal part of the prefixation zone and is absent in bacteroid-containing cells. Hence, in both symbioses, *cg12* expression is linked to the presence of infection threads in cells, not infection *per se*. As mentioned in section 6.1., *cg12* is not expressed in arbuscule-containing cells, which underlines the difference in the internalization processes for bacteria and fungi in plant cells.

At any rate, in both legumes and Casuarinaceae, at the level of transcription factors, there is a distinction between cells through which infection threads pass, including root-hair cells and root cortical cells that exist before signal exchange with the microsymbiont, and cells that either have been filled or are in the process of filling with branching infection threads or bacteroids (all these cells having been formed after signal exchange with the microsymbiont). The former cells show *cg12* expression but no expression of (leg-)hemoglobin, whereas the latter cells, in both legumes and Casuarinaceae, show (leg-)hemoglobin expression and, in Casuarinaceae, also express *cg12*.

Consistent with the link between infection-thread growth and cell cycling, in *M. truncatula* (*i.e.*, in a heterologous system) weak *cg12* promoter-*GFP* expression is also found in the root pericycle at the nodule-induction site, and in the nodule primordium cells that have not been entered by an infection thread, *i.e.*, in dividing cells and cells that are preparing to divide (S. Svistoonoff, L. Laplaze, C. Franche and D. Bogusz, unpublished data). Infection thread-containing cell-specific signal transduction seems to be derived from cell cycle-dependent signalling. In *Allocasuarina verticillata*, which is closely related to the source of the *cg12* promoter, *Casuarina glauca, cg12* promoter-*GFP* or -*GUS* are expressed exclusively in infection thread-containing cells (Svistoonoff *et al.,* 2003).

An unusual leghemoglobin gene (*Vflb29*) in *Vicia faba* has a promoter that was active specifically and exclusively in rhizobium bacteroid-containing and in arbuscule-containing cells (Vieweg *et al.,* 2004), so indicating some conservation of transcription factors between bacterium- and fungus-infected cells that cannot be related exclusively to the mechanism of infection-thread growth. Unfortunately, the activity of the *Vflb29* promoter has not been examined in actinorhizal plants.

Another data set indicates differences in the signal-transduction pathways involved in cell differentiation in nodules from Rosid I subgroups. Rubisco activase transcription is induced in *Datisca* nodules, but the transcript is not spliced correctly and does not leave the nucleus (Okubara *et al.,* 1999). Thus, the signal-transduction pathway that leads to the induction of rubisco activase transcription is active in *Datisca* nodules, implying that components of leaf light- or sugar-dependent signal-transduction pathways may have been recruited for the control of nodule development. However, no rubisco activase message could be detected in nodules of *Alnus glutinosa* (K. Pawlowski, unpublished data). Because differences in the regulation of a basic photosynthetic gene as conserved as rubisco activase are not likely between *Alnus* and *Datisca,* the induction of rubisco activase transcription in *Datisca,* but not *Alnus*, nodules seems to indicate that different signal-transduction pathways have been recruited for the control of nodule development in actinorhizal higher Hamamelidae and Cucurbitales, respectively. These data would support the idea that, based on the common precondition acquired by the Rosid I ancestor, the nodulation syndrome evolved independently in the different symbiotic subgroups of Rosid I (Soltis *et al.,* 1995; Swensen, 1996). On the other hand, an overlap between light signal-transduction pathways and the regulation of nodulation has been shown in legumes (Nishimura *et al.,* 2002), which are likely to be more closely related to actinorhizal Hamamelidae than to Cucurbitales (Pawlowski *et al.,* 2003). A homolog of the transcription factor HY5, which regulates photomorphogenesis and lateral root development in *Arabidopsis*, regulates photomorphogenesis and nodulation in *Lotus japonicus* (Nishimura *et al.,* 2002).

6.3. Structural Diversity of Nodules

Regarding the indeterminate and determinate growth pattern of legume nodules (Figure 2), it is obvious that the latter represents a derivative development (Doyle, 1998). Because perennial nodules require an indeterminate growth pattern, it is not surprising that, in spite of the structural diversity of actinorhizal nodules (Figure 3), a determinate nodule type never developed on woody host plants.

The main difference between legume- and actinorhizal-nodule development is the induction of cell divisions leading to the formation of the nodule primordium in the cortex (legumes) *versus* the pericycle (actinorhizal plants and *Parasponia*). It has been proposed that the stem-like organisation of legume nodules is a result of their induction in the root cortex instead of the pericycle, and is due to a predisposition of legumes to form lateral-root storage organs, a tendency that is not present in other symbiotic plant families (Joshi *et al.,* 1991). In some cases, legumes have been found to form intermediates between lateral roots and nodules (Hirsch, 1992). Furthermore, the necessity to obtain a vascular connection between nodule and root means that, in legumes also, divisions in the root pericycle are induced at an early time point in nodule development (Gualtieri and Bisseling, 2000), giving credibility to the argument that both legume and actinorhizal nodules are modified lateral roots.

As mentioned above, cortical cell divisions are also induced during intercellular infection of *Parasponia* and during intercellular infection of actinorhizal higher Hamamelidae and lead to the formation of the so-called prenodules, not to nodule primordia. Also, one inter-cellularly infected actinorhizal plant shows a putative relic of prenodule formation, in that cortical cells start dividing upon inter-cellular colonization of the root cortex, but do not become infected (*Ceanothus griseus*; see Liu and Berry, 1991a). Because prenodule cells show the same differentiation as their counterparts in mature nodule lobes, it has been proposed that prenodule-like structures were the origin of all root-nodule symbioses (Laplaze *et al.,* 2000a). In order to achieve both efficient removal of the fixed-N and an efficient supply of carbon sources by the host, a vascular connection of the symbiotic organ would have evolved later with different localization of the vascular system relative to the infected cells in legumes *versus* other plant families (Figure 3). In a similar vein, Kistner and Parniske (2002) have suggested that the evolution of N_2-fixing plants started with the ancestor of the symbiotic subclades of Rosid I (called "FaFaCuRo" for Fabales, Fagales, Cucurbitales and Rosales) acquiring the ability to internalize bacterial microsymbionts in infection threads, *i.e.*, that an infection-thread symbiosis was the origin of root-nodule symbioses (Figure 4). This hypothesis is supported by an electron-microscopy finding of bacterial cells in infection threads in roots of the non-nodulating legume tree *Gleditsia* sp. (Faria *et al.*, 1999). In all extant root-nodule symbioses examined, bacterial nitrogen fixation is only found in infected cells that arose by division after signal exchange with the microsymbiont, this infection of *Gleditisa* is probably consistent with this fact because the infected areas of the roots were macroscopically visible as small bumps that are probably indicative of cell division.

6.4. Microbial Signal Compounds Trigger Cell Cycling

Based on the assumption that the induction of incomplete cell cycling, namely endoreduplication, is required for infection-thread growth, the induction of nodule primordium formation - complete cell cycling - does not represent a complete new achievement apart from the internalization of the bacterial microsymbionts in

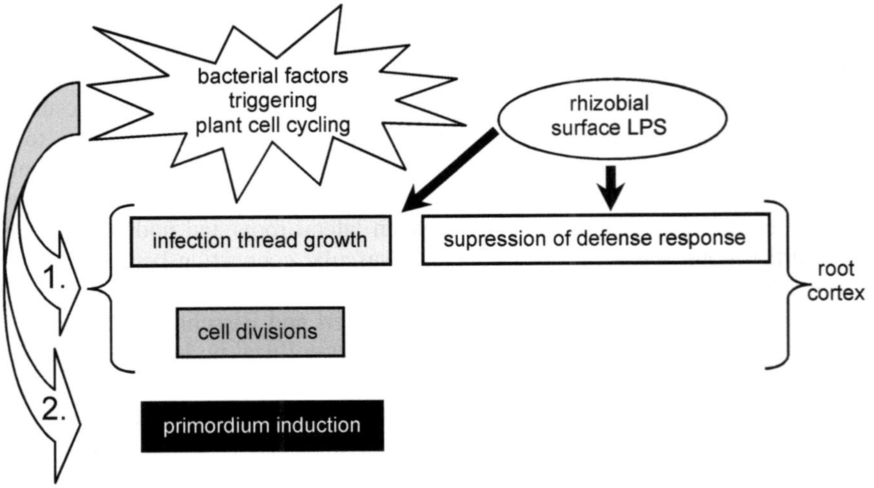

Figure 4. Manipulation of plant development by symbiotic bacteria (hypothetical). Triggering plant cell cycling, Nod factors could initially (1.) have caused infection thread growth and cell division, depending on the amount of receptors in the corresponding cells (indicated by different shades of grey). These events would have led to the formation of prenodule-like structures. Later, (2.) the induction of an organ primordium (i.e., sustained cell divisions) in root cortex and pericycle (legumes) or in the root pericycle (actinorhizal plants) could have been added, leading to nodule formation. In legumes, rhizobial surface lipopolysaccharides (LPS) are responsible for the suppression of the plant defence response and are required for infection-thread formation. It is not clear whether, and if yes, which, Frankia *surface molecules have similar functions.*

infection threads. On the other hand, the example of *Parasponia* shows that induction of cell division can cause ruptures in the plant root epidermis by which bacteria can enter the root cortex (Bender *et al.,* 1987). Thus, this might be interpreted to mean that the primary event was for bacteria to acquire the ability to manipulate plant cell division, with infection-thread formation being a later adjustment. However, the example of *Chamaecytisus* indicates that the induction of cortical cell division, to cause ruptures in the root epidermis by which bacteria enter the plant, is likely to represent an evolved mechanism derived from an earlier intra-cellular infection pathway (Vega-Hernández *et al.,* 2001).

It has been suggested that rhizobia with wide host ranges, such as NGR234, were basal in legume nodulation (Broughton *et al.,* 2000; Perret *et al.,* 2000) and that more specific associations only evolved later. Nevertheless, the difference between host specificity ranges of extant *Frankia* strains and rhizobia, as well as the obvious lack of chemical similarity between Nod factors of both groups of bacteria (Cérémonie *et al.,* 1999), implies that *Frankia* Nod factors do not use the same type of receptor as rhizobial Nod factors. A receptor-like kinase (RLK) containing three leucine-rich repeats (LRRs) and an unusual apoplastic domain is involved in the early stages of both AM and rhizobial symbioses and might represent the plant receptor of rhizobial Nod factors (called NORK or SYMRK; Endre *et al.,* 2002; Kistner and Parniske, 2002; Stracke *et al.,* 2002). RLKs differ in their apoplastic domains (Becraft, 2002), and no RLK with an apoplastic domain like that of NORK/SYMRK has been found outside of the legume family (Kistner and Parniske, 2002). RLKs like NORK/SYMRK, though without the unusual apoplastic domain and instead with more LRRs, *e.g.*, CLAVATA1 and the receptors of systemin and sulfokinins, bind proteinaceous effectors (Matsubayashi *et al.,* 2002; Scheer and Ryan, 2002; Trotochaud *et al.,* 2000). On the other hand, SERK-like RLKs have five LRRs and an apoplastic proline-rich domain, and their LRRs seem to serve only for multimerization, not for receptor binding (Shah *et al.,* 2001). Thus, it is not clear whether the LRRs of NORK/SYMRK have anything to do with effector binding, and what type of signal molecule is bound by NORK/SYMRK. It has been suggested that the acquisition of a NORK/SYMRK-like RLK by exon shuffling occurred at the beginning of the evolution of plants able to enter N_2-fixing root nodule symbioses (Kistner and Parniske, 2002).

Although rhizobial Nod factors and AM fungi signal substances share part of a signal transduction chain (Stougaard, 2001), it is not assumed that AM signal substances are structurally similar to Nod factors. The host range of AM fungi would argue against that. It is quite possible that AM signal factors as well as Nod factors bind to different receptors, thereby causing the production of a signal molecule that is recognized by NORK/SYMRK.

As mentioned above, some legumes and at least one actinorhizal plant, *Gymnostoma*, can form myconodules to host AM fungi (Béreau and Garbaye, 1994; Duhoux *et al.,* 2001). Also, the myconodules of legume trees do not have a legume nodule-like structure and the myconodules of *Gymnostoma* do not show prenodules. Thus, even if the formation of a lateral root-like structure is induced by AM fungal signal factors, its ontogenesis will not follow the model of bacterially induced nodules. However, *Penicillium nodositatum,* which is not an AM fungus, can use the infection pathway of *Frankia* to induced parasitic myconodules (Sequerra *et al.,* 1994; 1995). This can only be explained by *P. nodositatum* being able to produce signal molecules similar to those of *Frankia.* It also seems to imply that the formation of infection thread-like structures through which *Frankia* enters the plant roots does not require *Frankia* surface lipopolysaccharides, whereas legume infection-thread formation does require rhizobial LPS, a conclusion that seems convincing in the context of the diversity of *Frankia*- and *Frankia*-like strains that can be found in actinorhizal nodules (see chapter 2 in this volume). Interestingly, rhizobial LPS plays a role in the suppression of an anti-pathogen response of the plant against rhizobia (Albus *et al.,* 2001). Thus, although there seem to be similarities in the way rhizobia, *Frankia* and AM fungi infect plants, there seem to be striking differences in the way they suppress the anti-pathogen response.

REFERENCES

Alazard, D., and Duhoux, E. (1990). Development of stem nodules in a tropical forage legume, *Aeschynomene afraspera. J. Exp. Bot., 230,* 1199-1206.

Albus, U., Baier, R., Holst, O., Pühler, A., and Niehaus, K. (2001). Suppression of an elicitor-induced oxidative burst reaction in *Medicago sativa* cell cultures by *Sinorhizobium meliloti* lipopolysaccharides. *New Phytol., 151,* 597-606.

Andersson, C. R., Llewellyn, D. J., Peacock, W. J., and Dennis, E. S. (1997). Cell-specific expression of the promoters of two nonlegume hemoglobin genes in a transgenic legume, *Lotus corniculatus. Plant Physiol., 113,* 45-57.

Appleby, C. A. (1984). Leghemoglobin and *Rhizobium* respiration. *Annu. Rev.Plant Physiol., 35,* 443-478.

Appleby, C. A., Tjepkema, J. D., and Trinick, M. J. (1983). Hemoglobin in a nonleguminous plant, *Parasponia*: Possible genetic origin and function in nitrogen fixation. *Science, 220,* 951-953.

Barker, S. J., and Tagu, D. (2000). The roles of auxins and cytokinins in mycorrhizal symbioses. *J. Plant Growth Regul., 19,* 144–154.

Beckwith, J., Tjepkema, J. D., Cashon, R. E., Schwintzer, C. R., and Tisa, L. S. (2002). Hemoglobin in five genetically diverse *Frankia* strains. *Can. J. Microbiol., 48,* 1048-1055.

Becraft, P. W. (2002). Receptor kinase signaling in plant development. *Annu. Rev. Cell Dev. Biol.,* 18, 163-192.

Bender, G. L., Goydych, W., Rolfe, R. G., and Nyudu, M. (1987). The role of *Rhizobium* conserved and host specific nodulation genes in the infection of the non-legume *Parasponia andersonii. Mol. Gen. Genet., 210,* 299-306.

Benhamou, N., and Asselin, A. (1989). Attempted localization of a substrate for chitinases in plant cells reveals abundant N-acetyl-D-glucosamine residues in secondary walls. *Biol. Cell, 67,* 341-350.

Benson, D. R., and Clawson, M. L. (2000). Evolution of the actinorhizal plant symbiosis. In E. W. Triplett (Ed.), *Prokaryotic nitrogen fixation: A model system for analysis of a biological process* (pp. 207-224). Symondham, UK: Horizon Scientific Press.

Béreau, M., and Garbaye, J. (1994). First observations on the root morphology and symbioses of 21 major tree species in the primary tropical rain forest of French Guyana. *Ann. Sci. Forest., 51,* 407-416.

Berg, R. H. (1999). *Frankia* forms infection threads. *Can. J. Bot., 77,* 1327-1333.

Berg, R. H., Liu, L., Dawson, J. O., Savka, M. A., and Farrand, S. K. (1992). Induction of pseudoactinorhizae by the plant pathogen *Agrobacterium rhizogenes. Plant Physiol., 98,* 777-779.

Berg, R. H., and McDowell, L. (1987). Endophyte differentiation in *Casuarina* actinorhizae. *Protoplasma, 136,* 104-117.

Bergman, B., Johansson, C., and Söderback, E. (1992). The *Nostoc-Gunnera* symbiosis. *New Phytol., 122,* 379-400.

Berry, A. M., and Sunell, L. A. (1990). The infection process and nodule development. In C. R. Schwintzer, and J. D. Tjepkema (Eds.), *The biology of* Frankia *and actinorhizal plants* (pp. 61-82). San Diego, CA: Academic Press.

Berta, G., Fusconi, A., Sampo, S., Lingua, G., Perticone, S., and Repetto, O. (2000). Polyploidy in tomato roots as affected by arbuscular mycorrhizal colonization. *Plant Soil, 226,* 37–44.

Boogerd, F. C., and van Rossum, D. (1997). Nodulation of groundnut by *Bradyrhizobium*: A simple infection process by crack entry. *FEMS Microbiol. Rev., 21,* 5-27.

Brewin, N. J. (1991). Development of the legume root nodule. *Annu. Rev. Cell Biol., 7,* 191-226

Broughton, W. J., Jabbouri, S., and Perret, X. (2000). Keys to symbiotic harmony. *J. Bacteriol., 182,* 5641-5652.

Carlson, R. W., Price, N. P. J., and Stacey, G. (1994). The biosynthesis of rhizobial lipo-oligosaccharide nodulation signal molecules. *Mol. Plant-Microbe Interact., 7,* 684-695.

Cérémonie, H., Debelle, F., and Fernandez, M. P. (1999). Structural and functional comparison of *Frankia* root hair deforming factor and rhizobia Nod factor. *Can. J. Bot., 77,* 1293-1301.

Chandler, M. R. (1978). Some observations on infection of *Arachis hypogaea* L. by *Rhizobium. J. Exp. Bot., 29,* 749-755.

Chandler, M. R., Date, R. A., and Roughley, R. J. (1982). Infection and root-nodule development in *Stylosanthes* species by *Rhizobium. J. Exp. Bot., 33,* 47-57.

Chen, W-M., Laevens, S., Lee, T. M., Coenye, T., De Vos, P., *et al.* (2001). *Ralstonia taiwanensis sp. nov., isolated* from root nodules of *Mimosa* species and sputum of a cystic fibrosis patient. *Int. J. Syst. Evol. Micr., 51,* 1729-1735.

Clawson, M.L., Bourret, A., and Benson, D.R. (2004). Assessing the phylogeny of *Frankia*-actinorhizal plant nitrogen-fixing root nodule symbioses with *Frankia* 16S rRNA and glutamine synthetase gene sequences. *Mol. Phylogenet. Evol., 31,* 131-138.

Cooper, J. B., and Long, S. R. (1994). Morphogenetic rescue of *Rhizobium meliloti* nodulation mutants by trans-Zeatin secretion. *Plant Cell, 6,* 215-225.

Corby, H. D. L. (1988). Types of rhizobial nodule and their distribution among the Leguminosae. *Kirkia, 13,* 53-123.

de Bruijn, F. J. (1989). The unusual symbiosis between the diazotrophic stem-nodulating bacterium *Azorhizobium caulinodans* ORS571 and its host, the tropical legume *Sesbania rostrata.* In T. Kosuge and E. W. Nester (Eds.), *Plant-microbe interactions 3* (pp. 457-504). New York, NY: McMillan Publishing Company.

de Faria, S. M., Hay, G. T., and Sprent, J. I. (1988). Entry of rhizobia into roots of *Mimosa scabrella* Bentham occurs between epidermal cells. *J. Gen. Microbiol., 134,* 2291-2296.

de Faria, S. M., de Lima, H. C., and Olivares, F. L. (1999). Nodulação em espécies florestais, especifidade hospedeira e implicações na sistemática de Leguminosae. In J. O. Siqueira, F. M. S. Moreiram, A. S. Lopes, L. R. G. Guilherme, V. Faquin, A. E. Furtin Neto, and J. G. Carvalho (Eds.), *Inter-relação fertilidade, biologia do solo e nutrição de plantas. (Soil fertility, soil biology and plant nutrition interrelationships)* (p. 667–686). Universidade Federal da Lauras Brazil: Sociedade Brasileira de Ciênca do Solo.

de Jong, A. J., Hendricks, T., Meijer, E. A., Penning, M., Lo Schiavo, F., *et al.* (1995). Transient reduction in secreted 32 kD chitinase prevents somatic embryogenesis in the carrot (*Daucus carota* L.) variant ts11. *Dev. Genet., 16,* 332-343.

DeYoung, B. J, and Clark, S. E. (2001). Signaling through the CLAVATA1 receptor complex. *Plant Mol. Biol., 46,* 505-513.

Dean, R. M., Rivers, R. L., Zeidel, M. L., and Roberts, D. M. (1999). Purification and functional reconstitution of soybean nodulin 26. An aquaporin with water and glycerol transport properties. *Biochem., 38,* 347-353.

Ditengou, F. A., and Lapeyrie, F. (2000). Hypaphorine from the ectomycorrhizal fungus *Pisolithus tinctorius* counteracts activities of indole-3-acetic acid and ethylene but not synthetic auxins in eucalypt seedlings. *Mol. Plant-Microbe Interact., 13,* 151-158.

Ditengou, F. A., Beguiristain, T., and Lapeyrie, F. (2000). Root hair elongation is inhibited by hypaphorine, the indole alkaloid from the ectomycorrhizal fungus *Pisolithus tinctorius,* and restored by indole-3-acetic acid. *Planta, 211,* 722-728.

Domon, J. M., Neutelings, G., Roger, D., David, A., and David, H. (2000). A basic chitinase-like protein secreted by embryogenic tissues of *Pinus caribaea* acts on arabinogalactan proteins extracted from the same cell lines. *J. Plant Physiol., 156,* 33-39.

Doyle, J. J. (1998). Phylogenetic perspectives on nodulation: Evolving views of plants and symbiotic bacteria. *Trends Plant Sci., 3,* 473-478.

Duc, G., Trouvelot, A., Gianinazzi-Pearson, V., and Gianinazzi, S. (1989). First report of non-mycorrhizal plant mutants (Myc^-) otained in pea (*Pisum sativum* L.) and fababean (*Vicia faba* L.). *Plant Sci., 60,* 215-222.

Duhoux, E., Rinaudo, G., Diem, H. G., Auguy, F., Fernandez, D., *et al.* (2001). Angiosperm *Gymnostoma* trees produce root nodules colonized by arbuscular mycorrhizal fungi related to *Glomus. New Phytol., 149,* 115-125.

Dyachok, J. V., Wiweger, M., Kenne, L., and von Arnold, S. (2002). Endogenous Nod-factor-like signal molecules promote early somatic embryo development in Norway spruce. *Plant Physiol., 128,* 523-533.

Ellis, J., Dodds, P., and Pryor, T. (2000). The generation of plant disease resistance gene specificities. *Trends Plant Sci., 5,* 373-379.

Endre, G., Kereszt, A., Kevei, Z., Mihacea, S., Kalo, P., and Kiss, G. B. (2002). A receptor kinase gene regulating symbiotic nodule development. *Nature, 417,* 962-966.

Fernandez-Lopez, M., Goormachtig, S., Gao, M., D'Haeze, W., van Montagu, M., and Holsters, M. (1998). Ethylene-mediated phenotypic plasticity in root nodule development on *Sesbania rostrata. Proc. Natl. Acad. Sci. U.S.A., 95,* 12724-12728.

Franche, C., Diouf, D., Laplaze, L., Auguy, F., Frutz, T., *et al.* (1998). Soybean (*lbc3*), *Parasponia,* and *Trema* hemoglobin gene promoters retain symbiotic and nonsymbiotic specificity in transgenic Casuarinaceae: Implications for hemoglobin gene evolution and root nodule symbioses. *Mol. Plant-Microbe Interact.,* 11, 887–894.

Friml, J., and Palme, K. (2002). Polar auxin transport – old questions and new concepts? *Plant Mol. Biol.,* 49, 273-284.

Gualtieri, G., and Bisseling, T. (2000). The evolution of nodulation. *Plant Mol. Biol., 42,* 181-194.

Guinel, F. C., and Geil, R. D. (2002). A model for the development of the rhizobial and arbuscular mycorrhizal symbioses in legumes and its use to understand the roles of ethylene in the establishment of these two symbioses. *Can. J. Bot., 80,* 695-720.

Hafeez, F., Akkermans, A. D. L., and Chaudhary, A. H. (1984). Observations on the ultrastructure of *Frankia* sp. in root nodules of *Datisca cannabina* L. *Plant Soil, 79,* 383-402.

Harrison, M. J. (1999). Molecular and cellular aspects of the arbuscular mycorrhizal symbiosis. *Annu. Rev. Plant Physiol. Plant Mol. Biol., 50,* 361-389.

Haukka, K., Lindstrom, K., and Young, J. P. W. (1998). Three phylogenetic groups of *nodA* and *nifH* genes in *Sinorhizobium* and *Mesorhizobium* isolates from leguminous trees growing in Africa and Latin America. *Appl. Environ. Microbiol., 64,* 419-426.

Heidstra, R., Geurts, R., Franssen, H., Spaink, H. P., van Kammen, A., and Bisseling, T. (1994). Root hair deformation activity of nodulation factors and their fate on *Vicia sativa*. *Plant Physiol., 105,* 787-797.

Hirsch, A. M. (1992). Developmental biology of legume nodulation. *New Phytol., 122,* 211-237.

Hirsch, A. M. (1999). Role of lectins (and rhizobial exopolysaccharides) in legume nodulation. *Curr. Opin. Plant Biol., 2,* 320-326.

Hirsch, A. M., Bhuvaneswari, T. V., Torrey, J. G., and Bisseling, T. (1989). Early nodulin genes are induced in alfalfa root outgrowths elicited by auxin transport inhibitors. *Proc. Natl. Acad. Sci. U.S.A., 86,* 1244-1248.

Hirsch, P. (1996). Population dynamics of indigenous and genetically modified rhizobia in the field. *New Phytol.,* 133, 159-171.

Huss-Danell, K., Sverrisson, H., Hahlin, A. S., and Danell, K. (1999). Occurrence of *Alnus*-infective *Frankia* and *Trifolium*-infective *Rhizobium* in circumpolar soils. *Arct. Antarct. Alp. Res., 31,* 400-406.

Jacobsen-Lyon, K., Jensen, E. Ø., Jorgensen, J., Marker, K. A., Peacock, W. J., and Dennis, E. S. (1995). Symbiotic and non-symbiotic hemoglobin genes of *Casuarina glauca*. *Plant Cell, 7,* 213-222.

James, E. K., Sprent, J. I., Sutherland, J. M., McInroy, S. G., and Minchin, F. R. (1992). The structure of nitrogen fixing root nodules on the aquatic mimosoid legume *Neptunia plena*. *Ann. Bot., 69,* 173-180.

James, E. K., and Sprent, J. I. (1999). Development of N_2-fixing nodules on the wetland legume *Lotus uliginosus* exposed to conditions of flooding. *New Phytol., 142,* 219-231.

Joshi, P. A., Caetano-Anolles, G., Graham, E. T., and Gresshoff, P. M. (1991). Ontogeny and ultrastructure of spontaneous nodules in alfalfa (*Medicago sativa*). *Protoplasma 162,* 1-11.

Kijne, J. W. (1992). The *Rhizobium* infection process. In G. Stacey, R. H. Burris, and H. J. Evans (Eds.), *Biological nitrogen fixation* (pp. 349-398). New York, NY: Chapman and Hall.

Kistner, C., and Parniske, M. (2002). Evolution of signal transduction in intracellular symbiosis. *Trends Plant Sci., 7,* 511-518.

Knowlton, S., Berry, A., and Torrey, J. G. (1980). Evidence that associated soil bacteria may influence root hair infection of actinorhizal plants by *Frankia*. *Can. J. Microbiol., 26,* 971-977.

Lancelle, S. A., and Torrey, J. G. (1984). Early development of *Rhizobium*-induced root nodules of *Parasponia rigida*. I. Infection and early nodule initiation. *Protoplasma, 123,* 26-37.

Lancelle, S. A., and Torrey, J. G. (1985). Early development of *Rhizobium*-induced root nodules of *Parasponia rigida*. II. Nodule morphogenesis and symbiotic development. *Can. J. Bot., 63,* 25-35.

Landsmann, J., Llewellyn, D., Dennis, E. S., and Peacock, W. J. (1988). Organ regulated expression of the *Parasponia andersonii* haemoglobin gene in transgenic tobacco plants. *Mol. Gen. Genet., 214,* 68-73.

Laplaze, L., Duhoux, E., Franche, C., Frutz, T., Svistoonoff, S., *et al.* (2000a). *Casuarina glauca* prenodule cells display the same differentiation as the corresponding nodule cells. *Mol. Plant-Microbe Interact., 13,* 107-112.

Laplaze, L., Ribeiro, A., Franche, C., Duhoux, E., Auguy, F., *et al.* (2000b). Characterization of a *Casuarina glauca* nodule-specific subtilisin-like protease gene, a homologue of *Alnus glutinosa ag12*. *Mol. Plant-Microbe Interact., 13,* 113-117.

Lee, C. A. (1996). Pathogenicity islands and the evolution of bacterial pathogens. *Infect. Agents Dis., 5,* 1-7.

Liu, Q., and Berry A. M. (1991a). The infection process and nodule initiation in the *Frankia-Ceanothus* root nodule symbiosis: A structural and histochemical study. *Protoplasma, 163,* 82-92.

Liu, Q., and Berry A. M. (1991b). Localization and characterization of pectic polysaccharides in roots and root nodules of *Ceanothus* spp. during intercellular infection by *Frankia. Protoplasma, 164,* 93-101.

Martin, F., Duplessis, S., Ditengou, F., Lagrange, H., Voiblet, C., and Lapeyrie, F. (2001). Developmental cross talking in the ectomycorrhizal symbiosis: signals and communication genes. *New Phytol., 151, 145-154.*

Mathesius, U. (2001). Flavonoids induced in cells undergoing nodule organogenesis in white clover are regulators of auxin breakdown by peroxidase. *J. Exp. Bot., 52,* 419-426.

Mathesius, U., Bayliss, C., Weinman, J. J., Schlaman, H. R. M., Spaink, H. P., *et al.* (1998b). Flavonoids synthesized in cortical cells during nodule initiation are early developmental markers in white clover. *Mol. Plant-Microbe Interact., 11,* 1223-1232.

Mathesius, U., Schlaman, H. R. M., Spaink, H. P., Sautter, C., Rolfe, B. G., and Djordjevic, M. A. (1998a). Auxin transport inhibition precedes root nodule formation in white clover roots and is regulated by flavonoids and derivatives of chitin oligosaccharides. *Plant J.,* 14, 23-34.

Matsubayashi, Y., Ogawa, M., Morita, A., and Sakagami, Y. (2002). An LRR receptor kinase involved in perception of a peptide plant hormone, phytosulfokine. *Science, 296,* 1470-1472.

Matthysse, A. G., and McMahan, S. (2001). The effect of the *Agrobacterium tumefaciens attR* mutation on attachment and root colonization differs between legumes and other dicots. *Appl. Environ. Microbiol., 67,* 1070-1075.

Maunuksela, L., Zepp, K., Koivula, T., Zeyer, J., Haahtela, K., and Hahn, D. (1999). Analysis of *Frankia* populations in three soils devoid of actinorhizal plants. *FEMS Microbiol. Ecol., 28,* 11-21.

Mergaert, P., van Montagu, M., and Holsters, M. (1997). Molecular mechanisms of Nod factor diversity. *Mol. Microbiol., 25,* 811-817.

Miller, I. M., and Baker, D. D. (1986). The initiation development and structure of root nodules in *Elaeagnus angustifolia* (Elaeagnaceae). *Protoplasma, 128,* 107-119

Minchin, F. R. (1997). Regulation of oxygen diffusion in legume nodules. *Soil Biol. Biochem., 29,* 881-888.

Mirza, M. S., Hahn, D., and Akkermans, A. D. L. (1992). Isolation and characterization of *Frankia* strains from *Coriaria nepalensis. Syst. Appl. Microbiol., 15, 289-295.*

Mirza, M. S., Pawlowski, K., Hafeez, F. Y., Chaudhary, A. H., and Akkermans, A. D. L. (1994). Ultrastructure of the endophyte and localization of *nifH* transcripts in root nodules of *Coriaria nepalensis* Wall. by *in situ* hybridization. *New Phytol., 126,* 131-136.

Mort, A. J., and Grover, P. B., Jr. (1988). Characterization of root hair cell walls as potential barriers to the infection of plants by bacteria. The carbohydrate component. *Plant Physiol., 86,* 638-641.

Moulin, L., Munive, A., Dreyfus, B., and Boivin-Masson, C. (2001). Nodulation of legumes by members of the β-subclass of Proteobacteria. *Nature, 411,* 948-950.

Naisbitt, T., James, E. K., and Sprent, J. I. (1992). The evolutionary significance of the legume genus *Chamaecrista,* as determined by nodule structure. *New Phytol., 122,* 487-492.

Ndoye, I., de Billy, F., Vasse, J., Dreyfus B., and Truchet, G. (1994). Root nodulation of *Sesbania rostrata. J. Bacteriol., 176,* 1060-1068.

Newcomb, W. (1981). Nodule morphogenesis and differentiation. *Int. Rev. Cytol., 13 (Suppl.),* S247-S297.

Newcomb, W., and Pankhurst, C. E. (1982). Fine structure of actinorhizal nodules of *Coriaria arborea* (Coriariaceae). *New Zeal. J. Bot., 20,* 93-103.

Nishimura, R., Ohmori, M., Fujita, H., and Kawaguchi, M. (2002). A *Lotus* basic leucine zipper protein with a RING-finger motif negatively regulates the developmental program of nodulation. *Proc. Natl. Acad. Sci. U.S.A., 99,* 15206-15210.

Normand, P., and Lalonde, M. (1982). Evaluation of *Frankia* strains isolated from provenances of two *Alnus* species. *Can. J. Microbiol., 28,* 1133-1142.

Okubara, P. A., Pawlowski, K., Murphy, T. M., and Berry, A. M. (1999). Symbiotic root nodules of the actinorhizal plant *Datisca glomerata* express rubisco activase mRNA. *Plant Physiol., 120,* 411-420.

Parniske, M. (2000). Intracellular accommodation of microbes by plants: A common developmental program for symbiosis and disease? *Curr. Opin. Plant Biol., 3,* 320-328.

Passarinho, P. A., van Hengel, A. J., Fransz, P. F., and de Vries, S. C. (2001). Expression pattern of the *Arabidopsis thaliana AtEP3/AtchitIV* endochitinase gene. *Planta, 212,* 556-567.

Pawlowski, K., Swensen, S., Guan, C., Hadri, A.-E., Berry, A. M., and Bisseling, T. (2003). Distinct patterns of symbiosis-related gene expression in actinorhizal nodules from different plant families. *Mol. Plant-Microbe Interact., 16,* 796-807.

Perret, X., Staehelin, C., and Broughton, W. J. (2000). Molecular basis of symbiotic promiscuity. *Microbiol. Mol. Biol. Rev., 64,* 180-201.

Pommer, E. H. (1956). Beiträge zur anatomie und biologie der wurzelknöllchen von *Alnus glutinosa* (L.) Gaertn. *Flora, 143,* 603-634.

Pueppke, S. G. (1996). The genetic and biochemical basis of nodulation of legumes by rhizobia. *Crit. Rev. Biotechnol. 16,* 1-51.

Pueppke, S. G., and Broughton, W. J. (1999). *Rhizobium sp.* strain NGR234 and *R. fredii* USDA257 share exceptionally broad, nested host ranges. *Mol. Plant-Microbe Interact., 12,* 293-318.

Racette, S., and Torrey, J. G. (1989). Root nodule initiation in *Gymnostoma* (Casuarinaceae) and *Shephardia* (Elaeagnaceae) induced by *Frankia* strain HFPGpI1. *Can. J. Bot., 67,* 2873-2879.

Ramirez-Saad, H., Janse, J. D., and Akkermans, A. D. L. (1998). Root nodules of *Ceanothus caeruleus* contain both the N_2-fixing *Frankia* endophyte and a phylogenetically related Nod^-/Fix^- actinomycete. *Can. J. Microbiol., 44,* 140-148.

Reboutier, D., Bianchi, M., Brault, M., Roux, C., Dauphin, A., *et al.* (2002). The indolic compound hypaphorine produced by ectomycorrhizal fungus interferes with auxin action and evokes early responses in nonhost *Arabidopsis thaliana. Mol. Plant-Microbe Interact., 15,* 932-938.

Remy, W., Taylor, T. N., Hass, H., and Kerp, H. (1994). Four hundred-million-year-old vesicular arbuscular mycorrhizae. *Proc. Natl. Acad. Sci. U.S.A.*, 91, 11841–11843.

Ribeiro, A., Akkermans, A. D. L., van Kammen, A., Bisseling, T., and Pawlowski, K. (1995). A nodule-specific gene encoding a subtilisin-like protease is expressed in early stages of actinorhizal nodule development. *Plant Cell, 7,* 785-794.

Rivas, R., Velázquez, E., Willems, A., Vizcaíno, N., Subba-Rao, N. S., *et al.* (2002). A new species of *Devosia* that forms a unique nitrogen-fixing root-nodule symbiosis with the aquatic legume Neptunia natans. *Appl. Environ. Microbiol., 68,* 5217-5222.

Rodriguez-Barrueco, C., and Bermudez de Castro, F. (1973). Cytokinin induced pseudonodules on *Alnus glutinosa. Physiol. Plant., 29, 277-280.*

Rolfe, B. G., and Shine, J. (1984). *Rhizobium*-Leguminosae symbiosis: The bacterial point of view. In D. P. S. Verma and T. H. Hohn (Eds.), *Genes involved in microbe-plant interactions* (pp. 95-128). New York, NY: Springer-Verlag.

Schaede, R. (1940). Über die symbionten in den knöllchen der erle und des sanddornes und die cytologischen verhältnisse in ihnen. *Planta, 19,* 389-416.

Scheer, J. M., and Ryan, Jr., C. A. (2002). The systemin receptor SR160 from *Lycopersicon peruvianum* is a member of the LRR receptor kinase family. *Proc. Natl. Acad. Sci. U.S.A., 99,* 9585-9590.

Schmidt, J., Röhrig, H., John, M., Wieneke, U., Stacey, G., *et al.* (1993). Alteration of plant growth and development by *Rhizobium nodA* and *nodB* genes involved in the synthesis of oligosaccharide signal molecules *Plant J., 4,* 651-658.

Searle, I. R., Men, A. E., Laniya, T. S., Buzas, D. M., Iturbe-Ormaetxe, I., *et al.* (2002). Long-distance signaling in nodulation directed by a CLAVATA1-like receptor kinase. *Science, 299,* 109-112.

Sequerra, J., Capellano, A., Faure-Raynard, M., and Moiroud, A (1994). Root hair infection process and myconodules formation of *Alnus incana* by *Penicillium nodositatum. Can. J. Bot., 72,* 955-962.

Sequerra, J., Capellano, A., Gianinazzi-Pearson, V., and Moiroud, A. (1995). Ultrastructure of cortical root cells of *Alnus incana* infected by *Penicillium nodositatum. New Phytol., 130,* 545-555.

Shah, K., Gadella, Jr., T. W., van Erp, H., Hecht, V., and de Vries, S. C. (2001). Subcellular localization and oligomerization of the *Arabidopsis thaliana* somatic embryogenesis receptor kinase 1 protein. *J. Mol. Biol., 309,* 641–655.

Silvester, W. B., and Harris, S. L. (1989). Nodule structure and nitrogenase activity of *Coriaria arborea* in response to varying oxygen partial pressure. *Plant Soil, 118,* 97-110.

Silvester, W. B., Whitbeck, J., Silvester, J. K., and Torrey, J. G. (1988). Growth, nodule, morphology and nitrogenase activity of *Myrica gale* grown with roots at various oxygen levels. *Can. J. Bot., 66,* 1762-1771.

Soltis, D. E., Soltis, P. S., Morgan, D. R., Swensen, S. M., Mullin, B. C., *et al.* (1995). Chloroplast gene sequence data suggest a single origin of the predisposition for symbiotic nitrogen fixation in angiosperms. *Proc. Natl. Acad. Sci. U.S.A., 92,* 2647-2651.

Spaink, H. P. (2000). Root nodulation and infection factors produced by rhizobial bacteria. *Annu. Rev. Microbiol., 54,* 257-288.

Sprent, J. I. (1995). Legume trees and shrubs in the tropics: N_2 fixation in perspective. *Soil Biol. Biochem., 27,* 401-407.

Sprent, J. I. (2001). *Nodulation of legumes*. London, UK: Kew Royal Botanic Gardens.

Sprent, J. I., and Scott, R. (1979). The nitrogen economy of *Myrica gale* and its possible significance for the afforestation of peat soils. In J. C. Gordon, C. T. Wheeler, and D. A. Perry (Eds.), *Symbiotic nitrogen fixation in the management of temperate forests* (pp. 234-242). Corvallis, OR: Forestry Research Laboratory, Oregon State University.

Stokkermans, T. J. W., and Peters, N. K. (1994). *Bradyrhizobium elkanii* lipooligosaccharide signals induce complete nodule structures on *Glycine soja* Siebold et Zucc. *Planta, 193,* 413-420.

Stougaard, J. (2001). Genetics and genomics of root symbiosis. *Curr. Opin. Plant Biol., 4,* 328-335.

Stracke, S., Kistner, C., Yoshida, S., Mulder, L., Sato, S., *et al.* (2002). A plant receptor-like kinase required for both bacterial and fungal symbiosis. *Nature, 417,* 959-962.

Subba-Rao, N. S., Mateos, P. F., Baker, D., Pankratz, H. S., Palma, J., *et al.* (1995). The unique root-nodule symbiosis between *Rhizobium* and the aquatic legume, *Neptunia natans* (L. f.) Druce. *Planta, 196,* 311-320.

Sullivan, J. T., and Ronson, C. W. (1998). Evolution of rhizobia by acquisition of a 500-kb symbiosis island that integrates into a phe-tRNA gene. *Proc. Natl. Acad. Sci. U.S.A., 95,* 5145-5149.

Sullivan, J. T., Eardly, B. D., van Berkum, P., and Ronson, C. W. (1996). Four unnamed species of nonsymbiotic rhizobia isolated from the rhizosphere of *Lotus corniculatus. Appl. Environ. Microbiol., 62,* 2818-2825.

Sullivan, J. T., Patrick, H. N., Lowther, W. L., Scott, D. B., and Ronson, C. W. (1995). Nodulating strains of *Rhizobium loti* arise through chromosomal symbiotic gene transfer in the environment. *Proc. Natl. Acad. Sci. U.S.A., 92,* 8985-8989.

Svistoonoff, S., Laplaze, L., Auguy, F., Runions, J., Duponnois, R., *et al.* (2003a). *cg12* expression is specifically linked to infection of root hairs and cortical cells during *Casuarina glauca* and *Allocasuarina verticillata* actinorhizal nodule development. *Mol. Plant-Microbe Interact., 16,* 600-607.

Swensen, S. M. (1996). The evolution of actinorhizal symbioses - evidence for multiple origins of the symbiotic association. *Am. J. Bot., 83,* 1503-1512.

Tagu, D., Lapeyrie, F., and Martin, F. (2002). The ectomycorrhizal symbiosis: genetics and development. *Plant Soil, 244,* 97-105.

Timmers, A. C. J., Soupène, E., Auriac, M.-C., de Billy, F., Vasse, J., *et al.* (2000). Saprophytic intracellular rhizobia in alfalfa nodules. *Mol. Plant-Microbe Interact., 13,* 1204-1213.

Tjepkema, J. D., and Cartica, R. J. (1982). Diffusion limitation of oxygen uptake and nitrogenase activity in the root nodules of *Parasponia rigida* Merr. and Perry. *Plant Physiol., 69,* 728-733.

Torrey, J. G. (1976). Initiation and development of root nodules of *Casuarina* (Casuarinaceae). *Am. J. Bot., 63,* 335–344.

Torrey, J. G. (1990). Cross-inoculation groups within *Frankia*. In C. R. Schwintzer and J. D. Tjepkema (Eds.), *The biology of* Frankia *and actinorhizal plants* (pp. 83-106). New York, NY: Academic Press, Inc.

Torrey, J. G., and Callaham, D. (1979). Early nodule development in *Myrica gale. Bot. Gaz., 140 (suppl.),* S10-S14.

Trinick, M. J., and Galbraith, J. (1980). The *Rhizobium* requirements of the nonlegume *Parasponia andersonii* in relationship to the cross inoculation group concept of legumes. *New Phytol., 86,* 17-26.

Trinick, M. J., Goodchild, D. J., and Miller, C. (1989). Localization of bacteria and hemoglobin in root nodules of *Parasponia andersonii* containing both *Bradyrhizobium* strains and *Rhizobium leguminosarum* biovar *trifolii. Appl. Environ. Microbiol., 55,* 2046-2055.

Trotochaud, A. E., Jeong, S., and Clark, S. E. (2000). CLAVATA3, a multimeric ligand for the CLAVATA1 receptor-kinase. *Science, 289,* 613-617.

Truchet, G., Roche, P., Lerouge, P., Vasse, J., Camut, S., *et al.* (1991). Sulfated lipo-oligosaccharide signals of *Rhizobium meliloti* elicit root nodule organogenesis in alfalfa. *Nature, 351,* 670-673.

van Berkum, P., and Eardly, B. D. (2002). The aquatic budding bacterium *Blastobacter denitrificans* is a nitrogen-fixing symbiont of *Aeschynomene indica. Appl. Environ. Microbiol., 68,* 1132-1136.

van Brussel, A. A. N., Bakhuizen, R., van Spronsen, P. C., Spaink, H. P., Tak, T., and Lugtenberg, B. J. J. (1992). Induction of pre-infection thread structures in the leguminous host plant by mitogenic lipo-oligosaccharides of *Rhizobium*. *Science, 257,* 70-71.

van Spronsen, P. C., Bakhuizen, R., van Brussel, A. A., and Kijne, J. W. (1994). Cell wall degradation during infection thread formation by the root nodule bacterium *Rhizobium leguminosarum* is a two-step process. *Eur. J. Cell Biol., 64,* 88-94.

Vandamme, P., Goris, J., Chen, W.-M., de Vos, P., and Willems, A. (2002). *Burkholderia tuberum sp. nov.* and *Burkholderia phymatum sp. nov.,* nodulate the roots of tropical legumes. *Syst. Appl. Microbiol., 25,* 507-512.

Vasse, J., de Billy, F., Camut, S., and Truchet, G. (1990). Correlation between ultrastructural differentiation of bacteroids and nitrogen fixation in alfalfa nodules. *J. Bacteriol., 172,* 4295-4306.

Vega-Hernández, M. C., Perez-Galdona, R., Dazzo, F. B., Jarabo-Lorenzo, A., Alfayate, M. C., and Leon-Barrios, M. (2001). Novel infection process in the indeterminate root nodule symbiosis between *Chamaecytisus proliferus* (tagasaste) and *Bradyrhizobium* sp. *New Phytol., 150,* 707-721.

Vieweg, M. F., Frühling, M., Quandt, H.-J., Heim, U., Bäumlein, H., *et al.* (2004). The promoter of the *Vicia faba* L. leghemoglobin gene *VfLb29* is specifically activated in the infected cells of root nodules and in the arbuscule-containing cells of mycorrhizal roots from different legume and non-legume plants. *Mol. Plant-Microbe Interact., 17,* 62-69.

Vikman, P.-Å. (1992). The symbiotic vesicle is a major site for respiration in *Frankia* from *Alnus incana* root nodules. *Can. J. Microbiol., 38,* 779-784.

Wolters, D. J., van Dijk, C., and Akkermans, A. D. L., and Woldendorp, J. W. (1999). Ineffective *Frankia* and host resistance in natural populations of *Alnus glutinosa* (L.) Gaertn. *Acta Oecol., 20,* 71-79.

Woomer, P., Singleton, P. W., and Bohlool, B. B. (1988). Ecological indicators of native rhizobia in tropical soils. *Appl. Environ. Microbiol., 54,* 1112-1116.

Yang, W.-C., de Blank, C., Meskiene, I., Hirt, H., Bakker, J., *et al.* (1994). *Rhizobium* Nod factors reactivate the cell cycle during infection and nodule primordium formation, but the cycle is only completed in primordium formation. *Plant Cell, 6,* 1415-1426.

Young, J. M., Kuykendall, L. D., Martinez-Romero, E., Kerr, A., and Sawada, H. (2001). A revision of *Rhizobium* Frank 1889, with an emended description of the genus, and the inclusion of all species of *Agrobacterium* Conn 1942 and *Allorhizobium undicola* de Lajudie et al. 1998 as new combinations: *Rhizobium radiobacter, R. rhizogenes, R. rubi, R. undicola and R. vitis. Int. J. Syst. Evol. Microbiol., 51,* 89-103.

Zhong, R., Kays, S. J., Schroeder, B. P., and Ye, Z. H. (2002). Mutation of a chitinase-like gene causes ectopic deposition of lignin, aberrant cell shapes, and overproduction of ethylene. *Plant Cell, 14,* 165-179.

Chapter 11

PROSPECTS FOR THE STUDY OF A UBIQUITOUS ACTINOMYCETE, *FRANKIA*, AND ITS HOST PLANTS

P. NORMAND[1] AND B. C. MULLIN[2]
[1]*Department of Microbial Ecology CNRS, University Claude Bernard, Lyon, France.*
[2]*Department of Botany, The University of Tennessee, Knoxville, TN 37996-1100, U.S.A.*

1. INTRODUCTION

Since the publication of "The Biology of *Frankia* and Actinorhizal Plants" in 1990 (Schwintzer and Tjepkema, 1990), progress in the field of actinorhizal symbiosis has been made on a number of fronts. The development and application of molecular tools (see chapter 2 of this volume) has made it possible to assess *Frankia*'s presence and diversity not only in isolated *Frankia*, but also in uncultured strains and in the environment outside the plant. These techniques have also led to the recognition of four distinct clades within the genus *Frankia* (Normand *et al.*, 1996) and the identification of *Acidothermus* (and not *Geodermatophilus*) as the nearest neighbor of *Frankia* (Maréchal *et al.*, 2000). These two organisms share a common soil habitat and the uncommon morphological feature of multilocular sporangia. The application of molecular analysis to the study of relationships among host plants has led to the recognition of a clade within the Rosid I clade that appears to be predisposed to root-nodule symbioses (Soltis *et al.*, 1995; see chapters 4 and 10 in this volume). Further phylogenetic analysis of actinorhizal genera (Swensen, 1996) suggests that symbiosis is likely to have arisen independently at least four times among actinorhizal genera. Both host and microsymbiont phylogenetic trees are, for the most part, incongruent and do not support co-evolution of host and microsymbiont. In the Casuarinaceae, for instance, *Casuarina* spp. are nodulated by Group-1 strains, whereas *Gymnostoma* spp. are nodulated by Group-3 strains

K. Pawlowski and W. E. Newton (eds.), Nitrogen-fixing Actinorhizal Symbioses, 289–303.
© 2008 *Springer*.

(Navarro *et al.,* 1997); moreover, the closely related Group-4 *Frankia* strains nodulate distantly related host plants (Normand *et al.,* 1996).

Some progress has been made on understanding the physiology of *Frankia* and its host plants, both alone and in symbiosis (see chapter 7 of this volume), but little is known about the exchange of metabolites either between *Frankia* hyphae and vesicles or between *Frankia* and its host plants in symbiosis. One major advance has shown that the vesicle envelope is composed of hopanoid lipids (Berry *et al.,* 1993). Another advance has identified a number of host-plant genes that are expressed only in nodules (see chapter 9 in this volume), but the function and precise role in symbiosis has not yet been established for any of them.

2. DEVELOPMENT AND USE OF MOLECULAR TOOLS FOR GENETIC ANALYSIS OF SYMBIOSIS

Much of the research to date on actinorhizal symbioses has been descriptive in nature. More significant advances in understanding the biology of the symbiosis will only come from research with a stronger experimental component. An enormous impediment to moving beyond descriptive work is the lack of a genetic system for *Frankia*. For example, hemoglobins have recently been discovered in *Frankia* (Beckwith *et al.,* 2002), but their contribution to nodule physiology has not been established and will be difficult to establish in the absence of mutants. Also critical to advancing an understanding of symbiosis is the further development of molecular tools for the study of host-plant genes.

2.1. Frankia

2.1.1. Genome Analysis

The complete genome sequencing of two *Frankia* strains, ACN14a, an *Alnus*-infective isolate that is considered by us to be a model for genetics and physiology, and CcI3, a *Casuarina*-infective isolate, will provide both a wealth of new information about the *Frankia* genome as well as the information necessary for the design of probes and primers. With sequence data in hand, the genome of *Frankia* will be analyzed from several major points of view.

(i). Inter-strain Comparison. Given the ecological importance of *Frankia,* it is likely that several strains will be sequenced, ideally one from each of the four existing clades or clusters. The resulting data will allow comparisons to be made in an effort to correlate host spectrum with the presence and absence of specific genes and gene regions.

(ii). The family tree of *Frankia.* A surprise here is that its closest phylogenetic neighbor tends toward an extremophylic mode of existence. Close relatives include *Acidothermus*, *Sporichthya*, and *Geodermatophilus*, the most prominent of which is *Acidothermus*, a genus that thrives in acidic thermal environments and has cellulase activity (Mohagheghi *et al.,* 1986). The emphasis put on sequencing genomes of extremophiles should result in the genomic sequence of *Acidothermus* becoming available in the near future. Of particular interest will be the analysis of gene regions that encode two physiological traits shared by *Acidothermus* and *Frankia*, namely, cellulase activity and hopanoid biosynthesis. Hopanoids are prominent in extremophiles, such as *Alicyclobacillus* and *Zymomonas* (Vincent *et al.*, 2003), and are found in *Frankia* (Berry *et al.*, 1993) and *Acidothermus* (Maréchal *et al.*, 2000).

(iii). Comparisons with the other Root-symbiotic Microorganisms. Comparisons of *Frankia* with *Sinorhizobium meliloti* and *Bradyrhizobium japonicum* should indicate the importance, regulation, and position of genes related to symbiosis, such as *nif* (nitrogen fixation) and *nod* (nodulation) genes, and genes involved in coping with stress exerted by the host plant on its symbionts, such as *sod* (superoxide dismutase) genes (Santos *et al.*, 2001). Because *nod* genes have not been detected in *Frankia* (Ceremonie *et al.*, 1998), genomic sequencing will reveal whether indeed *nod* homologs are present.

(iv). Comparisons with Producers of Biologically Active Molecules. The genome of *Frankia* will be compared to that of *Streptomyces avermitilis* (Omura *et al.*, 2001) and *S. coelicolor* (Bentley *et al.*, 2002), two species of a genus that produces several bioactive molecules. It will also be compared to bacteria, such as *Pseudomonas putida*, which are capable of metabolizing aromatic compounds (Nelson *et al.*, 2002). *Frankia* is an actinomycete, a group of microorganisms whose members are recognized as being prolific producers of antibiotics. *Frankia* pro duce a benzoanthracene quinone (Gerber and Lechevalier, 1984), Frankiamide (a macrocycle) (Klika *et al.*, 2001), and phenylacetic acid, a compound described as antifungal and antibacterial (Hammad *et al.*, 2003). All of these molecules are based on aromatic rings, and the ability to metabolize aromatic compounds will be correlated with the effect of some aromatic compounds on *Frankia* morphology (Perradin *et al.*, 1983). A purpose of these genome comparisons is to look for the presence of predicted or unique metabolic pathways. Study of the chromosome should also help determine if *Frankia*'s genome is either circular, as for most bacteria, or linear, as in *Streptomyces* with unstable amplifiable regions in the extremities (Leblond and Decaris, 1998).

(v). The Codon-usage Pattern of *Frankia*. It is likely that the genome of every living organism has evolved a codon-usage pattern that contributes to optimal gene-expression levels in the environments in which that organism finds itself. In many bacteria, the highly expressed protein synthesis and processing genes exhibit codon-use frequencies that differ from the average gene in the genome. Other genes may be predicted to be highly expressed (PHX) on the basis of a similar a codon-usage pattern and have the potential to provide insight into the life history of an organism (Karlin and Mrázek, 2000). Take for example, the bacteria *Listeria*. Among the PHX genes identified on the basis of codon usage are genes necessary for pathogenicity in animals and for the production of cellulase, which is required for growth as a saprophyte. The high level of expression of these genes was confirmed experimentally and was consistent with the known reservoirs of *Listeria*, e.g., in ruminant animals and in silage (J. Mrázek, personal communication). It may be possible to apply codon-use frequency as a predictor of the level of expression of genes within the *Frankia* genome, a procedure which would allow prediction of such things as the relative importance of various carbon sources, metabolic pathways, etc., in the life history of *Frankia*, and perhaps even the relative importance of a symbiotic *versus* saprophytic life style. This approach will also be correlated with comparative genomics to understand the origin of genetic regions present in one neighbor and absent in the other. Codon usage, as well as the presence of insertion sequences, will permit an assessment of the importance of lateral transfer in the evolution of the *Frankia* genome.

(vi). Predictions of Metabolite Exchange Sites. With genomic sequence in hand, it may be possible to predict locations in metabolic pathways where metabolite exchange is occurring between symbiont and host. This could result from either the discovery of genes that code for specific export (or import) proteins or by the absence of (or predicted low expression of) a gene for an enzyme in a particular pathway. A practical result from this kind of analysis will be the identification of PHX genes that might reveal the need for a particular metabolite(s), which could lead to improved culturing methods and thus make the genetic manipulation of *Frankia* an easier prospect. It should also be possible to detect the presence of specific restriction-modification systems, which are known to have hindered transformation of other taxa (Koksharova and Wolk, 2002), and, therefore, could lead to the development of vectors devoid of sequences targeted by these systems.

2.1.2. Proteomics

Proteomics is the study of the protein-expression patterns of cells, tissues, and organisms and how protein patterns are altered in response to changing environmental conditions. The genome sequence of *Frankia* can reveal its genetic potential but it does not provide direct evidence of protein expression. Such fundamental knowledge is essential, if there is to be any hope of understanding the physiological responses that occur in *Frankia* upon establishment of symbiosis.

In a study designed to determine the earliest detectable changes in protein expression in *Frankia* strain ACN14a in response to root exudates of its *Alnus glutinosa* symbiotic partner, five up-regulated proteins were detected out of a few hundred unchanged proteins (Hammad *et al.,* 2001). Terminal sequencing of these five proteins identified them as the stress-response proteins, superoxide dismutase (SodF), heat shock protein (Hsp), tellurite-resistance protein (Ter), and bacterioferritin comigrating protein (Bcp), all of which indicated that, in this case, the symbiotic partner entered into its relationship armed and did not lower its defenses immediately. This kind of response has also been shown in other plant symbiotic associations. A *Sinorhizobium meliloti sod* mutant, for instance, cannot establish functional nodules (Santos *et al.,* 2001), indicating that, for some reason, either a stress response may be necessary for proper communication with a host plant or in order to become a symbiont, a bacterium must be able to withstand an array of plant-defense responses to which most pathogens will succumb.

A proteomics approach can be extended to different physiological situations, in particular, to understand what happens in various stages of symbiosis. *Frankia* diazovesicles can be isolated and compared to saprophytic cells at different time periods and different growth conditions to see how the genetic program leading to diazovesicle formation is implemented. Different variations on this theme can be considered, for instance, how the same strain modifies its developmental program in response to its different hosts, *e.g.*, to different *Alnus* species or *Myrica* species, and more specifically how a given strain will react to different compounds synthesized by a single host plant.

2.1.3. Genetic Manipulation of Frankia

Classic approaches to genetic manipulation, which work on other microbes and are based either on transformation with naked plasmid DNA molecules or on conjugation with broad host-range plasmids, have been unsuccessful with *Frankia.* The presence of plasmids in various *Frankia* strains cannot be predicted by phylogenetic analysis because plasmids are present in some strains and absent from phylogenetically related strains, suggesting the existence of a transfer mechanism. *Frankia* plasmids range in size from 8kb to 190kb (Simonet *et al.,* 1986). These plasmids have been characterized and three of them sequenced completely (John *et al.,* 2001; Lavire *et al.,* 2001; Xu *et al.,* 2002). The 8-kb plasmid present in several strains has a classic structure with Rep (plasmid replication) and Par (partitioning for stable plasmid maintenance) genes. These native plasmids were used to construct hybrid plasmids that were tested without success for their ability to transform *Frankia* or *Streptomyces*. All of these plasmids were shown to enter cells by electroporation (Cournoyer and Normand, 1992), but were not maintained in the culture.

Genome-sequence analysis should help in the development of vectors and transponsons by revealing: (i) the existence of restriction-modification systems; (ii) the importance of lateral gene transfer; (iii) the presence of insertion sequences, transposons, lysogenic phages and plasmids; and (iv) the presence of antibiotic-resistance genes. In parallel, basic work in physiology should be continued using different conditions known to promote DNA penetration and reduce DNA modification reactions. Identification of a *recA* mutant might make genetic manipulation more feasible.

2.2. Genetic Manipulation of Host Plants

2.2.1. Plant Transformation

Perhaps the most significant recent progress in actinorhizal plant research has been the development of transformation systems for actinorhizal trees (see chapter 9 in this volume). This accomplishment is celebrated in the beautiful prize-winning image of a transgenic *Allocasuarina* shoot in cross section as captured by Sergio Svistoonoff and featured in Nature Reviews Molecular Cell Biology (2003). In this image, the shoot is expressing *GFP*, driven by the CaMV 35S promoter, in the phloem. Using *Casuarina glauca* and *Allocasuarina verticillata*, and *Agrobacterium tumefaciens* and *A. rhizogenes*, the demonstrated transformation has allowed the study of promoter function through the use of promoter-reporter constructs. Promoter analysis in transgenic *C. glauca,* as well as in the closely related species *A. verticillata*, revealed that transgenic nodules expressing *GUS* driven by the

CaMV-35S promoter, exhibited GUS expression only in the vascular and cortical tissues and not in *Frankia*-infected cells (Laplaze *et al.,* 2000; Smouni *et al.,* 2002). In most plants, the CaMV-35S promoter acts as a constitutive promoter, so it is a significant finding that this promoter cannot be used for studies in actinorhizal nodules if expression in infected cells is desired.

Datisca glomerata is the only herbaceous actinorhizal species and as such may present the best hope for genetic analysis of host-plant functions in symbiosis. It sets seed within a few months of planting and it is, therefore, possible to make crosses and carry out segregation analysis within a reasonable time, and also to conduct mutagenesis and screening for symbiotic mutants. There are, as yet, no reports of a pure culture *Frankia* isolate from *Datisca*; however, analysis of nodule inhabitants by PCR amplification and sequencing of rRNA (Normand *et al.,* 1996) has demonstrated that *Datisca* is nodulated by Group II strains (see Figure 5 of chapter 4 in this volume). To date, there are no Group II strains in culture and it is not known how these strains differ from the more easily isolated *Frankia* strains from Group I and III. Efforts must continue to obtain *Datisca* isolates. Meanwhile, the advantages of using of *Datisca* as a model system for studying the intercellular infection pathway of actinorhizal nodule development must be weighed against the disadvantages of not having a well-characterized pure culture isolate.

2.2.2. Host-plant Gene Expression in Response to Nodulation

There are 24 genes known to be expressed in actinorhizal nodules and their locations within the nodule are also known (see chapter 9 in this volume). These genes were identified by numerous methods, including differential screening of nodule and root libraries, subtractive hybridization, heterologous probing, PCR, and sequencing of expressed sequence tags (ESTs). It is likely that, in the near future, hundreds more EST sequences will be available from a variety of actinorhizal plants, and putative nodule-specific ESTs will be determined by digital subtraction of root ESTs. With completion of both the *Arabidopsis* and rice genome sequences, and the availability of sequences from *Medicago* and *Lotus*, it is likely that many of the actinorhizal ESTs will be identified by BLAST searches. However, the identification of an EST must be followed by functional analyses in order to understand its role in nodule initiation, development, and function. ESTs will be of value in developing hypotheses, but the precise roles for the proteins that they encode will need to be determined by genetic and biochemical methods.

Another potential line of research will be the study of *Alnus* in symbiosis with an ectomycorhizal fungus. ESTs of *Alnus* obtained upon establishment of symbiosis with *Frankia* will be compared to those obtained upon establishment of symbiosis with an ectomycorhizal fungus. It will also be of interest to study the physiology of mutants, eventually mutants targeted toward homologues of those recently described in *Medicago* (Catoira *et al*., 2000).

2.2.3. Gene-silencing Systems in Host Plants

Chemical, fast neutron, T-DNA, and transposon-induced plant mutants have been critical in studying the role of specific genes in nodule formation and nitrogen fixation in legumes. Where these techniques have not worked, then decreasing the levels of the relevant gene product by transformation with either antisense or sense (to induce co-suppression) constructs has been useful. In such studies, the choice of the promoter driving the construct is critical as was demonstrated by two reports of antisense suppression of NADH-dependent glutamate synthase (NADH-GOGAT) in *Medicago sativa* L. (Cordoba *et al.,* 2003; Schoenbeck, 2000). An alternative to antisense and co-suppression as a means of post-transcriptional gene silencing (PTGS) is the use of small interfering RNA (siRNA or RNAi). The siRNA technologies are gaining in popularity because they result in higher and more predictable levels of silencing than either antisense or co-suppression, although all three methods are thought to be based on the same fundamental principles (reviewed by Waterhouse and Helliwell, 2003).

A variety of siRNA vectors have been constructed and are available for use. One such vector, pHANNIBAL, provides a way for a gene or gene fragment to be directionally cloned to make sense and antisense arms, separated by an intron (Wesley *et al.,* 2001). The arms form a hairpin loop that is optimized for effective gene silencing in plants. The intron hairpin loop can be sub-cloned into any binary vector and expressed in transgenic plants driven by the promoter of choice. More recently, it has been reported that the hairpin loop, which initiates the silencing response, can be heterologous to the sequence to be silenced, making it possible to clone any sequence of interest into a vector that already contains a generic stem loop (Brummell *et al.,* 2003). Furthermore, siRNA silencing is systemic, spreading from cell to cell, thus making it possible to achieve silencing in cells and tissues for which a promoter may not be available (Klahre *et al.,* 2002). Perhaps even more important for any actinorhizal species for which transformation systems have not been developed, is the report that silencing may be achieved biolistically (Klahre *et al.,* 2002). This would allow PTGS to be initiated at any stage in the life cycle of a plant without the need for stable transformation. The introduced PTGS signal is propagated in the plant and can spread systemically.

3. PROSPECTS FOR FUTURE UTILIZATION OF ACTINORHIZAL PLANTS

In addition to the uses of actinorhizal plants discussed in chapter 8 of this volume, actinorhizal plants may have potential for wider use in phytoremediation. There is a tremendous need for the development of efficient and cost-effective strategies for remediation of metal-contaminated sites around the world. No single plant species has been identified as the ideal candidate phytoremediator, although a number of plants have been identified which express individual components of an ideal phytoremediator (Clemens *et al.,* 2002).

Actinorhizal plants are thought to have evolved as pioneer species, able to thrive on disturbed sites, such as those resulting from glaciations or volcanic eruptions. They are able to thrive in such environments in part due to their association with *Frankia*. These same disturbed sites, where actinorhizals may have evolved, in addition to being low in both soil-N and organic matter, are likely to have contained higher natural levels of biologically available heavy metals than would be expected in mature soils. It is, therefore, likely that along with nitrogen fixation, these plants evolved as yet unidentified mechanisms to ameliorate metal toxicity. Both microcosm (Kairesalo *et al.,* 1999) and field studies (Dickinson, 2000) have shown that *A. glutinosa* is capable of accumulating heavy metals in foliar tissue within the range of 600-800μg metal/gm dry wt. with no obvious detrimental effects. Some *Frankia* strains are tolerant of millimolar levels of Ni (Wheeler *et al.,* 2001) as well as of other metals (Richards *et al.,* 2002), which is not surprising considering that host and symbiont may have evolved in the same metal-rich environments. Actinorhizal *Alnus* is a fast-growing, long-living, high biomass-producing phreatophyte with an extensive fibrous root system and has already been recognized as having moderate phytoremediation potential. It may be possible to enhance the phytoremediation potential of this and other actinorhizal species by either selection for or engineering of desirable traits, such as the expression of metal-binding proteins as described below.

As described in chapter of 9 of this volume, *ag*Nt84 and *ag*164 are two glycine- and histidine-rich nodulin genes expressed in the infection zone of *Alnus* nodules (Pawlowski *et al.,* 1997). Expression of the proteins encoded by these genes in *E. coli* allowed them to be characterized. They were found to bind multiple atoms of Ni^{2+}, Zn^{2+}, Co^{2+}, Cu^{2+}, Cd^{2+} and Hg^{2+} (Gupta *et al.,* 2002). Although their function in symbioses is not known, these proteins and their homologues (or paralogues) in other actinorhizal plants, are strong candidates for possible inclusion into phytoremediation strategies. Named metallohistins due to their ability to bind metals through histidine residues, these proteins are small and stable and are targeted extra-cellularly with apparent high affinity for the cell-wall matrix. They have a high heavy metal-binding capacity and the ability to bind metals both in solution and by exchange on a solid matrix. When either expressed at elevated levels in actinorhizal plants or expressed in heterologous systems, their potential is high for contributing to the advancement of phytoremediation technology. In summary, there must be a continued and expanded effort to explore uses for actinorhizal plants in agroforestry, land reclamation, and phytoremediation.

4. QUESTIONS STILL TO BE ADDRESSED

4.1. Evolutionary Questions

4.1.1. What is the Genetic Basis that Predisposes Plants in the Rosid 1 Clade to Enter into a Nitrogen-fixing Root-nodule Symbiosis?

All known plant families with N_2-fixing root-nodule symbioses group together on the basis of molecular data into a single N_2-fixing clade within Rosid 1. This clade is interspersed with non-N_2-fixing plants (see Figure 1 in chapter 4 of this volume). Within this clade, symbiosis arose more than once, at least once with rhizobia and once with *Frankia*, and likely arose multiple times (see chapters 4 and 10 in this volume). It has been suggested that plants within this clade have a predisposition to host a N_2-fixing symbiont (Soltis *et al.,* 1995).

4.1.2. Are Homologous Genes That Are Expressed in the Nodules of Different Aactinorhizal Species Orthologous or Paralogous?

As discussed in chapters 4 and 10 of this volume, it is likely that nodulation arose more than once among the actinorhizal genera and, in fact, may have arisen independently four or more times (Swenson, 1996). Independent origins of the capability to form nodules would require that nodule-expressed genes be recruited independently for functioning within the nodule. Each gene recruited would be selected from the available members of a gene family. Between any two independent origins, either orthologous or paralogous genes could be recruited. It is the promoters of genes and changes in signal-transduction pathways that would lead to recruitment for function in a new organ, such as the nodule (see chapter 10 in this volume). Examples of orthologous- and paralogous-gene recruitment are found in the plant globin genes. The symbiotic hemoglobin of *Parasponia* is paralogous with that of legumes, but the symbiotic hemoglobin of *Casuarina* belongs to the same clade as symbiotic leghemoglobins (Hunt *et al.,* 2001) and is, therefore, orthologous. Expression of paralogous genes in the nodules of different species of actinorhizal plants would support the hypothesis that nodules have arisen independently in these species. On the other hand, if for example, *Myrica*, *Casuarina*, *Alloccasuarina*, *Gymnostoma*, and *Alnus* expressed orthologous genes within the nodule, it would support the hypothesis that they were all descendants of a common nodulated ancestor.

4.1.3. Are Group-2 Frankia *Strains Incapable of Saprophytic Growth in Any Environment or Have The Proper Conditions for Culturing Them Not Been Found?*

Uncultured endophytes in root nodules of *Purshia tridentata*, *Coriaria arborea*, *C. plumose*, and *Datisca cannabina* cluster with endophytes in *Dryas* and *Ceanothus* on the basis of both 16S-ribosomal sequences and an insertion sequence in domain III of the 23S-RNA gene (see chapter 2). No isolates have been made from this group of host plants and no cultures are available of these Group-2 *Frankia* strains. Improvements have been made in the culturing of isolates from *Frankia* Groups 1 and 3 (Schwencke and Carú, 2001) and it is possible that, with further manipulation of culture conditions, isolates will be obtained from Group-2 *Frankia*.

4.1.4. Why Don't Group-3 Strains Nodulate Ceanothus *When They Do Nodulate Other Members of the Rhamnaceae?*

Ceanothus, a North American member of the Rhamnaceae, is nodulated by *Frankia* Group-2 strains, which also nodulate North American plants within the Rosaceae. The remainder of the actinorhizal plants within the Rhamnaceae are South American and New Zealand genera that are nodulated by Group-3 strains. Why?

4.2. Physiological Questions

As discussed in chapters 6 and 7, many questions related to carbon and nitrogen metabolism and host-symbiont signaling remain unanswered. These include:

- What are the components of host-symbiont signaling pathways and how do they operate?
- What carbon and nitrogen compounds are supplied to symbiotic *Frankia*?
- What are the nitrogen compounds leaving the vesicle and the hyphae?
- What are the metabolic roles of infected and uninfected cells within the nodule?
- What metabolic pathways are used, in what compartments do these pathways operate, and how are they regulated?
- How is the carbon flow to and from the nodules regulated?
- What determines the pattern of nitrogenous compounds in the xylem?

4.3. Ecological Questions

4.3.1. What are the Environmental Factors that Determine Strain Selection?

Even though two host plants can be infected by the same group of strains in cross-inoculation tests, the strains that they select from a mixed soil population can differ markedly, which indicates that there are environmental factors that determine strain selection. Where several actinorhizal plant species occur together, each may be nodulated by a distinct group of *Frankia*.

4.4. What is Life like for Frankia *Outside the Nodule*?

The presence of *Frankia* in a variety of locations outside the nodule and not in association with host plants has been revealed by molecular techniques. In addition to the presence of *Frankia* in the soil near host plants, *Frankia* has been found in soils devoid of host plants as well as in unusual non-soil environments, revealing its ubiquitous nature. On the basis of 16S-RNA sequencing, *Frankia* was identified on biodeteriorating wall paintings, one in Germany and one in Austria (Gurtner *et al.*, 2000; Rölleke *et al.*, 1996) as well as on historic glass (Rölleke *et al.*, 1999). Other closely related actinomycetes have been cultured from sandstone (Urzi and Realini, 1998) and other rock surfaces (Eppard *et al.*, 1996).

A three-way comparison of the genomes of *Frankia* that cannot infect any host, cultured *Frankia* that can infect, and unculturable *Frankia* within nodules might reveal what portion of the genome has evolved to facilitate life within the nodule and what portion has evolved to accommodate an independent plant-free existence. The demonstration that there were uncultivated close neighbors of *Frankia*, which were thriving in the rhizosphere of *Alnus viridis crispa* (Normand and Chapelon, 1997) but not in root nodules, complicates the question of what is a *bone fide Frankia* and what is not. Microbial genera generally do not have distinct boundaries due to genetic exchanges and shared biotopes. Thus, the situation with *Frankia* may become comparable to that of the complex of genera in the *Rhizobium-Agrobacterium* clade (Gaunt *et al.*, 2001) as more extensive rhizosphere sampling is performed and as the nodules of actinorhizal species growing in unexplored environments are studied.

5. PROSPECTS FOR EXTENDING SYMBIOSES BEYOND CURRENT HOST RANGE

Frankia is a strong candidate for extending symbioses to new host species (Mullin, 1993; Swenson and Mullin, 1997). No other known diazotropic bacteria possess the combination of characteristics listed below, which together may ultimately facilitate the development of new symbiotic or associative N_2-fixing relationships. In addition to working toward the development of nodule-based symbioses with *Frankia*, it is equally worthwhile to explore avenues that might lead to associative nitrogen fixation in the rhizosphere of new hosts. If *Frankia* can live in biofilms in deteriorating paintings, could it not be induced to live in a root-based biofilm, perhaps in combination with a mycorrhizal partner? Maybe it is already and we just have not looked for it in the right places using the right tools.

- *Frankia* has an intrinsic O_2-protection mechanism, the diazovesicle, and therefore need not rely on host-plant tissue for the protection of nitrogenase.
- The presence of hemoglobin in *Frankia* suggests that there may be an endogenous O_2-control mechanism in addition to the vesicle envelope.
- Host plants are phylogenetically diverse, indicating that *Frankia* can adapt to a wide range of genetic backgrounds and can be accommodated by a range of nodule structures.
- *Frankia* appears to be ubiquitous in the environment and so must be adaptable to a wide range of environmental conditions.
- Root nodules form by modifying a currently existing structure, the lateral root, in the root system of host plants.
- *Frankia* is not released into the cytoplasm of the host plant and so may be less likely to elicit a pathogenic response in a non-natural host plant.
- Strains are available which can infect either *via* root hair or by inter-cellular penetration.

6. CONCLUSIONS

The future holds great promise for the advancement of knowledge about *Frankia* and its host plants, with at least one complete *Frankia* genome being sequenced and many person-hours devoted to the analysis of sequence data. The most meaningful progress, however, will be made by the conception, clever design, and implementation of experiments that will ultimately lead to a greater understanding of how *Frankia* communicates with host plants to establish a productive symbiosis and how, in turn, host plants adjust their growth pattern and metabolism to accommodate *Frankia*. The development of ever more sophisticated molecular tools and methods of analysis will facilitate the process, but ultimately it is the creativity of the investigator and his/her intuition and observational skills that will determine the rate of progress and the magnitude and impact of the advances that are made.

Much remains to be learned about the ecology of actinorhizal plants and their response to the changing global environment. With worldwide decreasing water and land resources and the need to both reclaim and remediate depleted and contaminated soils, increased attention should be paid to the study and use of actinorhizal plants. Every effort should be made to utilize actinorhizal plants where appropriate and to work for their improvement through selection of superior plant material and co-selection for an optimal symbiotic partner.

ACKNOWLEDGMENTS

P. Normand would like to acknowledge Joelle Marechal for helpful discussions and B. C. Mullin would like to acknowledge the helpful comments of Nathan Whitfield as well as support from the USDA National Research Initiative-Competitive Grants Program.

REFERENCES

Beckwith, J., Tjepkema, J. D., Cashon, R. E., Schwintzer, C. R., and Tisa, L. S. (2002). Hemoglobin in five genetically diverse *Frankia* strains. *Can. J. Microbiol., 48,* 1048-1055.

Bentley, S. D., Chater, K. F., Cerdeno-Tarraga, A. M., Challis, G. L., Thomson, N. R., *et al.* (2002). Complete genome sequence of the model actinomycete *Streptomyces coelicolor* A3(2). *Nature, 417,* 141-147.

Berry, A. M., Harriot, O. T., Moreau, R. A., Osman, S. F., Benson, D. R., and Jones, A. D. (1993). Hopanoid lipids compose the *Frankia* vesicle envelope, presumptive barrier of oxygen diffusion to nitrogenase. *Proc. Natl. Acad. Sci. U.S.A., 90,* 6091-6094.

Brummell, D. A., Balint-Kurti, P. J., Harpster, M. H., Palys, J. M., Oeller, P. W., and Gutterson, N. (2003). Inverted repeat of a heterologous 3'-untranslated region for high-efficiency, high-throughput gene silencing. *Plant J., 33,* 793-800.

Catoira, R., Galera, C., de Billy, F., Penmetsa, R. V., Journet, E. P., *et al.* (2000). Four genes of *Medicago truncatula* controlling components of a nod factor transduction pathway. *Plant Cell, 12,* 1647-1666.

Clemens, S., Palmgren, M. G., and Krämer, U. (2002). A long way ahead: Understanding and engineering plant metal accumulation. *Trends Plant Sci., 7,* 309-315.

Ceremonie, H., Cournoyer, B., Maillet, F., Normand, P., and Fernandez, M. P. (1998). Genetic complementation of rhizobial nod mutants with *Frankia* DNA: Artifact or reality? *Mol. Gen. Genet., 260,* 115-119.

Cordoba, E., Shishkova, S., Vance, C. P., and Hernandez, G. (2003). Antisense inhibition of NADH glutamate synthase impairs carbon/nitrogen assimilation in nodules of alfalfa (*Medicago sativa* L.). *Plant J., 33*(6), 1037-1049.

Cournoyer, B., and Normand, P. (1992). Electropermeabilization of *Frankia* intact cells to plasmid DNA. *Acta Oecol., 13,* 369-378.

Dickinson, N. M. (2000). Strategies for sustainable woodland on contaminated soils. *Chemosphere, 41,* 259-263.

Duhoux, E., Rinaudo, G., Diem, H. G., Auguy, F., Fernandez, D., *et al.* (2001). Angiosperm *Gymnostoma* trees produce root nodules colonized by arbuscular mycorrhizal fungi related to *Glomus. New Phytol., 149,* 115-125.

Eppard, M., Krumbein, W. E., Koch, C., Rhiel, E., Staley, J. T., and Stackebrandt, E. (1996). Morphological, physiological, and molecular characterization of actinomycetes isolated from dry soil, rocks, and monument surfaces. *Arch. Microbiol., 166,* 12-22.

Gaunt, M. W., Turner, S. L., Rigottier-Gois, L., Lloyd-Macgilp, S. A., and Young, J. P. W. (2001). Phylogenies of *atpD* and *recA* support the small subunit rRNA-based classification of rhizobia. *Int. J. Syst. Evol. Microbiol., 51,* 2037-2048.

Gerber, N., and Lechevalier, M. (1984). Novel benzo [a] napthacene quinones from an actinomycete, *Frankia* G-2 (ORS020604). *Can. J. Chem., 62,* 2818-2821.

Gurtner, C., Heyrman, J., Piñar, G., Lubitz, W., Swings, J., and Rölleke, S. (2000). Comparative analyses of the bacterial diversity on two different biodeteriorated wall paintings by DGGE and 16S rDNA sequence analysis. *Int. Biodeter. Biodegr., 46,* 229-239.

Gupta, R. K., Dobritsa, S. V., Stiles, C. A., Essington, M. E., Liu, Z. Y., *et al.* (2002). Metallohistins: A new class of plant metal-binding proteins. *J. Protein Chem., 21*, 529-536.

Hammad, Y., Nalin, R., Marechal, J., Fiasson, K., Pepin, R., *et al.* (2003). The possible role of phenyl acetic acid (PAA) on alder nodulation by *Frankia*. *Plant Soil, 254,* 193-205.

Hammad, Y., Marechal, J., Cournoyer, B., Normand, P., and Domenach, A. M. (2001). Modification of the protein expression pattern induced in the nitrogen-fixing actinomycete *Frankia* sp. strain ACN14a-tsr by root exudates of its symbiotic host *Alnus glutinosa* and cloning of the *sodF* gene. *Can. J. Microbiol., 47*, 541-547.

Huguet, V., Batzli, J. M., Zimpfer, J. F., Normand, P., Dawson, J. O., and Fernandez, M. P. (2001). Diversity and specificity of *Frankia* strains in nodules of sympatric *Myrica gale, Alnus incana, and Shepherdia canadensis* determined by *rrs* gene polymorphism. *Appl. Environ. Microbiol., 67*, 2116-2122.

Hunt, P. W., Watts, R. A., Trevaskis, B., Llewelyn, D. J., Burnell, J., *et al.* (2001). Expression and evolution of functionally distinct haemoglobin genes in plants. *Plant Mol. Biol., 47,* 677-692.

John, T. R., Rice, J. M., and Johnson, J. D. (2001). Analysis of pFQ12, a 22.4-kb *Frankia* plasmid. *Can. J. Microbiol., 47*, 608-617.

Kairesalo, T., Haapa, A., Salkinoja-Salonen, M., and Salminen, J. (1999). Uptake and translocation of nickel from soil by *Alnus glutinosa* L. and the impacts on plant photosynthesis and growth. Meeting abstract: *Plant species useful for phytoremediation and wastewater treatment*, Lisbon, Portugal, November 18-20, 1999.

Karlin, S., and Mrázek, J. (2000). Predicted highly expressed genes of diverse prokaryotic genomes. *J. Bacteriol., 182*, 5238-5250.

Klahre, U., Crété, P., Leuenberger, S. A., Iglesias, V. A., and Meins, Jr., F. (2002). High molecular weight RNAs and small interfering RNAs induce systemic posttranscriptional gene silencing in plants. *Proc. Natl. Acad. Sci. U.S.A., 99,* 11981-11986.

Klika, K. D., Haansuu, J. P., Ovcharenko, V. V., Haahtela, K. K., Vuorela, P. M., and Pihlaja, K. (2001). Frankiamide, a highly unusual macrocycle containing the imide and orthoamide functionalities from the symbiotic actinomycete *Frankia*. *J. Org. Chem., 66*, 4065-4068.

Koksharova, O. A., and Wolk, C. P. (2002). Genetic tools for cyanobacteria. *Appl. Microbiol. Biotechnol., 58*, 123-137.

Laplaze, L., Bon, M.-C., Sy, M. O., Smouni, A., Allonneau, C., *et al.* (2000). Molecular biology of tropical nitrogen-fixing trees in the Casuarinaceae family. In S. M. Jain, and S. C. Minocha (Eds.), *Molecular biology of woody plants, Vol. 1* (pp. 269-285.). Dordrecht, The Netherlands: Kluwer Academic Publishers.

Lavire, C., Louis, D., Perriere, G., Briolay, J., Normand, P., and Cournoyer, B. (2001). Analysis of pFQ31, a 8551-bp cryptic plasmid from the symbiotic nitrogen-fixing actinomycete *Frankia*. *FEMS Microbiol. Lett., 197,* 111-116.

Leblond, P., Fischer, G., Francou, F. X., Berger, F., Guerineau, M., and Decaris, B. (1996). The unstable region of *Streptomyces ambofaciens* includes 210 kb terminal inverted repeats flanking the extremities of the linear chromosomal DNA. *Mol. Microbiol., 19*, 261-271.

Leblond, P., and Decaris, B. (1998). Chromosome geometry and intraspecific genetic polymorphism in Gram-positive bacteria revealed by pulsed-field gel electrophoresis. *Electrophoresis, 19*, 582-588.

Maréchal, J., Clement, B., Nalin, R., Gandon, C., Orso, S., *et al.* (2000). A *recA* gene phylogenetic analysis confirms the close proximity of *Frankia* to *Acidothermus*. *Int. J. Syst. Evol. Microbiol., 50,* 781-785.

Mohagheghi, A., Grohmann, K., Himmel, M., Leighton, L., and Updegraff, D. M. (1986). Isolation and characterization of *Acidothermus cellulolyticus* gen. nov., sp. nov., a new genus of thermophilic, acidophilic, cellulolytic bacteria. *Int. J. Syst. Bacteriol., 36,* 435-443.

Mullin, B. C. (1993). Symbiosis with *Frankia*. In G. S. Khush and J. Bennett, (Eds.). *Nodulation and nitrogen fixation in rice: Potential and prospects* (pp. 67-75). Los Baños, The Philippines: International Rice Research Institute.

Navarro, E., Nalin, R., Gauthier, D., and Normand, P. (1997). The nodular microsymbionts of *Gymnostoma* spp. are *Elaeagnus*-infective strains. *Appl. Environ. Microbiol., 63,* 1610-1616.

Nelson, K. E., Weinel, C., Paulsen, I. T., Dodson, R. J., Hilbert, H., *et al.* (2002). Complete genome sequence and comparative analysis of the metabolically versatile *Pseudomonas putida* KT2440. *Environ. Microbiol., 4*, 799-808.

Normand, P., and Chapelon, C. (1997). Direct characterization of *Frankia* and of close phyletic neighbors from an *Alnus viridis* rhizosphere. *Physiol. Plant., 99,* 722-731.

Normand, P., Orso, S., Cournoyer, B., Jeannin, P., Chapelon, C., *et al.* (1996). Molecular phylogeny of the genus *Frankia* and related genera and emendation of Family Frankiaceae. *Int. J. Syst. Bacteriol., 46,* 1-9.

Omura, S., Ikeda, H., Ishikawa, J., Hanamoto, A., Takahashi, C., *et al.* (2001). Genome sequence of an industrial microorganism *Streptomyces avermitilis*: Deducing the ability of producing secondary metabolites. *Proc. Natl. Acad. Sci. U.S.A., 98,* 12215-12220.

Pawlowski, K., Twigg, P., Dobritsa, S., Guan, C. H., and Mullin, B. C. (1997). A nodule-specific gene family from *Alnus glutinosa* encodes glycine- and histidine-rich proteins expressed in the early stages of actinorhizal nodule development. *Mol. Plant-Microbe Interact., 10,* 656-664.

Perradin, Y., Mottet, M., and Lalonde, M. (1983). Influence of phenolics on *in vitro* growth of *Frankia* strains. *Can. J. Bot., 61,* 2807-2814.

Richards, J. W., Krumholz, G. D., Chval, M. S., and Tisa, L. S. (2002). Heavy metal resistance patterns of *Frankia* strains. *Appl. Environ. Microbiol., 68,* 923-927.

Rölleke, S., Gurtner, C., Drewello, U., Drewello, R., Lubitz, W., and Weissmann, R. (1999). Analysis of bacterial communities on historical glass by denaturing gradient gel electrophoresis of PCR-amplified gene fragments coding for 16S rRNA. *J. Microbiol. Methods, 36,* 107-114.

Rölleke, S., Muyzer, G., Wawer, C., Wanner, G., and Lubitz, W. (1996). Identification of bacteria in a biodegraded wall painting by denaturing gradient gel electrophoresis of PCR-amplifed gene fragments coding for 16S rRNA. *Appl. Environ. Microbiol., 62,* 2059-2065.

Santos, R., Herouart, D., Puppo, A., and Touati, D. (2000). Critical protective role of bacterial superoxide dismutase in *Rhizobium*-legume symbiosis. *Mol. Microbiol., 38,* 750-759.

Santos, R., Franza, T., Laporte, M. L., Sauvage, C., Touati, D., and Expert, D. (2001). Essential role of superoxide dismutase on the pathogenicity of *Erwinia chrysanthemi* strain 3937. *Mol. Plant-Microbe Interact., 14,* 758-767.

Schoenbeck, M. A., Temple, S. J., Trepp, G. B., Blumenthal, J. M., Samac, D. A., *et al.* (2000). Decreased NADH glutamate synthase activity in nodules and flowers of alfalfa (*Medicago sativa* L.) transformed with antisense glutamate synthase transgene. *J. Exp. Bot., 51,* 29-39.

Schwencke, J., and Carú, M. (2001). Advances in actinorhizal symbiosis: Host plant-*Frankia* interactions, biology, and applications in arid land reclamation. A Review. *Arid Land Res. Manag., 15,* 285-327.

Schwintzer, C. R., and Tjepkema, J. D. (1990). *The biology of* Frankia *and actinorhizal plants.* San Diego, CA: Academic Press.

Simonet, P., Haurat, J., Normand, P., Bardin, R., and Moiroud, A. (1986). Localization of *nif* genes on a large plasmid in *Frankia* sp. strain ULQ0132105009. *Mol. Gen. Genet., 204,* 492-495.

Smouni, L., Laplaze, L., Bogusz, D., Guermache, F., Auguy, F., *et al.* (2002). The 35S promoter is not constitutively expressed in the transgenic tropical actinorhizal tree, *Casuarina glauca. Functl. Plant Biol., 29,* 649-656.

Soltis, D. E., Soltis, P. S., Morgan, D. R., Swensen, S. M., Mullin, B. C., *et al.* (1995). Chloroplast gene sequence data suggest a single origin of the predisposition for symbiotic nitrogen fixation in angiosperms. *Proc. Natl. Acad. Sci. U.S A., 92,* 2647-2651.

Swensen, S. M. (1996). The evolution of actinorhizal symbioses: Evidence for multiple origins of the symbiotic association. *Am. J. Bot., 83,* 1503-1512.

Swensen, S. M., and Mullin, B. C. (1997). The impact of molecular systematics on hypotheses for the evolution of root nodule symbioses and implications for expanding symbioses to new host plant genera. *Plant Soil, 194,* 185-192.

Svistoonoff, S. (2003). *Nature Rev. Mol. Cell Biology, 4,* 262.

Urzi, C., and Realini, M. (1998). Colour changes of Noto's calcareous sandstone as related to its colonisation by microorganisms. *Int. Biodeter. Biodegr., 42,* 45-54.

Vincent, S. P., Sinay, P., and Rohmer, M. (2003). Composite hopanoid biosynthesis in *Zymomonas mobilis*: N-acetyl-D-glucosamine as precursor for the cyclopentane ring linked to bacteriohopanetetrol. *Chem. Commun.*, 782-783.

Waterhouse, P. M., and Helliwell, C. A. (2003). Exploring plant genomes by RNA-induced gene silencing. *Nature Rev. Genet., 4,* 29-38.

Wesley, S. W., Helliwell, C. A., Smith, N. A., Wang, M. B., Rouse, D. T., *et al.* (2001). Construct design for efficient, effective and high-throughput gene silencing in plants. *Plant J., 27,* 581-590.

Wheeler, C. T., Hughes, L. T., Oldroyd, J., and Pulford, I. D. (2001). Effects of nickel on *Frankia* and its symbiosis with *Alnus glutinosa* (L.) Gaertn. *Plant Soil, 231,* 81-90.
Xu, X., Kong, R., de Bruijn, F. J., He, S. Y., Murry, M. A., *et al.* (2002). DNA sequence and genetic characterization of plasmid pFQ11 from *Frankia alni* strain CpI1. *FEMS Microbiol. Lett., 207,* 103-107.

SUBJECT INDEX

Nitrogen Fixation: Origins, Applications, and Research Progress

1. B.E. Smith, R.L. Richards and W.E. Newton (eds.): *Catalysts for Nitrogen Fixation.* Nitrogenases, Relevant Chemical Models and Commercial Processes. 2004
ISBN 1-4020-2508-4
2. W. Klipp, B. Masepohl, J.R. Gallon and W.E. Newton (eds.): *Genetics and Regulation of Nitrogen Fixation in Free-Living Bacteria.* 2004 ISBN 1-4020-2178-X
3. R. Palacios and W.E. Newton (eds.): *Genomes and Genomics of Nitrogen-fixing Organisms.* 2005 ISBN 1-4020-3053-3
4. D. Werner and W.E. Newton (eds.): *Nitrogen Fixation in Agriculture, Forestry, Ecology, and the Environment.* 2005 ISBN 1-4020-3542-X
5. C. Elmerich and W.E. Newton (eds.): *Associative and Endophytic Nitrogen-fixing Bacteria and Cyanobacterial Associations.* 2007 ISBN 1-4020-3541-1
6. K. Pawlowski and W.E. Newton (eds.): *Nitrogen-fixing Actinorhizal Symbioses.* 2008
ISBN 978-1-4020-3540-1

MIX
Papier aus verantwortungsvollen Quellen
Paper from responsible sources
FSC® C105338

If you have any concerns about our products,
you can contact us on
ProductSafety@springernature.com

In case Publisher is established outside the EU,
the EU authorized representative is:
Springer Nature Customer Service Center GmbH
Europaplatz 3, 69115 Heidelberg, Germany

Printed by Libri Plureos GmbH
in Hamburg, Germany